EL NIÑO

# EL NIÑO
## Historical and Paleoclimatic Aspects of the Southern Oscillation

Edited by

### HENRY F. DIAZ

*Meteorologist, Environmental Research Laboratories, National Oceanic and Atmospheric Administration, Boulder, Colorado*

*and*

### VERA MARKGRAF

*Institute of Arctic and Alpine Research, University of Colorado, Boulder, Colorado*

CAMBRIDGE
UNIVERSITY PRESS

Published by the Press Syndicate of the University of Cambridge
The Pitt Building, Trumpington Street, Cambridge CB2 1RP
40 West 20th Street, New York, NY 10011–4211, USA
10 Stamford Road, Oakleigh, Victoria 3166, Australia

First published 1992

Printed in Great Britain at the University Press, Cambridge

*A catalogue record for this book is available from the British Library*

*Library of Congress cataloguing in publication data*

El Niño: historical and paleoclimatic aspects of the southern
oscillation / edited by Henry F. Diaz and Vera Markgraf.
        p.    cm.
    ISBN 0-521-43042-9
    1. El Niño Current.  2. Southern Oscillation.  I. Diaz, Henry F.
II. Markgraf, Vera.
GC296.8.E4N56   1992
551.47'5--dc20   92-24682   CIP

ISBN 0 521 43042 9     hardback

Dedication to

# BILL QUINN

*whose foresight and meticulous work in extending back the recent El Niño record helped spark a surge of interest in long-term aspects of ENSO*

# Contents

# Contributors

R. Y. ANDERSON
Department of Geology
University of New Mexico
Albuquerque, NM 87131
U.S.A.

J. L. BETANCOURT
U.S. Geological Survey
1675 West Anklam Rd
Tucson, AZ 85705
U.S.A.

G. W. BRANSTATOR
National Center for Atmospheric
Research
P.O. Box 3000
Boulder, CO 80307
U.S.A.

D. R. CAYAN
Climate Group A-024
Scripps Institution of Oceanography
UCSD
La Jolla, CA 92093

M. K. CLEAVELAND
Department of Geography
University of Arkansas
Fayetteville, AR 72701
U.S.A.

J. E. COLE
INSTAAR
Campus Box 450
University of Colorado
Boulder, CO 80309
U.S.A.

E. R. COOK
Lamont-Doherty Geological Observatory
Palisades, NY 10964
U.S.A.

R. D. D'ARRIGO
Lamont-Doherty Geological Observatory
Palisades, NY 10964
U.S.A.

H. F. DIAZ
NOAA/ERL
325 Broadway
Boulder, CO 80303
U.S.A.

D. B. ENFIELD
NOAA/AOML/PHOD
4301 Rickenbacker Causeway
Miami, FL 33149
U.S.A.

R. G. FAIRBANKS
Lamont-Doherty Geological Observatory

Palisades, NY 10964
U.S.A.

G. C. JACOBY
Lamont-Doherty Geological Observatory
Palisades, NY 10964
U.S.A.

T. C. JOHNSON
Duke Marine Laboratory
Beaufort, NC 28516
U.S.A.

G. N. KILADIS
CIRES
University of Colorado
Campus Box 216
Boulder, CO 80309
U.S.A.

A. P. KERSHAW
Department of Geography
Monash University
Clayton, Victoria 3168
Australia

J. M. LOUGH
Australian Institute of Marine Science
P.M.B. 3
Townsville, M.C., Queensland 4810
Australia

V. MARKGRAF
INSTAAR
Campus Box 450
University of Colorado
Boulder, CO 80309
U.S.A.

M. S. McGLONE
DSIR – Land Resources
Private Bag
Christchurch
New Zealand

G. A. MEEHL
National Center for Atmospheric
  Research
P.O. Box 3000
Boulder, CO 80307
U.S.A.

D. M. MEKO
Laboratory of Tree-Ring Research

University of Arizona
Tucson, AZ 85721
U.S.A.

J. MICHAELSEN
Department of Geography
University of California – Santa Barbara
Santa Barbara, CA 93106
U.S.A.

M. MOORE
Reef Research Group
Museum of Paleontology
University of California
Berkeley, CA 94720
U.S.A.

E. MOSLEY-THOMPSON
Byrd Institute of Polar Studies
The Ohio State University
Columbus, OH 43210
U.S.A.

N. NICHOLLS
Bureau of Meteorology Research Centre
GPO Box 1289K
Melbourne, Victoria 3001
Australia

R. S. PULWARTY
CIRES
University of Colorado
Campus Box 216
Boulder, CO 80309
U.S.A.

W. H. QUINN
College of Oceanography
Oregon State University
Corvallis, OR 97331-5503
U.S.A.

D. W. STAHLE
Department of Geography
University of Arkansas
Fayetteville, AR 72701
U.S.A.

G. D. SHARP
NOAA/COAP
2560 Garden Rd
Monterey, CA 93940
U.S.A.

*Contributors*

G. T. SHEN
Department of Oceanography
University of Washington
Seattle, WA 98195
U.S.A.

A. SOUTAR
Scripps Institution of Oceanography
Mail Code A-014
La Jolla, CA 92037
U.S.A.

T. W. SWETNAM
Laboratory of Tree-Ring Research
University of Arizona

Tucson, AZ 85721
U.S.A.

L. G. THOMPSON
Byrd Institute of Polar Studies
The Ohio State University
Columbus, OH 43210
U.S.A.

P. A. THOMPSON
Department of Decision and
   Information Science
University of Florida
Gainesville, FL 32611
U.S.A.

# Acknowledgment

The editors wish to thank all of the contributing authors for their dedication and effort to this project; we hope that they are as proud of the outcome as we are.

We wish to thank our reviewers for generously giving of their time and expertise to each of the manuscripts. Chapters were generally reviewed by two specialists in the field, as well as the editors. We would like, specially, to single out the assistance of Ms. Kathleen Salzberg, our technical editor, for her painstaking and able help throughout the editorial process, and Ms. Grace Norman for expert help in preparing the final manuscripts.

We also thank the National Geophysical Data Center of the National Oceanic and Atmospheric Administration for their financial support, which made it possible to hold the workshop that engendered this volume, and who provided the additional resources that enabled the editors to accomplish the task of producing this book.

The diagram shown in the book cover is based on information provided by W. H. Quinn and C. F. Ropelewski, and the editors gratefully acknowledge their contribution.

# 1

# Introduction

HENRY F. DIAZ

*NOAA/ERL, 325 Broadway, Boulder, Colorado 80303, U.S.A.*

VERA MARKGRAF

*Institute of Arctic and Alpine Research, University of Colorado, Boulder, Colorado 80309, U.S.A.*

Significant progress has been made over the last decade in understanding the mechanisms and global manifestations of the weather and climate anomalies referred to as El Niño/Southern Oscillation (ENSO). The ENSO phenomenon constitutes the largest single source of interannual climatic variability on a global scale, and because its effects are wide-ranging and often severe, it has attracted the attention of many scientists worldwide.

In broad outline, the large-scale sea-level pressure 'seesaw' across the tropical Pacific Ocean which defines the Southern Oscillation, and the anomalous oceanographic and atmospheric conditions which occur periodically along the upwelling zone of the eastern equatorial Pacific and along the coast of southern Ecuador and Peru, known as 'El Niño,' are manifestations of slowly evolving, coupled ocean-atmosphere processes that give rise to characteristic responses in the atmosphere and the ocean.

A concerted effort to monitor the principal atmospheric and oceanographic variables that make up the ENSO has led to an improved understanding of the phenomenon and an emergent ability to predict the development of recent ENSO events with a few months lead time. Because the global climatic patterns associated with the extreme modes of the ENSO cycle are quite different, there has been some concern that one or the other pattern (either the warm 'El Niño' phase, or its opposite, cold 'La Niña' phase) could become much more frequent in the future as a result of changes in the Earth's climate brought about by increases in atmospheric concentrations of so-called greenhouse gases. Such a possibility has spurred interest in determining whether the ENSO system has undergone low-frequency variations in the past. This has led to efforts to extend the modern ENSO record by making use of historical records of climatic anomalies known to be associated with the ENSO, such as failure of the Indian monsoon and unusually heavy rainfall and flooding in the normally arid coastal region of northern Peru. These studies have resulted in the extension of the chronology of the warm or El Niño phase of the Southern Oscillation by several centuries. More recently, the annual record of Nile River floods which extends back in time to the 7th century AD has been used to develop a chronology of ENSO variability in the western portion of the core ENSO region. That work is documented in Section B of this book, which deals with the use of historical records in ENSO reconstruction.

Analyses of high-resolution proxy records (tree-rings, ice cores, sea corals, varved sediments) have made it possible to extend the ENSO chronology farther back in time, thereby providing a longer temporal basis to evaluate ENSO changes in the time domain. They also furnish additional geographical 'anchors' that are useful in evaluating the spatial consistency of ENSO patterns through time. The results of this work, which are detailed in Section C of this volume help us to learn more about the century to millennial-scale characteristics of the ENSO phenomenon. Finally, lower resolution proxy records such as flood deposits and time series derived from paleoecological changes recorded in the sedimentary record from terrestrial and coastal marine environments provide information that may help us understand changes in the long-term behavior of ENSO, and perhaps provide clues regarding possible forcing mechanisms.

This book is an outgrowth of a workshop that was held in Boulder, Colorado, in May 1990 to examine some of the proxy evidence of ENSO-related climatic variability. Although the strongest manifestations of the ENSO are found in the tropical Pacific and Indian oceans and some of the adjacent coastal regions, atmospheric and oceanic teleconnections outside the tropics have also been well documented. It was felt that a comprehensive review of worldwide ENSO-sensitive paleoclimate indices was needed. It was also felt that the spatial and temporal resolution of many of the proxy records was such that it would be profitable to have paleoclimatologists present their findings to an audience that included not only people within their own general discipline, but also to oceanographers and atmospheric scientists involved in ENSO research. The stated goals of the workshop were to facilitate the exchange of information among different workers in the different paleoclimate fields and to promote greater interaction among those researchers working with the field observations (palynologists, dendroclimatologists, geomorphologists, etc.) and atmospheric scientists and oceanographers doing ENSO research. Our goal with this book is to bring together a variety of related topics under the common theme of ENSO variability. The techniques discussed in the separate chapters range from the use of written historical records which document the climatic effects known to be modulated by the different phases of the ENSO phenomenon, to analysis of tree-ring records, ice cores, tropical coral records, sedimentary and pollen records from lacustrine and near-coastal marine environments.

The book is divided thematically as well as by the temporal resolution of the ENSO records themselves. In Section A, *ENSO in the modern record*, which immediately follows this introduction, the observational basis for describing the phenomenon and some of the modeling work applicable to understanding the physical mechanisms of the ENSO are presented. Section B, *Use of historical records in ENSO reconstruction*, presents the results of studies aimed at reconstructing past ENSO occurrences from the historical record. Several chapters in this section examine the evidence for changes in the frequency and intensity of ENSO events (mostly with regard to its warm El Niño phase) and one describes some plausible linkages between the ENSO-modulated climate, which is manifested in the form of enhanced climatic variability and the vegetation and fauna in the Australasian region. The period of time covered here extends over roughly the last one-and-a-half millennia. In Section C, *Paleoclimate reconstructions of ENSO from tree-ring records*, we examine how tree-ring records are used to 'predict' the occurrence of ENSO events prior to available instrumental records from the statistical relationships between climatic anomalies forced by the ENSO phenomenon, and its effect on tree growth. Section D, *Records from ice cores and corals*, describes the use of other types of paleoindicators. Coral atolls along the equatorial Pacific provide several geochemical measures that can be related to environmental changes associated with changes in the ENSO cycle. The use of ice cores retrieved from mountain glaciers in areas affected by atmospheric circulation changes induced by ENSO are described here, and its utility to reconstruct ENSO events (signals) at annual resolution for the last several centuries is evaluated by comparison with ENSO-sensitive tree ring records from the southwestern United States and with ENSO reconstructions based on historical records.

The last section, titled *Low-resolution paleoclimate reconstruction of ENSO: marine and terrestrial proxy indicators* (Section E), examines the basis for interpreting lower resolution proxy records in the context of high frequency phenomena such as ENSO. The studies describe how low frequency variations in the vegetation assemblages, fire histories, bioproductivity, sedimentary records of fisheries abundances, etc. that can be inferred from such records can be used to interpret the possible state of the ENSO system during those times. In this and previous chapters, the possible connection between solar forcing variations and changes in the expression of the ENSO signal is also explored. The last chapter, *Synthesis and future prospects*, attempts to bring together the key findings and ideas presented in the book and discusses future prospects for interdisciplinary research aimed at reconstructing climatic variability at year to century and longer time scales.

It is hoped that this book will serve to stimulate further work both in the area of constructing additional reliable proxy chronologies from these and other ENSO-sensitive regions, as well as to stimulate the development of new, innovative ways to analyse existing proxy and instrumental records. A conviction that was shared by many people at the workshop was that improved reliability of individual ENSO indices might be achieved by pooling together as many of the proxy series as is feasible. This should also lead to increased confidence in the interpretation of specific features in the reconstructed record. If this were possible, one might also be able to ascertain with some degree of confidence whether there have been any significant changes in the spatial structure of ENSO during the past several hundred years, and whether ENSO has undergone changes in its frequency characteristics during the late Holocene.

We feel that the ample talent and multidisciplinary expertise that has been brought to bear on this important scientific question will have a broad audience, and that it will be useful to people involved not only on scientific research issues related to ENSO, but also to those working on the socioeconomic impacts associated with ENSO variability.

# ENSO in the modern record

# 2

# Atmospheric teleconnections associated with the extreme phases of the Southern Oscillation

HENRY F. DIAZ

*Environmental Research Laboratories, NOAA, Boulder, Colorado 80303, U.S.A.*

GEORGE N. KILADIS

*Cooperative Institute for Research in Environmental Sciences, University of Colorado, Boulder, Colorado 80309, U.S.A.*

## Abstract

An overview is presented of the principal climatic characteristics associated with the development of warm and cold phases of the ocean-atmospheric pheno-menon known as El Niño/Southern Oscillation (ENSO), and of the most salient large-scale teleconnection features related to those extremes. Besides giving the reader some appreciation of the typical climatic patterns in different parts of the globe during the extreme ENSO phases, we have made an effort to illustrate some of the event-to-event variability inherent in various climatic indices associated with this phenomenon.

ENSO is not a stationary system; there are substantial differences between events that are reflected in a variety of ENSO indices. It is shown that even for a particular set of ENSO measures, the association among such indices may vary with time. It is important to keep this mind when analysing long-term associa-tions with individual proxy variables of ENSO activity, such as tropical corals or glacier varves.

## Introduction

Elements of the global-scale climate phenomenon now referred to as the El Niño/Southern Oscillation (ENSO) began to be noted (though not by that name) toward the end of the 19th century (see Nicholls 1992, this volume). Starting with the observations of Sir Charles Todd, the South Australian Government Observer, published in Australia in the late 1880s, and culminating in a series of papers by Sir Gilbert Walker and collaborators (Walker 1923; Walker and Bliss

1932), knowledge of these large-scale atmospheric pressure changes and related fluctuations in rainfall in the tropics gradually accumulated. It was Walker who gave the name 'Southern Oscillation' (SO) to the sea-level pressure 'seesaw' that he documented in the tropical Pacific and Indian oceans.

Associated with this large-scale fluctuation of atmospheric mass (see Trenberth 1976; van Loon and Madden 1981; Trenberth and Shea 1987) are marked changes in tropical Pacific sea surface temperature (SST), particularly along the eastern equatorial Pacific upwelling zone and off the coast of Peru (see Philander 1990). One of the primary effects of the evolution of tropical Pacific SSTs and related changes in air-sea interaction is found in the strong interannual variation of rainfall throughout the global tropics. In fact, Nicholls (1988) found that the interannual variance of mean annual precipitation in the core region of the El Niño/Southern Oscillation is larger than in areas less affected by the phenomenon.

After the work of Walker in the 1920s and 1930s, studies on the general characteristics and mechanisms of remote atmospheric and oceanic responses (teleconnections) associated with sea-level pressure (SLP) and SST fluctuations in the equatorial Pacific waned. Interest in the Southern Oscillation and the El Niño was re-established in the late 1950s, beginning with the work by Berlage (1957) and following the 1972 El Niño, which brought about considerable economic hardship to Peru and other countries around the world (see overviews by Julian and Chervin 1978; Rasmusson and Carpenter 1982; Enfield 1989).

Beginning with the work of J. Bjerknes (1966, 1969), our understanding of the physical mechanisms responsible for the observed large-scale associations of seasonal and longer time scale climatic variations linked to the ENSO has grown manyfold. Modern refinements to our knowledge of the workings of ENSO, starting with Troup (1965), have added considerably to our understanding of this dominant mode of atmospheric and upper ocean variability. In particular, the synthesis provided by Rasmusson and Carpenter (1982) of the composite structure of ENSO in the tropics and the analysis of the planetary scale associations provided by Horel and Wallace (1981) and van Loon and Madden (1981), van Loon and Shea (1987), and Rasmusson and Wallace (1983), among others, have underscored the pervasive influence of this dynamical system. More recently, Ropelewski and Halpert (1987, 1989) and Kiladis and Diaz (1989) have detailed the spatial and temporal variability of surface temperature and precipitation throughout the globe associated with ENSO.

The reader is also referred to the recently published book by Glantz et al. (1991), which addresses the subject of ENSO-induced teleconnections linking climate anomalies throughout the globe, for a review of the societal impacts associated with this phenomenon, as well as a review of the physical mechanisms operating during ENSO.

In the following pages, we will first describe the typical evolution of atmospheric features associated with ENSO during its life cycle. We will then focus on an important issue that bears on the reliability of paleo-ENSO reconstructions, namely, how consistent from one event to another the remote atmospheric and oceanic responses are to ENSO. In this regard, our emphasis will be on those regions where analysis of ENSO-sensitive proxy records are discussed in other sections of this book. We note that various other observational aspects of ENSO are also examined in the chapters in this book by Enfield (1992), Nicholls (1992), and Quinn (1992).

## Typical evolution and climatic features of ENSO

In this section, we review the general climatic characteristics of the 'typical' ENSO event. Here, we use 'ENSO' to refer to the general system that comprises both warm (the 'El Niño' phase) and cold (so-called 'La Niña' episodes) sea surface temperature extremes of Walker's Southern Oscillation. The signals discussed here are based primarily on studies using a composite of several events (Rasmusson and Carpenter 1982; Ropelewski and Halpert 1986, 1987, 1989; Kiladis and van Loon 1988; Kiladis and Diaz 1989). However, we wish to emphasize that individual events may differ markedly from the mean pattern; the next section discusses some of these differences among ENSO events and gives some examples of the range of variability to be expected from one event to another.

By far the best-defined ENSO teleconnection is associated with sea-level pressure (SLP), and it was this signal that led Hildebrandson (1897), Lockyer (1906), and Walker (1923) to detect and define the SO. Essentially, the pressure signal of the SO is observed as a 'seesaw' between the southeastern tropical Pacific and the Australian-Indonesian region, such that when pressure is below normal in one region it tends to be above normal in the other on time scales ranging from monthly to annual (see Fig. 2.1). A measure of the state of the SO is nowadays defined by the Southern Oscillation Index (SOI), which is based on the standardized SLP difference between Tahiti and Darwin, Australia. Located near the core regions of the SO, these stations have relatively long periods of record, so that the SOI can be calculated back to the year 1882 (see Ropelewski and Jones 1987).

Early workers on the SO were aware that extremes in the pressure seesaw were associated with marked climatic anomalies in both the tropics and the subtropics. For example, Walker (1923) established that high pressure over the Australasian region was accompanied by drought over India and Australia, and cool, wet winters over the southeastern United States.

While the pressure oscillation associated with the SO has been known for about a century, it was not until the late 1950s and the 1960s that the connection between the SO and SST was demonstrated (Berlage 1957; Ichiye and Petersen 1963; Bjerknes 1966, 1969). Generally, SST along the western coast of South America and along the equator in the central and eastern Pacific is anomalously cold for its latitude. This is due to strong oceanic upwelling associated with trade winds along the equatorial regions and equatorward flow along the South

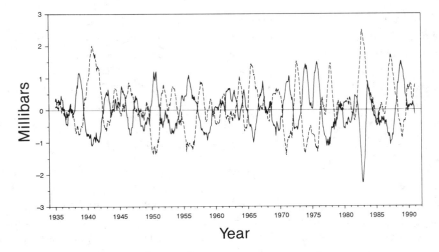

Fig. 2.1  Nine-month running mean of monthly sea-level pressure anomalies at Darwin, Australia (dashed line) and at Tahiti (solid line). Period of record is January 1935–May 1991. Correlation = −0.75.

American coast (see Enfield 1989). These cool SSTs stabilize the lower atmosphere, inhibiting precipitation and giving rise to the hyperarid climate of coastal Peru. This aridity extends well into the central Pacific, occupying a wedge-shaped region that extends as far west as the dateline and south-eastward from this general vicinity to about the latitude of Santiago, Chile, along the South American coast. This region is referred to as the Pacific 'dry zone.'

Every few years this aridity is broken by periodic heavy rainfall episodes lasting several months, associated with a dramatic increase of equatorial Pacific SST. Along the Peruvian coast this phenomenon has been known as El Niño because of its general occurrence near Christmas time (see Wyrtki 1975; Enfield 1989). The coastal plains of Ecuador and northern Peru are particularly suscepti-ble to flooding during El Niño, although regions farther south and the islands along the equatorial upwelling zone can also see spectacular rainfall increases. In essence, the weakening of the South Pacific High is accompanied by a decrease in the strength of the trade winds, weakening the oceanic upwelling and causing SSTs to rise. This warming of the ocean surface causes increased evaporation and heating of the troposphere, and decreased atmospheric stability, creating con-ditions favorable for convection and rainfall. The process tends to be self-sustaining because the development of organized areas of convection near the equator will tend to further weaken the trade winds within and to the west of the convective area, thereby causing SSTs in that area to remain anomalously warm. Thus, forcing mechanisms of these so-called warm ENSO events are due to a complex ocean-atmosphere instability dependent on the coupling of the two media (see Rasmusson and Wallace 1983; Philander 1990).

The increase of precipitation over the eastern tropical Pacific during warm ENSO events actually represents part of a major reorganization of the tropical rainfall patterns. On average, the wettest part of the tropics extends from equatorial regions of Indonesia eastward to New Guinea and the western Pacific, where the rainfall is prevalent throughout the year. During northern summer the zone of heavy rainfall expands northwestward from Indonesia into India and southeast Asia, and is known as the South Asian summer monsoon. During northern winter these regions become dry, while monsoon rainfall migrates southward to give northern Australia and the southwestern tropical Pacific its rainy season. In general, these monsoon regions experience drier than normal conditions during warm events, so that in some sense, these events represent an eastward shift in the main regions of convection from the eastern Indian Ocean/western Pacific to the central and eastern tropical Pacific. It should be emphasized that warm events are not accompanied by a complete failure in monsoon rainfall, but a significant reduction in the usually heavy rains that these regions rely on for their intensive agricultural economies (see Kiladis and Sinha 1991).

An intriguing feature of ENSO is its apparent 'phase locking' to the annual cycle. Both warm and cold events show a propensity to develop during the March through May period, and last at least until a year later, but frequently longer (Rasmusson and Carpenter 1982; Bradley et al. 1987). In the following discussion, the year in which a warm (or cold) event develops will be referred to as 'year 0' of that event. While equatorial Pacific SST anomalies tend to peak in northern summer or fall of year 0, there is often a secondary peak along the coast of Peru observable in the so-called 'mature phase' during the northern winter season (Rasmusson and Carpenter 1982). It is during northern winter (at the end of year 0 and into year +1) that ENSO teleconnections are most evident in the extratropics.

Another aspect of ENSO, overlooked by most workers until the mid-1980s, was the state opposite to El Niño, the so-called La Niña or cold event (van Loon and Shea 1985; Philander 1990). During this phase of the SO, the climatological conditions of heavy rainfall over the monsoon regions and relatively cool SST over the eastern equatorial Pacific appear to be amplified. Moreover, the global-scale climatic anomalies during cold events are for the most part opposite in sign to those seen during warm events (van Loon and Shea 1985; Yarnal and Diaz 1986; Bradley et al. 1987; Lau and Sheu 1988; Kiladis and Diaz 1989; Ropelewski and Halpert 1989).

It now appears that warm and cold events represent but two poles of the ENSO 'cycle,' with a pronounced tendency for the system to go from one extreme to the other in adjacent years (see e.g., van Loon and Shea 1985; Meehl 1987; Kiladis and Diaz 1989; Rasmusson et al. 1990). Thus, one extreme of the SO seems to set the conditions in the ocean/atmosphere system for the transition to the opposite extreme in the following year. This is reflected in the fact that warm and cold events are frequently seen to follow each other. While this 'biennial

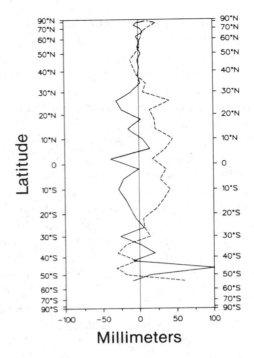

Fig. 2.2   Profile of zonally averaged annual precipitation anomalies (in millimeters) over global land areas during warm (solid line) and cold (dashed line) ENSO events. The years corresponding to each set are listed in Table 2.1. Reference period is 1951–1970. (Data are from Eischeid et al. 1991.)

tendency' is by no means common to all ENSO events, it clearly shows up in composite analyses of global ENSO signals as an opposition in the sign of seasonal temperature and precipitation anomalies between year − 1 (i.e, the year prior to the development of the event) and year 0 for both cold and warm ENSO events, even if the preceding year is not defined as an 'event' year by the usual definitions.

As we noted earlier, rainfall is strongly enhanced over many tropical land areas during the cold SO phase and diminished during the warm phase. This feature is illustrated in Figure 2.2, which shows the zonally averaged profile of annual precipitation departures over land areas from a 1951–1970 reference mean. Interestingly, in large areas of the continents of both hemispheres the cold and warm event patterns are reversed, particularly over the Southern Hemisphere. We emphasize that this figure does not represent the global ENSO signals since a large portion of the precipitation variability associated with this phenomenon is concentrated over eastern equatorial Pacific, where there are little or no land data.

In the following section, we review the main climatic signals observed during ENSO extremes. We will discuss the strongest and most consistent anomalies observed during the warm event phase of the oscillation. As we have shown above, the anomalies in these regions during the ENSO cold phase tend to be opposite in sign.

## ENSO Teleconnections

### Precipitation signals

As was mentioned above, large positive anomalies in precipitation are typically observed during warm events along coastal Peru and in the central and eastern equatorial Pacific, related to the increase of SST there. The bulk of the rainfall over the Pacific tends to be concentrated in two relatively narrow zones which are related to the convergence of the trade wind regimes in the tropical Pacific. At about 8°N to 15°N, the Intertropical Convergence Zone (ITCZ) comprises a west-to-east oriented band of convective rainfall extending across the entire Pacific from the Philippines to the Central American coast. South of the equator, the South Pacific Convergence Zone (SPCZ) is a somewhat broader region of heavy rains extending southeastward from New Guinea towards the Polynesian region. Unlike the ITCZ, the SPCZ is interrupted on its eastern margin by the relatively cold SST of the South Pacific dry zone (see Kiladis et al. 1989). Although both convergence zones exist throughout the year, they are notably stronger during their respective summer seasons.

During warm ENSO events, both the ITCZ and the SPCZ shift equatorward, and appear to become merged over their western portions near the dateline (Kiladis and van Loon 1988). This results in wetter than normal conditions along the equator, with anomalously dry conditions in the usual positions of these convergence zones, such as the Caroline Islands of the western Pacific and the South Pacific islands of Fiji and New Caledonia. Farther west, drier than normal conditions affect a broad region of the tropics bordering the eastern Indian Ocean, such as Australia (see Nicholls 1992, this volume), Indonesia, and the South Asian monsoon. Figures 2.3 and 2.4 illustrate the extent and locations of the areas of strongest response to the ENSO, as well as the season of the year when the strongest anomalies are typically experienced. It is noteworthy that some of the most notable failures of the Indian monsoon have occurred coincident with warm ENSO events (Rasmusson and Carpenter 1983; Kiladis and Sinha 1991), with flood years over India also tending to coincide with cold events (Parthasarathy and Pant 1985; Parthasarathy et al. 1988) (Fig. 2.3a). In contrast to India, Sri Lanka is one small region of Asia which actually experiences wet conditions during warm events (Fig. 2.3b); this occurs in phase with a tendency for heavy rainfall over the Indian Ocean during the normal September through November rainy season in this region (Kiladis and Diaz 1989).

Over Africa, the influence of the SO is spatially somewhat more restricted. During southern summer and fall, in the early phase of a developing warm event, there is a significant tendency for precipitation to be above normal over southeastern Africa, affecting a large area including Zimbabwe, Mozambique,

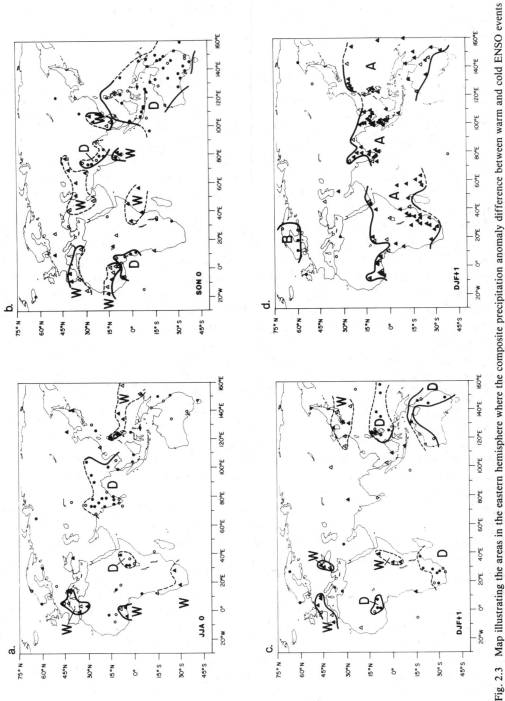

Fig. 2.3    Map illustrating the areas in the eastern hemisphere where the composite precipitation anomaly difference between warm and cold ENSO events (warm minus cold) is statistically significant at the 1% (solid symbols) or 5% (open symbols) level. The anomaly sign indicated in these panels is representative of warm events ("W" for wetter and "D" for drier than normal conditions); the opposite sign is representative of cold events. Panels: (a) JJA, (b) SON, and (c) DJF+1 seasons. Panel (d) illustrates the corresponding surface temperature anomalies of DJF+1 season. Signif... are symbols as above. "A" stands for

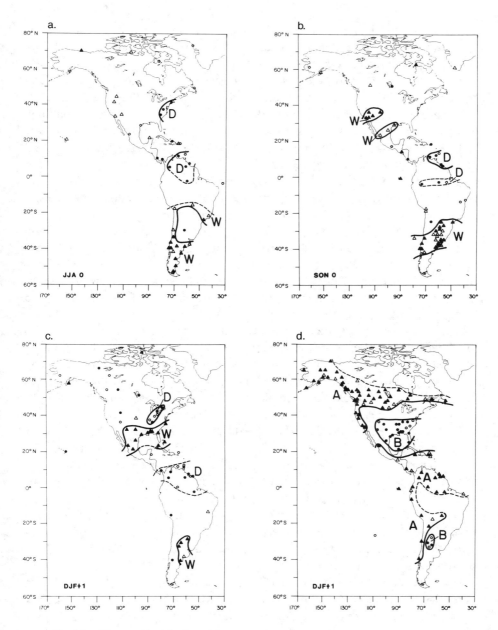

Fig. 2.4 Same as Figure 2.3, except for the American sector. (From Kiladis and Diaz 1989.)

and South Africa. Following the mature phase of the warm event, precipitation tends to be lower than normal over southern Africa (Fig. 2.3c). The tendency for drought during year +1 (late summer and fall) in this region leads to an early termination of the southern rainy season in association with warm ENSO events,

and an extension of the rainy season following the development of cold events.
This results in a marked biennial tendency for rainfall fluctuations in this area
(see Nicholson and Entekhabi 1986).

During the northern summer season in year 0, the area encompassing the
highlands of Ethiopia experiences drier than normal conditions (see Fig. 2.3a).
This feature has been used by Quinn (1992, this volume) to develop a chronology
of possible ENSO warm events by considering the low Nile River flow years
recorded in Egypt. The principal source of the Nile is located in the highland
region of Ethiopia, and is related to the occurrence of the summer rainy season
(Griffiths 1972; Leroux 1983). There is a weak indication that dry conditions are
favored over the Sahel region north of the equator in the summer rainy season
during warm events (Fig. 2.3a; see also, Janowiak 1988).

Over the American sector (Figs. 2.4a-c), the main tropical precipitation
signals, apart from the El Niño flooding mentioned previously, are manifested
in South America to the east of the Andes. Once the rains have set in along the
equatorial Pacific and over Ecuador and northern Peru, anomalously dry conditions then affect the Amazon region of Brazil northward to the Caribbean
through the latter half of the year. Farther south, however, from the coast of central Chile eastward into Argentina and Uruguay, a more active storm track leads
to higher precipitation from winter (JJA 0) to summer (DJF + 1). The opposite
occurs during cold events. The two strongest droughts of this century in Uruguay
occurred in connection with cold ENSO events. Southern Oscillation-related
precipitation and atmospheric circulation anomalies over South America are
discussed in greater detail by Aceituno (1988, 1989).

The strongest ENSO precipitation signal over North America affects the Gulf
coast region of the United States and parts of northern Mexico, Texas, and the
Caribbean islands, where wetter than normal conditions are found during winter
(Fig. 2.4c). This signal is one of the most consistent extratropical teleconnections
associated with the SO and it was evident to Walker (1923). In the following
spring, wet conditions tend to prevail over the southern High Plains area (see
Kiladis and Diaz 1989). These signals are related to a strengthening of the subtropical jet over the Gulf of Mexico during warm events, associated with an
active southern storm track to its north.

There are other areas over western Canada and to the south of the Great Lakes
where drier than normal conditions are suggested. However, these signals are not
sufficiently reliable from one event to the next to be useful for forecasting purposes. A rather more consistent signal is found in Hawaii, where a lack of winter
'Kona type' storms during warm events is responsible for diminished rainfall.

### Temperature signals

By far the most impressive temperature signal associated with ENSO is the strong
tendency for the global tropics as a whole to be warmer (colder) than normal during warm (cold) events. Newell and Weare (1976) first demonstrated that the sign

of the tropospheric temperature anomaly follows that of SST in the tropical eastern Pacific, with the latter leading by 3 to 6 mo (see also Angell 1981). This was verified for land surface temperature by Bradley et. al. (1987). Figures 2.3d and 2.4d depict the areas where statistically significant signals are found during DJF + 1 of ENSO events. The maps for the following season (MAM + 1) are similar (see Kiladis and Diaz 1989). The great extent of the tropical warmth during these seasons is striking, as is its extent into the subtropics over southern Africa, eastern Asia, and South America. This warmth also affects the islands of the tropical Pacific, as shown by Kiladis and van Loon (1988), and is evident in SST anomalies over tropical portions of the Indian and Atlantic oceans. Thus, the tropical warming indeed appears to be global in extent (see Spencer and Christy 1990).

The global temperature signal is further illustrated in Figure 2.5, which shows the zonally averaged surface temperature anomaly by five-degree latitude bands during warm and cold ENSO events (years given in Table 2.1). The 12-mo period from April of year 0 to the following March was used to composite the anomalies for each type event. This figure shows both the temperature anomalies over the

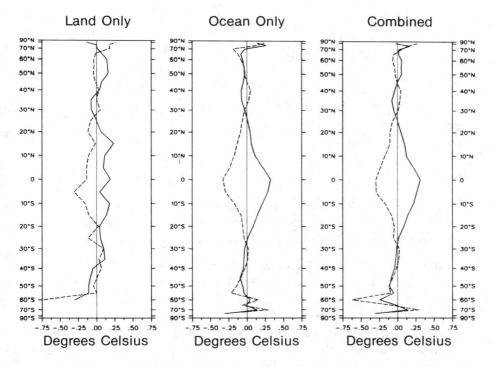

Fig. 2.5 Profile of zonally averaged annual temperature anomalies (°C) over both land and ocean areas during warm (solid line) and cold (dashed line) ENSO events. The years corresponding to each set are listed in Table 2.1. Reference period is 1951–1970 for the land data, 1951–1980 for the ocean data. (Land data are from Jones et al. 1986a,b, ocean data are from Bottomley et al. 1990.)

Table 2.1 *Years associated with cold and warm ENSO events based on instrumental records[a]*

| Warm event years | Cold event years |
|---|---|
| 1877 | 1886 |
| 1880 | 1889 |
| 1884 | 1892 |
| 1888 | 1898 |
| 1891 | 1903 |
| 1896 | 1906 |
| 1899 | 1908 |
| 1902 | 1916 |
| 1904 | 1920 |
| 1911 | 1924 |
| 1913 | 1928 |
| 1918 | 1931 |
| 1923 | 1938 |
| 1925 | 1942 |
| 1930 | 1949 |
| 1932 | 1954 |
| 1939 | 1964 |
| 1951 | 1970 |
| 1953 | 1973 |
| 1957 | 1975 |
| 1963 | 1988 |
| 1965 | Total = 21 cold (La Niña) events |
| 1969 | |
| 1972 | |
| 1976 | |
| 1982 | |
| 1986 | |
| Total = 27 warm (El Niño) events | |

[a]Based on classification methodology described in Kiladis and Diaz (1989).

land areas, as well as over the oceans and their combined values. It is easy to understand the mechanisms associated with the warming of the troposphere taking place in the regions of the tropical Pacific and west coast of South America directly affected by the rise in SST. Over the more remote regions, the process leading to warming must be less direct and are presumably related to changes in net radiation. For example, over Australia, the Indian Ocean, and eastern South America the drier than normal conditions described earlier are accompanied by a decrease in cloudiness, implying an increase in solar radiation and subsidence warming.

The largest extratropical temperature signals related to ENSO occur over North America during winter (DJF + 1); this is illustrated in Figure 2.4d. The

# Region A DJF + 1 Temperature Departures

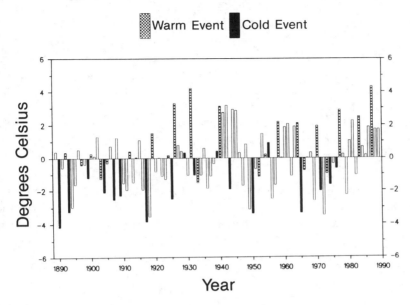

# Region B DJF + 1 Temperature Departures

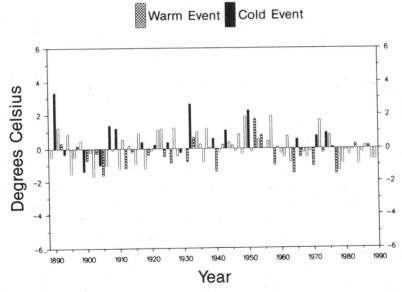

Fig. 2.6   Time series of winter temperature anomalies (in °C) averaged over (a) the region marked "A" (above normal), and (b) the region marked "B" (below normal) in Figure 2.4d. Warm and cold ENSO events are highlighted by hatching and shading, respectively.

reason for this sensitivity is that the atmospheric circulation over the North Pacific is strongly affected by ENSO through changes in the tropical rainfall patterns (see e.g., van Loon and Madden 1981; Horel and Wallace 1981; Hamilton 1988; Tribbia 1991). Meehl and Branstator (1992, this volume) discuss the results of some modeling experiments that bear on the question of paleo-climatic interpretations of ENSO-sensitive proxy records, as well as implications for future patterns if the climate changes in response to increased atmospheric greenhouse gas concentrations.

One of the most consistently observed changes associated with extremes in the SO relates to a strengthening of the Asian jet stream in the central North Pacific during warm events, which leads to stronger and more frequent westerly flow into the North American west coast and across central Canada (Emery and Hamilton 1985; Yarnal and Diaz 1986). Figure 2.6a gives the temperature anomalies for the North American region depicted in Figure 2.4d showing above normal winter season temperatures in northern North America. The increased influx of mild Pacific air results in much above normal seasonal temperatures from Alaska southward along the west coast and inland across the whole of southern Canada. Conversely, during cold events a weakening of the zonal circulation leads to an increase in the amount of cold arctic air moving southward across the continent. These changes are strong enough to cause a mean difference of nearly 4°C between warm and cold event winters at locations such as Winnipeg and Edmonton, Canada (see, Kiladis and Diaz 1989).

The more active storm track that develops during warm events over the northern Gulf of Mexico leads to below normal temperatures over a large portion of the southeastern United States (see Fig. 2.6b). A mean difference of about 2°C between warm and cold events occurs in this area. The magnitude of the climate anomalies in both of these North American regions is large enough to be useful for seasonal forecasting purposes. These anomalies typically extend into the spring season, although they are not as strong or consistent as those present during winter.

### Differences among ENSO events

The composite mean patterns of surface ocean variability associated with warm ENSO events since World War II were detailed in Rasmusson and Carpenter (1982). However, even in those regions where ENSO has its strongest signal (e.g., Australasia and Peru–Ecuador coast), there are occasional ambiguities regarding the occurrence of ENSO events not only in the pre-instrumental historical record (see Quinn et al. 1987; Allan 1991), but also since the advent of regular instrumental measurements (Deser and Wallace 1987; Wright et al. 1988). Deser and Wallace (1987) illustrated some of differences in the relationship between some basic measures of the El Niño phenomenon (e.g., SSTs along the Peru and Ecuador coast, compared to those along the eastern equatorial Pacific), while Wright et al. (1988) examined differences in the association among various indices of the El Niño and the SO. Trenberth and Shea (1987) and Elliott and Angell (1988) focused on the time variation of the most important ENSO indices, and examined some of the changes that have occurred in the spatial and temporal structure of the SO during the instrumental period. They showed that ENSO relationships are not stationary through time, and that the presence of statistical associations between particular ENSO indices at some specified period of time does not guarantee that those relationships will be applicable to that same degree during some other time interval. Below, we show some examples of the inherent variability in some fundamental measures of ENSO as well as in some of its large-scale teleconnections.

One of the characteristic features in the evolution of warm surface water along the central equatorial Pacific during warm events is the eastward displacement of the warm water pool that is normally found to the west of the international dateline (see Fu et al. 1986). Figure 2.7 illustrates this east-to-west movement of the warm water area over the past 40 yr. In a climatological sense, the location of the convective maximum in the equatorial Pacific tends to be found in the general vicinity of the warm water pool which is roughly delimited by the 28.5°C isotherm. Note that during cold event years, the warm water pool retreats west of 170°E. Figure 2.8 depicts the eastward movement of the 28.5°C isotherm along the equator during the last 10 warm ENSO events. It can be seen that there are substantial differences in the individual profiles, both with regard to the initial timing of the eastward warm water displacement, as well as with respect

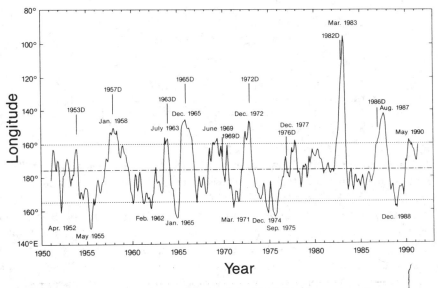

Fig. 2.7   Time variation of the longitude of the 28.5 °C isotherm along the equator (4°N–4°S) for the period January 1951–June 1991. Values are 3-month running means. Mean longitude for this period is 175.6 °W. Extreme months are identified, with minima in the curve associated with cold ENSO events, maxima with warm events.

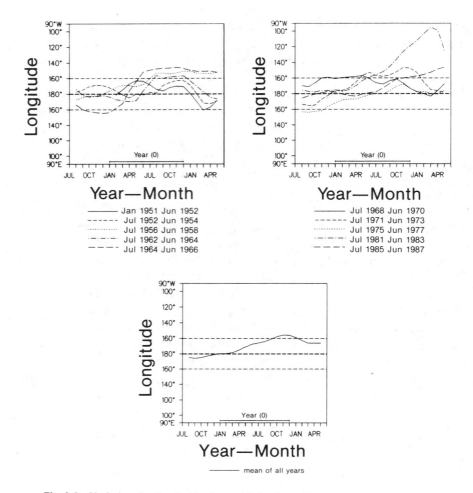

Fig. 2.8   Variation in the longitude position of warm SSTs along the equator (4 °N–4 °S) during the last 10 warm ENSO events. Curves show the monthly position of the 28.5 °C isotherm from the summer of year −1 to the spring of year +1 of each warm event.

to its final retreat (end of the warm event). Differences in the general pattern of SST anomalies throughout the tropical Pacific associated with different warm events have also been documented (Fu et al. 1986).

It is the *tropical* atmospheric response to the SST field during ENSO that determines the extratropical response of the general circulation (see Tribbia 1991, for a review). These SST anomaly differences among warm (and cold) events result in differences in the thermodynamic forcing of the tropical atmosphere, which in turn can lead to variability in the extratropical response from one event to the next.

Ropelewski and Halpert (1986, 1987, 1989) and Kiladis and Diaz (1989) have identified those areas of the world where the tropical and extratropical climatic

Table 2.2 *Percentage of warm and cold ENSO events winters in Regions A and B of Fig. 4d with the sign and magnitude of the seasonal temperature anomaly as indicated*

| | Region A | | Region B | |
|---|---|---|---|---|
| **Warm Events** | | | | |
| No. of events with pos. anom. | 16/23(70%) | No. of events with neg. anom. | 16/23(70%) | |
| Pos. anom. $>0.5\sigma$ | 10/23(43%) | Neg. anom. $< -0.5\sigma$ | 12/23(52%) | |
| Mean standardized deviation | $0.64\sigma$ | Mean standardized deviation | $-0.49\sigma$ | |
| **Cold Events** | | | | |
| No. of events with neg. anom. | 15/19(79%) | No. of events with pos. anom. | 14/19(74%) | |
| Neg. anom. $< -0.5\sigma$ | 13/19(68%) | Pos. anom. $>0.5\sigma$ | 10/19(53%) | |
| Mean standardized deviation | $-0.84\sigma$ | Mean standardized deviation | $0.65\sigma$ | |

response to ENSO is statistically significant. These authors also provided a measure of the variability in mean temperature and precipitation response among the different events for these areas. Figure 2.6 illustrates the nature of this variability for two areas in North America (see Fig. 2.4d) where there is a significant seasonal response to the SO (see also previous discussion). The composite shows a distinct difference in the winter temperature response for these two regions during warm and cold ENSO events. Table 2.2 further quantifies this association showing, for these two regions in North America, the percentage of warm and cold events exhibiting the expected sign of the seasonal temperature anomaly for the warm and cold event years, and secondly, the percentage of years in which the anomaly exceeded the expected sign, based on the composite mean, by half a standard deviation from the long-term mean. The strength of these associations is one reason they are used in seasonal forecasting. Nevertheless, as shown in Figure 2.6 and Table 2.2, there are occasions when even the sign of the anomaly is opposite to its expected value, in some cases by a half a standard deviation or more.

The foregoing discussion was meant to provide some indication of the internal variability of the tropical ENSO cycle and its remote manifestations. It remains true, however, that over very large regions, for example the global tropics and even hemispheric and global-scale averages, the influence of the ENSO on the global climate system comprises the largest documented source of interannual variability in the troposphere. It seems plausible that low-frequency changes in the ENSO phenomenon would be associated with decadal and longer term climatic variability as well. This potential linkage between ENSO and long-term climate variations appears to be one area where ENSO-sensitive paleoclimate indicators could help to interpret decadal to century-scale variability.

## Concluding remarks

The principal elements of the El Niño/Southern Oscillation phenomenon involve a large-scale redistribution of mass, heat, and momentum within the ocean-atmosphere system, which results in large deviations from climatology of monthly, seasonal, and annual values of temperature and precipitation in many parts of the world. We have examined the seasonal climatic response to SO extremes experienced over those tropical and extratropical regions where the signal is strongest.

From the point of view of ENSO reconstructions, it is important to be able to select those areas with suitable proxy records where the characteristic regional climatic response to the ENSO cycle is strongest and most consistent. In most instances, the ENSO reconstructions described in this book come from areas where the climatic sensitivity to ENSO fluctuations is relatively high (the tropics, the southern United States and northern Mexico). In particular, the series developed by Quinn from records of Nile River streamflow is a very useful extension of his El Niño chronology (see Quinn 1992, this volume). Nevertheless, one should bear in mind that these two widely separated areas (the Ethiopian Highlands and the northwest coast of South America), although both affected by ENSO, are also affected by many other factors, so that a match of the years listed by Quinn (Quinn et al. 1987; Quinn 1992, this volume) result in a correlation which accounts for only about 20% of their mutual variance (see also Diaz and Pulwarty, 1992, this volume). The contemporary relationship between Ethiopian Highlands summer rainfall and South American El Niño indices is rather modest.

It is interesting that, despite the many 'filters' that are applied in the cascade of physical interactions from events in the central equatorial Pacific to climatic fluctuations recorded in the paleoclimatic series in other regions, one is still able to extract useful information from these proxy records.

The work on tropical corals described in the contribution by Cole et al. 1992, this volume) will help delineate ENSO events along the core ENSO areas to the west of the Ecuador-Peru coast, thus providing an association between ENSO-related activity near the South American coast and the Horn of Africa. The work on tropical ice cores by Thompson and his colleagues (see, Thompson et al. 1992, this volume) will continue to add important information on regional aspects of ENSO teleconnections, and, together with the tree-ring work being done by a number of investigators in this country and abroad, should provide important clues regarding the low frequency behavior of ENSO.

*Acknowledgments* We thank Jon Eischeid of CIRES, University of Colorado at Boulder for his assistance in data analysis and graphics display. We also thank H. van Loon and C. Deser for suggesting ways to improve the manuscript. Work on this study was partially funded through a NOAA Climate and Global Change grant.

# References

ACEITUNO, P., 1988: On the functioning of the Southern Oscillation in the South American sector. Part I: Surface climate. *Monthly Weather Review*, 116: 505–524.

ACEITUNO, P., 1989: On the functioning of the Southern Oscillation in the South American sector. Part II: Upper-air circulation. *Journal of Climate*, 2: 341–355.

ALLAN, R. J., 1991: Australasia. *In* Glantz, M. H., Katz, R. W. and Nicholls, N. (eds.), *Teleconnections Linking Wordwide Climate Anomalies*. Cambridge: Cambridge University Press, 73–120.

ANGELL, J. K., 1981: Comparison of variations in atmospheric quantities with sea surface temperature variations in the equatorial eastern Pacific. *Monthly Weather Review*, 109: 230–243.

BERLAGE, H. P., 1957: Fluctuations in the general atmospheric circulation of more than one year, their nature and prognostic value. *Koninklijk Nederlands Meteorologisch Institut Mededlingen en Verhandelingen*, 69. 152 pp.

BJERKNES, J., 1966: A possible response of the atmospheric Hadley circulation to equatorial anomalies of ocean temperature. *Tellus*, 18: 820–829.

BJERKNES, J., 1969: Atmospheric teleconnections from the equatorial Pacific. *Monthly Weather Review*, 97: 163–172.

BOTTOMLEY, M., FOLLAND, C. K., HSIUNG, J., NEWELL, R. E., and PARKER, D. E., 1990: *Global Ocean Surface Temperature Atlas*. UK Meteorological Office and Massachusetts Institute of Technology.

BRADLEY, R. S., DIAZ, H. F., KILADIS, G. N., and EISCHEID, J. K., 1987: ENSO signal in continental temperature and precipitation records. *Nature*, 327: 497–501.

COLE, J. E., SHEN, G. T., FAIRBANKS, R. G., and MOORE, M., 1992: Coral monitors of El Niño/Southern Oscillation dynamics across the equatorial Pacific. *In* Diaz, H. F. and Markgraf, V. (eds.), *El Niño: Historical and Paleoclimatic Aspects of the Southern Oscillation*. Cambridge: Cambridge University Press, 349–375.

DESER, C. and WALLACE, J. M., 1987: El Niño events and their relation to the Southern Oscillation: 1925–1986. *Journal of Geophysical Research*, 92: 14,189–14,196.

DIAZ, H. F. and PULWARTY, R. S., 1992: A comparison of Southern Oscillation and El Niño signals in the tropics. *In* Diaz, H. F. and Markgraf, V. (eds.), *El Niño: Historical and Paleoclimatic Aspects of the Southern Oscillation*. Cambridge: Cambridge University Press, 175–192.

EISCHEID, J. K., DIAZ, H. F., BRADLEY R. S., and JONES, P. D., 1991: A Comprehensive precipitation data set for global land areas. U.S. Department of Energy Technical Report TR051. 82 pp.

ELLIOTT, W. P. and ANGELL, J. K., 1988: Evidence for changes in Southern Oscillation relationships during the last 100 years. *Journal of Climate*, 1: 729–737.

EMERY, W. J. and HAMILTON, K., 1985: Atmospheric forcing of interannual variability in the northeast Pacific: Connections with El Niño. *Journal of Geophysical Research*, 90: 857–868.

ENFIELD, D. B., 1989: El Niño, past and present. *Reviews of Geophysics*, 27: 159–187.

ENFIELD, D. B., 1992: Historical and prehistorical overview of El Niño/Southern Oscillation. *In* Diaz, H. F. and Markgraf, V. (eds.), *El Niño: Historical and Paleoclimatic Aspects of the Southern Oscillation*. Cambridge: Cambridge University Press, 95–117.

F U, C., D I A Z, H. F., and F L E T C H E R, J. O, 1986: Characteristics of the response of sea surface temperature in the central Pacific associated with warm episodes of the Southern Oscillation. *Monthly Weather Review*, 114: 1716–1738.

G L A N T Z, M. H., K A T Z, R. W., and N I C H O L L S, N. (eds.), 1991: *Teleconnections Linking Worldwide Climate Anomalies*. Cambridge: Cambridge University Press. 535 pp.

G R I F F I T H S, J. F., (ed.), 1972: *Climates of Africa*. World Survey of Climatology, Vol. 10. Amsterdam: Elsevier. 604 pp.

H A M I L T O N, K., 1988: A detailed examination of the extratropical response to tropical El Niño/Southern Oscillation events. *Journal of Climatology*, 8: 67–86.

H I L D E B R A N D S S O N, H. H., 1897: Quelques recherches sur les centre d'action de l'atmosphère. *Kungliga Svenska Vetenskapsakademiens Handlingar*, 29. 33 pp.

H O R E L, J. D. and W A L L A C E, J. M., 1981: Planetary-scale atmospheric phenomena associated with the Southern Oscillation. *Monthly Weather Review*, 109: 813–829.

I C H I Y E, T. and P E T E R S E N, J. R., 1963: Anomalous rainfall of the 1957–58 winter in the equatorial Central Pacific arid area. *Journal of Meteorological Society of Japan*, 41: 172–182.

J A N O W I A K, J. E., 1988: An investigation of interannual rainfall variability in Africa. *Journal of Climate*, 1: 240–255.

J O N E S, P. D., R A P E R, S. C. B., B R A D L E Y, R. S., D I A Z, H. F., K E L L Y, P. M., and W I G L E Y, T. M. L. 1986a: Northern Hemisphere surface air temperature variations: 1851–1984. *Journal of Climate and Applied Meteorology*, 25: 161–179.

J O N E S, P. D., R A P E R, S. C. B., and W I G L E Y, T. M. L., 1986b: Southern Hemisphere surface air temperature variations: 1851–1984. *Journal of Climate and Applied Meteorology*, 25: 1213–1230.

J U L I A N, P. R. and C H E R V I N, R. M., 1978: A study of the Southern Oscillation and Walker circulation phenomena. *Monthly Weather Review*, 106: 1433–1451.

K I L A D I S, G. N. and D I A Z, H. F., 1989: Global climatic anomalies associated with extremes of the Southern Oscillation. *Journal of Climate*, 2: 1069–1090.

K I L A D I S, G. N. and S I N H A, S. K., 1991: ENSO, monsoon and drought in India. *In* Glantz, M. H., Katz, R. W. and Nicholls, N. (eds.), *Teleconnections Linking Worldwide Climate Anomalies*. Cambridge: Cambridge University Press, 431–458.

K I L A D I S, G. N. and van L O O N, H., 1988: The Southern Oscillation. Part VII: Meteorological anomalies over the Indian and Pacific sectors associated with the extremes of the oscillation. *Monthly Weather Review*, 116: 120–136.

K I L A D I S, G. N., von S T O R C H, H., and van L O O N, H., 1989: Origin of the South Pacific Convergence Zone. *Journal of Climate*, 2: 1185–1195.

L A U, K.-M. and S H E U, P. J., 1988: Annual cycle, quasi-biennial oscillation, and Southern Oscillation in global precipitation. *Journal of Geophysical Research*, 93: 10,975–10,988.

L E R O U X, M., 1983: *The Climate of Tropical Africa Atlas*. Paris: Champion. 24 pp., 250 plates.

L O C K Y E R, S. N., 1906: Barometric variations of long duration over large areas. *Proceedings of the Royal Society of London*, 78: 43–60.

M E E H L, G. A., 1987: The annual cycle and interannual variability in the tropical Pacific and Indian Ocean regions. *Monthly Weather Review*, 115: 27–50.

M E E H L, G. A. and B R A N S T A T O R, G. W., 1992: Coupled climate model simulation of El Niño/Southern Oscillation: Implications for paleoclimate. *In* Diaz, H. F. and Markgraf, V. (eds.), *El Niño: Historical and Paleoclimatic Aspects of the Southern Oscillation*. Cambridge: Cambridge University Press, 69–91.

NEWELL, R. E. and WEARE, B. C., 1976: Factors governing tropospheric mean temperature. *Nature*, 194: 1413–1414.

NICHOLLS, N., 1988: El Niño-Southern Oscillation and rainfall variability. *Journal of Climate*, 1: 418–421.

NICHOLLS, N., 1992: Historical El Niño/Southern Oscillation variability in the Australasian region. *In* Diaz, H. F. and Markgraf, V. (eds.), *El Niño: Historical and Paleoclimatic Aspects of the Southern Oscillation*. Cambridge: Cambridge University Press, 151–173.

NICHOLSON, S. E. and ENTEKHABI, D., 1986: The quasi-periodic behavior of rainfall variability in Africa and its relationship to the Southern Oscillation. *Archiv für Meteorologie, Geophysik und Bioklimatologie*, A34: 311–348.

PARTHASARATHY, B. and PANT, G. B., 1985: Seasonal relationships between Indian summer monsoon rainfall and the Southern Oscillation. *Journal of Climatology*, 5: 369–378.

PARTHASARATHY, B., DIAZ H. F., and EISCHEID, J. K., 1988: Prediction of all-India summer monsoon rainfall with regional and large-scale parameters. *Journal of Geophysical Research*, 93: 5341–5350.

PHILANDER, S. G., 1990: *El Niño, La Niña, and the Southern Oscillation*. San Diego: Academic Press. 293 pp.

QUINN, W. H., 1992: A study of Southern Oscillation-related climatic activity for A.D. 622–1990 incorporating Nile River flood data. *In* Diaz, H. F. and Markgraf, V. (eds.), *El Niño: Historical and Paleoclimatic Aspects of the Southern Oscillation*. Cambridge: Cambridge University Press, 119–149.

QUINN, W. H., NEAL, V. T., and ANTUNEZ de MAYOLO, S. E., 1987: El Niño over the past four and a half centuries. *Journal of Geophysical Research*, 92: 14,449–14,461.

RASMUSSON, E. M. and CARPENTER, T. H., 1982: Variations in tropical sea surface temperature and surface wind fields associated with the Southern Oscillation/El Niño. *Monthly Weather Review*, 110: 354–384.

RASMUSSON, E. M. and CARPENTER, T. H., 1983: The relationship between eastern equatorial Pacific sea surface temperatures and rainfall over India and Sri Lanka. *Monthly Weather Review*, 111: 517–528.

RASMUSSON, E. M. and WALLACE, J. M., 1983: Meteorological aspects of the El Niño/Southern Oscillation. *Science*, 222: 1195–1202.

RASMUSSON, E. M., WONG, X., and ROPELEWSKI, C. F., 1990: The biennial component of ENSO variability. *Journal of Marine Systems*, 1: 71–96.

ROPELEWSKI, C. F. and HALPERT, M. S., 1986: North American precipitation and temperature patterns associated with the El Niño/Southern Oscillation (ENSO). *Monthly Weather Review*, 114: 2352–2362.

ROPELEWSKI, C. F. and HALPERT, M. S., 1987: Global and regional scale precipitation patterns associated with El Niño/Southern Oscillation. *Monthly Weather Review*, 115: 1606–1626.

ROPELEWSKI, C. F. and HALPERT, M. S., 1989: Precipitation patterns associated with the high-index phase of the Southern Oscillation. *Journal of Climate*, 2: 268–284.

ROPELEWSKI, C. F. and JONES, P. D., 1987: An extension of the Tahiti-Darwin Southern Oscillation Index. *Monthly Weather Review*, 115: 2161–2165.

SPENCER, R. W. and CHRISTY, J. R., 1990: Global atmospheric temperature monitoring with satellite microwave measurements: Method and results 1979–84. *Journal of Climate*, 3: 1111–1128.

THOMPSON, L. G., MOSLEY-THOMPSON, E., and THOMPSON, P. A., 1992: Reconstructing interannual climate variability from tropical and subtropical ice-core records. *In* Diaz, H. F. and Markgraf, V. (eds.), *El Niño: Historical and Paleo-climatic Aspects of the Southern Oscillation.* Cambridge: Cambridge University Press, 295–321.

TRENBERTH, K. E., 1976: Spatial and temporal variations of the Southern Oscillation. *Quarterly Journal of the Royal Meteorological Society*, 102: 639–653.

TRENBERTH, K. E. and SHEA, D. J., 1987: On the evolution of the Southern Oscillation. *Monthly Weather Review*, 115: 3078–3096.

TRIBBIA, J. J., 1991: The rudimentary theory of atmospheric teleconnections associated (with ENSO. *In* Glantz, M. H., Katz, R. W. and Nicholls, N. (eds.), *Teleconnections Linking Worldwide Climate Anomalies.* Cambridge: Cambridge University Press, 285–308.

TROUP, A. J., 1965: The Southern Oscillation. *Quarterly Journal of the Royal Meteorological Society*, 91: 490–506.

van LOON, H. and MADDEN, R. A., 1981: The Southern Oscillation. Part I: Global associations with pressure and temperature in northern winter. *Monthly Weather Review*, 109: 1150–1162.

van LOON, H. and SHEA, D. J., 1985: The Southern Oscillation. Part IV: The precursors south of 15°S to the extremes of the oscillation. *Monthly Weather Review*, 113: 2063–2074.

van LOON, H. and SHEA, D. J., 1987: The Southern Oscillation. Part VI: Anomalies of sea (level pressure on the Southern Hemisphere and of Pacific sea surface temperature during the development of a warm event. *Monthly Weather Review*, 115: 370–379.

WALKER, G. T., 1923: Correlation in seasonal variations of weather. Part VIII: A preliminary study of world weather. *Memoirs of the Indian Meteorological Department*, 24: 75–131.

WALKER, G. T. and BLISS, E. W., 1932: World Weather V. *Memoirs of the Royal Meteorological Society*, 4: 53–84.

WRIGHT, P. B., WALLACE, J. M., MITCHELL, T. P., and DESER, C., 1988: Correlation structure (of the El Niño/Southern Oscillation phenomenon. *Journal of Climate*, 1: 609–625.

WYRTKI, K., 1975: El Niño – The dynamic response of the equatorial Pacific Ocean to atmospheric forcing. *Journal of Physical Oceanography*, 5: 572–584.

YARNAL, B. and DIAZ, H. F., 1986: Relationships between extremes of the Southern Oscillation and the winter climate of the Anglo-American Pacific coast. *Journal of Climatology*, 6: 197–219.

# 3

# El Niño/Southern Oscillation and streamflow in the western United States

DANIEL R. CAYAN

*U.S. Geological Survey, Climate Research Division Scripps Institution of Oceanography, La Jolla, California 92093-0224, U.S.A.*

ROBERT H. WEBB

*U.S. Geological Survey, Desert Laboratory 1675 W. Anklam Rd., Tucson, Arizona 85745, U.S.A.*

## Abstract

The response of western United States precipitation, snow water content (SWC), and streamflow to El Niño/Southern Oscillation (ENSO) conditions was examined. Analyses were conducted over time scales from a season to a day, using several decades of instrumental records. The results reinforce and clarify the association with precipitation found in previous studies. Seasonal SWC and streamflow tend to be enhanced in the southwestern United States and diminished in the northwestern United States during the mature Northern Hemisphere winter El Niño phase of ENSO. Opposite behavior occurs during the La Niña phase of ENSO.

An analysis of the behavior of daily precipitation during the El Niño, La Niña, and 'other year' phases of ENSO reveals further detail. Here we examined differences in the distribution of daily precipitation, stratified seasonally, over the Yellowstone River, Montana, representing the interior northwestern United States, and the Salt River, Arizona, representing the southwestern United States. Both the frequency and the amount of precipitation exhibited significant, opposite-tending changes during the warm and cool phases of ENSO. At Yellowstone River, there is a noticeable reduction in the amount and the frequency of occurrence of early winter daily precipitation during El Niño and a corresponding increase during La Niña. At Salt River, there is an increase in the amount and particularly the frequency of early winter and late winter/spring precipitation during El Niño and a corresponding decrease during La Niña.

Typical precipitation-producing atmospheric circulations were examined via composites of Northern Hemisphere 700-mb height associated with days having appreciable precipitation during El Niño vs. La Niña. A consistent picture emerged for Yellowstone and Salt River. While the synoptic-scale anomaly features of El Niño and La Niña daily precipitation were similar during early winter, there were significant differences during late winter/spring. The El Niño pattern shows evidence of a North Pacific basinwide activated storm track imbedded in strong westerlies. The La Niña pattern appears to be more confined to the eastern North Pacific and regions downstream, with a high to the west in the Gulf of Alaska vicinity and a low (often a cutoff low) along the West Coast.

Flood frequency in the southwestern United States is affected by ENSO, but this effect varies spatially, and the climatic causes are unclear. The largest floods in Arizona have occurred during El Niño years, and certain rivers, such as the Santa Cruz River, have an increased flood frequency during or just after El Niño conditions. However, rivers strongly affected by winter storms and snowmelt, such as the Salt River, may be affected more by storms occurring during years other than ENSO. One reason for the complexity of the response of streams to ENSO is that there are at least three different types of storms which generate floods. These include frontal systems, which occur predominantly in winter, and monsoonal storms and dissipating tropical cyclones, which usually occur in summer or early fall. The types of storms that generate flooding are affected by El Niño conditions, but the effects are not consistent among El Niño years. Thus, the occurrence of El Niño conditions alone is not sufficient to explain increased flood frequency; instead, large floods appear to occur during a small subset of years associated with El Niño conditions.

## Introduction

Since the El Niño/Southern Oscillation (ENSO) is probably the foremost climate variability feature in the short time scale range of a season to a few years, we wish to investigate associated variability in surface hydrological conditions. In this case we focus on behavior of the water supply and flooding over the western United States.

Several previous studies have examined the effect of ENSO on seasonal climate in the extratropics. Specifically, these have shown that during the warm 'mature' phase of El Niño, cool-season precipitation tends to be less in the Pacific Northwest and greater in the southwestern United States (Douglas and Englehart 1981; Ropelewski and Halpert 1986; Andrade and Sellers 1988; Cayan and Peterson 1989; Redmond and Koch 1991.) Previous studies (e.g., Kiladis and Diaz 1989) have shown that opposite effects tend to occur in most regions, including western North America, during the cold phase of the tropical Pacific, which is often called La Niña. While they are weaker than ENSO-related effects elsewhere on the globe (Ropelewski and Halpert 1986), these western North America connections are statistically significant (Kiladis and Diaz 1989), and are as strong as other prominent large-scale climate associations in the Northern Hemisphere (Cayan and Peterson 1989). Furthermore, these connections may have a practical value in that the low frequency nature of tropical ENSO variations introduces a time lead, permitting some predictive capability at a season or more in advance. Finally, despite the problems in statistical reliability, the great impact of ENSO connections during certain events, such as 1983, cannot be ignored.

The reason for the ENSO connection lies in the atmospheric circulation, where upper level wind shifts change the track and intensity of storms. Theoretical studies suggest that these are responses to anomalous heating in the tropics, induced by the altered sea surface temperatures and lower level wind convergence. During the winter mature phase of El Niño, a tendency for a stronger, more persistent high pressure ridge over western Canada is symptomatic of fewer storms in the Pacific Northwest, and a diversion of these storms into the Alaskan coast. A tendency for low pressure activity in the south also occurs during these episodes, leading to heavier than normal precipitation in the southwestern United States. Unfortunately, the connection is complicated in that there is not a unique pattern associated with El Niño in the North Pacific/North American sector, and work by Fu et al. (1986) and Livezey and Mo (1987) suggest that two or three patterns may be possible.

The intent of this paper is to examine closer the relationship of precipitation, snowpack, streamflow, and flood events in the western United States to ENSO. We examine the El Niño influence on the surface hydrology in terms of both seasonal and shorter term daily fluctuations, including extreme events represented by floods in the southwestern United States.

## Data

Data employed in this study include instrumental records of atmospheric, surface meteorological, streamflow, and snow water content variables. Time scales employed include daily through seasonal averaged data.

The atmospheric circulation is represented by twice-daily analyses of the 700-mb height over the Northern Hemisphere, gridded onto a 5°C latitude-longitude grid, available since 1947. Anomalies were computed from long-term means of the daily average (0Z and 12Z analyses averaged together) over the 30-yr period 1950–1979. The height of the 700-mb pressure surface, which is typically 3 km aloft, provides a good representation of the mid-tropospheric circulation. A longer record of the atmospheric circulation is provided by Northern Hemisphere sea-level pressure (SLP) analyses which are available in monthly mean form beginning in 1899 (Trenberth and Paolino 1980). When used here, SLP anomalies were computed from the long-term mean over the entire record (1899–1988).

The state of the tropical Pacific is represented by the Southern Oscillation Index (SOI), an index of the surface pressure gradient across the tropical Pacific basin. When SOI is negative, the tropical Pacific is usually in its warm (El Niño) state, and when it is positive, the tropical Pacific is usually in its cool ('La Niña') state. The form of the SOI used here is the standardized anomaly difference between Tahiti and Darwin as employed by the NOAA Climate Analysis Center (Ropelewski and Jones 1987). Seasonal average values are employed here. Also, to group events into different year types, years of El Niño and La Niña, as identified by Quinn et al. (1987) and Rasmusson and Carpenter (1982) are listed in Table 3.1.

To explore further relationships to atmospheric circulation over the central North Pacific, abbreviated CNP, a sea-level pressure (SLP) index, was constructed (Cayan and Peterson 1989). This is the average of SLP anomalies over a broad region of the central North Pacific: 35–55°N and 170°E–150°W. To characterize the magnitude of CNP during strong and weak central North Pacific low cases, the average CNP for the 25 strong CNP cases is less than $-7$ mb, while it exceeds $+7$ mb for the 25 weak cases. In both composites, the strongest anomaly center is found at about (50°N, 165°W). This index is similar to the well-known Pacific-North American (PNA) index, the most important winter

Table 3.1 *El Niño and La Niña years*[a]

| El Niño | La Niña |
|---------|---------|
| 1901 | 1904 |
| 1903 | 1907 |
| 1906 | 1909 |
| 1912 | 1917 |
| 1915 | 1921 |
| 1919 | 1925 |
| 1924 | 1929 |
| 1926 | 1932 |
| 1931 | 1939 |
| 1933 | 1943 |
| 1940 | 1950 |
| 1941 | 1955 |
| 1942 | 1965 |
| 1947 | 1971 |
| 1952 | 1974 |
| 1954 | 1976 |
| 1958 | 1989 |
| 1964 | |
| 1966 | |
| 1970 | |
| 1973 | |
| 1977 | |
| 1978 | |
| 1983 | |

[a]Years listed are water years (October of previous calendar year through September of listed calender year).

circulation mode in the central North Pacific (Davis 1978; Barnston and Livezey 1987; Namias et al. 1988). The SLP data for deriving CNP is available since 1899 while the upper level geopotential height data to derive the PNA index does not begin until 1947. This simple index is well related to PNA in the cool months; correlations between CNP and PNA are $-0.69$, $-0.90$, and $-0.75$ in fall, winter, and spring (Cayan and Peterson 1989).

The streams employed were selected to provide reliable records over the longest possible period. Monthly average streamflow from 61 stations over western North America and Hawaii were obtained from the U.S. Geological Survey and Canadian archives (U.S. Department of Interior 1975). The 61 streams include 52 from the western coterminous United States, 3 from British Columbia, 4 from Alaska, and 2 from Hawaii. Most of the streamflow records begin between 1900 and the 1930s; the longest series, Spokane River at Spokane, Washington, begins in 1891. Most of the streams employed have records updated through 1986, and several have records through 1988. These latter include the Yellowstone River, Corwin Springs, Montana, and the Salt River, Roosevelt, Arizona, which are used for more detailed analyses with ENSO.

For several decades, the USDA Soil Conservation Service (SCS) has archived snow observations at several hundred mountain snow courses in the western United States. Snow water content (SWC) is affected by both accumulation and ablation of snow, but at many locations in the West, there is maximum SWC in spring near the beginning of April. This study employs a set of approximately 400 SWC records for the period 1950 to 1989. Snow courses are usually located at relatively high elevation sites, but because of the variety of topography in the West, a broad range of elevations are represented, from about 500 m to over 3000 m. Many snow courses have observations taken at the beginning and sometimes the middle of each of the months with substantial snow cover from January through May. For purposes of this study, we use only those taken on the beginning of the month.

To examine the ENSO connection to individual watersheds, daily precipitation at two individual river basins was utilized. These basins, representing the northwestern United States and the southwestern United States were the Yellowstone River upstream of the Corwin Springs, Montana gauge, and the Salt River, upstream of the Roosevelt, Arizona gauge. Basin average precipitation for Salt River and Yellowstone River was constructed from a weighted average of precipitation from six stations within or nearby the watershed. The stations (Table 3.2) were selected because they had several decades of record and showed no apparent artificially caused discontinuities or trends. In deriving the basin average precipitation, a scheme was used to weigh contributions from stations which contained sometimes quite different precipitation statistics. The idea of this scheme was to average the standardized precipitation anomaly, multiply this standardized anomaly times the 'overall' basin standard deviation, and add this anomaly to the 'overall' basin mean for that day. On days when all stations recorded zero precipitation, the basin average precipitation was zero. On days

Table 3.2 *Precipitation stations for individual watersheds*

| Station | Elevation (m) | Period |
|---|---|---|
| *Yellowstone River* | | |
| Hebgen Dam, Montana | 1973 | 1948–1989 |
| Island Park, Idaho | 1912 | 1937–1989 |
| Lake Yellowstone, Wyoming | 2369 | 1948–1989 |
| Tower Falls, Wyoming | 1912 | 1948–1989 |
| West Yellowstone, Montana | 2030 | 1924–1989 |
| Yellowstone Park, Wyoming | 1890 | 1948–1989 |
| *Salt River* | | |
| Buckeye, Arizona | 265 | 1893–1989 |
| Clifton, Arizona | 1055 | 1893–1989 |
| McNary, Arizona | 2231 | 1933–1989 |
| Miami, Arizona | 1085 | 1914–1989 |
| Roosevelt 1 WNW, Arizona | 674 | 1905–1989 |
| Springerville, Arizona | 2152 | 1911–1989 |

when at least one station recorded precipitation, the basin average precipitation was constructed by averaging the standardized precipitation anomaly times an 'overall' basin standard deviation, plus the average daily precipitation for the basin. The overall basin standard deviation, $\sigma_T$, was computed using

$$\sigma_T \left[ \frac{1}{n} \sum_{i-1} \sigma_i^2 \right]^{1/2}$$

where $\sigma_i^2$ is the daily variance at station $i$. The basin average precipitation was equal to the mean of the daily precipitation at the $n$ stations; $n$ ranged from 3 to 6 stations. To reduce sampling variations of the daily means and standard deviations from finite record estimations, they were smoothed using a 15-d Gaussian filter (the smoothed standard deviation was computed from the smoothed daily variance).

To examine the broad-scale precipitation during individual events, gridded daily precipitation over the coterminous United States (Roads and Maisel 1991) was employed. This set was constructed from 831 stations from the United States cooperative station network, and provided data from 1950 through 1988. The stations, a subset of the National Climatic Data Center United States Historical Climatology Network (Quinlan et al. 1987), were selected because they were relatively insulated from urban effects and had a nearly complete record over the 1950–1988 period. The station precipitation was transformed to a 2.5° grid by averaging all nonmissing daily station values contained within the 2.5° latitude by 2.5° longitude region of each grid point. Note that for the present purpose, these data are used to discover the *qualitative* nature of the spatial distribution of precipitation associated with certain types of events.

## ENSO vs. seasonal hydrological variations

Both SWC and streamflow are important components of the surface hydrology and the water supply. They are cumulative measures of the surface climate, responding to behavior over the wet season in the western United States. Streamflow has the added benefit of providing a spatial average over a watershed. In many respects SWC and streamflow are superior to the precipitation in measuring the water supply in the West, since they represent the amount which is available. Streamflow also records flood events, which will be discussed later for southern Arizona watersheds. Cross correlations between seasonal precipitation, SWC, and streamflow at Yellowstone River and Salt River are shown in Table 3.3. At Yellowstone River, these components are all correlated at about 0.7 or greater. At the more arid Salt River region, the correlations of November–March precipitation vs. 1 February SWC and water year (October–September) streamflow are 0.7 to 0.8, but the correlation with 1 April SWC is only about 0.4. The time series of SWC and streamflow at the two locations is shown in Figure 3.1, along with an indication of years of El Niño and La Niña.

A broad-scale effect of ENSO on surface hydrological variations is evident from correlations of seasonal anomalies of SOI with observations of 1 April SWC and mean December–August streamflow. The 1 April SWC and (December–August) streamflow measures are cumulative seasonal indices representative of several months of climatic activity. To examine behavior concurrent with, and subsequent to the tropical conditions, correlations are calculated using both winter and preceding summer SOI. For comparison with mid-latitude circulation influences, correlations with seasonal CNP (winter and the preceding fall) are also presented. Summer CNP is not well related to the following winter atmospheric circulation, so this relationship is not shown.

The correlations between SOI and SWC, and SOI and streamflow are similar in pattern and magnitude. The correlation of SOI during summer and winter vs.

Table 3.3. *Cross correlations, Nov–Mar precipitation 1 Feb SWC, 1 Apr SWC, and water year streamflow*

|  | 1 Feb SWC | 1 Apr SWC | Water year streamflow |
|---|---|---|---|
| *Yellowstone River* |  |  |  |
| Nov-Mar Precipitation | 0.83 | 0.71 | 0.74 |
| 1 Feb SWC |  | 0.78 | 0.78 |
| 1 Apr SWC |  |  | 0.67 |
| *Salt River* |  |  |  |
| Nov-Mar Precipitation | 0.66 | 0.42 | 0.84 |
| 1 Feb SWC |  | 0.25 | 0.58 |
| 1 Apr SWC |  |  | 0.56 |

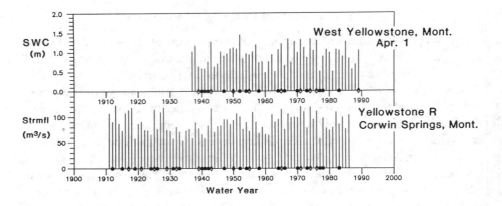

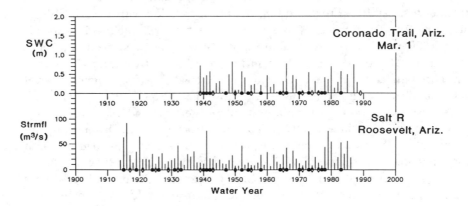

Fig. 3.1  Seasonal snow water content (SWC) and water year mean streamflow, Yellowstone River, Corwin Springs, Montana, and Salt River Roosevelt, Arizona, El Niño and La Niña years are denoted by solid circles and open diamonds, respectively.

the subsequent 1 April snow course SWC over the West is shown in Figure 3.2. Corresponding patterns of the correlations between streamflow and SOI are shown in Figure 3.3, along with comparable patterns for CNP. For SOI, correlations are positive (low SWC during winters of El Niño) over most of the Northwest and negative over a broad region of the Southwest. Strongest positive correlations, with magnitudes about 0.5, are found in patches over Idaho, Montana, and Wyoming. Maximum negative correlations occur on a broad band over the Southwest, having values of approximately −0.4. Streamflow anomalies in the Southwest from southern California through Arizona and southern Colorado tend to be positive (wet) while those in the Pacific Northwest are negative (dry) during the warm eastern tropical Pacific phase of ENSO. SWC anomalies from available snow courses in the Southwest also tend to be enhanced during El Niño. These relationships are consistent with previous studies of monthly and seasonal precipitation (Kiladis and Diaz 1989). For the moment

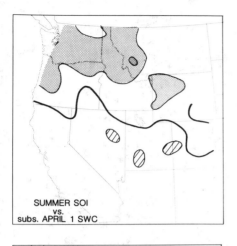 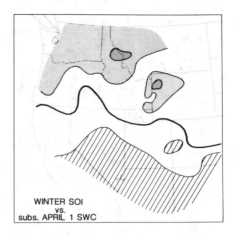

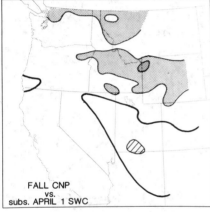 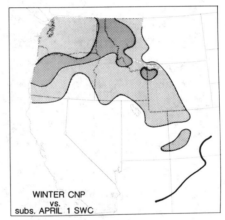

Fig. 3.2 Cross correlations: SOI (above) and CNP (below) vs. subsequent 1 April snow water content (SWC) from western U.S. snow courses. Two seasons SOI and CNP are used: summer or fall preceding, and winter preceding the subsequent (subs.) 1 April SWC. Stippling and hatching indicates correlations $\geq 0.3$, $0.5$ and $\leq -0.3$, $-0.5$, respectively.

assuming a symmetric relationship between the two ENSO phases, negative SOI (the warm state of the tropical Pacific) is associated with reduced snowpack and streamflow in the Northwest and enhanced snowpack and streamflow in the Southwest. Since the Northwest tends to exhibit warmer winter temperatures (Ropeleski and Halpert 1986) as well as drier than normal conditions during El Niño, the anomalous temperature may also contribute to the reduced snowpack (Redmond and Koch 1991).

While the correlations of SOI vs. SWC and streamflow are somewhat weak (maximum correlations between these seasonal variables are about $\pm 0.5$) there are potentially useful lag relationships. The relatively slow time evolution of the

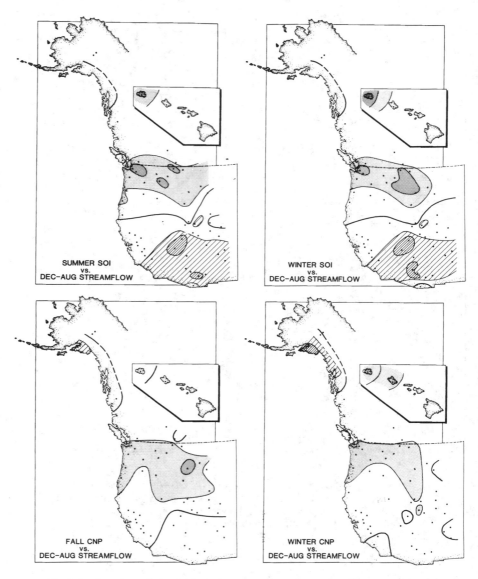

Fig. 3.3 Cross correlations SOI (above) and CNP (below) vs. December–August streamflow of several stream gauges in western North America and Hawaii. Two seasons' SOI and CNP are used; summer or fall preceding, and winter during the December–August streamflow period. Stippling and hatching indicates correlations $\geqslant 0.3$, 0.5 and $\leqslant -0.3$, $-0.5$, respectively. Dots indicate stream gauge locations. (Modified from Cayan and Peterson, 1989.)

SOI is sufficient to produce the same pattern and nearly the same correlation magnitudes for summer SOI leading as for winter SOI contemporary values with the surface hydrological variables.

The 'mature' winter phase (Rasmusson and Carpenter 1982) of ENSO is associated with low pressure in the central North Pacific in winter. The SOI vs.

streamflow anomaly correlation pattern is similar to that of CNP and PNA, but with two differences. Comparing the SOI and the CNP seasonal correlations, there is a strong similarity with the patterns in the Northwest but not in the Southwest. Correlations with CNP indicate virtually no relationship in the Southwest. The SOI-CNP similarity in the Northwest is not surprising, since, as mentioned, the deep Aleutian low phase of the PNA (the negative phase of the CNP) often occurs during the Northern Hemisphere winter mature phase of El Niño. The connection between SOI and CNP and SWC and streamflow in the Pacific Northwest can be translated in terms of the large scale circulation, the carrying current for North Pacific storms. When the central Pacific low is well developed, North Pacific winter storms tend to be carried northward toward northern British Columbia and Alaska, making that region wet while the northwestern United States is dry. The negative SLP anomalies in the central North Pacific correspond to enhanced storminess south of the Aleutian Islands, and a high pressure anomaly ridge downstream over the Northwest. This flow pattern steers the storm track toward the north into the Alaskan coast instead of into the Pacific coast farther south (see Klein [1957] for storm tracks). In the subtropical North Pacific, there is an anomalous westerly (west-to-east) flow in association with the gradient between the deep central Pacific low to the north and anomalous high pressure to the south. This flow anomaly reduces the trade winds and diminishes precipitation and streamflow on the eastern coastal slopes of the Hawaiian Islands. Conversely, when the Aleutian low is weak, there is a tendency for a more active low to the east in the Gulf of Alaska, so that heavier than normal precipitation, higher SWC, and larger streamflow anomalies tend to occur.

The second difference between SOI and CNP is that along the Alaskan coast, correlations of SOI and streamflow are weaker than with CNP. Nonetheless, the negative correlations with SOI may be meaningful, since Yarnal and Diaz (1986) reported significant correlations between ENSO and coastal Alaskan precipitation. Much of the weakness of this relationship arises from the variability in the longitude of the deepened central North Pacific low during winter that is associated with El Niño (Peterson et al. 1986, Fig. 8). A more westerly position of this trough favors the winter storm track moving into the Alaskan coast, while a more easterly position encourages storms along the West Coast farther to the south. Further insight into the types of Northern Hemisphere circulation patterns that appear with ENSO is provided by Fu et al. (1986) and Livezey and Mo (1987) who suggest that the configuration of tropical Pacific heating may help to determine the pattern of extratropical response.

It is noteworthy that parts of the western United States are *not* significantly correlated to either SOI or CNP. Concerning the deep Aleutian Low phase of SOI (warm equatorial Pacific) and CNP, the relative weakness of the southern portion of the teleconnection downstream over the West Coast means that the tendency for high pressure (lack of storminess) is not very reliable. Hence, there is not a strong CNP relationship over California and the Southwest. For SOI, the statistical relationship begins to reverse in southern California, with a

tendency for above normal streamflow during El Niño, a relationship observed
in seasonal precipitation that has been noted by Schonher and Nicholson (1989).
This is an example where the statistics are weak even though there have been
individual El Niño cases such as in 1982/83 and 1940/41 when the stormy North
Pacific El Niño effect spread throughout much of the West. Statistically, the circulation pattern that is best correlated with anomalous precipitation and
streamflow in California has negative geopotential or SLP anomalies just offshore, centered at about (40°N, 130°W) (Klein and Bloom 1987; Cayan and
Peterson 1989). The winter 700-mb height teleconnection pattern centered at this
origin (Namias 1981) illustrates this connection. In comparison to the Aleutian
low teleconnection in the central North Pacific, the California pattern is more

(a)

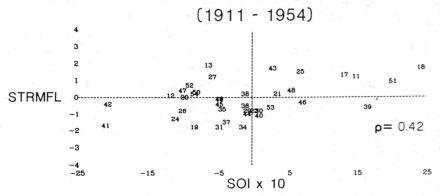

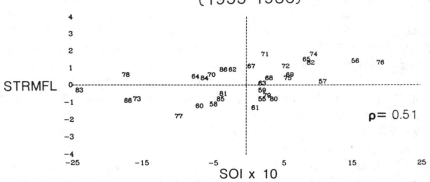

Fig. 3.4 (*a*) Summer SOI leading standardized December–August streamflow for
Yellowstone River. Record is divided into two sections, 1911–1954 (top) and 1955–1986
(bottom). Year of each data point is labeled on plots as year-1900. Correlation (rho)
is indicated for each period. (b) Same as (*a*) but for Salt River, whose two sections are
1914–1954 (top) and 1955–1985 (bottom).

regionally confined, and not well related to anomalies in the Aleutian Low region. For coastal British Columbia, a similar offshore regional atmospheric circulation pattern is associated with streamflow anomalies (Cayan and Peterson 1989). These more regional circulation connections explain the lack of a statistical connection between streamflow in these areas and strong or weak atmospheric circulation in the central North Pacific. Stated differently, California and British Columbia are close to the node of the atmospheric long-wave pattern that emanates from the central North Pacific, and small variations in the position of the remote circulation anomaly center can yield both positive and negative precipitation variations.

There is an important distinction between the results with streamflow and the results of Ropelewski and Halpert (1986). These authors found some evidence that the El Niño phase of ENSO was associated with lighter than average

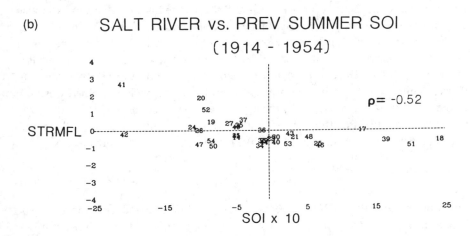

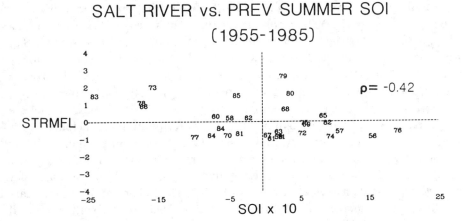

Fig. 3.4 (*b*)

precipitation in the Pacific Northwest, but this tendency was not strong enough for them to consider it to be reliable. The weakness in this relationship appears to arise, in part, from variations in the timing of the dry conditions between fall and spring of the mature El Niño period. However, the snow courses considered and the streamflow in basins with higher elevations are not as sensitive to these timing changes as is the precipitation, since SWC is cumulative and snowmelt constitutes a significant portion of the streamflow. Heavy Southwest runoff appears to result from active southerly-displaced middle-latitude storms that tap subtropical Pacific moisture, which is often transported by the subtropical jet stream. Heightened activity occurs over several months from fall through spring, as discussed by Douglas and Englehart (1981), Ropelewski and Halpert (1986), and Andrade and Sellers (1988), and as indicated by the frequency of daily precipitation discussed below. Increased precipitation in the Southwest occurs during late fall through spring in the El Niño phase of the Southern Oscillation, which would contribute to above-normal SWC in the higher elevations and also enhanced streamflow. To investigate the temporal consistency of the relationship between SOI and streamflow, the relationship was examined for two halves of the data period (before and after 1955). Figure 3.4 shows a pair of scatter plots of summer SOI vs. December–August streamflow at Yellowstone River and Salt River. While the correlations between SOI and streamflow in the Northwest and in the Southwest are not strong, these plots are fairly consistent from the early to the later period. For the earlier and later halves of the data, the Yellowstone and Salt River correlations are 0.42, 0.51 and −0.52, −0.42, respectively.

## Daily precipitation

For more detail of the ENSO effects exposed by the seasonal relationships, the behavior of daily precipitation during different phases of the ENSO cycle was investigated. Two individual watersheds were studied, representing the core regions of the positive and negative SOI-SWC correlation patterns discussed above. These were the Yellowstone River, near the Corwin Springs, Montana gauge in the heart of the El Niño winter dry region in the Northwest, and the Salt River near the Roosevelt, Arizona gauge in the El Niño winter wet region in the Southwest. The daily precipitation is broken into five successively heavier classes: all days with equal or greater than 1, 5, 10, 20, and 30 mm. In this categorization, the higher precipitation categories overlap each of those that are lower; e.g., all the days with 5 mm and greater precipitation are included in the 1 mm and greater category, and so on. Diaz (1991) has examined changes in different categories of United States regional precipitation according to longer period wet and dry regimes. The present analysis takes a similar approach, but here we explicitly considered the effect of ENSO. The El Niño and La Niña years employed in this analysis are shown in Table 3.1. The three 'seasons' employed were fall (August, September, and October, abbreviated ASO), early winter (November, December, and January, abbreviated NDJ), and late winter/early spring (February, March, and April, abbreviated FMA). These seasons are centered on the mature northern winter phase of the warm or cold event, or 'year 1' of Rasmusson and Carpenter (1982) and Kiladis and Diaz (1989). For example, for the El Niño with mature phase in winter of 1973, NDJ is November and December 1972 and January 1973; FMA is February, March, and April 1973. Two fall seasons were included, designated ASO(0) and ASO(1), which in the example would be August, September, October 1972, and August, September, October 1973, respectively. To avoid mid-summer behavior within the fall aggregate, the ASO definition was from 15 August through 31 October.

First, consider Yellowstone River in the positive SWC and streamflow vs. SOI correlation region of the northwestern United States. The number of days with measurable precipitation, the amount of precipitation per day on days with precipitation, and the frequency of different categories of precipitation during El Niño, La Niña, and 'other' years, stratified by season is presented in Table 3.4. For Yellowstone River, the greatest difference between El Niño, La Niña, and 'other' year precipitation frequency is for the NDJ period. Considering daily precipitation of all amounts (greater than 1 mm), the mean precipitation over all precipitation days for the 18 El Niño years, the 12 La Niña years, and the 36 'other' years was 3.5, 4.4, and 3.7 mm d$^{-1}$, respectively. The frequency of NDJ daily precipitation of all amounts was lowest for El Niño (42.0%), greatest for La Niña (52.2%), and intermediate for 'other years' (47.5%). A higher frequency of precipitation during La Niña and lower frequency during El Niño prevailed for each successive category, especially for precipitation categories greater than 10 mm. The most impressive differences in precipitation frequency occurred in the higher daily precipitation categories. For example, days with 10 mm and greater precipitation were three times as likely to occur during La Niña (5.8%) as they did during El Niño (1.8%), and nearly twice as likely during 'other years' (3.1%). Although there are few days upon which to base a comparison, the only two days in the 64-yr period with greater than 30 mm were in La Niña years.

The other seasons at Yellowstone River showed less disparity in their El Niño, La Niña, and 'other year' frequencies, although FMA and ASO(0) exhibited a higher frequency of precipitation during La Niña than during El Niño. 'Other years' had nearly the same or a slightly higher frequency of most categories of precipitation as the La Niña years for these seasons, however.

To test the statistical significance of these differences, the following Monte Carlo exercise was performed. Sets of years equal in number to the available period of record were randomly chosen from the record. This random mix of years, whose numbers were the same as the years in the El Niño, La Niña, and 'other' year categories, was repeated 100 times, and each time the precipitation frequency statistics were computed. To preserve the annual distribution of precipitation, the daily precipitation within each year was retained unscrambled. The likelihood of achieving the observed El Niño, La Niña, and 'other' year statistics was judged by comparing the observed frequency or amount of daily precipitation (broken into the various precipitation amount bins) with that from

Table 3.4 *Daily precipitation statistics Yellowstone River region (1924–1988)*

| | ASO (0) | | | NDJ | | | FMA | | | ASO (+1) | | |
|---|---|---|---|---|---|---|---|---|---|---|---|---|
| | E[a] | L[a] | O[a] | E | L | O | E | L | O | E | L | O |
| No. days | 1460 | 843 | 2737 | 1653 | 1063 | 3213 | 1683 | 1046 | 3070 | 1437 | 913 | 2620 |
| No. days w/Ppt | 406 | 259 | 835 | 695 | 555 | 1525 | 645 | 456 | 1343 | 395 | 261 | 844 |
| Mean daily Ppt[b] | 4.0 | 4.8 | 4.2 | 3.5 | 4.4 | 3.7 | 3.5 | 3.6 | 3.7 | 4.4 | 4.1 | 4.2 |
| Ppt ≥ 1 mm[c] | 27.81 | 30.72 | 30.51 | 42.04 | 52.21 | 47.46 | 38.32 | 43.59 | 43.75 | 27.49 | 28.59 | 31.38 |
| ≥ 5 mm | 8.22 | 10.91 | 9.68 | 11.74 | 17.12 | 12.82 | 9.39 | 11.66 | 11.56 | 9.46 | 9.42 | 9.48 |
| ≥ 10 mm | 2.67 | 4.15 | 3.22 | 1.81 | 5.83 | 3.14 | 2.14 | 2.77 | 3.29 | 3.55 | 2.63 | 3.23 |
| ≥ 20 mm | 0.21 | 0.71 | 0.29 | 0.12 | 0.66 | 0.19 | 0.18 | 0.29 | 0.23 | 0.28 | 0.11 | 0.45 |
| ≥ 30 mm | 0.00 | 0.12 | 0.04 | 0.00 | 0.19 | 0.00 | 0.06 | 0.00 | 0.00 | 0.07 | 0.00 | 0.04 |

[a]E, L, and O designate seasons occuring within El Niño, La Niña, or "other" years.

[b]Mean daily precipitation for days with measurable precipitation; days with no precipitation are not included.

[c]Entries under each precipitation (ppt) category are the percentage of all days during period with ppt amount equal or exceeding that category's threshold. Lower threshhold categories include days entering higher threshhold categories (e.g., days with 5 mm and greater are included in the 1 mm and greater category).

the 100 randomly generated sets of the three categories of years. The rank of the frequency or amount of the observed precipitation defined its percentile placement within the sorted (lowest to highest) stack of the frequency or amount statistics from the randomly generated sets.

First considering Yellowstone River, to assess the significance of these El Niño vs. La Niña differences, observed precipitation frequency and amounts for each category were compared with those from the stack of 100 Monte Carlo runs. At Yellowstone River, the NDJ results had interesting differences in comparison with the Monte Carlo series. For all precipitation days greater than 1 mm, the difference in precipitation frequency was not extremely unusual, but the difference in the amount of precipitation was quite significant. The *frequency* of El Niño precipitation days was ranked 14, and the La Niña precipitation frequency ranked 69. That is, 86 of the shuffled NDJ series had a frequency of precipitation higher than that of El Niño, and 31 of them had a frequency greater than that of La Niña. (Remember the rank is the percentile order from 1–100 of the sorted [lowest to highest] Monte Carlo precipitation series.) The *amount* of precipitation on all NDJ El Niño precipitation days was ranked 2, and on all La Niña precipitation days ranked 99. That is, 98 of the shuffled winter series had precipitation amounts greater than that of El Niño, and only 1 had an amount greater than that of La Niña. Thus a rainy day during La Niña had significantly heavier precipitation than a rainy day during El Niño. This difference in amounts is related to the frequency of larger precipitation amounts during El Niño versus La Niña years. While the larger amounts contribute a much smaller sample, the differences in the higher amount frequencies appear to be considerably more unusual than those of the all-precipitation-amounts frequencies. For days with 10 mm or more precipitation, the observed El Niño frequency ranked 0 (lowest), and the La Niña frequency ranked 100th (highest), among the 100 Monte Carlo series; that is, none of the Monte Carlo year series had frequencies as extreme. Results for the other seasons at Yellowstone River were not as significant (Table 3.4); i.e. the frequency and amount of daily precipitation during El Niño and La Niña differed little more than one would expect by chance.

Second, consider the Salt River, in the heart of the negative SWC and streamflow vs. SOI correlation region of the southwestern United States (Table 3.5). For Salt River, the greatest difference between El Niño, La Niña, and 'other year' precipitation frequencies was during FMA, and to a lesser extent, during NDJ. For FMA, considering all daily events, the mean precipitation showed opposite differences as those for Yellowstone River. Over all FMA precipitation days for the 24 El Niño years, the 16 La Niña years, and the 47 'other' years, the mean precipitation was 4.8, 4.3, and 4.6 mm d$^{-1}$, respectively. Correspondingly, the frequency of daily precipitation of any amount was highest during El Niño (24.8%), lowest during La Niña (14.9%), and intermediate during other years (21.2%). As they were at Yellowstone River, the differences were accentuated for higher precipitation amounts. At Salt River, days in FMA with precipitation equal or greater than 10 mm were about twice as likely during El

Table 3.5 *Daily precipitation statistics Salt River region (1901–1988)*

| | ASO (0) | | | NDJ | | | FMA | | | ASO (+1) | | |
|---|---|---|---|---|---|---|---|---|---|---|---|---|
| | E[a] | L[a] | O[a] | E | L | O | E | L | O | E | L | O |
| No. days | 1925 | 1232 | 3619 | 2219 | 1523 | 4288 | 2250 | 1440 | 4230 | 1925 | 1232 | 3619 |
| No. days w/Ppt | 560 | 329 | 1049 | 495 | 273 | 914 | 557 | 214 | 895 | 537 | 381 | 1020 |
| Mean daily Ppt (mm)[b] | 4.3 | 4.4 | 4.2 | 5.9 | 5.2 | 5.5 | 4.8 | 4.3 | 4.6 | 4.2 | 4.3 | 4.3 |
| Ppt ≥ 1 mm[c] | 29.09 | 26.70 | 28.99 | 22.13 | 17.93 | 21.32 | 24.76 | 14.86 | 21.16 | 27.90 | 30.93 | 28.18 |
| ≥ 5 mm | 9.09 | 7.79 | 8.93 | 9.37 | 7.35 | 8.26 | 8.36 | 4.58 | 7.26 | 8.10 | 9.74 | 8.79 |
| ≥ 10 mm | 2.81 | 3.33 | 2.68 | 4.24 | 2.69 | 3.73 | 3.42 | 1.74 | 2.79 | 2.91 | 2.92 | 2.76 |
| ≥ 20 mm | 0.42 | 0.65 | 0.47 | 1.17 | 0.66 | 0.77 | 0.58 | 0.28 | 0.40 | 0.57 | 0.49 | 0.44 |
| ≥ 30 mm | 0.10 | 0.16 | 0.17 | 0.23 | 0.13 | 0.19 | 0.22 | 0.07 | 0.09 | 0.10 | 0.16 | 0.17 |

[a] E, L, and O designate seasons occuring within El Niño, La Niña, or 'other' years.

[b] Mean daily precipitation for days with measurable precipitation; days with no precipitation are not included.

[c] Entries under each precipitation (ppt) category are the percentage of all days during period with ppt amount equal or exceeding that category's threshold. Lower threshhold categories include days entering higher threshhold categories (e.g., days with 5 mm and greater are included in the 1 mm and greater category).

Niño (3.4%) than during La Niña (1.7%) and about 1.2 times as likely during El Niño than during other year FMAs (2.8%). Days with 30 mm and greater precipitation occurred about 3 times more frequently during El Niño than during La Niña and more than two times more frequently than during 'other years.'

The Salt River NDJ differences were not as distinct as those for FMA. This was largely due to the frequency of occurrence of lighter precipitation days, which was more similar across the El Niño, La Niña, and 'other year' types. Days with less than 10 mm still occurred most frequently during El Niño (18.1%), least during La Niña (15.2%), and intermediately during other NDJs (17.6%), but these differences were fairly bland. (The frequency of days with less than 10 mm is obtained by subtracting the frequency of greater than or equal 10 mm days from the frequency of greater than or equal 1 mm days.) For the larger daily precipitation categories, the relative differences were greater. Daily precipitation for NDJ greater than 10 mm occurred 1.6 times as often during El Niño (4.2%) as during La Niña (2.7%) and also more frequently during El Niño than during other years (3.7%).

To determine the significance of these Salt River precipitation frequency and amount differences among the three categories of years, they were each compared with the stack of 100 Monte Carlo precipitation frequency calculations. Compared to the Monte Carlo series, the Salt River FMA results were highly significant. Considering all precipitation days of any amount (greater than 1 mm), the *frequency* of precipitation during El Niño was ranked 100, and the frequency during La Niña ranked 1. That is, none of the shuffled Salt River FMA series had a frequency of precipitation as high as that which occurred during El Niño years or as low as that which occurred during La Niña years. Differences in FMA daily precipitation *amounts* were not as remarkable, however. For FMA precipitation days of any amount (greater than 1 mm), the mean daily observed precipitation during El Niño ranked 78 and the mean daily observed precipitation during La Niña ranked 28 in the stack of 100 Monte Carlo sets. While drawn from a much smaller sample of days, the frequency of larger precipitation amounts during El Niño and La Niña was quite significant. For example, considering only days with precipitation of 10 mm or more, the observed El Niño frequency ranked 95th, and the La Niña frequency ranked 7th.

Results for NDJ at Salt River were not as significant, but were nonetheless consistent with those for FMA. For all precipitation days greater than 1 mm, the frequency of El Niño precipitation was ranked 70th, and the La Niña precipitation was ranked 20th. That is, of 100 Monte Carlo series, 30 NDJs had a higher precipitation frequency than the NDJ history for El Niño, while 20 had a lower NDJ precipitation frequency than the NDJ history for La Niña. For all NDJ precipitation days greater than 1 mm, the mean amount observed on El Niño precipitation days was ranked 78 and the mean amount observed on La Niña precipitation days was ranked 28. Considering only days with 10 mm or greater precipitation, the El Niño frequency ranked 85th and the La Niña frequency ranked 4th.

## YELLOWSTONE RIVER STORMS, EL NIÑO YEARS
### FALL/EARLY WINTER (NDJ)

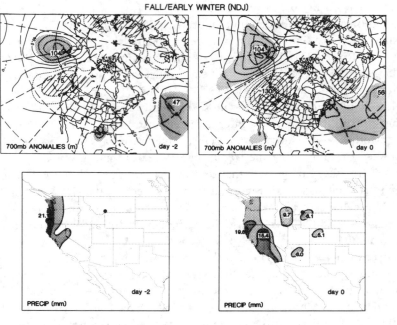

## YELLOWSTONE RIVER STORMS, LA NIÑA YEARS
### FALL/EARLY WINTER (NDJ)

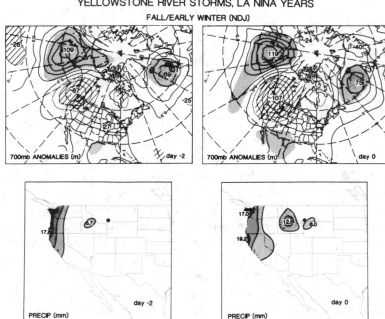

Fig. 3.5   A composite 700-mb height anomaly (m) and gridded precipitation anomaly (mm) for day −2 and day 0 of heavy (≥10 mm) daily precipitation events at Yellowstone River during November/December/January (NDJ). El Niño year cases are above and La Niña year cases below. 700-mb heights are from 12 00 GMT gridded

## Daily atmospheric circulation

The Yellowstone River and Salt River daily precipitation show that the El Niño and La Niña phases of ENSO tend to produce opposite behavior of precipitation frequency and amounts. For further insight into the causes of daily precipitation associated with ENSO, we examined the atmospheric circulation and the spatial distribution of precipitation during heavy precipitation events of Yellowstone River and Salt River. The events considered were grouped by season (fall/early winter, or NDJ, and late winter/spring, or FMA) and by ENSO year type (El Niño or La Niña). The composites were based upon days with equal or greater than 10 mm of precipitation at the six stations of each basin. For groups with an ample number of heavy precipitation days, the threshold was raised to 15 mm of precipitation. These days were selected from the 1947–1988 period. The composite contained as few as 7 cases and as many as 24 cases. In an attempt to provide an independent set, consecutive days which happened to exceed the 10 mm threshold were eliminated by only including the first day of the episode. Both 700-mb height anomalies over the western part of the Northern Hemisphere and gridded precipitation over the western United States are mapped for these days. To provide an indication of the development of the precipitation event, these fields are shown for two days beforehand (day $-2$) as well as the day of the composite event. The statistical significance of the 700-mb height anomalies is indicated by a two-tailed $t$-test. Values exceeding the null hypothesis (no difference in pattern from the long-term mean) at the 95% level of confidence are shaded. Since by chance, it might be expected that a certain number (50 or less) of the 360 grid points comprising this field would contain a 700-mb anomaly composite exceeding the 95% confidence threshold, the composite anomalies were tested for field significance (Livezey and Chen 1983). Each of the maps in the following set of analyses contains more grid points with significant composite anomalies than would be expected by chance (95% of the time) using a Monte Carlo procedure. Moreover, these maps are meaningful because the anomalies form physically reasonable patterns associated with the regional precipitation considered.

As indicated by the 95% confidence level (relative to the null hypothesis) shaded regions on the 700-mb anomaly composites, the patterns involved in the precipitation events are quite distinct from a random sample. By inference from the shading pattern, the El Niño vs. La Niña maps are nearly indistinguishable for the NDJ sets but they are quite different for the FMA sets. As will be seen, this difference between NDJ and FMA occurs for both Yellowstone River and Salt River precipitation patterns.

Figure 3.5 shows the composite 700-mb height anomaly and precipitation distribution for fall/early winter at Yellowstone River during El Niño and La

---

Fig. 3.5 (*cont.*)   analysis from days entering the composite. Regions of composite anomalies significant at the 95% confidence level are indicated by shading. Shading of precipitation indicates regions with equal or greater than 5, 10, and 15 mm.

Niña. The atmospheric circulation patterns for El Niño and La Niña are very similar in the North Pacific–western North America sector. Both have well defined patterns, with a strong positive anomalies to the west over the North Pacific and negative anomalies over the eastern North Pacific and western North America. The positive anomalies are centered in the Aleutian Island–Bering Sea region and the negative anomalies are centered in Washington State. This pattern induces a southwesterly flow into the central Rockies region that includes Yellowstone River. The precipitation maps indicate that the heavy precipitation associated with this pattern usually develops out of a system which first appears over coastal Washington, and propagates eastward into the Rockies. The precipitation is regional in scope, as evidenced by the several grid points having precipitation exceeding 5 mm (note that the precipitation on the grid point maps may appear spotty because some stations fall below the threshold chosen for the shading.) There is a systematic development over at least 2 d, as indicated by the day − 2 composite 700-mb height anomalies: note the already strong positive anomaly center upstream in the Aleutian Islands. In fact, this development can be traced back at least 4 d beforehand (maps not shown). The greatest dissimilarity in the El Niño vs. La Niña patterns occurs downstream over the North Atlantic south of Greenland, where the El Niño circulation has a strong negative anomaly and the La Niña circulation has a positive anomaly.

The late winter spring circulation patterns for Yellowstone River (Fig. 3.6) offer much greater contrast. The El Niño pattern exhibits negative 700-mb height anomalies across the entire North Pacific basin north of 30°, culminating in a negative anomaly center just west of Washington State. Strong southwesterly wind anomalies again occur over the central Rockies. In the Pacific sector, the La Niña pattern has an anomaly structure which is somewhat more spatially confined than that of El Niño. Negative anomalies of this pattern are located north of the Bering Sea and in the Gulf of Alaska, and extend over much of the southern Canada and the northern coterminous United States. Over the North Pacific, the lack of distinct anomalies reflects a lack of consistency in the pattern there; evidently there are several upstream circulation patterns contributing to La Niña precipitation. In common with all the other composites this late winter/spring pattern has negative 700-mb height anomalies over the West and anomalous southwesterly flow into the central Rockies. Here, the negative anomalies are shifted farther east than for the other cases, however, with a center over Idaho. Downstream, another interesting feature of this pattern is the couplet of strong negative anomalies south of Greenland and positive anomalies over Scandinavia. Inspection of composite maps subsequent to this event (maps not shown) indicate that this downstream anomaly continues to grow, with centers exceeding 200 m. (Apparently in some cases, rain in Montana is a good predictor of European weather conditions, although this suggestion should be examined using a larger sample.)

Even though the Salt River basin is more than 10° latitude south of the Yellowstone region, the Salt River 700-mb composite patterns during heavy

YELLOWSTONE RIVER STORMS, EL NIÑO YEARS

LATE WINTER/SPRING (FMA)

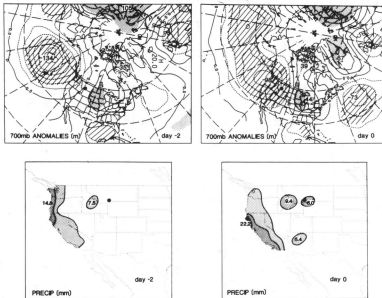

YELLOWSTONE RIVER STORMS, LA NIÑA YEARS

LATE WINTER/SPRING (FMA)

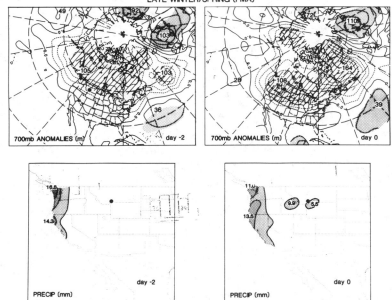

Fig. 3.6   Same as Figure 3.5, but for FMAs of El Niño years, Yellowstone River.

precipitation days have a resemblance to those for Yellowstone. For Salt River, the El Niño and La Niña heavy precipitation patterns are much alike in NDJ (Fig. 3.7) but quite different in late winter/spring. The El Niño and La Niña fall/early winter patterns are both examples of the development of sharp troughs along the West Coast, probably including a number of 'cut-off lows' (see discussion on flood events below). In both composites, the event is preceded on day −2 by the development of large positive anomalies (a high pressure ridge) upstream in the Gulf of Alaska, as well as a significant negative anomaly region (a trough) stationed just off the northern California coast. By the day of the event, the negative anomaly has strengthened and moved southeastward to a position near southern California, and a strong positive anomaly just downstream over the eastern United States, centered over the Ohio Valley. Regionally, in the transition from day −2 to day 0, strong southwesterly flow develops over southern Arizona, and the precipitation spreads from coastal Washington, Oregon, and northern California to southern California and Arizona.

The late winter spring (FMA) patterns at Salt River are quite different between El Niño and La Niña (Fig. 3.8). The La Niña pattern has local negative 700-mb height anomalies and a strong positive anomaly in the eastern Gulf of Alaska. This pattern is somewhat similar to the 'sharp trough' pattern observed in NDJ, with a strong positive anomaly to the northwest and a negative anomaly (cutoff low) centered at low latitudes (about 30°S, just offshore of Baja California, Mexico). This positive/negative anomaly couplet develops out of a strong negative anomaly far upstream to the south of Kamchatka on day −2. The gridded precipitation shows a drastic change from day −2 to day 0 of the La Niña event, with significant precipitation in western Washington on day −2 and then an isolated region of heavy precipitation in Arizona on day 0.

In contrast to the La Niña pattern, the El Niño late winter/spring pattern is a North Pacific basin wide negative anomaly which extends into the Southwest. This pattern represents an expanded circumpolar vortex, with strengthened westerly winds centered about 30°N, and positive 700-mb height anomalies far to the north over Alaska, central Canada, and south of Greenland in the North Atlantic. The movement of this system onshore from day −2 to day 0 is apparent from the two 700-mb anomaly patterns as well as from the precipitation maps, which indicate heavy precipitation in California on day −2 spreading to Arizona and southern New Mexico by day 0.

The limited sample size in these composites may allow the resulting anomaly field to be heavily biased by an unusual pattern during one particular year. One reviewer cautioned that the impressive FMA anomaly patterns for the Yellowstone River and Salt River El Niño composites might have been largely produced by the strong El Niño of winter 1983. For Yellowstone River, only one of the 12 cases included in the composite was from 1983, so that this pattern is not dominated by the winter of 1983. The Salt River FMA composite consisted of 18 individual storm cases, of which 4 storms were from 1983, so that may have had a significant impact. However, we tested this by removing these four cases

SALT RIVER STORMS, EL NIÑO YEARS

FALL/ EARLY WINTER (NDJ)

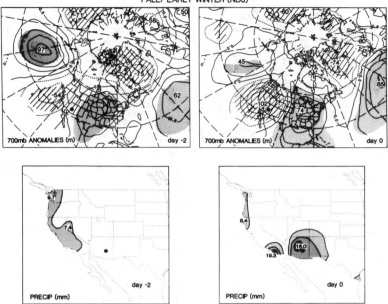

SALT RIVER STORMS, LA NIÑA YEARS

FALL/EARLY WINTER (NDJ)

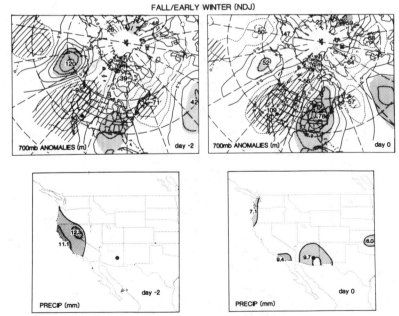

Fig. 3.7   Same as Figure 3.5, but for NDJs of El Niño years, Salt River.

SALT RIVER STORMS, EL NIÑO YEARS

LATE WINTER/SPRING (FMA)

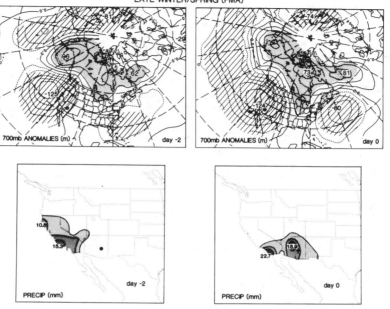

SALT RIVER STORMS, LA NIÑA YEARS

LATE WINTER/SPRING (FMA)

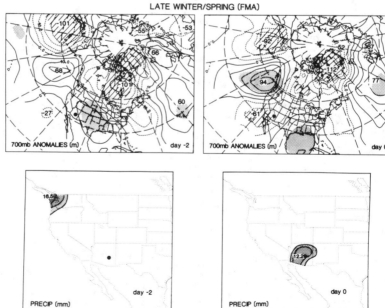

Fig. 3.8   Same as Figure 3.5, but for FMAs of El Niño years, Salt River.

and re-computing the composite. The resultant map, while not quite as strong as the original, had the same pattern, so it appears that the FMA vs. NDJ composite differences are robust.

### El Niño and flood frequency

In the southwestern United States, the statistically significant negative relations between SOI and precipitation and SOI and streamflow suggest that El Niño and large floods should be related. During El Niño, atmospheric circulation is unusual and anomalously warm water in the eastern Pacific Ocean provides a source for large amounts of precipitable moisture. Therefore, the hydroclimatology of storms that cause floods should be different during El Niño conditions. For example, the hydrograph and associated precipitation of exceptionally large floods on the Salt River near Roosevelt, Arizona, during the strong El Niño of

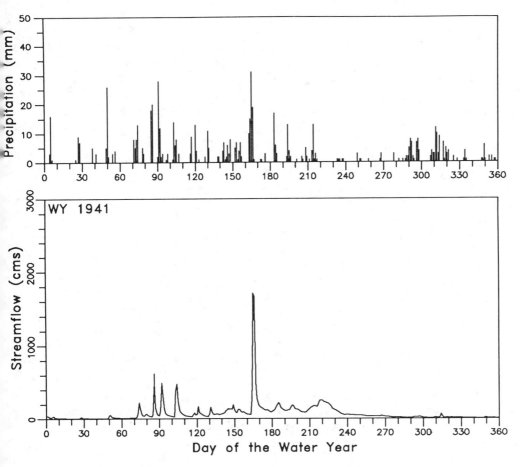

Fig. 3.9  Daily streamflow at Salt River ($m^3 \ s^{-1}$) and daily precipitation (mm) at Roosevelt, Arizona for water year (WY, October–September) 1941.

1941 is shown in Figure 3.9. Most of the largest floods on large rivers in Arizona have occurred during or within 6 mo following cessation of El Niño conditions (Table 3.6). Although most of the years in which large floods occurred have been rated at strong or very strong El Niño conditions (Quinn et al. 1987), others occurred under moderate or weak El Niño conditions (Table 3.6). Many of these rivers have headwaters in Colorado, Utah, or New Mexico, which underscores the large-scale nature of the anomalous circulation patterns, as illustrated in the previous section.

Flood frequency in Arizona is particularly sensitive to El Niño conditions because floods can occur in three distinct hydroclimatic seasons (Hirschboeck

Table 3.6 *Recorded large floods on drainages larger than 3600 km$^2$ in Arizona during or within 6 months after El Niño conditions.*[a]

| Year | Month | El Niño severity[b] | River gauging station | Peak discharge (m$^3$/s) | Rank of flood[c] |
|------|-------|--------------------|-----------------------|--------------------------|------------------|
| 1862 | Jan? | W/M? | Colorado River near Topock | 11300 | 1 |
| 1862 | Jan | W/M? | Virgin River at Littlefield | Unknown | 1* |
| 1868 | Sep | M | Gila River at Kelvin | Unknown | 2* |
| 1884 | Jul | S+ | Colorado River near Grand Canyon | 8500 | 1 |
| 1891 | Feb | VS | Salt River near Roosevelt | Unknown | 1* |
| 1891 | Feb | VS | Bill Williams River near Planet | 5670 | 1 |
| 1891 | Feb | VS | Verde River below Tangle Creek | 4250 | 1 |
| 1905 | Nov | W/M | San Francisco River at Clifton | 1840 | 3 |
| 1905 | Nov | W/M | Gila River at Kelvin | 5380 | 1 |
| 1923 | Sep | M | Little Colorado River at Grand Falls | 3400 | 1 |
| 1926 | Oct | VS | Paria River at Lees Ferry | 456 | 1 |
| 1926 | Sep | VS | San Pedro River at Charleston | 2780 | 1 |
| 1941 | Mar | S | Salt River near Roosevelt | 3310 | 1 |
| 1966 | Dec | M+ | Virgin River at Littlefield | 1000 | 1[d] |
| 1972 | Oct | S | Gila River at head of Safford Valley | 2330 | 3 |
| 1972 | Oct | S | San Francisco River at Clifton | 1810 | 3 |
| 1977 | Oct | M | Santa Cruz River at Tucson | 651 | 2 |
| 1983 | Oct | S | Gila River at head of Safford Valley | 3740 | 1 |
| 1983 | Oct | S | Santa Cruz River at Tucson | 1490 | 1 |
| 1983 | Oct | S | San Francisco at Clifton | 2580 | 1 |

[a]Data from U.S. Geological Survey records.
[b]Strength of El Niño events is from Quinn et al. (1987), except for 1862 (Quinn, pers. comm., 1990). VS, very strong; S, strong; M, moderate; W/M, weak to moderate; + / – indicates an event between strength ratings.
[c]A rank of 1 is the largest flood of record. * indicates uncertainty because the flood was not part of the systematic gauging record but probably is of the rank listed.
[d]A larger flood in 1990 was caused by a dam failure.

1985, 1987). Floods commonly result from frontal storms in winter (November–March); Figure 3.7a and 3.8a give composite 700-mb height anomalies that are typically associated with this type of storm. During summer (July–August), advection of low-level moisture from the Gulf of Mexico and eastern North Pacific Ocean into the Southwest causes what is termed the Arizona 'monsoon,' which is characterized by local to mesoscale thunderstorms. However, summer thunderstorms do not cause the largest floods in Arizona (Table 3.6), except in small drainages (less than about 259 km$^2$). There have been large floods in fall (September–October), as moisture from dissipating tropical cyclones may be advected into the Southwest. This source has caused some of the largest floods in Arizona, particularly between 1972 and 1983 (Table 3.6). For the flood-frequency analysis discussed below, we have redefined the water year, which the U.S. Geological Survey conventionally defines as 1 October to 30 September, as 1 November to 31 October to be in accord with the hydroclimatological seasons.

Webb and Betancourt (1992) investigated the effects of low-frequency climatic variability on flood frequency of the Santa Cruz River in southern Arizona. Six of the ten largest floods at Tucson occurred during or within 6 mo following

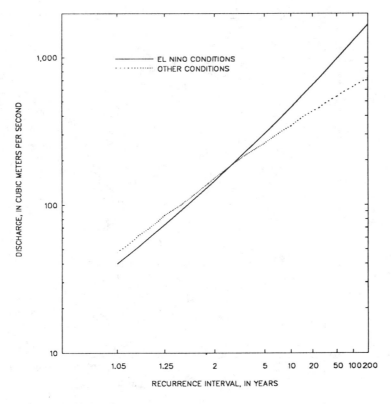

Fig. 3.10   Flood frequency for El Niño and other years for the Santa Cruz River at Tucson, Arizona, 1914–1986. (Modified from Webb and Betancourt 1992.)

cessation of El Niño conditions. Flood frequency was greater during El Niño
conditions at recurrence intervals of greater than 10 yr (Fig. 3.10). The difference
in flood frequency between El Niño and other conditions is caused primarily by
a significant increase in variance of size of floods during El Niño conditions
(Webb and Betancourt 1992); the mean annual flood was unchanged.

Flood frequency of the Salt River (Fig. 3.1) illustrates some of the problems
associated with flood frequency in relation to El Niño and La Niña conditions.
The two largest floods on the Salt River occurred during the El Niño years of
1891 and 1941 (Table 3.6). Although floods during El Niño years are nearly twice
as large as floods during La Niña years at all recurrence intervals (Fig. 3.11),
floods during other years are much larger than either at recurrence intervals
greater than about 5 yr. Statistically, the differences among the different sets of
years occurs in the skew coefficient, which is − 0.44 for La Niña conditions, 0.20
for El Niño conditions, and 0.42 for other conditions. The differences in skew
coefficients account for differences in the curvature of the flood-frequency relations (log-Pearson type III distribution); the more positive skew coefficients,
such as calculated for El Niño and other years, cause an upward curvature in the

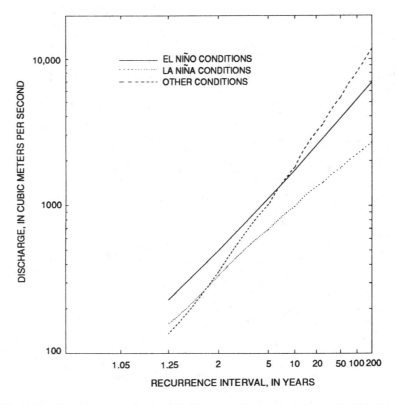

Fig. 3.11   Flood frequency for the Salt River near Roosevelt, Arizona, for El Niño,
La Niña, and other years, 1925–1988.

relations (Fig. 3.11). Although it would appear from Figure 3.11 that El Niño has a minor effect on flood frequency of the Salt River, the results may be deceptive and influenced by lag effects of large floods which occurred in 1927, 1937, 1952, 1960, 1978, 1979, and 1980, none of which was an El Niño or La Niña year, but they were years with some months of negative SOI.

As this example shows, fundamental knowledge of hydroclimatic controls on flood frequency may not be gained from simple analysis of El Niño and La Niña conditions. Instead, the flood-generating mechanism, and the effect that El Niño may have on the mechanism, are of greater importance. Hirschboeck (1985, 1987) analysed flood-generating storm types in the Gila River basin of southern Arizona, and Webb and Betancourt (1992) combined many of Hirschboeck's storm types into the three types: frontal systems, monsoonal storms, and dissipating tropical cyclones. A fourth type of storm, cutoff low-pressure systems, occurs in fall to spring, but we had insufficient data to adequately account for floods caused by this type of storm. Nonetheless, cutoff low-pressure systems are considered here because of their influence as a steering mechanism for dissipating tropical cyclones in fall (Smith 1986). Analysis of the influence of El Niño on these storm types would result in a more direct examination of the causal mechanism of floods in the southwestern United States.

## Storm types and floods affected by El Niño

In general, several climatic anomalies occur during El Niño conditions that affect storm types for the southwestern United States. As discussed earlier in this chapter, winter storms increase in severity and the amount of winter precipitation is increased during years with El Niño conditions. Not surprisingly, this leads to increased frequency of winter floods in some El Niño years, particularly on rivers in Arizona with extensive drainage areas at elevations greater than 2000 m. However, the effects of ENSO on other types of storms that may cause floods are more subtle and involve changes in sea-surface temperatures and atmospheric circulation.

During El Niño conditions, the anomalously warm pool of water in the eastern equatorial Pacific Ocean provides a source of precipitable moisture for advection into the extratropical latitudes. Also, the strength of the westerlies tends to increase, providing a steering mechanism for disturbances to move into the southwestern United States. Carleton et al. (1990) discuss in detail some of the atmospheric circulation mechanisms associated with variability of summer monsoon rainfall in the southwest United States and their relationship to ENSO phases.

The relation between El Niño conditions and tropical cyclones in the eastern North Pacific Ocean provides an example of the complex and interrelated changes associated with ENSO. The greatest incidence of tropical cyclones affecting Arizona occurs during or within 6 mo of cessation of El Niño conditions (Smith 1986; Webb and Betancourt 1992; Hereford and Webb in press). Most tropical cyclones in the eastern North Pacific move in a westerly or northwesterly direction; ones that affect the continental United States must have sufficient longevity and concurrently favorable steering winds to recurve towards the northeast.

Generation of tropical cyclones in the eastern North Pacific Ocean is slightly suppressed during El Niño conditions and to a lesser extent during La Niña conditions (Table 3.7). Also, the number of days with hurricane conditions is also lower during El Niño and La Niña conditions (Table 3.7). These results are similar to previous findings for tropical cyclones in the Australian region (Nicholls 1985) and Atlantic Ocean hurricanes for El Niño conditions (Gray 1984). Reasons for the suppression of tropical cyclones in the eastern North Pacific may be increased vertical shearing in the atmosphere caused by increased upper-level westerlies and the southward shift of the ITCZ in those years.

As in the Atlantic Ocean, the number of tropical cyclones in the eastern North Pacific Ocean is not much below average for El Niño years between 1965 and 1990. Although the lowest annual numbers of tropical cyclones occurred in the El Niño years of 1969 and 1977, the second and third largest annual number of tropical cyclones occurred during and just after El Niño conditions in 1982/83 (Fig. 3.12). The above-average number of tropical cyclones between 1982 and 1987 (Fig. 3.12), and consequent increase in hurricane days, reduces the statistical certainty of decreased tropical cyclone activity during El Niño conditions by increasing the interannual variability.

The latitude and longitude of the origin point for tropical cyclones is an important factor in whether the storm may recurve towards the northeast and affect the continental United States. Storms that form too far south and/or west of the Pacific coast of Mexico or other Central American countries must travel farther to reach a position where they could affect the continental United States and are likely to be steered away from the mainland. During July–September of El Niño years, the average point of origin of tropical cyclones tends to be slightly west

Table 3.7 *Numbers of tropical cyclones and duration of hurricanes in the eastern North Pacific Ocean, 1965–1988, for El Niño, La Niña, and 'other' years*

| | Number of years | Mean | Standard deviation |
|---|---|---|---|
| Number of tropical storms and hurricanes per year | | | |
| El Niño years | 9 | 13.8 | 4.4 |
| La Niña years | 4 | 14.8 | 2.8 |
| 'Other' years | 11 | 16.8 | 3.1 |
| Hurricane days per year | | | |
| El Niño years | 9 | 60 | 36 |
| La Niña years | 4 | 51 | 19 |
| 'Other' years | 11 | 72 | 26 |

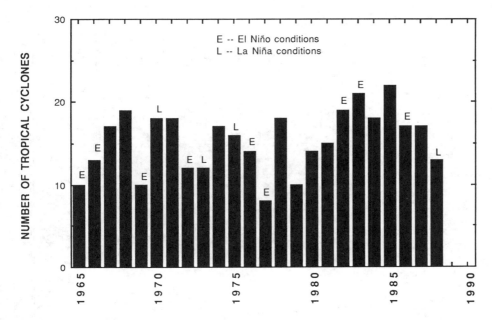

Fig. 3.12 Tropical cyclone occurrence in the eastern North Pacific Ocean, 1965–1988.

or southwest of the point of origin of tropical cyclones in other years (Table 3.8). Again, interannual variability precludes a conclusion of statistically significant changes, but the origin point appears to be shifting towards the pool of warmer water that tends to form off the west coast of Mexico during El Niño conditions. A shifting of the point of origin for tropical cyclones, albeit weak, has also been reported for the South Pacific (Revell and Goulter 1986).

During average El Niño years, fewer tropical cyclones are generated at points farther from the continental United States, and hurricanes that develop from these cyclones have a shorter duration. However, these averages are contrary to the fact that more tropical cyclones dissipate over the southwestern United Stated during or within 6 mo of cessation of El Niño conditions. El Niño-induced changes in the steering mechanism of most recurving tropical cyclones may explain why more tropical cyclones tend to recurve in years in which fewer tropical cyclones are generated.

Cutoff low-pressure systems interact with many tropical cyclones to cause recurvature and advect low-level moisture over the continent (Smith 1986). These upper-atmospheric cyclones often form during the transition from meridional flow to zonal flow and can be defined by examination of 500-mb daily-weather maps (Webb and Betancourt 1992). As Table 3.9 shows, the average number of cutoff low-pressure systems that occurred during El Niño and other years are 4.07 and 3.50, respectively, for fall months. Although large interannual variability precludes statistical significance (Table 3.9), it suggests that the probability

Table 3.8 Locations of origin and numbers of eastern North Pacific Ocean tropical cyclones during El Niño, La Niña and 'other' years, 1949–1989.

| Month | | El Niño | | La Niña | | Other years | |
|---|---|---|---|---|---|---|---|
| | | Latitude | Longitude | Latitude | Longitude | Latitude | Longitude |
| May | Number | 5 | 5 | 0 | 0 | 5 | 5 |
| | Mean | 10.68 | 103.3 | – | – | 12.3 | 103.9 |
| | Variance | 4.44 | 46.8 | – | – | 4.0 | 76.0 |
| June | Number | 16 | 16 | 12 | 12 | 39 | 39 |
| | Mean | 12.98 | 100.99 | 12.8 | 100.4 | 13.1 | 103.2 |
| | Variance | 4.5 | 35.69 | 2.4 | 54.0 | 5.4 | 46.8 |
| July | Number | 27 | 27 | 19 | 19 | 65 | 65 |
| | Mean | 13.64 | 109.77 | 13.9 | 102.8 | 13.7 | 107.3 |
| | Variance | 8.57 | 126.64 | 5.2 | 48.8 | 6.9 | 68.9 |
| August | Number | 39 | 39 | 18 | 18 | 66 | 66 |
| | Mean | 13.63 | 112.69 | 13.9 | 108.6 | 14.1 | 112.5 |
| | Variance | 12.75 | 181.5 | 7.7 | 129.5 | 9.0 | 209.7 |
| September | Number | 54 | 54 | 22 | 22 | 60 | 60 |
| | Mean | 14.27 | 113.77 | 16.9 | 113.0 | 14.9 | 111.0 |
| | Variance | 8.57 | 352.3 | 22.5 | 413.9 | 10.2 | 377.7 |
| October | Number | 30 | 30 | 9 | 9 | 40 | 40 |
| | Mean | 13.47 | 106.96 | 12.6 | 105.0 | 14.1 | 108.8 |
| | Variance | 6.56 | 121.31 | 1.1 | 67.8 | 9.6 | 102.0 |

Table 3.9 *Number of cutoff low-pressure systems affecting the southwestern United States in fall and winter, 1945–1988, for El Niño, La Niña, and 'other' years.*

| | Number of years | Fall Mean | (Sep–Oct) Standard deviation | Winter Mean | (Nov–Mar) Standard deviation |
|---|---|---|---|---|---|
| El Niño years | 15 | 4.07 | 1.62 | 8.67 | 2.29 |
| La Niña years | 7 | 3.71 | 1.60 | 7.33 | 3.39 |
| Other years | 22 | 3.50 | 1.82 | 8.27 | 3.63 |

of occurrence a favorable steering mechanism may be slightly larger during El Niño conditions.

The example of the complex effect of El Niño conditions on tropical cyclones illustrates the complicated effects ENSO also has on southwest United States flood frequency. Under typical conditions, fewer cyclones may be generated with a shorter life expectancy, but the southward extension of the westerlies may increase the potential for recurvature. Moreover, the potential for a favorable steering mechanism, which typically is a cutoff low-pressure system, is also higher during average El Niño conditions. Nevertheless, this coincidence of favorable factors does not occur in all years with El Niño conditions; hence, flood frequency may be affected only in a small, as yet undefined subset of years with El Niño conditions.

## Conclusions

Seasonal snow water content and streamflow are enhanced in the southwestern United States and diminished in the northwestern United States during the mature Northern Hemisphere winter phase of El Niño, and vice-versa during the corresponding phase of La Niña. There is also some evidence for increased streamflow along the Alaskan coast during El Niño. This is consistent with relationships between ENSO and precipitation, discussed by several previous authors. Areas such as California and British Columbia are not reliably connected to ENSO variations, but have been influenced by ENSO conditions in certain extreme cases. There is apparently some predictive value in the ENSO connection to streamflow in the southwestern and northwestern United States, albeit small in terms of variance accounted for. Correlations with the summer Southern Oscillation Index (SOI) leading the following December–August stream discharge have magnitudes of approximately 0.4, and are nearly as strong as those between winter SOI and December–August stream discharge. The reliability of these seasonal correlations appears to hold up from one period to the next; this was demonstrated by splitting the sample into halves and comparing the SOI vs. discharge correlations from the two for Yellowstone River in Montana, and Salt River in Arizona.

While the above correlations are weak, they are statistically significant at a fairly high level of confidence. Ropelewski and Halpert (1986) discounted the ENSO precipitation relationship in the Northwest and Southwest because these relationships were not strongly repeatable from one El Niño or La Niña event to the next. However, the distribution of daily precipitation, stratified seasonally, during El Niño years and during La Niña years appears to be quite distinct. This is demonstrated by randomly choosing sets of years corresponding to the number of years of El Niño and the number of years of La Niña from observed records in the heart of ENSO connections in the northwest and southwest United States. In this Monte Carlo exercise, the lowered frequency of precipitation observed during El Niño in November, December, and January (NDJ) and heightened frequency of precipitation observed during La Niña NDJs in the Yellowstone River, Montana region is unlikely to occur by chance. This difference between El Niño and La Niña is even more apparent in the frequency of heavy precipitation amounts (equal or greater than 10 mm d$^{-1}$) at Yellowstone River. In NDJ, the La Niña precipitation occurs about three times as often as during El Niño. Other years have a precipitation frequency intermediate between El Niño and La Niña. Differences are greater for high precipitation amounts than low precipitation amounts. Similar conclusions are drawn for Salt River, Arizona from analogous comparison between the precipitation frequency of the observed record and the statistics of sets from randomly generated groups of years. However, at Salt River, precipitation during El Niño occurs more frequently, and during La Niña less frequently than expected by chance. At Salt River, NDJ and especially February, March, and April (FMA) are involved – the winter rainy season is lengthened during El Niño. For Salt River heavy precipitation amounts (equal or greater than 10 mm d$^{-1}$) in FMA, the El Niño heavy precipitation occurs about twice as often as during La Niña.

There is some evidence that both the frequency and amount of precipitation changes during ENSO. At Yellowstone River, the NDJ El Niño vs. La Niña reduction in amount of all precipitation days is about as unusual as their El Niño vs. La Niña reduction in frequency. At Salt River, the FMA El Niño vs. La Niña increase in amount of all precipitation days is not as remarkable as their El Niño vs. La Niña increase in frequency, but clearly both phases of ENSO (El Niño and La Niña) have noticeable effects on the precipitation in both regions. 'Other years' have precipitation statistics which tend to fall in-between the El Niño and La Niña frequencies and amounts.

In NDJ, atmospheric patterns creating heavy precipitation during El Niño are very similar to those for La Niña both in interior Northwest (Yellowstone River) and in the Southwest (Salt River). However, in FMA, there is evidence that the synoptic-scale precipitation systems differ between El Niño and La Niña. The El Niño pattern shows evidence of a North Pacific basin-wide activated storm track imbedded in strong westerlies. The La Niña pattern is more confined to the eastern North Pacific and regions downstream, with a high to the west in the Gulf of Alaska vicinity and a low (often a cutoff low) along the West Coast. This same kind of contrast appears for both Yellowstone River and for Salt River heavy FMA precipitation patterns.

It is emphasized that the daily circulation patterns shown in the composites correspond to basin average precipitation events equal or greater than 10 mm $d^{-1}$. Using this threshold, we have included only approximately 7% of the precipitation days at Yellowstone River and approximately 15% of the precipitation days at Salt River. The sample size of these composites ranged from 7 to 24 individual maps. As in all composites, there is a danger that there may be important differences masked by the averaging. For instance, an important issue that this analysis has not addressed is the amount of rain vs. snow, which may affect the water supply in the mountainous western United States. Since some storms within a particular group may be warm and some may be cool, further scrutiny of the cases included in each composite would be useful. Also, further tests should be performed to examine the circulation patterns associated with lighter precipitation events. Finally, since the weather during the between-storm periods is probably also a factor in the surface hydrology (via snow melt, evaporation, etc.), it would also be useful to look at differences between El Niño and La Niña dry spell weather patterns.

Flood frequency in the southwestern United States is affected by the El Niño/Southern Oscillation, but the significance of the effect varies spatially and the climatic causes are less clear. Many of the largest floods in Arizona have occurred during El Niño years. Also, certain rivers have an increased flood frequency during or within 6 mo after cessation of El Niño conditions. However, rivers strongly affected by winter storms and snowmelt may be affected more by storms that occur in years other than El Niño. This may be explained partially by lag effects after El Niño occurrence, or by floods occurring during winters with conditions of persistent, slightly negative SOI conditions that do not develop into full-fledged El Niños, such as the winter of 1978/79.

The types of storms that generate flooding are affected by El Niño conditions, but the effects are not consistent among El Niño years. Winter storms during some El Niño years may be enhanced. Although slightly fewer tropical cyclones are generated during El Niño and La Niña years, the number of dissipating tropical cyclones that affect the southwestern United States is greatest during or following El Niño conditions. Similarly, the number of cutoff low-pressure systems affecting the southwestern United States is higher during El Niño years, but the increase is not statistically significant. The occurrence of El Niño conditions per se is not sufficient to explain increased flood frequency; instead, flood frequency may be enhanced only during a small subset of El Niño events.

*Acknowledgments* We thank Mike Dettinger, Henry Diaz, Art Douglas, and Katie Hirschboeck for many useful comments. We thank Editors Diaz and Markgraf for organizing the Paleoclimatic-ENSO Workshop and for encouragement and patience in seeing this manuscript through to publication. Larry Riddle and Emelia Bainto processed data, carried out analyses, and plotted results.

Marguerette Schultz drafted several figures and Jean Seifert word processed the manuscript. Part of the work by DRC was supported by the NOAA Experimental Climate Forecast Center and by the University of California Water Resources Center under Grant W-768.

## References

ANDRADE, E. R. and SELLERS, W. D., 1988: El Niño and its effect on precipitation in Arizona and western New Mexico. *Journal of Climate*, 8: 403–410.

BARNSTON, A. G. and LIVEZEY, R. E., 1987: Classification, seasonality and persistence of low-frequency atmospheric circulation patterns. *Monthly Weather Review*, 115: 1083–1126.

CARLETON, A. M., CARPENTER, D. A., and WESER, P. J., 1990: Mechanisms of interannual variability of the southwest United States summer rainfall maximum. *Journal of Climate*, 3: 999–1015.

CAYAN, D. R. and PETERSON, D. H., 1989: The influence of North Pacific atmospheric circulation on streamflow in the west. *In* Peterson, D. H. (ed.), *Aspects of Climate Variability in the Pacific and the Western Americas*. Geophysical Monograph 55. Washington, D.C.: American Geophysical Union, 375–397.

DAVIS, R. E., 1978: Predictability of sea level pressure anomalies over the North Pacific Ocean. *Journal of Physical Oceanography*, 8: 233–246.

DIAZ, H. F., 1991: Some characteristics of wet and dry regimes in the contiguous United States: Implications for climate change detection efforts. *In* Schlesinger, M. E. (ed.), *Greenhouse-Gas-Induced Climatic Change: A Critical Appraisal of Simulations and Observations*. Amsterdam: Elsevier, 269–296.

DOUGLAS, A. V. and ENGLEHART, P. J., 1981: On a statistical relationship between rainfall in the central equatorial Pacific and subsequent winter precipitation in Florida. *Monthly Weather Review*, 109: 2377–2382.

FU, C., DIAZ, H. F., and FLETCHER, J. O., 1986: Characteristics of the response of sea-surface temperature in the central Pacific associated with warm episodes of the Southern Oscillation. *Monthly Weather Review*, 114: 1716–1738.

GRAY, W. M., 1984: Atlantic seasonal hurricane frequency. Part I: El Niño and 30 mb quasi-biennial oscillation influences. *Monthly Weather Review*, 112: 1649–1668.

HEREFORD, R. and WEBB, R. H., in press: Historic variation in warm-season rainfall on the Colorado Plateau, U.S.A. *Climatic Change*.

HIRSCHBOECK, K. K., 1985: Hydroclimatology of flow events in the Gila River Basin, central and southern Arizona. Ph.D. dissertation, University of Arizona, 335 pp.

HIRSCHBOECK, K. K., 1987: Hydroclimatically defined mixed distributions in partial duration flood series. *In* V. P. Singh (ed.), *Hydrologic Frequency Modeling*, Dordrecht: Reidel, 199–212.

KILADIS, G. N. and DIAZ, H. F., 1989: Global climatic anomalies associated with extremes in the Southern Oscillation. *Journal of Climate*, 2: 1069–1090.

KLEIN, W. H., 1957: Principal tracks and mean frequencies of cyclones and anti-cyclones in the Northern Hemisphere. U.S. Weather Bureau, Washington, D.C. *Research Paper* 40. 60 pp.

KLEIN, W. H. and BLOOM, H. J., 1987: Specification of monthly precipitation over the United States from the surrounding 700 mb height field. *Monthly Weather Review*, 115: 2118–2132.

LIVEZEY, R. E. and CHEN, W. Y., 1983: Statistical field significance and its determination by Monte-Carlo techniques. *Monthly Weather Review*, 111: 46–59.

LIVEZEY, R. E. and MO, K. C., 1987: Tropical-extratropical teleconnections during the northern hemisphere winter, Part II: Relationships between monthly mean northern hemisphere circulation patterns and proxies for tropical convection. *Monthly Weather Review*, 115: 3115–3132.

NAMIAS, J., 1981: Teleconnections of 700 mb Height Anomalies for the Northern Hemisphere. CalCOFI Atlas No. 29. Fleminger, Marine Life Research Program, Scripps Institution of Oceanography, University of California, San Diego, La Jolla, California.

NAMIAS, J., YUAN, X., and CAYAN, D. R., 1988: Persistence of North Pacific sea surface temperature and atmospheric flow patterns. *Journal of Climate*, 1: 682–703.

NICHOLLS, N., 1985: Predictability of interannual variations of Australian seasonal tropical cyclone activity. *Monthly Weather Review*, 113: 1144–1149.

PETERSON, D. H., CAYAN, D. R., and FESTA, J. F., 1986: Interannual variability in biogeochemistry of partially mixed estuaries dissolved sillate cycles in northern San Francisco Bay. *In* Wolfe, D. A. (ed.), *Estuarine Variability*. New York: Academic Press; 123–138.

QUINLAN, F. T., KARL, T. R., and WILLIAMS, C. N., JR., 1987: United States Historical Climatology Network (HCN) serial temperature and precipitation data. Prepared by T. A. Boden, Carbon Dioxide Information Analysis Center, Environmental Sciences Division, Oak Ridge National Laboratory, Oak Ridge, Tennessee.

QUINN, W. H., NEAL, V. T., and ANTUNEZ de MAYOLO, S. E., 1987: El Niño occurrences over the past four and a half centuries. *Journal of Geophysical Research*, 92: 14,449–14,461.

RASMUSSON, E. M. and CARPENTER, T. H., 1982: Variations in tropical sea surface temperature and surface wind fields associated with the Southern Oscillation/El Niño. *Monthly Weather Review*, 110: 354–384.

REDMOND, K. and KOCH, R., 1991: ENSO vs. surface climate variability in the western United States. *Water Resources Research*, 27: 2381–2399.

REVELL, C. G. and GOULTER, S. W., 1986: South Pacific tropical cyclones and the Southern Oscillation. *Monthly Weather Review*, 114: 1138–1145.

ROADS, J. O. and MAISEL, T. N., 1991: Evaluation of the National Meteorological Center's medium range forecast model precipitation forecasts. *Weather and Forecasting*, 6: 123–132.

ROPELEWSKI, C. F. and HALPERT, M. S., 1986: North American precipitation and temperature patterns associated with the El Niño/Southern Oscillation (ENSO). *Monthly Weather Review*, 114: 2352–2362.

ROPELEWSKI, C. F. and JONES, P. D., 1987: An extension of the Tahiti-Darwin Southern Oscillation index. *Monthly Weather Review*, 115: 2161–2165.

SCHONHER, T. and NICHOLSON, S. E., 1989: The relationship between California rainfall and ENSO events. *Journal of Climate*, 2: 1258–1269.

SMITH, WALTER, 1986: The effects of eastern North Pacific tropical cyclones on the southwestern United States. Salt Lake City, Utah, *NOAA Technical Memorandum*, NWS WR-197. 229 pp.

TRENBERTH, K. E. and PAOLINO, D. A., 1980: The northern hemisphere sea-level pressure data set trends, errors, and discontinuities. *Monthly Weather Review*, 108: 855–872.

U.S. Department of the Interior, Geological Survey, 1975: Index stations and selected large-river streamgaging stations in the west. *Water Resources Review*, p. 11.

WEBB, R. H. and BETANCOURT, J. L., 1992: Climatic variability and flood frequency of the Santa Cruz River, Pima County, Arizona. *U.S. Geological Survey, Water-Supply Paper*, 2379. 40 pp.

YARNAL, B. and DIAZ, H. F., 1986: Relationships between extremes of the Southern Oscillation and the winter climate of the Anglo-American Pacific Coast. *Journal of Climate*, 6: 197–219.

# 4

# Coupled climate model simulation of El Niño/Southern Oscillation: implications for paleoclimate

GERALD A. MEEHL AND GRANT W. BRANSTATOR

*National Center for Atmospheric Research, Boulder, Colorado 80307–3000, U.S.A.*

## Abstract

Paleoclimatic reconstructions of El Niño/Southern Oscillation (ENSO) generally assume a constant extratropical response to ENSO tropical sea surface temperature (SST) anomalies through time. Results shown here from a coupled general circulation model (GCM) indicate that this assumption may not always be valid. A global, coarse-grid coupled ocean-atmosphere GCM is used to study climate sensitivity associated with a significant change of external forcing (a doubling of atmospheric carbon dioxide ($CO_2$), or $2 \times CO_2$). This increase of $CO_2$ in the model could be considered analogous to other changes in external forcing that have occurred in the paleoclimatic record. The result is a significant alteration of the atmosphere-ocean circulation. In the model with doubled $CO_2$ ($2 \times CO_2$), ENSO continues to function in the tropics in much the same way as it does contemporarily (in relative terms) but with mean SSTs in the tropical eastern Pacific higher by about 1°C. However, the changed mean climate in the extratropics with doubled $CO_2$ is associated with altered extratropical teleconnection patterns in a composite $2 \times CO_2$ El Niño, compared to a composite present-day ($1 \times CO_2$) El Niño. These altered teleconnection patterns are thought to be associated with the changed extratropical circulation due to the modified external forcing associated with increased $CO_2$. We test this hypothesis with two alternate model configurations: (1) an atmospheric GCM coupled to a simple mixed-layer ocean with prescribed ENSO SST anomalies in the tropical eastern Pacific and (2) an atmospheric GCM with idealized tropical heating anomalies. All the model results show that a change in the equilibrium state in the extratropics associated with altered external forcing of the climate system causes a systematic change in the extratropical teleconnections from ENSO events. The implication for paleoclimate from these experiments is that a change in past climate (e.g., due to a change in orbital parameters and resulting adjustment of snow-ice extent and atmospheric circulation) would produce virtually unaltered ENSO events in the tropics but a different response of the climate system in the extratropics than presently observed. This could be associated with a different signature of ENSO in the paleoclimatic record in the extratropics compared to present.

## Introduction

El Niño/Southern Oscillation (ENSO) is a global mode of weather and climate variability accompanied by extreme regional weather events (droughts, floods, etc.) that have significant signatures in the paleoclimate record. Climate change induced by a change in external forcing in the atmosphere could, conceivably, affect the dynamics and thermodynamics associated with ENSO in ways that are not intuitively obvious. Consequently, this necessitates the use of computer climate models to determine what effects, if any, a change in external forcing could have on ENSO. Only recently have such global coupled ocean-atmosphere general circulation models (GCMs) been capable of internally generating some aspects of ENSO phenomena (Sperber et al. 1987; Philander et al. 1989; Meehl 1990). In these coarse-grid models, only a subset of the processes thought to be taking place in observed ENSOs is simulated.

For example, Meehl (1990) shows that processes in the eastern Pacific associated with a modulation of the seasonal cycle in that region are present in ENSOs simulated in a global coupled GCM developed at the National Center for Atmospheric Research. In that model, features of ENSO in the western Pacific are not well simulated. Anomalous warm water associated with the inception of ENSO in the eastern Pacific in the model appears off the South American coast in northern spring and moves toward the date line. During the mature phase of ENSO events in the coupled model (December–January–February, or DJF), going from the year of the inception of the event to the year following (commonly perceived as the time of year when teleconnections to the extratropics in the Northern Hemisphere are strongest), positive sea surface temperature (SST) anomalies are established near the date line. Associated with those SST anomalies in the equatorial Pacific are relatively low sea-level pressure (SLP) in the eastern Pacific and relatively high SLP over southern Asia and the western Pacific, the signal of the Southern Oscillation (SO). In the extratropics, the effects on the Northern Hemisphere atmospheric circulation (referred to as teleconnections) resemble those during observed ENSO events and include a deepened Aleutian low and a weak ridge over the western United States.

These extratropical teleconnections are typically assumed to remain more or less constant by-products of ENSO events over time when performing paleoclimatic reconstructions and identifying ENSO signals in the proxy record.

However, long-term fluctuations of climate (a result of changes in external forcing or large-amplitude low-frequency inherent climate variability) could lead to altered patterns of extratropical teleconnections that could have important implications for ENSO in the extratropical paleoclimatic record. The purpose of this paper is to explore the effects on ENSO of a significantly different external forcing and the consequential alteration of ENSO teleconnections associated with the resulting change of atmospheric circulation in the extratropics. We will analyse results from a coupled ocean-atmosphere GCM and expand on those results with two other simpler model configurations.

## Model description and experimental design

The coupled model includes a global, spectral atmospheric GCM with rhomboidal 15 (R15) resolution (about 4.5° latitude by 7.5° longitude), realistic geography, nine layers in the vertical, and parameterized land-surface processes. The global ocean GCM has a latitude-longitude resolution of 5°, realistic geography, bottom topography, four layers, and a simple thermodynamic sea-ice formulation. Washington and Meehl (1989) provide details of the atmosphere and ocean models, as well as the thermodynamic sea-ice formulation (Fig. 4.1).

The other model configurations described herein involve variants of the atmospheric GCM with different ocean surface formulations. In the second configuration, we couple the atmospheric GCM to a simple mixed-layer ocean. In a third configuration used for idealized heating experiments, we run the

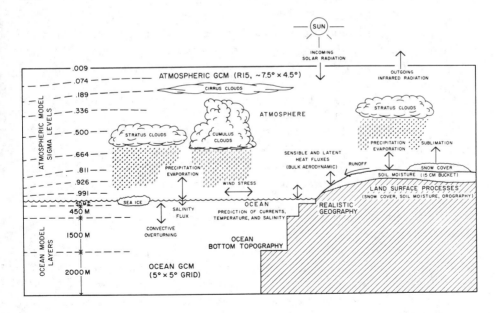

Fig. 4.1 Schematic diagram of the components of the coupled ocean-atmosphere GCM.

atmospheric GCM in perpetual January mode with observed January SSTs and prescribed heating anomalies, as described by Branstator (1990).

The results of the coupled atmosphere-ocean GCM are from two 30-yr experiments (Washington and Meehl 1989) – one is run with present amounts of carbon dioxide ($CO_2$) ($1 \times CO_2$) and one with instantaneously doubled $CO_2$ ($2 \times CO_2$). Only the results after year 20 are available for time-series analysis. Some results are derived from the last 10 yr of each integration (years 21–30). Recently, the experiments were extended 5 more years, and some results are for the last 15 yr (years 21–35).

In the second configuration, an ENSO anomaly is specified in a version of the model with the atmosphere coupled to a simple slab (nondynamic) mixed-layer 50-m deep ocean (mixed-layer model here). This model has been run to equilibrium in $1 \times CO_2$ and $2 \times CO_2$ experiments (Meehl and Washington 1990). The ENSO SST anomaly was placed in roughly the same position in the tropical Pacific (175°W–125°W, 10°N–10°S) as the comparable SST anomaly generated in the coupled model during the mature phase DJF season. We started the mixed-layer model in July with this anomaly and ran it to the following 1 March for the $1 \times CO_2$ and $2 \times CO_2$ cases. We ran eight integrations for each case starting on consecutive days in July. The SST anomaly relative to the respective mean SST in the tropical Pacific was about $+1°$ averaged over the area of the specified anomaly, with values ranging from roughly $+0.5$ to $+2.0$. The amplitude of these anomalies is comparable to that of previous specified El Niño experiments with the atmospheric model (Blackmon et al. 1983; Meehl and Albrecht 1991) and is about twice those of the coupled model. However, these latter two studies specified SST anomalies only over the tropical eastern Pacific, whereas the coupled model and mixed-layer model configurations allow SST elsewhere to respond. The omission of extratropical ENSO SST anomalies was thought to contribute to some of the model errors in reproducing the observed extratropical response in those earlier experiments (Meehl and Albrecht 1991). In the coupled model, the global pattern of SST anomalies associated with ENSO is surprisingly well simulated (Meehl 1990).

The third configuration is designed to test the effects of idealized tropospheric heating (assumed to be associated with an ENSO tropical SST anomaly) in the atmospheric GCM run with observed January SSTs in a perpetual January mode. An imposed steady heat source specified in the model acts directly on the atmosphere, thereby eliminating the question of efficiency of SST anomalies in transferring heating effects associated with enhanced convection to the atmosphere (Branstator 1990). The heat source is maximum at a specified point at the equator and decreases linearly in all directions from that point to 1500 km. The heating is sinusoidal in the vertical with wavelength 2 in sigma coordinates (such that heating is maximized in the midtroposphere) with a peak value of $2.5 \text{ K d}^{-1}$ in the midtroposphere. We ran the model for 600 d and computed averages for the last 500 d of simulation. These averages are differenced from a 6000-d control run with no heating anomalies.

### ENSO simulated in the coupled model

The SO, of which ENSO is a manifestation, is a seesaw of atmospheric pressure between the Southeast Asian/Australian region and the tropical eastern Pacific. The characteristic opposition of sign of SLP anomalies associated with the SO is evident for observational data from Tahiti and Darwin (Fig. 4.2a). A similar oscillation is evident in the coupled model (Fig. 4.2b) if area-averaged SLP anomalies are computed for an area north of Australia (Indonesia) and an area in the tropical eastern Pacific (NINO3).

Meehl (1990) describes the structure and evolution of the SO in the coupled model involving interacting sets of processes in the atmosphere and ocean. Composite anomalies for ENSO events or warm events (low phase of the SO in the model, that is, low SLP in the eastern Pacific, high in the western Pacific) are reproduced from Meehl (1990) and shown in Fig. 4.3 for the northern summer season just as an ENSO event has begun. In Fig. 4.3a, warm-water anomalies have become established in the eastern equatorial Pacific with cool water to the west. Associated with these SST anomalies are low SLP over the warm water in the tropical eastern Pacific (Fig. 4.3b) and high SLP to the west, extending to the

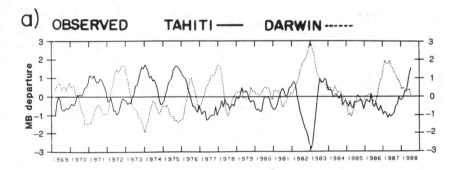

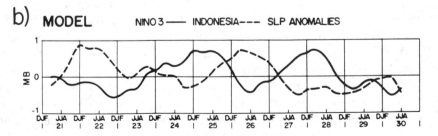

Fig. 4.2   (a) SLP anomalies for Tahiti (solid line) in the eastern tropical Pacific and Darwin (dotted line) in north Australia; and (b) area-averaged SLP anomalies from the coupled model for the NINO3 area in the eastern equatorial Pacific (10 °N–10 °S, 90 °W–150 °W, solid line) and for Indonesia (10 °N–13 °S, 110 °E–155 °E, dashed line).

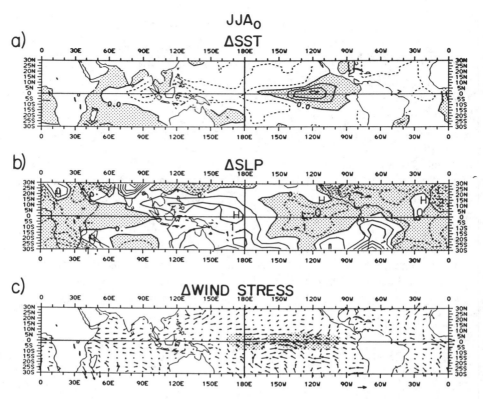

Fig. 4.3   ENSO events in the coupled model usually begin in northern spring (March-April-May, or MAM). For the season June-July-August (JJA) shown here, ENSO warm-event composites minus cold-event composites (two of each event from years 21–30 in the model) give the sign of the anomalies associated with an ENSO event in the tropics. (a) SST anomalies, positive differences are stippled; (b) SLP anomalies, negative differences are stippled; and (c) surface wind-stress vector anomaly differences with stippled areas representing suppressed upwelling between 10 °N and 10 °S. Scaling arrow at bottom right is 0.3 dyn cm$^{-2}$ (0.03 N m$^{-2}$).

region of the Indian monsoon. These SLP anomalies cause near-equatorial ageostrophic surface wind-stress anomalies just to the west of the positive SST anomalies (Fig. 4.3c) and a suppression of upwelling in the ocean (stippled area in Fig. 4.3c). The coupled sets of anomalies then propagate to the west as the event evolves with the seasonal cycle.

Figure 4.4 shows the movement of the coupled anomalies associated with ENSO. Time-longitude plots of composite events are shown for SST differences and $u$-component wind-stress differences (Fig. 4.4a, b) from the coupled model and for observed events (Fig. 4.4c, d for $u$-component surface wind). The establishment of sustained positive SST anomalies in the tropical eastern Pacific in the model occurs in the northern spring-summer season (horizontal line in Fig. 4.4). Westerly wind-stress anomalies lie just to the west of the warm water, and the entire set of anomalies moves to the west.

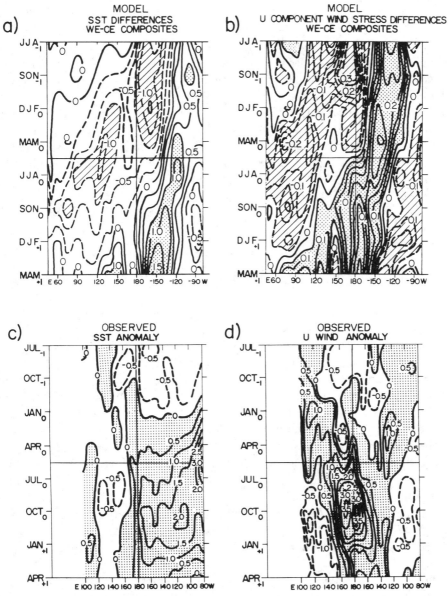

Fig. 4.4 Time-longitude plots showing the evolution of composite ENSO (warm) events; vertical line is the date line, horizontal line is the northern spring to northern summer transition (MAM) for the onset of an event. (a) warm-minus-cold-event composite SST differences (°C) from the coupled model averaged from 10 °N–10 °S. Stippled areas are anomalies greater than +0.75 °C, hatching less than −0.75 °C; (b) warm-minus-cold-event composite *u*-component (east-west) wind stress differences averaged from 10 °N–10 °S; stippling for anomalies greater than +0.5 dyn cm$^{-2}$ (+0.05 N m$^{-2}$); (c) observed SST anomalies along the equator for composite of 1957, 1965, and 1972 ENSO events (after Rasmusson et al. 1986); stippling indicates positive SST anomalies; and (d) same as (c) except for *u*-component surface wind anomalies; stippling indicates positive (westerly) wind anomalies. (From Meehl 1990.)

For the observations in Fig. 4.4c, d, seasonal timing and movement of SST and $u$-wind anomalies are similar to the model in the eastern Pacific. However, a large signal of westerly wind anomalies starts in the western Pacific the year before the event and moves east to meet the anomalies propagating west from the eastern Pacific. Small-amplitude positive SST anomalies are associated with the $u$-wind anomalies in the western Pacific in the observed composites. These become larger near the date line in the northern spring of the year of an event. Comparing Fig. 4.4a and b with Fig. 4.4c and d, we find that the coupled model is not capturing the processes taking place in the western Pacific. Meehl (1990) concludes that the coupled model is simulating a subset of coupled instabilities associated with ENSO. Yet, Meehl (1990) shows that the ENSO events internally generated by the coupled model are associated with global patterns of anomalies that resemble in many respects those anomalies observed during ENSO.

### ENSO events with altered external forcing

Figure 4.5 depicts the model interannual variability in the tropical Pacific for years 21–35. Area averages of SST anomalies for the NINO3 area (90°W–150°W, 10°N–10°S) for $1 \times CO_2$ (Fig. 4.5a) and $2 \times CO_2$ (Fig. 4.5b) indicate that area-averaged SSTs in this region oscillate about their respective means. The $2 \times CO_2$ case shows a slight upward trend, thus reflecting the slow model approach to equilibrium after the instantaneously doubling of $CO_2$ at the beginning of the experiment (Washington and Meehl 1989). The trend is considerably less than the interannual fluctuations. The amplitude of the area-averaged SST anomalies with respect to the $1 \times CO_2$ and $2 \times CO_2$ means is about 0.5°. The significance of changes in amplitude of the SST oscillation on longer time scales is difficult to assess because of the short period shown here and inherent low-frequency variability. This issue is being explored as longer time series from the coupled model become available. Geographical plots of the model-generated SST anomalies show maximum values on the order of 1° (Meehl 1990). This is somewhat less than observed ENSO SST anomalies (Rasmusson and Carpenter 1982). Significantly, the $2 \times CO_2$ SSTs are warmer in the mean by about a degree and *SSTs continue to oscillate about the warmer mean in the $2 \times CO_2$ case*.

Figure 4.5c, d shows SLP anomalies for the NINO3 area and for an area in the western tropical Pacific (Indonesia, 110°E–155°E, 10°N–13°S) for $1 \times CO_2$ and $2 \times CO_2$. Both cases show the opposition of sign of anomalies characteristic of the SO (van Loon and Madden 1981) and noted in Fig. 4.2. The oscillation of SLP is coincident with the SST oscillation with relatively lower SLP in the NINO3 area coinciding with higher SST anomalies. This is consistent with the observed associations of SLP and SST in the tropical eastern Pacific (Lindzen and Nigam 1987). The mean SLP for the $2 \times CO_2$ case is relatively lower than the $1 \times CO_2$ case associated with the warmer mean SSTs caused by the increase of $CO_2$.

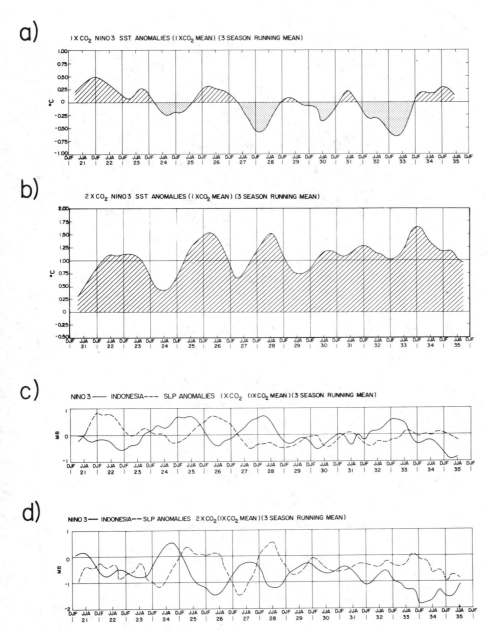

Fig. 4.5 Area-averaged SST anomalies for the NINO3 area in the tropical eastern Pacific (10 °N–10 °S, 90 °W–150 °W) for the coupled model experiment (a) $1 \times CO_2$ seasons minus the $1 \times CO_2$ long-term mean; (b) $2 \times CO_2$ seasons minus the $1 \times CO_2$ long-term mean; hatching indicates positive anomalies; stippling is negative anomalies relative to the $1 \times CO_2$ mean. Area-averaged SLP anomalies for NINO3 (solid line) and Indonesian area (dashed line) for (c) $1 \times CO_2$ seasonal values minus the $1 \times CO_2$ long-term mean, and (d) $2 \times CO_2$ seasonal values minus the $1 \times CO_2$ long-term mean.

a)
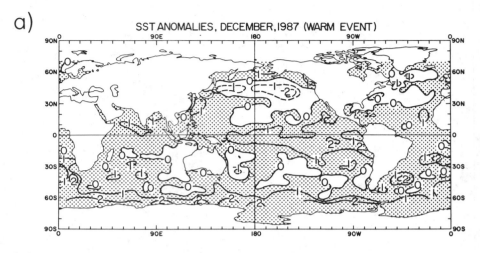
SST ANOMALIES, DECEMBER, 1987 (WARM EVENT)

b)
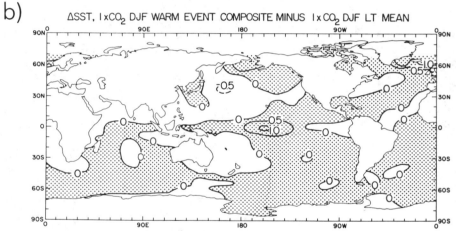
ΔSST, 1×CO₂ DJF WARM EVENT COMPOSITE MINUS 1×CO₂ DJF LT MEAN

c)
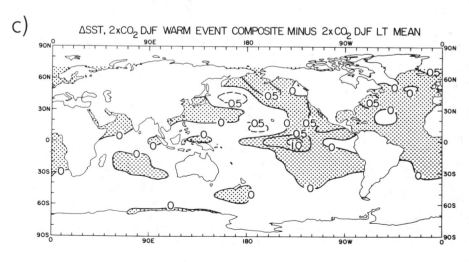
ΔSST, 2×CO₂ DJF WARM EVENT COMPOSITE MINUS 2×CO₂ DJF LT MEAN

We define ENSO events in this coupled model as an area-averaged SST anomaly in the NINO3 region greater than $+0.25°C$ and SO Index (derived from the SLP anomalies in Fig. 2.3.5) less than $-0.5$ during the year of initiation of an ENSO event (Meehl 1990). For the $1 \times CO_2$ experiment, in the 15-yr period from years 21–35 there are four such DJF ENSO event seasons (years 22–23, 23–24, 26–27, and 34–35). Four DJF ENSO event seasons can be identified in the $2 \times CO_2$ experiment defined in relation to the mean $2 \times CO_2$ climate (years 25–26, 28–29, 31–32, and 33–34).

We compute composites for the four warm-event DJF seasons in each of the two experiments. From the averages of the $1 \times CO_2$ and $2 \times CO_2$ warm events, we subtract the $1 \times CO_2$ and $2 \times CO_2$ 15-yr mean northern winter SSTs. These differences show the changes in ENSO phenomena, compared to the respective $1 \times CO_2$ and $2 \times CO_2$ mean climates. Fig. 4.6a shows the observed SST anomalies for the recent warm event of December 1987. The typical observed warm-event pattern of positive SST anomalies in the central equatorial Pacific and negative SST anomalies to the north and southwest are present in the $1 \times CO_2$ and $2 \times CO_2$ composites (Fig. 4.6). The amplitude of the warm anomaly in the tropical eastern Pacific is about half of the observed anomaly and is limited to a narrower longitudinal extent (Fig. 4.4). Even though the mean SST has warmed in the $2 \times CO_2$ case (Fig. 4.5b), the amplitude and pattern of the SST anomalies for the $2 \times CO_2$ ENSO events (Fig. 4.6c) are similar to those for the $1 \times CO_2$ ENSO events (Fig. 4.6b).

Peak SST anomaly values for the composite $1 \times CO_2$ ENSO events are located near 160°W, and around 140°W for the $2 \times CO_2$ ENSO events. This relatively small shift in longitudinal position should not have a large effect on the pattern of extratropical teleconnections in the model (Geisler et al. 1985).

Figure 4.7 depicts SLP anomalies for the same season. The correspondence is good between the amplitude and pattern of the SLP anomalies in the tropics in the $1 \times CO_2$ and $2 \times CO_2$ ENSOs, with relatively low SLP over the warm water in the eastern Pacific and high SLP in the western Pacific and Indian Ocean, as observed. Since the positive SST anomalies do not extend all the way to South America from the central equatorial Pacific (the narrower longitudinal extent noted in Figs. 4.4 and 4.6), the negative SLP anomalies are also of smaller amplitude in the far eastern equatorial Pacific (Fig. 4.7b) compared to the observed values (Fig. 4.7a). The extratropical pattern in the $2 \times CO_2$ warm events is markedly different from both the observed and the $1 \times CO_2$ composites. Relatively high pressure dominates most of the high northern latitudes

Fig. 4.6 (a) Observed warm-event (ENSO event) SST anomalies for December 1987 from Climate Analysis Center; solid lines (stippling) are positive values, dashed lines are negative anomalies; (b) SST anomalies from the coupled model for $1 \times CO_2$ warm-event composite minus $1 \times CO_2$ 15-yr mean, DJF season for mature phase (end of year zero-beginning of year plus one), four events from a 15-yr period; (c) same as (b) except for $2 \times CO_2$ warm event composites minus $2 \times CO_2$ 15-yr mean.

a)

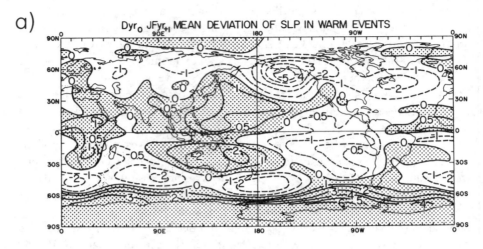

b)

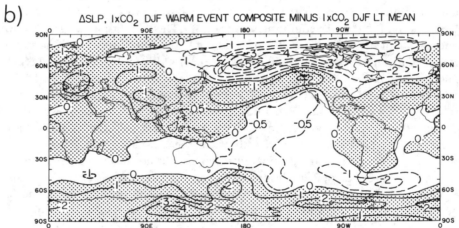

c)

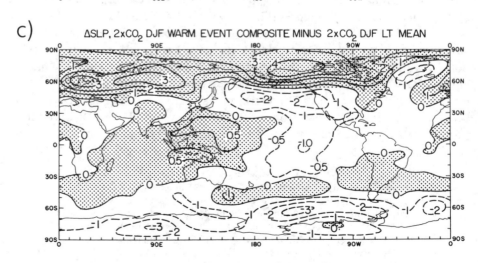

in the $2 \times CO_2$ case compared to low pressure in the observed and $1 \times CO_2$ case, whereas relatively low pressure appears near 60–70°S in the $2 \times CO_2$ case (Fig. 4.7c) where positive SLP anomalies had appeared in the observed and $1 \times CO_2$ case (Fig. 4.7a, b).

The changes in lower tropospheric temperature that accompany the SLP anomalies are considerable. Temperature differences at 850 mb for the $1 \times CO_2$ ENSO events and $2 \times CO_2$ ENSO events in Fig. 4.8 show patterns that are considerably different in the extratropics of both hemispheres. For example, over North America in the $1 \times CO_2$ ENSO events, positive temperature anomalies are evident over Alaska and most of Canada, while over the eastern United States, there are negative anomalies (similar to observed composites). In the $2 \times CO_2$ ENSO composites from the coupled model, the pattern is reversed, with negative temperature anomalies over Alaska and Canada and positive anomalies over the eastern United States. Notable changes are also evident over Europe, Central Asia, and Australia. The circulation patterns through the depth of the troposphere associated with the surface anomalies are consistent in that the extratropical anomalies are quite different in the $1 \times CO_2$ and $2 \times CO_2$ ENSO events. Height anomalies at 200 mb (Fig. 4.9) for the $1 \times CO_2$ and $2 \times CO_2$ events are similar in the tropics (higher heights at all longitudes indicative of a warming of the midtroposphere, as observed). But the extratropical anomalies in the $1 \times CO_2$ events (deepened Aleutian low, ridge over the western United States, and low over the eastern United States with positive height anomalies in the high latitudes are changed in the $2 \times CO_2$ events. These are characterized by more zonally uniform positive height anomalies in the Northern Hemisphere high latitudes and negative height anomalies in the Southern Hemisphere high latitudes, as well as large phase shifts in regional anomalies.

## Simulations with alternate model configurations

Because only four ENSO events are available for each of the $1 \times CO_2$ and $2 \times CO_2$ coupled model time series, the question arises as to whether the altered extratropical teleconnections in the $2 \times CO_2$ ENSO events are systematically different or statistically significant. If comparable signals can be generated for other model configurations with revised experimental schemes and different forcing, greater confidence can be placed in the coupled model results.

The first alternate model configuration is the atmospheric model coupled to the mixed-layer ocean described earlier. Most aspects of the $2 \times CO_2$ climate of the mixed layer are similar to those of the coupled model (Washington and Meehl

---

Fig. 4.7   (a) Observed SLP anomalies for DJF during mature phase of warm events minus long-term mean that does not include warm-event seasons from van Loon (1986); stippling indicates positive anomalies; (b) SLP anomalies from the coupled model for present-day ($1 \times CO_2$) warm-event composites minus $1 \times CO_2$ 15-yr mean for DJF during mature phase; (c) same as (b) except for $2 \times CO_2$ warm-event composites minus $2 \times CO_2$ 15-yr mean from the coupled model.

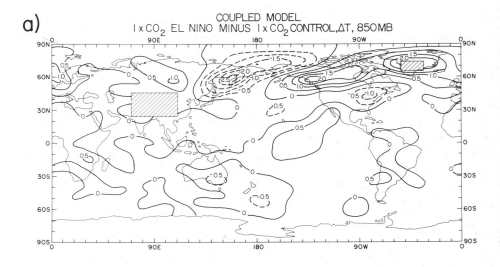

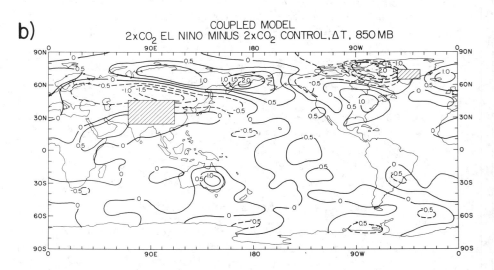

Fig. 4.8   Temperature anomalies in the lower troposphere (850 mb). (a) $1 \times CO_2$ El Niño composite (4 events) for northern winter (after the inception of an event) minus 15-yr $1 \times CO_2$ DJF average; and (b) same as (a) except for $2 \times CO_2$. Areas with no contours (hatching in the Northern Hemisphere and over Antarctica) are at elevations above the 850-mb level.

1991). The main differences occur around Antarctica and in the North Atlantic where the coupled model shows much smaller-amplitude changes of temperature due to ocean mixing. Figure 4.10 shows 200-mb height differences for ENSO SST anomaly ensemble averages (eight cases each) minus the long-term $1 \times CO_2$ and $2 \times CO_2$ averages for each of the $1 \times CO_2$ ENSO and $2 \times CO_2$ ENSO experiments. The area-averaged SST anomalies in the eastern equatorial Pacific

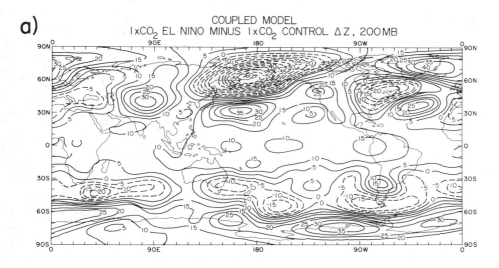

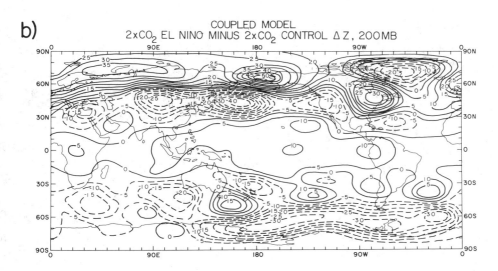

Fig. 4.9  Height differences at 200 mb from the coupled GCM. (a) composite of four ENSO events in the $1 \times CO_2$ case for northern winter after the inception of an event minus 15-yr DJF average of the $1 \times CO_2$ control; dashed lines are negative differences, solid lines are positive differences; and (b) same as (a) except for $2 \times CO_2$.

are more than double those in the coupled model. For the area 10°N–10°S, 180°W–120°W, the coupled model peak value of the ENSO SST anomaly is about $+1.0$°C; for the mixed-layer model it is slightly greater than $+2.0$°C. The mean background area-averaged SST is also greater in the mixed-layer model. The SST averaged for that area in the coupled model is 23.6°C for the $1 \times CO_2$ case and 28.7°C for the mixed-layer model, with the observed value in between at 26.5°C. Consequently, convection is weaker in the tropics in the coupled

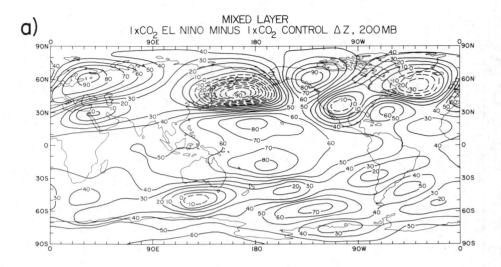

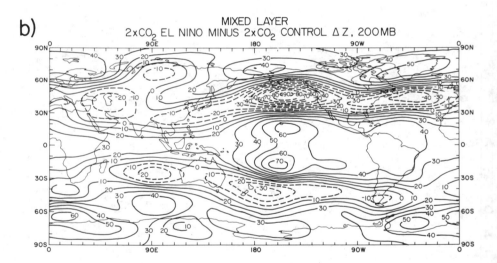

Fig. 4.10   Height differences at 200 mb from the atmospheric GCM coupled to the simple mixed-layer ocean. (a) Composite of 8 winter seasons with a specified ENSO SST anomaly in the equatorial eastern Pacific (all starting the previous July) minus an 8-yr DJF average from the $1 \times CO_2$ case in Meehl and Washington (1990); (b) same as (a) except for $2 \times CO_2$ differences greater than roughly 20 m in the tropics and about 30 m in the extratropics are significant at the 5% level.

model (Washington and Meehl 1989). However, precipitation anomalies are closely tied to tropical SST anomalies with enhanced precipitation associated with positive SST anomalies (Meehl 1990). Due to the nonlinear relationship between SST and evaporation, a positive SST anomaly in the mixed-layer model tropics is associated with relatively greater evaporation and absolute precipitation increase than in the coupled model.

The 200-mb height differences in Fig. 4.10 are larger in the mixed-layer model, particularly in the tropics, but the patterns are similar to those from the coupled model in Fig. 4.9. In the $1 \times CO_2$ ENSO case, we see a deepened Aleutian low, a ridge over western North America, and a low shifted southwest to central North America. In the Southern Hemisphere, changes are also evident with a band of positive height anomalies near $50°S$ in the Pacific in the $1 \times CO_2$ case with negative anomalies in the $2 \times CO_2$ case. For the $2 \times CO_2$ ENSO case (Fig. 4.10b), there is zonalization of height anomalies similar to that seen in Fig. 4.9b. Differences in the tropics greater than about 20 m and anomalies in the extratropics greater than roughly 30 m are significant at the 5% level. All of the large-amplitude features of the height anomalies in Fig. 4.10 are statistically significant. These results imply the existence of physical processes that significantly alter extratropical teleconnections with the change of external forcing supplied by increased $CO_2$ in the models.

Experiments with linear barotropic models have demonstrated that one way that the response to tropical heating associated with ENSO is altered is by changing the time mean atmospheric state (Branstator 1983; Simmons et al. 1983). This could be the reason that the midlatitude structure of ENSO events is altered in our $2 \times CO_2$ experiments, for the increase of $CO_2$ has changed the basic atmospheric state from that simulated in the $1 \times CO_2$ case (see Washington and Meehl [1989] for changes of mean climate for the coupled model; Meehl and Washington [1990] for the mixed-layer model; and Washington and Meehl [1991] for a discussion of the comparison of differences between the mixed-layer and coupled models). These changes involve the entire temperature structure of the troposphere, the strength of the mid-latitude westerlies, changes in strength and position of the long waves, etc.

Figure 4.11a illustrates this changed basic state from increased $CO_2$ by means of the streamfunction differences from the upper troposphere for a 5-yr average (years 26–30), $2 \times CO_2$ minus $1 \times CO_2$ for the coupled model. Averages for other periods show a similar pattern. The increase of $CO_2$ manifests itself in the circulation as alternating positive and negative streamfunction differences arching from the tropics into the midlatitudes in both hemispheres. This considerable change in the longwave pattern in the extratropics associated with the increased $CO_2$ would be expected to affect the teleconnections in the $2 \times CO_2$ El Niño events. The reason for the change in $CO_2$ being manifested by such a streamfunction change is under investigation. The fact that similar patterns are associated with an anomalous diabatic heat source in the tropical Pacific in Fig. 4.11b from Branstator (1990) suggests that augmented rainfall in the $2 \times CO_2$ experiment in the Indonesian region might be responsible. Besides modifications of the time mean extratropical state, other changes in the equilibrium state induced by external forcing can alter the structure of ENSO. Because extratropical transients are thought to provide important feedbacks in ENSO events (Held et al. 1989; Branstator 1990), modifications in the average storm tracks could also be important. Furthermore, changes in the mean thermal

# Δ STREAMFUNCTION

a)          2×CO₂ MINUS CONTROL

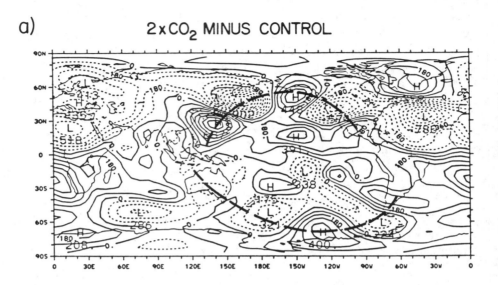

b)          PERPETUAL JANUARY

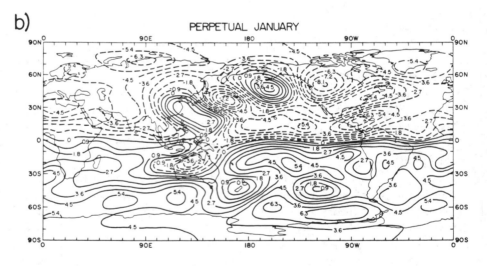

Fig. 4.11 (a) Streamfunction difference, $2 \times CO_2$ minus $1 \times CO_2$ (years 26–30), for the upper troposphere ($0.189\ \sigma$ in the model). Contour interval is $90 \times 10^4\ \mathrm{m}^2\ \mathrm{s}^{-1}$; arching patterns of anomalies stretching from the tropics to the midlatitudes in both hemispheres are indicated by thick dashed lines; and (b) streamfunction difference for a 500-d average of the atmospheric GCM with a specified tropospheric heating anomaly centered at equator, 135 °E (cross in circle) minus a 6000-day control for perpetual January with observed January mean SSTs, upper troposphere ($0.336\ \sigma$ in the model); contour interval is $0.9 \times 10^6\ \mathrm{m}^2\ \mathrm{s}^{-1}$.

and moisture distribution of the tropics could influence the structure of diabatic anomalies associated with ENSO. These effects are being studied in a subsequent set of experiments and will be reported elsewhere.

To test further the effect of a modified climate on extratropical teleconnections, we ran the third model configuration described earlier (atmospheric GCM with observed January SSTs in perpetual January mode) with prescribed tropical tropospheric heat sources. Since the pattern of streamfunction change in Fig. 4.11b was analogous to the mean change caused by doubled $CO_2$ in the coupled model in Fig. 4.11a, we call the climate state resulting from a diabatic source placed in the western tropical Pacific an analog for doubled $CO_2$ in Fig. 4.12a. If we place the same-magnitude diabatic heat source in the eastern equatorial Pacific (Fig. 4.12b), a pattern of streamfunction anomalies resembling those observed during an ENSO event occurs (deepened Aleutian low, etc.).

Therefore, we have an analog to the mean climate change from increased $CO_2$ (Fig. 4.12a) and an analog to the circulation anomalies from an ENSO event with present $CO_2$ (Fig. 4.12b). We can include both heating anomalies in the model simultaneously then to obtain an analog to the effects of an ENSO event taking place (heating anomaly in the eastern Pacific) with doubled $CO_2$ (heating anomaly in the western Pacific) (see result in Fig. 4.12c). As noted for the coupled and mixed-layer models, there is a systematic change of the teleconnection pattern compared to Fig. 4.12b. The anomalies in Fig. 4.12c have become more zonal in the Northern Hemisphere, and major zonal changes of sign have occurred in the high latitudes of the Southern Hemisphere. The consistency of results from the three different model configurations lends confidence to the conclusion that the change in climate due to the change in external forcing causes a systematic alteration of extratropical teleconnections associated with ENSO.

As noted earlier, we are investigating the mechanisms of these systematic changes to ascertain the role of various physical processes. Preliminary results indicate that changes in tropical precipitation anomalies and alterations in the midlatitude transient eddies associated with the altered atmospheric circulation (Held et al. 1989) are mainly responsible for the systematic changes in teleconnection patterns between the $1 \times CO_2$ and $2 \times CO_2$ ENSO events.

## Conclusions

To investigate the effects on ENSO of a significant change in external forcing (in this case a doubling of atmospheric $CO_2$), we have analysed results from several different model configurations. This change in external forcing could be analogous to circulation changes in the climate system in the paleoclimatic record. The climate equilibrium state (e.g., temperature structure, strength of the westerlies, storm-track structure, etc.) change significantly with doubled $CO_2$ in the model. The coupled ocean-atmosphere GCM shows that ENSO continues to function in the tropics with doubled $CO_2$, but with tropical SSTs increased by

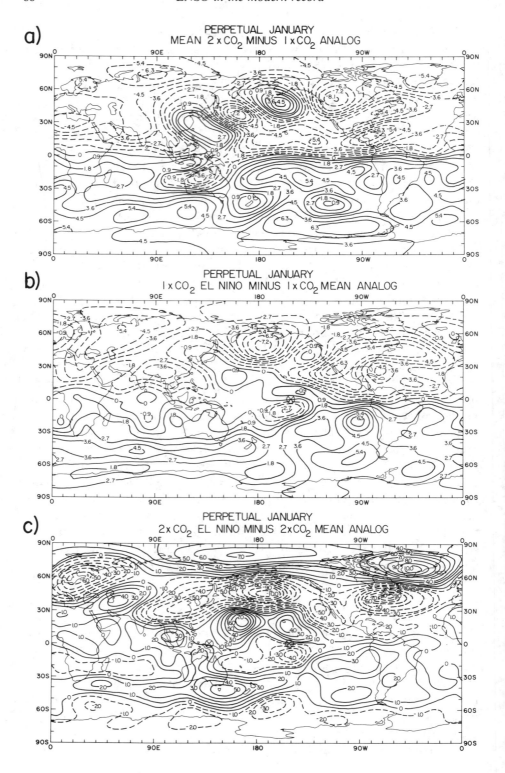

a) PERPETUAL JANUARY
MEAN 2 x CO₂ MINUS 1 x CO₂ ANALOG

b) PERPETUAL JANUARY
1 x CO₂ EL NINO MINUS 1 x CO₂ MEAN ANALOG

c) PERPETUAL JANUARY
2 x CO₂ EL NINO MINUS 2 x CO₂ MEAN ANALOG

about 1°C in the mean. However, extratropical teleconnections associated with ENSO events are altered with increased $CO_2$.

To determine if these changes in teleconnections are systematic and statistically significant, we make use of two other model configurations. One involves the atmospheric GCM coupled to a simple 50-m deep slab mixed-layer ocean (with no currents). Specified ENSO SST anomalies placed in this model in the eastern equatorial Pacific produce results similar to those obtained with the coupled model for $1 \times CO_2$ ENSOs and $2 \times CO_2$ ENSOs. That is, tropical ENSO effects are similar, but there is a systematic zonalization of SLP and 200-mb height anomalies in the Northern Hemisphere in the $2 \times CO_2$ ENSO events, as well as large phase shifts in regional anomalies. Zonal mean changes of sign of anomalies are also evident in the Southern Hemisphere.

A third model configuration with essentially the same atmospheric GCM is run with observed January SSTs in a perpetual January mode. In this model, we insert idealized heating anomalies in the tropics as analogs to (1) the change in climate basic state with $2 \times CO_2$ (a heat source in the equatorial western Pacific), (2) an ENSO event in the present-day climate (a heat source in the equatorial eastern Pacific), and (3) an ENSO event in a $2 \times CO_2$ climate (heat sources in both the equatorial western and eastern Pacific). Results from these experiments show a similar systematically different response in the extratropics in the case analogous to an ENSO event with increased $CO_2$ compared to the case analogous to an ENSO event with present amounts of $CO_2$. As in the other two model experiments, there is a more zonal response in the Northern Hemisphere with zonal changes of sign in the Southern Hemisphere and significant changes in the positions of local anomalies.

The consistency of these results from the various model configurations suggests that a change in the climate (in this case due to doubled $CO_2$) alters the pattern of extratropical anomalies associated with ENSO events, while tropical circulation anomalies remain fairly similar between basic states. The implication for analyses of extratropical signals from ENSO events in the paleoclimatic record is that we cannot assume that present-day ENSO composite anomaly patterns are valid for paleo-ENSO events.

*Acknowledgements* A portion of this study is supported by the Office of Health and Environmental Research of the U.S. Department of Energy as part of its

---

Fig. 4.12 (a) Streamfunction difference reproduced from Figure 4.11b for a 500-d average of the atmospheric GCM with a specified tropospheric heating anomaly centered at equator, 135 °E (cross in circle) minus a 6000-d control for perpetual January with observed January mean SSTs, upper troposphere (0.336 $\sigma$ in the model), taken to be an analog for the mean change in climate basic state in $2 \times CO_2$ minus $1 \times CO_2$; contour interval $0.9 \times 10^6$ m$^2$ s$^{-1}$; (b) same as (a) except for a heating anomaly placed at the equator, 150 °W as an analog to an ENSO event with $1 \times CO_2$ and (c) same as (a) except for heating anomalies placed at the equator and both 135 °E and 150 °W as an analog to an ENSO event with $2 \times CO_2$; contour interval $1.0 \times 10^6$ m$^2$ s$^{-1}$.

Carbon Dioxide Research Program. The National Center for Atmospheric Research is sponsored by the National Science Foundation. Lynda VerPlank ran and archived the coupled model and mixed-layer model results, Ann Modahl edited the text, and Suzanne Whitman drafted the figures.

# References

BLACKMON, M. L., GEISLER, J. E., and PITCHER, E. J., 1983: A general circulation model study of January climate anomaly patterns associated with interannual variation of equatorial Pacific sea surface temperatures. *Journal of the Atmospheric Sciences*, 40: 1410–1425.

BRANSTATOR, G. W., 1983: Horizontal energy propagation in a barotropic atmosphere with meridional and zonal structure. *Journal of the Atmospheric Sciences*, 40: 1689–1708.

BRANSTATOR, G. W., 1990: Mechanisms affecting the midlatitude atmospheric response to equatorial heating anomalies. *In: Proceedings of the TOGA International Scientific Conference, Honolulu, Hawaii, 16–20 July 1990*. Geneva: WMO, 211–216.

GEISLER, J. E., BLACKMON, M. L., BATES, G. T., and MUÑOZ, S., 1985: Sensitivity of January climate response to the magnitude and position of equatorial Pacific sea surface temperature anomalies. *Journal of the Atmospheric Sciences*, 42: 1037–1049.

HELD, I. M., LYONS, S. W., and NIGAM, S., 1989: Transients and the extratropical response to El Niño. *Journal of the Atmospheric Sciences*, 46: 163–174.

LINDZEN, R. S. and NIGAM, S., 1987: On the role of sea surface temperature gradients in forcing low level winds and convergence in the tropics. *Journal of the Atmospheric Sciences*, 44: 2418–2436.

MEEHL, G. A., 1990: Seasonal cycle forcing of El Niño-Southern Oscillation in a global, coupled ocean-atmosphere GCM. *Journal of Climate*, 3: 72–98.

MEEHL, G. A. and ALBRECHT, B. A., 1991: Response of a GCM with a hybrid convective scheme to a tropical Pacific sea-surface temperature anomaly. *Journal of Climate*, 4: 672–688.

MEEHL, G. A. and WASHINGTON, W. M., 1990: $CO_2$ climate sensitivity and snow-sea-ice albedo parameterization in an atmospheric GCM coupled to a mixed-layer ocean model. *Climatic Change*, 16: 283–306.

PHILANDER, S. G. H., LAU, N. C., PACANOWSKI, R. C., and NATH, M. J., 1989: Two different simulations of Southern Oscillation and El Niño with coupled ocean-atmosphere general circulation models. *Philosophical Transactions of the Royal Society*, A329: 167–178.

RASMUSSON, E. M. and CARPENTER, T. H., 1982: Variations in tropical sea surface temperature and surface wind fields associated with the Southern Oscillation/El Niño. *Monthly Weather Review*, 110: 354–384.

RASMUSSON, E. M., KOUSKY, V. E., and HALPERT, M. S., 1986: Interannual variability in the equatorial belt: Evolution and relationship to El Niño/Southern Oscillation. *In* Chagas, C. and Puppi, G. (ed.), *Persistent Meteo-Oceanographic Anomalies and Teleconnections*. Pontificiae Academiae Scientiarum Scripta Varia, 69. Rome, Italy, 17–57.

SIMMONS, A. J., WALLACE, J. M., and BRANSTATOR, G. W., 1983: Barotropic wave propagation and instability, and atmospheric teleconnection patterns. *Journal of the Atmospheric Sciences*, 40: 1363–1392.

SPERBER, K. R., HAMEED, S., GATES, W. L., and POTTER, G. L., 1987: Southern Oscillation simulated in a global climate model. *Nature*, 329: 140–142.

van LOON, H., 1986: The characteristics of sea level pressure and sea surface temperature during the development of a warm event in the Southern Oscillation. *In* Roads, J. O. (ed.), *Namias Symposium*, Scripps Institution of Oceanography Reference Series 86–17, Scripps Institution of Oceanography. La Jolla, California, 160–173.

van LOON, H. and MADDEN, R. A., 1981: The Southern Oscillation. Part I: Global associations with pressure and temperature in northern winter. *Monthly Weather Review*, 109: 1150–1162.

WASHINGTON, W. M. and MEEHL, G. A., 1989: Climate sensitivity due to increased $CO_2$: Experiments with a coupled atmosphere and ocean general circulation model. *Climate Dynamics*, 4: 1–38.

WASHINGTON, W. M. and MEEHL, G. A., 1991: Characteristics of coupled atmosphere – ocean $CO_2$ sensitivity experiments with different ocean formulations. *In* Schlesinger, M. E. (ed.), *Greenhouse-Gas-Induced Climatic Change: A Critical Appraisal of Simulations and Observations*. Amsterdam: Elsevier Scientific Publishers, 79–110.

# Use of historical records in ENSO reconstructions

# 5

# Historical and prehistorical overview of El Niño/Southern Oscillation

DAVID B. ENFIELD

*NOAA Atlantic Oceanographic and Meteorological Laboratory 4301 Rickenbacker Causeway, Miami, Florida 33149, U.S.A.*

## Abstract

This paper presents a thumbnail sketch of what El Niño/Southern Oscillation (ENSO) is, how it may have existed (or not) in previous epochs, and its relation to paleoclimatic studies. El Niño is a recurrent aperiodic, interannual (2–5 yr) warming of the tropical Pacific Ocean, with an associated atmospheric counterpart, the Southern Oscillation. The two synchronous phenomena appear to occur as a result of an unstable interaction between ocean and atmosphere and vacillate between two contrasting states. They are of very large spatial scale and involve remote interannual climatic perturbations beyond the equatorial Pacific region of intense interaction. These 'teleconnections' are one of the primary ways in which El Niño-like fluctuations during historical and prehistorical epochs have been preserved in surrogate records by biospheric and geological processes. Analyses of historical, proxy, and geological information suggest that (1) ENSO has existed intermittently, as today, over at least the last 5000 yr; (2) that its statistics are probably nonstationary during that period, although apparently not in relation to variations of the background climate, such as the Little Ice Age. Evidence as to the existence of ENSO prior to 5000 BP is mainly geological and as yet equivocal. Strategies for paleoclimatic research in this area are discussed in regard to several possible scenarios.

## Introduction

For many years paleoclimate research has progressed on a virtually independent path from the mainstream of El Niño/Southern Oscillation (ENSO) studies. The common ground has centered largely on global climate models, with little or no shared interests in observations. This has changed rather quickly in recent years, in response to at least two factors. On the one hand, our knowledge of ENSO has been tremendously enhanced, along with the realization that it has widely felt climatic impacts around the globe. On the other hand, there has been an accelerated interest in the problem of climate change on all time scales, from interannual to longer, due to the potential impact of human activities over the next century.

The recent interest in interdisciplinary 'paleo-ENSO' studies stems from the realization that short-term climate phenomena are often microcosms of those that operate on the time scales of centuries to millennia. Thus, the interpretation of paleoclimate records can benefit from the knowledge gained of climate processes observable in the 20th century, while the predictive schemes of modelers can potentially be evaluated by verifying hindcasts against proxy histories of previous climates. Since one of the most extensive and observable climate signals on time scales of less than a century is El Niño, it is natural that research on prehistoric ENSO variability has emerged as an important branch of paleoclimate studies. This marriage of convenience brings a burden of interdisciplinary communication, which I try to alleviate in this paper.

I will first provide an overview of El Niño and the Southern Oscillation, followed by a discussion of 'teleconnections,' ENSO impacts and their relevance to paleoclimate research. A review of recent research on recorded El Niño events since the Spanish conquest of Peru in the 16th century is then followed by a discussion of several representative paleoclimatic studies based on geological evidence and proxy data. Much of the discussion is necessarily brief, and the reader is referred to the review by Enfield (1989) for a more complete treatment of ENSO and references to the wider literature, or to the treatise by Philander (1990) for details on the ocean-atmosphere dynamics of ENSO. Diaz and Kiladis (1992, this volume) cover the subject of global teleconnections in more detail.

### What is 'El Niño'?

The 'El Niño' is an unusual warming of the normally cool near-surface waters off the west coast of South America. It was so named by the inhabitants of northern Peru in reference to the Christ-child, because it typically appears as an enhancement to the annual onset of a warm, southward setting current that occurs there around the Christmas season. The anomalous event typically lasts for about a year and the unusual oceanic conditions that accompany it include a 2 to 8°C anomalous warming of the near-surface isothermal layer, a deepening of the thermocline, a reversal of normal northward setting coastal currents, the reduction of the nutrient supply to the photic zone where phytoplankton grow, and a series of consequent disturbances in the coastal ecosystem. Historically, the most notorious impact of El Niño events off South America has been the drastic reduction in the stocks of anchovy and sardine as well as the bird populations and fisheries that depend on them (Fig. 5.1).

The warmth of the El Niño condition is actually typical of most other tropical zones, while the coolness of the Peru (or Humboldt) Current is anachronous at low latitudes. We now know that these productive waters are cooled by the wind-induced upwelling of nutrient-rich water from lower (and colder) layers. The upwelling-favorable winds may be briefly interrupted north of about 6°S during El Niño and thus may play a partial role in the warming. This is not the case farther south, however (Enfield 1981). There the thermocline deepens during El Niño, causing the upwelled water to be warmer, in spite of the persistence of upwelling-favorable winds. The local climatic disturbances associated with the warming include (1) unusual storminess and rainfall in coastal areas of Peru and Ecuador north of about 6°S, (2) increased rainfall over the Andean cordillera of northern and central Peru with resulting coastal flooding and erosion as far south as Lima, and (3) drought in the Andes of Southern Peru. More remote impacts include wet winters in the southwestern and southeastern U.S. and central Chile, droughts in northeastern Brasil, and ocean ecosystem anomalies in the high latitudes of the eastern Pacific.

Beginning with the work of Jacob Bjerknes in the 1960s, research has shown that the El Niño phenomenon of South America is actually a regional

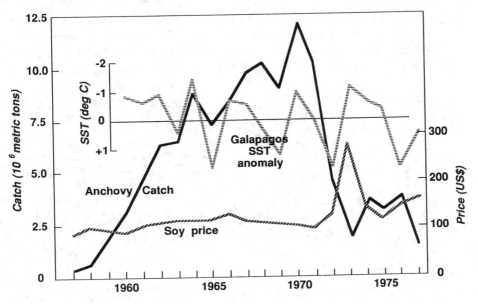

Fig. 5.1 The annual Galapagos sea surface temperature (SST) anomalies, Peru anchovy catch, and the cost of soybean meal illustrate the economic impact of the 1972/73 El Niño on world markets. The SST anomaly scale is inverted, with upward swings indicating enhanced upwelling and cooler conditions. The switch to less anchovy meal and more soy meal was permanent, motivating large-scale clearings of Amazon forest and North American wetlands for agriculture. (Redrawn from Barber 1988.)

manifestation of ocean-atmosphere interactions on a much larger scale. During El Niño, the equatorial zone and much of the tropical belt are anomalously warm across the breadth of the Pacific and are characterized by strong perturbations in the currents and deeper thermal structure as well.

The atmospheric counterpart to El Niño is the Southern Oscillation (SO), recognized by Walker (1923) and Walker and Bliss (1932) as a coherent variation of barometric pressures at interannual intervals. The low surface atmospheric pressure in regions dominated by tropical convection (ascending air) and rainfall, such as Indonesia, is inversely correlated with the high pressure in regions typified by subsidence (descending air) and dry conditions, such as the southeast Pacific (Fig. 5.2). Air is continually transferred at low levels – through the zonal trade wind circulations – from the subsidence regions to the convective regions. The air returns at upper tropospheric levels, completing a series of zonal cells around the globe that comprise the Walker Circulation (Bjerknes 1969; Flohn and Fleer 1975). In the 'high phase' of the SO, the southeast Pacific high pressure is higher than normal while the Indonesian trough (low pressure) is lower than normal. The increased pressure gradient between the two regions drives stronger Pacific trade winds and thus creates a greater mass exchange in the dominant Indo-Pacific Walker cell. During the low phase the pressure see-saw is reversed and the trades are weaker than normal.

The state of the SO pressure system is characterized by the Southern Oscillation Index (SOI), usually defined as the anomaly of pressure difference between Papeete (Tahiti) and Darwin (Australia). Meteorologists were unaware of the relation between the Southern Oscillation and El Niño until Berlage (1957) noted a strong correlation between SO anomalies and sea surface temperatures (SST)

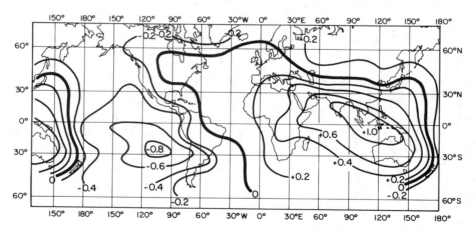

Fig. 5.2  The global distribution of the correlation coefficient between the barometric pressure variations at Djakarta, Indonesia, and those elsewhere. The negative correlation between the eastern and western hemispheres, with centers over Indonesia the southeast subtropical Pacific, characterizes the pressure seesaw associated with the SO, an interannual fluctuation in the strength of the Walker Circulation. (Redrawn from Berlage 1957.)

in Peru (Fig. 5.3), with the low phase of the SO coinciding with warm episodes (El Niño).

It was Bjerknes (1966a,b), who proposed a reasonable ocean-atmosphere interaction scenario that has since been embellished by extensive research. During the SO high phase, the system of westward tropical surface currents normally causes much of the upper layer water to accumulate in the western Pacific with a deep thermocline (150–200 m), leaving only a shallow layer and thermocline in the east (30–50 m). As Ekman (1905) had argued, the easterly (westward flowing) trade winds, combined with the earth's rotation, cause a steady movement of water away from the equator (Ekman transport) into both hemispheres, creating a horizontal divergence of surface water that in turn induces an equatorial upwelling of water from a few tens of meters below. The upwelling results in an elongated east-west tongue of cool surface water in the central and eastern equatorial Pacific Ocean, where the winds are strong and colder water is brought up from beneath the shallow thermocline.

Under the action of suddenly reduced trade winds (onset of the SO low phase), the accumulated warm water can return eastward toward the central Pacific, while equatorial upwelling ceases or weakens. As the cold tongue warms, atmospheric convective activity normally absent over the colder water spreads eastward toward the central Pacific, and the trade winds west of the new convective center are weakened or reversed to westerly. These changes in turn feed back onto the ocean and the anomalies grow in an unstable manner.

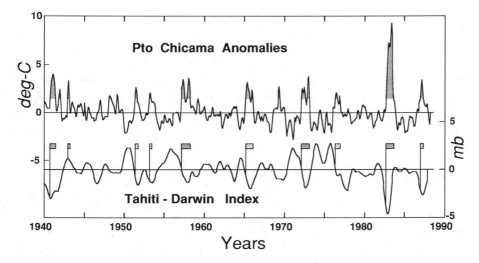

Fig. 5.3 Time series of sea surface temperature (SST) anomalies at Puerto Chicama, Peru (smoothed with a 3-mo running mean) and the Southern Oscillation Index (SOI) (smoothed with an 11-mo recursive filter). SST anomalies above one standard deviation are shaded in black. Small flags attached to the SOI curve signal the beginning (staff) and duration (flag) of the SST shading of each El Niño event. Solid (hatched) flags refer to S and VS (W and M) events from the compilation of Quinn et al. (1987).

Meanwhile, wave-like oceanic disturbances excited by the wind anomalies cause the thermocline to be depressed in the remote eastern Pacific, resulting in the El Niño off South America. Over most of the Peru coast the winds remain favorable for upwelling, forcing a slow drift northward in a shallow (10–30 m) surface layer, while most of the water column below moves anomalously poleward, bringing less productive equatorial and offshore waters to the upwelling zones (Smith 1983). Water continues to be upwelled from 50 to 100 m depth, but from a deepened and impoverished thermocline rather than a productive mixed layer (Huyer et al. 1987). The result: a breakdown of the normally rich coastal ecosystem, reproductive failures of some species, emigration of others from the region, mass mortalities of fish and birds, etc. (Barber et al. 1985).

Although this simplified picture conveys the essence of ENSO, it belies the many complexities of the phenomenon, its varied behavior from one event to another, and the degree of uncertainty that still exists. Is it accurate, for example, to say that El Niño begins in response to a well-defined anomaly in the trade winds that spontaneously occurs every 3 to 4 yr (the typical return interval between events)? Or is the time scale between onsets determined by an inherent cycle within the ocean, whose phase must be timed properly as a precondition for an oceanic response to wind forcing? Currently there are adherents to both points of view (see Enfield 1989).

## El Niño impacts

Understandably, most of El Niño research has been concentrated in the equatorial Pacific where the core instability and ocean-atmosphere interaction takes place. However, Bjerknes (1966b, 1969) also proposed that an unusually warm equatorial Pacific Ocean over a large zonal extent would create anomalous zonal and meridional SST gradients over large space scales, and that these gradients would provide an enhanced input of thermal energy to the direct atmospheric circulations, especially the meridional circulation of the atmospheric 'Hadley cells' in that quadrant of the globe. This would in turn increase the poleward flux of angular momentum to the winter hemisphere jet streams (Fig. 5.4) through a more efficient meridional circulation, ultimately strengthening the mid-latitude westerlies and affecting weather patterns downstream (to the east) of the original disturbance. In this way the El Niño warming in the equatorial Pacific can project 'teleconnected' climatic anomalies to remote regions of the globe. There is a statistical tendency for pressure and circulation anomalies to occur in certain patterns (Horel and Wallace 1981; Rasmusson and Wallace 1983) and for precipitation to be consistently higher or lower in certain regions (Ropelewski and Halpert 1987).

In a sense, any climatic anomaly removed from but correlated with the core interaction along the equator is a 'teleconnection' – a remote association – including the anomalies that occur along the South American coast and in

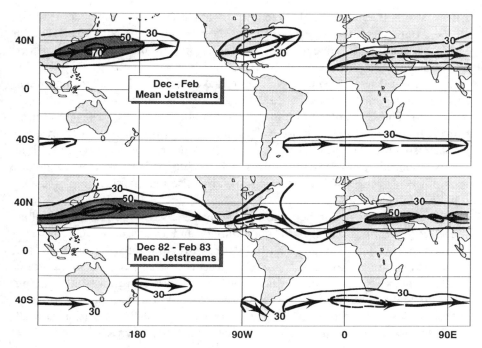

Fig. 5.4  Typical December–February jetstream pattern (top) and the anomalous (teleconnection) pattern during December 1982–February 1983 (bottom). Isotachs of wind speed are in m s$^{-1}$. (Redrawn from Rasmusson 1984).

Australia, two regions that lie at the opposite geographic and phenomenological extremes of the Southern Oscillation.

Public and scientific awareness of the atmospheric teleconnections proposed by Bjerknes was increased because of marked weather abnormalities that occurred in North America during the 1976/77 winter: an exaggerated circulation pattern brought a severe drought to the U.S. west coast while the east coast suffered the numbing effects of repeated snow storms and record cold (Namias 1978). Additional analyses of observations and atmospheric model results confirmed the teleconnecton principle and showed that the mid-latitude westerlies are indeed energized during El Niño, especially over the eastern Pacific and North America. The 1982/83 El Niño produced the strongest impacts ever observed, both locally in South America and worldwide (see Kiladis and Diaz 1986; Enfield 1989), but the teleconnection patterns during this event differed markedly from their predecessors. It has since become clear that the variation of teleconnection patterns about their statisitical 'mean' is quite large, and that, while the tendencies are significant, they account only for a small portion of the overall climatic variability around the globe (Namias and Cayan 1984).

Other climatic anomalies associated with the ENSO include droughts in Australia and northeast Brasil and excessive rainfall in the Great Basin of North

America and the southeast United States (Fig. 5.5; Ropelewski and Halpert 1987). Atmospheric teleconnections may reach as far as the Nile Basin (Quinn 1992, this volume), but due to their remoteness they are more difficult to detect and substantiate. However, the Nile catchment is influenced strongly by the Asian monsoon regime, and recent studies have established clear links between the Southern Oscillation and the strength of the summer monsoon (e.g., Rasmusson and Carpenter 1983; Parthasarathy and Pant 1984). The teleconnections at higher latitudes are most evident during the winter season of the respective hemisphere, as already discussed for North America. Similarly, central Chile has experienced devastating rainstorms in June–July of El Niño years, and correlations between the SOI and rainfall in Valparaiso and Santiago bear this out (Quinn and Neal 1983).

The paleoclimate researcher is interested in the ENSO core region as well as the teleconnections, because both create climatic and oceanic anomalies that may be potentially reflected in many naturally occurring surrogate records. Thus, for example, humic acids (organic runoff constituents derived from land vegetation) in the coral skeletons of the Great Barrier reef record centuries of alternating droughts and rainfall over northern Australia (Isdale 1984), while tree rings in the Great Basin of North America have recorded statistically significant rainfall surpluses associated with ENSO (Kay and Diaz 1985; Michaelsen 1989). The

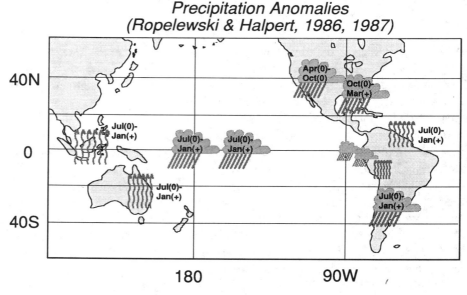

Fig. 5.5 Schematic showing teleconnection regions (Indo-Pacific and surrounding continental areas) with surplus or deficit of precipitation during various phases of a typical El Niño/Southern Oscillation cycle. ENSO phase is indicated in parentheses by a minus (antecedent year), zero (El Niño year) or plus (year following El Niño onset). (Based on Ropelewski and Halpert 1986, 1987.)

annually layered Quelccaya Ice Cap in the Andes of southern Peru reflects the occurrence of ENSO-related droughts (Thompson et al. 1984).

The ocean, as well as the atmosphere, can transmit anomalies to remote regions, but these are primarily confined to the equatorial Pacific and the higher latitudes of the eastern Pacific (Wooster and Fluharty 1985). The surrogate records for oceanic anomalies are seen mainly in the abundance, distribution, and characteristics of affected marine organisms. Examples include the cadmium uptake and other aspects of east Pacific corals, related mostly to ocean temperature (Shen et al. 1987, 1990; Glynn 1988), and the planktonic remains found in varved marine sediments near Guaymas, Mexico (Baumgartner et al. 1989). Other potential effects include the relative abundances of species recorded in fish scale deposits, the characteristics of relict mollusc shells, and the species composition of siliceous plankton deposits (see a review by Sharp 1992, this volume).

## Historical record

Many of the surrogate records (e.g., corals, tree rings) provide annual resolution (or better) over periods ranging from several centuries to a millennium or more. However, because ENSO episodes occur with widely varying characteristics and adhere only partially to a canonical pattern (see Enfield 1989), the teleconnections that produce surrogate records are not perfectly consistent from one event to another. Add to this the unrelated processes that are also recorded, plus stochastic noise, and it is easy to understand why these proxy time series are imperfect as chronologies of past El Niño occurrences. Usually, investigators must 'calibrate' their proxy indices against whatever knowledge we have of past events, as a means of establishing the representativeness of the proxy statistics.

The most comprehensive compilation of 'known' El Niño events is the anecdotal history by Quinn et al. (1987), who revised the earlier work of Quinn et al. (1978) and Hamilton and Garcia (1986), while extending the list to the early Spanish conquest when written records began. They classify El Niños as weak (W), moderate (M), strong (S), or very strong (VS) depending on the intensity and duration of climatic anomalies and their social and economic impacts as recorded by the conquistadores and the colonial inhabitants of Peru. In their work, an El Niño event is identified when confirming anecdotal evidence is found from several sources, e.g., reports of severe and prolonged desert flooding, massive crop failures, disease and insect infestations, and marine observations from ships' logs. A confidence index is attached to each event in proportion to the amount of evidence uncovered.

Use of the Quinn et al. (1987) data entails some caution. One must presume, of course, that there is more confidence in the identification of an event than in the assignment of an intensity category for it. Hence, studies based on the frequency of events in broad categories (e.g., 'strong' or 'weak') should have more validity than an analysis of intensities. Secondly, even if the anecdotal approach could classify perfectly the response to El Niño in Peru, it has no way of estimating conditions elsewhere in the Pacific. We know that the relationship between the broader ENSO phenomenon and its regional manifestations (such as El Niño) is not perfectly 1 : 1 (Deser and Wallace 1987). Hence, conclusions derived from the anecdotal data apply more precisely to the recurrence of the regional El Niño phenomenon than to the global scale ENSO system.

What the anecdotal data set shows is that El Niño events of all intensities have occurred most frequently at intervals of 3 to 4 yr, while the S events seldom occur less than 6 to 7 yr apart. Only eight VS events have occurred in nearly 5 centuries (the last one was 1982/83); none of these repeated at intervals of less than 20 yr. The first documented El Niño (though not by that name) was in 1525/26, as evidenced by the abundant vegetation encountered on the northern Peru desert and described in the campaign logs of Francisco Pizarro (Quinn et al. 1987).

It is important to understand the statistics of past El Niños, such as documented in the Quinn et al. (1987) chronology, because if proxy indices are to be credible, they must reproduce similar statistics over that 4- to 5-century period. For example, in addition to the basic statistics outlined above, what can be said of changes in the rate of occurrence of El Niño between the Little Ice Age (LIA, 16th/18th centuries) and the present, warmer climate? Enfield (1988) did a qualitative analysis of the Quinn et al. (1987) data, but could find no suggestion that El Niño had been more or less frequent several centuries ago than it is today; he did note, however, what seem to be century-scale oscillations in El Niño frequency. These were also seen by Anderson (1990, 1992, this volume), who remarked on their apparent phase relationship with alternating epochs of high or low solar variability.

A more quantitative analysis of the Quinn et al. (1987) data was done by Enfield and Cid (1991), which confirms the previous results. Stratifying the data according to event strength (all intensities; S, VS) and epochs (16th–18th centuries vs. 18th–20th centuries; high solar vs. low solar), they find no evidence of a change in onset intervals associated with the emergence from the LIA. The return intervals for strong events are (95% significance) longer during epochs of high solar variability (mode = 12.8 yr) than during those of low variability (mode = 8.5 yr). A similar result (90% significance) obtains for events of all intensities (Fig. 5.6). It is important to note that the differences in return intervals between epochs is significant to a reasonably high degree, but that the association with solar activity may be coincidental, because there are only a few realizations of the cycles (Fig. 5.7). The authors offer speculation on a possible linkage between solar variability and El Niño return intervals that involves the quasi-biennial oscillation (QBO) of the atmosphere. This possibility is suggested by recent studies showing a relation between the QBO and various tropospheric climatic variations (e.g., Labitzke and van Loon 1990).

These results provide an important lesson for paleoclimate studies: we should strive to understand changes in the short-term variability of climate as well as the secular drift of the longer term background climate. Civilization may be minimally affected by a gradual increase in temperature (several degrees) and sea

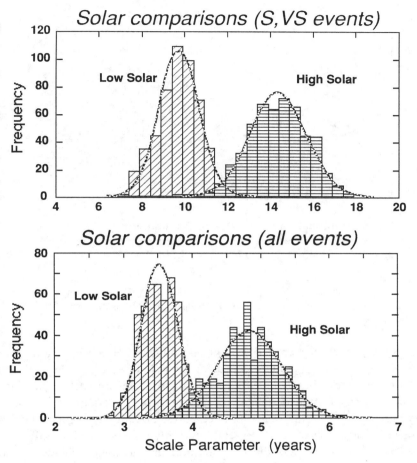

Fig. 5.6  Histograms and fitted normal distribution curves for the Weibull scale parameters estimated from the bootstrap samples taken from groups 4,5 (above) and 9,10 (below). The distributions overlap by 5% and 10%, respectively. (From Enfield and Cid 1991.)

level (10–30 cm) over the next century, as that time scale is probably adequate for social and technological responses to become effective. In contrast, a significant change in the frequency of extreme events can have a large impact on the ability to recover over short intervals, e.g., from periods of excessive storminess or droughts. Very little is known of how short-term climate variability has changed in the past, making this a potentially fruitful area for research.

### Prehistory of El Niño

Although there is some hope of extending the historical record backward past the Spanish conquest (Quinn 1992, this volume), this depends on less reliable

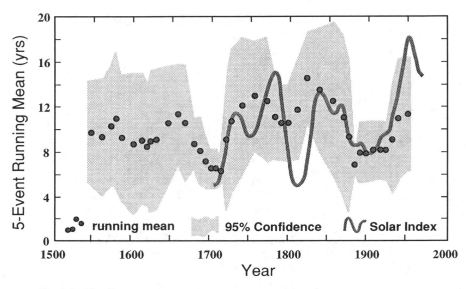

Fig. 5.7   The five-event running mean return interval for strong El Niño events, superimposed on the envelope of the 11-yr solar activity cycle. (From Enfield and Cid 1991.)

indicators such as those recorded through teleconnected effects in remote regions of the globe, e.g., the Egyptian records of the Nile River discharge (see Quinn 1992, this volume). Our ability to understand El Niño in earlier times must depend primarily on paleoclimatic records of two types: annually laminated proxy time series such as tree rings, ice cores, aquatic sediments, and coral sections (Baumgartner et al. 1989); and intermittent geological indicators, such as beach ridges, river deposits/erosion, and archaeological evidence. Both approaches have proven useful in detecting changes in the background climate. In addition, the former offer the possibility of reconstructing the statistics of El Niño in past epochs (Thompson et al. 1987), while the latter are better suited to telling us if El Niño occurrences affected other epochs at all, or if previous periods may have had a permanent 'El Niño-like' condition (DeVries 1987). The former are typically limited to time spans of a few millennia or less but have relatively complete and continuous information; the latter are much more sporadic and discontinuous but extend much farther into the past. Space does not permit an exhaustive review of the many papers that have appeared in recent years, hence I will limit myself to citing a few representative cases that are instructive. The reader is referred to other chapters of this volume for a more thorough treatment of this subject.

### Proxy time series

Research on El Niño using surrogate indices can potentially answer some intriguing questions that have relevance to modern ENSO investigations. Is ENSO an unstable oscillation between cold or warm tropical ocean states that can alternately become permanently El Niño-like (warm ocean), devoid of El Niño (cold ocean), or intermittent (warm El Niño episodes in an otherwise cold ocean) under different background climates? Or, is the present intermittent situation so robust as to defy the vagaries of large changes in wind patterns, sea level, and land area that must have accompanied the passage of Quaternary climatic epochs (at most having nonstationarities in the intensity and/or interval statistics)? The results of Enfield and Cid (1991, above) suggest the latter to be the case within the last half millennium, during the transition out of the LIA. No consensus on such questions has yet been reached, however, for prior periods. In fact, most proxy research is still struggling with the initial attempts at climate reconstruction, with verification against the more recent anecdotal and instrumental records being the primary objective.

An implicit assumption in the use of paleoclimatic records to study prehistoric ENSO activity is that the processes that connect the ENSO phenomenon and the surrogate time series remain invariant, or nonstationary through time. This allows the investigator to 'calibrate' the proxy series against modern instrumental records of ENSO so as to interpret past epochs when such records did not exist. Similar assumptions have been made in trying to relate precipitation or other weather manifestations with sunspot activity, and failed when tested against independent observations. Ramage (1983) gives examples of such relationships, which generally involve nonstationarities in the climate system. These may be the rule, rather than the exception, and argue for extreme caution in the pursuit of paleoclimatic information.

Caveats aside, an appropriate example of a paleoclimatic proxy record is the annual layering of ice accumulation, dust particles and oxygen isotope concentration in the Quelccaya Ice Cap of the southern Peruvian Andes. The variations preserved in ice cores correlate well with 20th century occurrences of El Niño (Fig. 5.8; Baumgartner et al. 1989) and clearly reflect overall climatic changes between the LIA and the present (Thompson and Mosley-Thompson 1989). As one of the more promising proxy records available, how well can the ice data reconstruct past El Niño activity?

Thompson et al. (1987) did an interesting experiment in attempting to statistically 'hindcast' El Niño since 1851. Simultaneous transfer function estimates were made (separately) of the SOI and Puerto Chicama SST (8°S, Peru coast). Four Quelccaya ice variables were used as predictors for SST along with the SOI, and the ice variables and SST were used as predictors for the SOI. The models were 'calibrated' using development samples for the period after 1935. The variance explained was 81% and 66%, respectively, for the SOI and SST models. Most of the model success for SST is due to the relationship between the SOI and SST, and the dependence on the ice variables is significant but not strong. In the model for SOI the ice variables have a fairly strong contribution to the explained variance. The models were then tested against verification samples for 1851–1935 by scoring the outputs in five categories of intensity from 'no event' to 'strong'

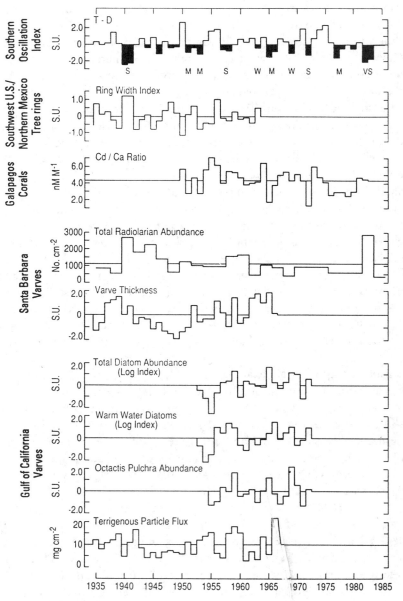

Fig. 5.8  Time series of proxy variables from four of the five natural systems plotted below the SOI (Tahiti minus Darwin). Negative values of the SOI are shaded and the well-defined El Niño episodes identified by letters corresponding to the subjective classification of El Niño strengths in Quinn et al. (1978, 1987): W (weak), M (moderate), S (strong), VS (very strong). The standard unit label on vertical axes indicates that values are plotted in units of standard deviation. Ticks on time axis correspond to month of July. (From Baumgartner et al. 1989, copyright by the American Geophysical Union.)

and comparing them against three similar categories in the Quinn et al. (1987) chronology. The calibration and hindcast series are shown in Figure 5.9.

I have attempted to summarize the Thompson et al. (1987) hindcast results in Table 5.1. There appear to be more hits (or near-hits) than misses in each category and the overall hit/miss score is 50/35. In a performance comparison by this author, the correlation coefficients and percent variance explained for the SOI against several variables are as follows (all values are significant at the 95% confidence level or better except for ice amount):

(1) Against Thompson et al. Quelccaya ice variables:

| Ice Amount | $^{18}O$ Isotope | Conductivity | Particles |
|---|---|---|---|
| −0.17 (3%) | −0.34 (12%) | −0.24 (6%) | 0.30 (9%) |

(2) Against SST at Pto Chicama, Peru:  −0.69 (48%)

Much of the hindcast skill in the statistical model is clearly due to the SOI–SST relationship and not to the ice variables. Neither way of looking at the ice data (correlations or discrete category comparisons) suggests that ENSO event identification can be done with much accuracy, using only the ice data from a single site as input. Thus, the proxy data may permit event detection with better than

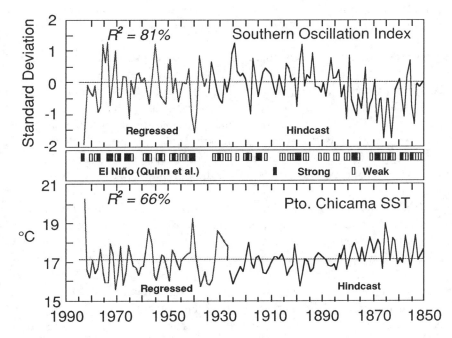

Fig. 5.9  Thompson et al. (1987) hindcast of SOI (top) and sea surface temperature (bottom) at Puerto Chicama, Peru, from 1851 (right) to 1983 (left), from a regression on Quelccaya ice variables. The lighter curve (left) is the regression based on the 1936–1983 developmental sample; the solid curve (right) are the hindcast 'predictions' for the 1851–1935 period. Weak and strong El Niño episodes found anecdotally from archival data are shown in the middle panel. (Redrawn from Thompson et al. 1987.)

Table 5.1 *Verification of the Quelccaya ice regression El Niño hindcast model*[a]

| Scored strength | Actual strength (Quinn et al.) | | | Total |
|---|---|---|---|---|
| | None | Weak | Strong | |
| None | 16 | 8 | 3 | 16/11 |
| None/Weak | 6 | 5 | 1 | 11/1 |
| Weak | 2 | 2 | 1 | 2/3 |
| Weak/Strong | 7 | 12 | 4 | 16/7 |
| Strong | 4 | 9 | 5 | 5/13 |
| Totals | 22/13 | 19/17 | 9/5 | 50/35 |
| Events only | **19/16** | **26/10** | **10/4** | **55/30** |

[a]The Quelccaya regression model of Thompson et al. (1987) hindcasts El Niño events for the period 1851–1935, according to the vertically arranged intensity categories on the left. The hindcast events (and non events) above the double line are arranged according to where they fall within the Quinn et al. (1987) chronology (categories horizontally at the top). Hindcasts within (outside) the boxed areas are summed to be 'hits' or near-hits (misses). The boldface numbers summarize the ratios of hits to misses. The same results without regard to intensity (events only) are summarized below the dashed line, assuming a 50/50 split of hindcast years in the None/Weak category.

coin-toss effectiveness, but many misses are likely to filter through any procedure used.

Like other proxy data, the Quelccaya time series are affected by multiple processes. In this case, meteorological phenomena in the high Andes are influenced by the Amazon and Atlantic climatic regimes as well as the Pacific El Niño, with a dose of stochastic noise thrown in for good measure. One can likely improve the reliability of El Niño identifications by prior statistical modeling and removal of the unwanted sources, but the miss rate is apt to remain unacceptably high for many purposes.

Another problem is that annual horizons in core records are sometimes lost or false horizons identified as years, producing imperfect chronologies. Since the El Niño only lasts about a year, a ± 1- to 2-yr uncertainty in the chronology can have a large effect on the results. Another proxy variable, [18]O isotopic composition in Pacific corals, also has skill in tracking most El Niño events, but likewise cannot do so unequivocally, for similar reasons (Druffel et al. 1990).

This difficulty in reliably identifying past El Niño events is a reality that must be accepted for most, if not all proxy-based investigations and a suitable research strategy must be adopted accordingly. One approach might be to improve chronologies and diversify response detection by combining two or more proxy sources of different types and/or from different regions (Michaelsen and Thompson 1992, this volume). Another might be to reconstruct an El Niño 'climate' in such a way that the number of predicted events verifies successfully, even if the hit/miss rate is imperfect. Thus, one might detect past changes in the event rate, provided the model skill is high enough.

We are warned, however (Michaelsen and Thompson 1992, this volume), that any attempt at reconstruction may be rendered invalid if there are nonstationarities in the relative importance of the various proxy variables that enter the model. Thus, the cadmium/calcium ratio in a massive coral at Urvina Bay in the Galapagos archipelago reflects the nutrient deficiency of the shallow waters during recent El Niño occurrences (Fig. 5.10; Shen et al. 1987; Cole 1992, this volume). However, when the proxy history is extended several centuries back using a San Cristobal coral farther east, the hybrid time series appears nonstationary (Shen, 1989, pers. comm.). Clearly, ensemble reconstructions must be done carefully, even when samples are taken from similar or neighboring environments.

Finally, we should note that remoteness from the El Niño core region does not necessarily mean that proxy hindcasts will be less skillful. For example, a tree-ring index developed by Michaelsen (1989) is surprisingly coherent with the SOI, explaining about half the variance in the SOI – almost as much as Puerto Chicama SST, above (see also the paper by Lough 1992, this volume). The Michaelsen (1989) index explains about half of the SOI variance, considerably more than any of the Quelccaya ice variables, even though the latter are much closer to the zone of maximum ocean anomalies.

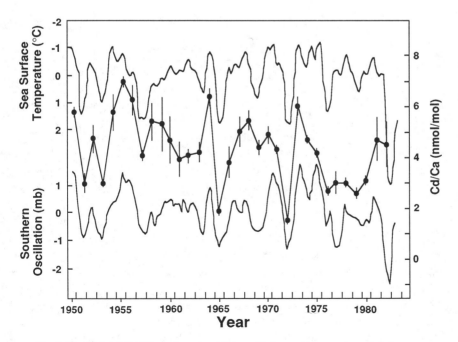

Fig. 5.10  Skeletal Cd/Ca mole ratios (right axis) from a core section of Galapagos coral compared against historical SST anomalies at Puerto Chicama, Peru, and the SOI (left axes). (Redrawn from Shen et al. 1987.)

While the tree-ring chronologies indeed appear promising, we must view them with extreme caution, because they are generally extratropical and remote from the center of action of ENSO, thus being more susceptible to nonstationarities in the teleconnection processes. In a coupled ocean-atmosphere GCM model with doubled $CO_2$, Meehl and Branstator (1992, this volume) found that ENSO activity continues to function in the tropics as it does now, but with mean eastern tropical Pacific SSTs about 1°C higher. The extratropical response to this situation differs from the present patterns, however, indicating a significant nonstationarity in the teleconnection process. Since past climates have easily differed from the present by similar amounts, one can imagine that the tree-ring response to past ENSO activity would be different and therefore misleading when interpreted against modern calibrations such as Michaelsen (1989). In spite of the difficulties in preparing reliable coral climatologies, they are probably well worth the effort because of their direct and local response to the intrinsic ENSO variables such as SST and precipitation.

The best results for any sort of retrospective study of El Niño, based on proxy data, will probably come from time-consuming but rewarding interdisciplinary studies that combine various kinds of proxy records from different regions. Such studies can improve chronologies through cross comparison of temporal signatures and hiatuses, represent more of the variables that bear on the ENSO process through multivariate analysis, and eliminate regional differences from one event to another by analysing broadly dispersed sites.

### Intermittent evidence

DeVries (1987) has summarized the efforts of geologists, archaeologists and paleontologists to detect El Niño manifestations over the last 34,000 yr. On the time scale of millennia, only ice cores hold the potential to resolve interannual variability at the subdecadal time scale, with 1500 BP being the most extensive core record to this point (Thompson et al. 1984). The dominant type of evidence beyond this are the intermittent events preserved in the geological record. Dating techniques do not permit chronologies with the accuracy of annually layered series and discontinuities in the records become common. Much of the evidence for prehistoric El Niño activity has come from the Holocene, and some from the earlier Pleistocene.

Studies of Holocene shell middens (garbage heaps) left by the early inhabitants along the north-central Peru coast (back to 10,000 BP) identify the interval around 5000 BP as an important date: sea level had stabilized at a high level following a period of rapid rise, and the shellfish diet of the indigenous coastal population changed from predominantly warm water to cold water species (Rollins et al. 1986). The colder environment was apparently accompanied by an onset of El Niño events, confirmed by a series of beach ridges deposited after large coastal floods (Sandweiss 1986). Similar evidence for such large events is found in coastal river deposits from the same region (Wells 1987). The suggestion from the middens is of a geologically rapid transition from a warm environment without El Niño episodes to the present cooler background climate interrupted by El Niños every few years. Although the evidence of El Niño activity in the last 5000 yr seems acceptable, DeVries (1987) points out that the rapidly rising Holocene sea renders the midden evidence only fragmentary and contradictory before that, and that alternate interpretations for the warm water species exist. As for the last 5 millennia, it seems likely that the beach ridges correspond to catastrophic events and can say little about the frequency of moderate and strong El Niños as we know them today.

DeVries (1987) cites clear indications of climate changes along the Peruvian littoral during the Pleistocene, with alternations between arid and savanna environments. He proposes that an early Pleistocene southward retreat of cold-water molluscs from northwest Peru could be accounted for by the onset of El Niño conditions. It is not as clear, however, how the ocean and atmosphere circulations had changed. Unfortunately, the ability to resolve El Niño variability becomes seriously degraded as the researchers delve farther into the past, and ambiguities in the interpretation of their findings become more commonplace. DeVries (1987) argues that to identify El Niño episodes in proxy records unambiguously the methods must resolve event durations of a few months to two years, and establish unequivocal anomalous warming along the Peru-Ecuador coast.

All that we can presently say with confidence is that El Niño-like environmental variations have occurred in the South American region over prehistoric time scales and are not unique to the period following the Spanish conquest. We cannot say whether the variability has been statistically stationary in the sense that oceanographers and meteorologists understand it, although there are some indications that the variability in previous periods may have been different in ways that numerical modelers would find interesting.

## Discussion

Many paleoclimatologists and most of the theoretical and descriptive oceanographers describing modern phenomena have been approaching the question of climate change from the perspective of gradual, secular changes in the background climate. At issue is the possibility of significant changes in ocean circulation, e.g., to or from a cooler ocean without El Niño, from or to a warmer ocean reminiscent of a permanent El Niño condition. Because of the oscillatory character of El Niño as we now know it, the more probable extremes would seem to be a cold ocean without El Niño versus a cold ocean with intermittent El Niño events, much as we have today. In view of some recent results using anecdotal and proxy data sets, we should perhaps entertain a third notion, i.e., that El Niño is very robust against changes in the background climate and has always occurred intermittently, or at least within the geologically 'recent' past during which the continents have occupied their present configuration (Pleistocene or even most of the Quaternary). One important objective of paleoclimatic research should therefore be aimed at resolving which of these several situations is the case.

Another goal, especially if the third scenario holds, should be to determine to what extent the statistics of El Niño variability have been and are sensitive to the background climate (an 'internal' factor or interaction between time scales) or to 'external' influences such as vulcanism and/or solar variability. Since the biosphere adapts less readily to rapid changes than to slow ones, this will be a particularly important research area in respect of variations that may occur over the next century, possibly due to human activities. Because El Niño/Southern Oscillation is the most energetic global-scale variation we know of on short time scales, an increase or decrease in either its intensity or frequency (be the causes internal or external) is likely to have a significant impact. Clearly, more frequent and/or intense flood/drought cycles (associated with ENSO) would be more devastating than gradual increments in temperature and precipitation.

As a final note I emphasize the importance of broadening the interaction between the paleoclimatic and oceanographic communities in attacking these questions. Ocean modelers, in particular, should attempt to understand the mechanisms behind changes in El Niño variability. On the one hand, it will help them to understand the basic nature of ENSO itself. On the other, it will greatly aid paleoclimatologists in fleshing out the sketchy picture that geological and proxy data provide.

*Acknowledgments* The preparation of this paper has been supported by the NOAA Atlantic Oceanographic and Meteorological Laboratory under a grant from the U.S. Tropical Ocean/Global Atmosphere (TOGA) program. Particularly useful have been past discussions of material presented here with W. H. Quinn, L. G. Thompson, G. T. Shen, and R. Anderson.

## References

ANDERSON, R. Y., 1990: Solar-cycle modulation of ENSO: A possible source of climatic change. *In* Betancourt J. L. and Mckay, A. M. (eds.), *Proceedings of the 6th Pacific Climate (PACLIM) Workshop,* California Department of Water Resources Technical Report, 23: 77–82.

ANDERSON, R. Y., 1992: Long-term changes in the frequency of occurrence of El Niño events. *In* Diaz, M. F. and Markgraf, V. (eds.), *El Niño: Historical and Paleoclimatic Aspects of the Southern Oscillation.* Cambridge: Cambridge University Press, 193–200.

BARBER, R. T., 1988: The ocean basin ecosystem. *In* Alberts, J. J. and Pomeroy, L. R. (eds.), *Concepts of Ecosystem Ecology.* New York: Springer–Verlag, 166–188.

BARBER, R. T., CHAVEZ, F. P., and KOGELSCHATZ, 1983: Biological effects of El Niño. *In* Vegas, M. (ed.), *Ciencia, Tecnologiía y Agresión Ambiental: El Fenómeno 'El Niño'.* Lima: CONCYTEC Press, 399–438.

BAUMGARTNER, T. R., MICHAELSEN, J., THOMPSON, L. G., SHEN, G. T., SOUTAR, A., and CASEY, R. E., 1989: The recording of interannual climatic change by high-resolution natural systems: Tree-rings, coral bands, glacial ice layers, and marine varves. *In* Peterson, D. W. (ed.), *Aspects of Climate Variability in the Pacific and the*

*Western Americas.* Geophysical Monograph, 55. Washington, D.C.: American Geophysical Union, 1–14.

BERLAGE, H. P., 1957: Fluctuations of the general atmospheric circulation of more than one year, their nature and prognostic value. *Koninklijk Nederlands Meteorologisch Instituut Mededlelingen en Verhandelingen*, 69.

BJERKNES, J., 1966a: Survey of El Niño 1957–58 in its relation to tropical Pacific meteorology. *Inter-American Tropical Tuna Commission Bulletin,* 12: 1–62.

BJERKNES, J., 1966b: A possible response of the atmospheric Hadley circulation to equatorial anomalies of ocean temperature. *Tellus,* 18: 820–829.

BJERKNES, J., 1969: Atmospheric teleconnections from the equatorial Pacific. *Monthly Weather Review,* 97: 163–172.

COLE, J. E., SHEN, G. T., FAIRBANKS, R. G., and MOORE, M., 1992: Coral monitors of El Niño/Southern Oscillation dynamics across the equatorial Pacific. *In* Diaz, H. F. and Markgraf, V. (eds.), *El Niño: Historical and Paleoclimatic Aspects of the Southern Oscillation.* Cambridge: Cambridge University Press, 349–375.

DESER, C. and WALLACE, J. M., 1987: El Niño events and their relation to the Southern Oscillation. *Journal of Geophysical Research,* 92: 14,189–14,196.

DE VRIES, T. J., 1987: A review of geological evidence for ancient El Niño activity in Peru. *Journal of Geophysical Research,* 92: 14,471–14,479.

DIAZ, H. F. and KILADIS, G. N., 1992: Atmospheric teleconnections associated with the extreme phases of the Southern Oscillation. *In* Diaz, H. F. and Markgraf, V. (eds.), *El Niño: Historical and Paleoclimatic Aspects of the Southern Oscillation.* Cambridge: Cambridge University Press, 7–28.

DRUFFEL, E. R. H., DUNBAR, R. B., WELLINGTON, G. M., and MINNIS, S. A., 1990: Reef-building corals and identification of ENSO warming episodes. *In* Glynn, P. W. (ed.), *Global Ecological Consequences of the 1982–83 El Niño/Southern Oscillation.* Oceanographic Series, 52. Amsterdam: Elsevier, 233–253.

EKMAN, V. W., 1905: On the influence of the earth's rotation on ocean currents. *Arkiv för Matematik, Astronomi och Fysik*, 2(11): 1–52.

ENFIELD, D. B., 1981: El Niño: Pacific eastern boundary response to interannual forcing. *In* Glantz, M. H. and Thompson, J. D. (eds.), *Resource Management and Environmental Uncertainty: Lessons from Coastal Upwelling Fisheries.* New York: Wiley, 213–254.

ENFIELD, D. B., 1988: Is El Niño becoming more common? *Oceanography Magazine,* 1: 23–27,59.

ENFIELD, D. B., 1989: El Niño, past and present. *Reviews of Geophysics,* 27: 159–187.

ENFIELD, D. B. and CID, S., L., 1991: Low-frequency changes in El Niño-Southern Oscillation. *Journal of Climate,* 4: 1137–1146.

FLOHN, H. and FLEER, H., 1975: Climatic teleconnections with the equatorial Pacific and the role of ocean/atmosphere coupling. *Atmosphere,* 13: 98–109.

GLYNN, P. W., 1988: El Niño-Southern Oscillation 1982–83: Nearshore population community and ecosystem responses. *Annual Review of Biology and Systematics,* 19: 309–345.

HAMILTON, K. and GARCIA, R. R., 1986: El Niño/Southern Oscillation events and their associated midlatitude teleconnections. *Bulletin of the American Meteorological Society,* 67: 1354–1361.

HOREL, J. D. and WALLACE, J. M., 1981: Planetary scale atmospheric phenomena associated with the Southern Oscillation. *Monthly Weather Review,* 109: 813–829.

HUYER, A., SMITH, R. L., and PALUSZKIEWICZ, T., 1987: Coastal upwelling off Peru during normal and El Niño times, 1981-1984. *Journal of Geophysical Research*, 92: 14,297-14,307.

ISDALE, P., 1984: Fluorescent bands in massive corals record centuries of coastal rainfall. *Nature*, 310: 578-579.

KAY, P. A. and DIAZ, H. F. (ed.), 1985: Problems and prospects for predicting Great Salt Lake levels. *Proceedings*. Salt Lake City: University of Utah. 309 pp.

KILADIS, G. N. and DIAZ, H. F., 1986: An analysis of the 1977-78 ENSO episode and comparison with 1982-83. *Monthly Weather Review*, 114: 1035-1047.

LABITZKE, K. and van LOON, H., 1990: Association between the 11-year solar cycle, the quasi-bienniela oscillation and the atmosphere: a summary of recent work. *Philosophical Transactions of the Royal Society of London*, A330: 577-589.

LOUGH, J. M., 1992: Southern Oscillation index reconstructed from western North American tree-ring chronologies. *In* Diaz, H. F. and Markgraf, V. (eds.), *El Niño: Historical and Paleoclimatic Aspects of the Southern Oscillation*. Cambridge: Cambridge University Press, 215-226.

MEEHL, G. A. and BRANSTATOR, G. W., 1992: Coupled climate model simulation of El Niño/Souther Oscillation: Implications for paleoclimate. *In* Diaz, H. F. and Markgraf, V. (eds.), *El Niño: Historical and Paleoclimatic Aspects of the Southern Oscillation*. Cambridge: Cambridge University Press, 69-91.

MICHAELSEN, J., 1989: Long-period fluctuations in El Niño amplitude and frequency reconstructed from tree-rings. *In* Peterson, D. W. (ed.), *Aspects of Climate Variability in the Pacific and the Western Americas*. Geophysical Monograph, 55. Washington, D.C.: American Geophysical Union, 69-74.

MICHAELSEN, J., and THOMPSON, L.G., 1992: A comparison of proxy records of El Niño/Southern Oscillation. *In* DIAZ, H. F. and MARKGRAF, V. (eds.), *El Niño: Historical and Paleoclimatic Aspects of the Southern Oscillation*. Cambridge: Cambridge University Press, 323-348.

NAMIAS, J., 1978: Multiple causes of the North American abnormal winter 1976-77. *Monthly Weather Review*, 106: 279-295.

NAMIAS, J. and CAYAN, D. R., 1984: El Niño: Implications for forecasting. *Oceanus*, 2: 41-47.

PARTHASARATHY, B. and PANT, G. P., 1984: The spatial and temporal relationships between the Indian summer monsoon rainfall and the Southern Oscillation. *Tellus*, 36A: 269-277.

PHILANDER, S. G. H., 1990: *El Niño, La Niña and the Southern Oscillation*. New York: Academic Press. 293 pp.

QUINN, W. H., 1992: A study of Southern Oscillation-related climatic activity for A.D. 622-1989 incorporating Nile River flood data. *In* Diaz, H. F. and Markgraf, V. (eds.), *El Niño: Historical and Paleoclimatic Aspects of the Southern Oscillation*. Cambridge: Cambridge University Press, 119-149.

QUINN, W. H., ZOPF, D. O., SHORT, K. S., and KUO YANG, R. T., 1978: Historical trends and statistics of the Southern Oscillation, El Niño, and Indonesian droughts. *Fisheries Bulletin*, 76: 663-678.

QUINN, W. H. and NEAL, V. T., 1983: Long-term variations in the Southern Oscillation, El Niño, and Chilean subtropical rainfall. *Fisheries Bulletin*, 81: 363-374.

QUINN, W. H., NEAL, V. T., and ANTUNEZ de MAYOLO, S.E., 1987: El Niño occurrences over the past four and a half centuries. *Journal of Geophysical Research*, 92: 14,449-14,461.

RASMUSSON, E. M., 1984: The ocean-atmosphere connection. *Oceanus*, 27: 5–13.

RASMUSSON, E. M. and CARPENTER, 1983: The relationshop between eastern equatorial Pacific sea surface temperatures and rainfall over India and Sri Lanka. *Monthly Weather Review*, 111: 517–528.

RASMUSSON, E. M. and WALLACE, J. M. 1983: Meteorological aspects of the El Niño/Southern Oscillation. *Science*, 222: 1195–1202.

ROLLINS, H. B., III, RICHARDSON, J. B., and SANDWEISS, D. H., 1986: The birth of El Niño: Geoarcheological evidence and implications. *Geoarcheology*, 1: 17–28.

ROPELEWSKI, C. F. and HALPERT, M. S., 1986: North American Precipitation and temperature patterns associated with the El Niño/Southern Oscillation (ENSO). *Monthly Weather Review*, 114: 2352–2362.

ROPELEWSKI, C. F. and HALPERT, M. S., 1987: Global and regional scale precipitation patterns associated with the El Niño/Southern Oscillation. *Monthly Weather Review*, 115: 1606–1626.

SANDWEISS, D. H., 1986: The beach ridges at Santa, Peru: El Niño, uplift and prehistory. *Geoarcheology*, 1: 17–28.

SHARP, G., 1992: Fishery catch records, El Niño/Southern Oscillation, and longer-term climate change as inferred from fish remains in marine sediments. *In* Diaz, H. F. and Markgraf, V. (eds.), *El Niño: Historical and Paleoclimatic Aspects of the Southern Oscillation*. Cambridge: Cambridge University Press, 379–417.

SHEN, G. T., BOYLE, E. A., and LEA, D. W., 1987: Cadmium in corals as a tracer of historical upwelling and industrial fallout. *Nature*, 328: 794–796.

SHEN, G. T. and SANFORD, C. L., 1990: Trace element indicators of climate variability in reef-building corals. *In* Glynn, P. W. (ed.), *Global Ecological Consequences of the 1982–83 El Niño/Southern Oscillation*. Oceanography Series 52. Amsterdam: Elsevier, 255–283.

SMITH, R. L., 1983: Peru coastal currents during El Niño: 1976 and 1982. *Science*, 221: 1397–1399.

THOMPSON, L. G., MOSLEY-THOMPSON, E., and MORALES-ARNAO, B., 1984: El Niño-Southern Oscillation events recorded in the stratigraphy of the tropical Quelccaya Ice Cap, Peru. *Science*, 226: 50–53.

THOMPSON, P. A., THOMPSON, L. G., and MOSLEY-THOMPSON, E., 1987: Hindcasts of El Niño events in the 19th century. Columbus: Ohio State University College of Business, *Working Paper Series* 87–124. 31 pp.

THOMPSON, L. G. and MOSLEY-THOMPSON, E., 1989: One-half millennium of tropical climate variability as recorded in the stratigraphy of the Quelccaya Ice Cap, Peru. *In* Peterson, D. W. (ed.), *Aspects of Climate Variability in the Pacific and the Western Americas*. Geophysical Monograph, 55. Washington, D.C.: American Geophysical Union, 15–31.

WALKER, G. T., 1923: World weather I. *Memoirs of the Indian Meteorological Department*, 24: 75–131.

WALKER, G. T. and BLISS, E. W., 1932: World weather V. *Memoirs of the Royal Meteorological Society*, 4: 53–84.

WELLS, L. E., 1987: An alluvial record of El Niño events from northern coastal Peru. *Journal of Geophysical Research*, 92: 14,463–14,470.

WOOSTER, W. S. and FLUHARTY, L. (eds.), 1985: *El Niño North: Niño Effects in the Eastern Subarctic Pacific Ocean*. Seattle: University of Washington Sea Grant. 312 pp.

# 6

# A study of Southern Oscillation-related climatic activity for A.D. 622–1900 incorporating Nile River flood data

WILLIAM H. QUINN

*College of Oceanography, Oregon State University, Corvallis, Oregon 97331-5503, U.S.A.*

## Abstract

A record of the recurring large-scale climatic fluctuation over the lower latitudes, which is usually referred to as the El Niño/Southern Oscillation (ENSO), is derived based on applicable data, reports, and historical information accumulated over the past five centuries. It is then used to further refine a previously derived record on its closely related regional feature, El Niño. In order to extend the record of low Southern Oscillation Index (SOI)-related climatic activity farther back in time, it was necessary to focus the investigation on the western extremity of the area that is primarily affected by this large-scale ocean-atmosphere climatic fluctuation associated with the Southern Oscillation (SO). Information over many centuries is available on droughts, floods, plagues, famines, and Nile River levels for that area (East Africa, the Middle East, and India). The yearly flood of the Nile River has been the basis of Egyptian agriculture for thousands of years; and prior to the continually expanded upstream regulatory facilities several decades ago, its maximum flow level at Cairo registered the effects of the summer monsoon (June–September) rainfall over the highlands of Ethiopia, with a low SOI relating to a below-normal flood level at Cairo. The large-scale feature (ENSO) which extends over the lower latitudes from East Africa eastward to Central and South America manifests itself as a 'seesaw' in ocean-atmosphere activity between the area in and surrounding the tropical Indian Ocean and the area in and surrounding most of the tropical Pacific Ocean. Here, the regional climatic changes activated by the large-scale ENSO over the western side of the 'seesaw' are demonstrated, and the stage is then set for extending the low (negative) SOI climatic record back to the early

7th century by using the annual maximum Nile River flood level record for Cairo. Findings concerning extreme changes, as well as extended period changes and very long-term variations (e.g., the Little Climatic Optimum and Little Ice Age) are discussed.

## Introduction

The basic premise for constructing a truly useful, long climatic record dictates that its foundation be based on a recurring large-scale physical phenomenon that is deeply rooted (yet readily recognizable) in the earth's ocean-atmosphere system. It also requires that its recurrence be verifiably continuous over the past several thousand years. The large-scale Southern Oscillation (SO)-related ocean-atmosphere climatic fluctuation (often referred to as the El Niño/Southern Oscillation or ENSO) appears to meet that requirement. This near-global feature (Fig. 6.1) is particularly prevalent over the tropics and much of the subtropics extending from East Africa eastward to Central and South America. It manifests itself as a 'seesaw' in ocean-atmosphere activity between the area in and surrounding the tropical Indian Ocean and the area in and surrounding most of the tropical Pacific Ocean. Atmospheric pressures, winds, sea surface temperatures (SSTs), air temperatures, rainfall amounts, and sea levels are the parameters associated with the shifting climatic features, and they are often used in indices to represent and evaluate the changes. Sea-level pressure differences between sites in the Indonesian equatorial low-pressure region and sites in the southeast Pacific subtropical high-pressure region have traditionally been used as SO indices (SOIs). Critical changes in the depth and location of the equatorial low and changes in strength and location of the southeast Pacific subtropical high are

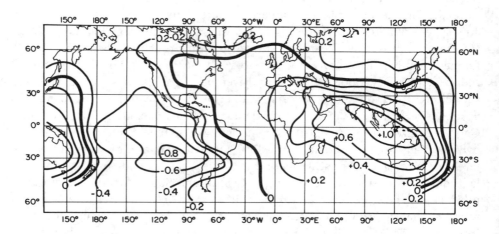

Fig. 6.1   Walker's Southern Oscillation, as shown in Berlage (1957). This map shows the worldwide distribution of correlations of annual pressure anomalies with simultaneous pressure anomalies in Djakarta, Indonesia.

reflected in SOIs. Quinn et al. (1978) show some of the pressure indices and pressure index components that have been utilized at various times to monitor the phase changes of the SO, depending on the availability of suitably located and reliable weather stations. When the low is deep and centered over Indonesia or farther to the west and the southeast Pacific subtropical high is strong, there is a high SOI; and when the equatorial low fills and/or moves to the east and the southeast Pacific subtropical high breaks down, there is a low SOI and large-scale climatic conditions occur that are typical of the ENSO and its regional manifestation, the El Niño (Ropelewski and Halpert 1987; Kiladis and Diaz 1989). Recurrence intervals in the long record generally range from 2 to 7 yr and on the average about 3.9 yr at moderate to very strong intensity levels.

The following sections are provided to justify the use of the very long Nile River flood level record at Cairo (which represents the summer monsoon rainfall amounts over the highlands of Ethiopia) to extend the record of SO-related climatic activity back into the 7th century A.D. It should be noted that the primary concentration over the whole record has been on the low SOI-related climatic activity. However, the high SOI-related features have been used to separate the low SOI (ENSO) events and sometimes to estimate event intensity based on the amplitude of phase shift from high to low values.

First, a large-scale ENSO record has been constructed, based on contents of literature sources in Quinn et al. (1987), Quinn and Neal (1992), and the references of this chapter. It was tentatively extended back through the beginning of the 16th century for the benefit of those working on Little Ice Age data. Then the regional El Niño record was refined to a minor extent by reconsidering all available information, current concepts, and time-limits established by the parent large-scale feature.

Next, it was essential to show the interrelationships between the large-scale ENSO and its regional features on both sides of the ocean-atmosphere 'seesaw.' The specific aim here was to justify the use of the very long annual maximum Nile River flood level record to extend the low SOI-related climatic record back to the early part of the 7th century A.D.

Prior to constructing the early record of weak annual maximum Nile River floods that occurred during the period A.D. 622–1522, the following are considered: (1) the ancient record on Nile River data; (2) the use of Nilometers; (3) background information on sources of data used to construct the record; (4) the use of lunar year dates by the Arabs, and the skip year introduced to relate the 33-year lunar period to its general 32-year equivalency on the solar calendar; (5) the use of the original cubit and finger measurement scales, as well as various levels of reference for the measurements (e.g., the floor of a Nilometer well, zero level on an established measurement scale, mean sea level of the Mediterranean at Alexandria). The above complications had to be dealt with as Table 6.6 was constructed.

In the latter part the unusually strong events, the extended periods of low Nile floods, near-decadal and longer climatic changes, and the very long-term climatic variations (e.g., the Little Climatic Optimum and Little Ice Age) are discussed in relation to this long record of inferred SO-related climatic activity. The recent warming over the tropical Pacific and subtropical South Pacific that set in during the latter half of 1976 and its ramifications is also mentioned.

## The ENSO and El Niño records

Most of our interests up until recently were primarily concentrated on the tropical and subtropical areas extending from western South America over to the Indonesian equatorial low pressure region and also on the regional El Niño (Quinn et al. 1978; Quinn 1980; Quinn and Neal 1983; Quinn et al. 1987). However, over recent years I became increasingly interested in extending the record of SO-related climatic activity much farther back in time and fully realized the essentiality of broadening my understanding of the large-scale ENSO developmental patterns and the various regional features they spawn. Therefore, it was necessary to develop a long record of the large-scale ENSO events, since it was noted several years ago that the onset, duration and strength of the ENSO can vary considerably from that of the regional El Niño; yet, one expects the El Niño to occur within the time-frame of its parent ENSO. Table 6.1 construc-

Table 6.1 *Years (Yr) in which large-scale El Niño/Southern Oscillation (ENSO) events occurred, with some years modified by E (early), M (mid), or L (late). Strengths (Str) are moderate (M), strong (S), or very strong (VS) with a + or − added for intermediate values. Recent weak (W), very weak (VW) ENSOs are given at end of table. Confidence (Conf) ratings (1–5) are estimates based on the quantity and quality of information afforded, ranging from minimal (1) to complete (5) (see text).*

| Yr | Str | Conf | Yr | Str | Conf | Yr | Str | Conf |
|---|---|---|---|---|---|---|---|---|
| 1497 | S | 2 | 1703–04 | S | 3 | 1854–55 | S | 5 |
| 1510 | M | 2 | 1707–09 | M | 2 | 1857–59 | M+ | 5 |
| 1518 | M+ | 3 | 1713–14 | M+ | 3 | 1860 | M | 3 |
| 1520 | M | 2 | 1715–16 | S+ | 3 | 1862 | M− | 2 |
| 1525–E26 | M | 2 | 1718 | M | 2 | 1864 | S+ | 5 |
| 1531–E32 | M | 2 | 1720 | M+ | 3 | 1865–66 | M+ | 4 |
| 1535 | M+ | 2 | 1723 | S | 4 | L1867–E69 | S+ | 5 |
| 1539–41 | S | 2 | 1725 | M | 2 | 1871 | M | 3 |
| 1544 | M+ | 3 | 1728 | M | 3 | 1873–74 | M+ | 5 |
| 1546–47 | S | 2 | 1731 | M+ | 2 | L1876–78 | VS | 5 |
| 1552–53 | S | 3 | 1734 | M | 2 | 1880–81 | M+ | 5 |
| 1558–E61 | S | 3 | 1737 | S | 3 | 1884–85 | M+ | 4 |
| 1565 | M+ | 2 | 1744 | M+ | 3 | L1887–E89 | S | 5 |
| 1567–68 | S+ | 3 | 1747–48 | S | 3 | 1891 | M | 5 |
| 1574 | S | 2 | 1751 | M+ | 2 | 1896–97 | M+ | 5 |
| 1578–E79 | S | 3 | 1754–55 | S | 2 | 1899–M1900 | VS | 5 |
| 1581–82 | M+ | 3 | 1761–62 | S | 3 | L1901–02 | S+ | 5 |

Table 6.1 (*cont.*)

| Yr | Str | Conf | Yr | Str | Conf | Yr | Str | Conf |
|---|---|---|---|---|---|---|---|---|
| 1585 | M | 2 | 1765–66 | M+ | 2 | 1904–05 | S | 5 |
| 1589–91 | S | 3 | 1768–69 | M+ | 4 | 1907 | M+ | 5 |
| 1596 | M | 2 | 1772–73 | M | 3 | 1911–12 | M+ | 5 |
| 1600–01 | S | 3 | 1776–E78 | M+ | 3 | M1913–M15 | S+ | 5 |
| 1604 | S | 3 | 1782–84 | VS | 5 | 1918–E20 | S+ | 5 |
| 1607–08 | S | 3 | 1785–86 | M+ | 3 | 1923 | M | 5 |
| 1614 | S | 3 | 1790–93 | VS | 5 | 1925–26 | S | 5 |
| 1618–19 | M | 3 | 1794–97 | M+ | 3 | L1929–E31 | M+ | 5 |
| 1621 | S | 3 | 1799 | M | 2 | 1932 | M+ | 5 |
| 1624 | M+ | 2 | 1802–04 | S+ | 5 | 1939 | M | 4 |
| 1630–31 | S+ | 3 | 1806–07 | M | 3 | 1940–41 | VS | 5 |
| 1635 | M | 3 | 1810 | M | 2 | 1943–44 | M | 5 |
| 1641 | S+ | 3 | 1812 | M+ | 3 | 1951–E52 | M+ | 5 |
| 1647 | M | 2 | 1814 | S | 3 | 1953 | M | 5 |
| 1650 | S+ | 3 | 1817 | M+ | 3 | 1957–58 | S | 5 |
| 1652 | M | 2 | 1819 | M+ | 3 | 1965–66 | S | 5 |
| 1655 | M | 2 | 1821 | M | 3 | M1968–69 | M− | 3 |
| 1661 | VS | 3 | 1824–25 | S | 5 | 1972–73 | S+ | 5 |
| 1671 | M+ | 2 | 1827–28 | S+ | 5 | 1976–77 | M | 5 |
| 1681 | S | 2 | 1830 | M | 3 | 1982–M83 | VS | 5 |
| 1683–84 | M+ | 2 | 1832–33 | S+ | 5 | M1986–87 | M | 5 |
| 1687–88 | S | 3 | 1835–36 | M | 3 | *Recent weaker ENSOs* | | |
| 1692 | M+ | 2 | 1837–39 | S | 5 | M1963–E64 | W | 3 |
| 1694–95 | VS | 2 | 1844–E46 | VS | 5 | M1974–E75 | VW | 5 |
| 1697 | M | 2 | 1850 | S | 5 | 1979–80 | W | 3 |
| 1701 | M | 3 | 1852–53 | M | 4 | E1990 | VW | 3 |

tion was based on the contents of references included in Quinn et al. (1987), Quinn and Neal (1992), and this chapter. It was extended back through the beginning of the 16th century by considering the tabular data and textual information of Toussoun (1925) and tables and plots of Popper (1951). El Nino/Southern Oscillation information was quite thin through much of the 1500s as a result of the Turkish conquest of Egypt in A.D. 1522; over this period, information depended heavily on the sources in Quinn et al. (1987), Quinn and Neal (1992), and related drought/famine reports in Elliot (1867), Walford (1879), and textual information in Toussoun (1925). Information sources improved in the late 1500s and 1600s and were quite good from 1700 on. Strengths of ENSOs were primarily based on the number of regional features activated and their respective intensities; however, for the first two centuries the values are quite qualitative and based on the degree of activity exhibited, similar to the strength scale of Quinn and Neal (1992). The lowest level recognized was moderate (M), the next level was strong (S) where the activity was highly significant, the extreme events were very strong (VS). A plus (+) entry was used to indicate intermediate strength estimates. The confidence figure ranging from 1 to 5 indicates the number of sources of evidence, with 5 indicating 5 or more sources.

Weak ENSOs were not included in the basic Table 6.1 since details over the distant past were insufficient for identification purposes. However, recent weak event occurrences are listed at the end of Table 6.1 since they may be of particular interest to scientists working on the coral reef problem which has drawn increasing attention over the past decade. Rising temperatures (along with decreasing sea-level pressures) were first reported in Quinn (1979) over the southeast Pacific. The temperature increase since mid-1976 over the tropics and lower subtropics was further discussed in Quinn and Neal (1984) and the extended effects can be seen in the generally lowered SOI values of Figure 6.2. Background information on the recent weak events is as follows:

(1) mid-1963 – early 1964, which is listed as such in Quinn et al. (1978), exhibits heavy equatorial Pacific precipitation as documented in Quinn

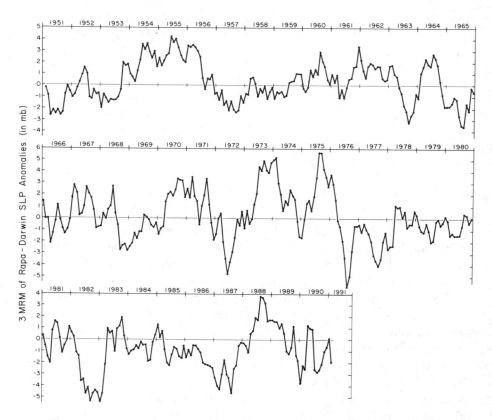

Fig. 6.2   Three-month running mean plot of anomalies of the difference in sea level atmospheric pressure (mb) Rapa Island (27 °37′S, 144 °20′W) and Darwin, Australia (12 °26′S, 130 °52′E). Anomalies are based on data for 1951–1988.

and Burt (1972), and shows an Indonesian east monsoon drought in Figure 6.2 of Flohn (1986);

(2) mid-1974 – early 1975 which is discussed in Patzert (1975), Wyrtki et al. (1976), Quinn et al. (1978), and Quinn (1980), and which also shows a deficient Indian summer monsoon in 1974 (Mooley and Parthasarathy 1984);

(3) 1979–1980, is discussed by Donguy et al. (1982) and Graham and White (1988), and it also shows a deficient summer monsoon rainfall over India in 1979 (Mooley and Parthasarathy 1984);

(4) early 1990 showed a very low SOI, above normal SSTs in the central equatorial Pacific, and very heavy rainfall at Tarawa for 1990. Both the 1979–1980 and the early 1990 periods showed adverse affects on several of the coral communities. It is expected that the sensitivity of the coral communities to the weak ENSO events over the past decade is a result of the general increase in tropical and subtropical SSTs over the past 14 to 15 yr.

Regional El Niño events did not occur during any of these weak ENSOs. Table 6.2 is a modified list of regional El Niño events as a result of further study, additional information obtained through this research, and identification with the large-scale ENSO developments. Strengths and confidence values on events, although modified in several cases, are in general, determined as they were in Quinn et al. (1987) and Quinn and Neal (1992).

## Justification for use of Nile River record

Subsequent work sought to find a way to extend this record of SO-related climatic activity farther back in time than the 16th century. Looking at the graphic representation of Walker's (1937) SO (Fig. 6.1), as shown by Berlage (1957: p. 59, Fig. 10), noting the areas primarily affected by the SO, and considering the seesaw-like relationship mentioned in the Introduction, it became readily apparent that the longest record could be obtained over the western extremity of that expanse significantly affected by the SO. There are many sources of information on that area (East Africa, the Middle East, and India) concerning floods, droughts, plagues, famines, and the Nile River flood levels. The yearly flood of the Nile has been the basis of Egyptian agriculture for several thousand years; and the flood level at Cairo, up until recent decades when upstream flow regulation facilities were established (Shahin 1985), registered the effect of the summer monsoon (June–September) rainfall over the highlands of Ethiopia. As early as 1908, it was stated in *The Imperial Gazetteer of India* (vol. 1, p. 127): 'it is now fully established that years of drought in western or northwestern India are almost invariably years of low Nile flood. The relation is further confirmed by the fact that years of heavier rain than usual in western India are also years of high Nile flood.' Although the vicissitudes of the

Table 6.2 *Years (Yr) in which regional El Niño events occurred, with some years modified by E (early), M (mid), or L (late). Strengths (Str) are moderate (M), strong (S), or very strong (VS) with a + or − added for intermediate values. Confidence (Conf) ratings (1–5) are based on the number of information sources with 5 indicating 5 or more. See text for details.*

| Yr | Str | Conf | Yr | Str | Conf | Yr | Str | Conf |
|---|---|---|---|---|---|---|---|---|
| 1525–E26 | M | 2 | 1704 | M | 2 | 1857–58 | M | 5 |
| 1531–E32 | M | 2 | 1707–09 | M/S | 3 | 1860 | M | 4 |
| 1535 | M+ | 2 | 1713 | M | 2 | 1862 | M− | 2 |
| 1539–41 | M/S | 3 | 1715–16 | S | 3 | 1864 | S | 5 |
| 1544 | M+ | 4 | 1718 | M+ | 2 | 1866 | M+ | 4 |
| 1546–47 | S | 4 | 1720 | VS | 5 | L1867–68 | M+ | 5 |
| 1552 | S | 3 | 1723 | M+ | 4 | 1871 | S+ | 5 |
| 1558–E61 | M/S | 4 | 1728 | VS | 5 | 1874 | M | 4 |
| 1565 | M+ | 2 | 1734 | M | 2 | 1877–78 | VS | 5 |
| 1567–68 | S+ | 5 | 1737 | S | 3 | 1880 | M | 3 |
| 1574 | S | 3 | 1744 | M+ | 3 | 1884 | S+ | 5 |
| 1578–E79 | VS | 5 | 1747 | S+ | 5 | L1887–E89 | M−/M+ | 4 |
| 1581–82 | M+ | 3 | 1751 | M+ | 3 | 1891 | VS | 5 |
| 1585 | M+ | 2 | 1754–55 | M | 2 | 1897 | M+ | 4 |
| 1589–91 | M/S | 3 | 1761 | S | 5 | 1899–E1900 | S | 5 |
| 1596 | M+ | 2 | 1765 | M | 2 | 1902 | M+ | 5 |
| 1600 | S | 3 | 1768 | M | 3 | 1904–05 | M− | 5 |
| 1604 | M+ | 3 | 1772 | M | 2 | 1907 | M+ | 4 |
| 1607–08 | S | 5 | 1776–E78 | S | 3 | 1910 | M+ | 4 |
| 1614 | S | 5 | 1782–83 | S | 3 | 1911–12 | M | 5 |
| 1618–19 | S | 4 | 1785–86 | M+ | 2 | 1914–E15 | M+ | 5 |
| 1621 | M+ | 2 | 1791 | VS | 5 | 1917 | M+ | 5 |
| 1624 | S+ | 4 | 1803–04 | S+ | 5 | 1918–19 | M | 5 |
| 1630 | M | 2 | 1806–07 | M | 3 | 1923 | M | 5 |
| 1635 | S | 3 | 1810 | M | 2 | 1925–26 | VS | 5 |
| 1641 | M | 2 | 1812 | M+ | 2 | L1930–E31 | M | 5 |
| 1647 | M+ | 3 | 1814 | S | 3 | 1932 | S | 5 |
| 1650 | M | 3 | 1817 | M+ | 4 | 1939 | M+ | 5 |
| 1652 | S+ | 3 | 1819 | M+ | 3 | L1940–41 | S | 5 |
| 1655 | M | 2 | 1821 | M | 4 | 1943 | M+ | 5 |
| 1661 | S | 2 | 1824 | M+ | 4 | 1951 | M− | 5 |
| 1671 | S | 3 | 1828 | VS | 5 | 1953 | M+ | 5 |
| 1681 | S | 2 | 1830 | M | 2 | 1957–58 | S | 5 |
| 1684 | M+ | 3 | 1832 | M+ | 5 | 1965 | M+ | 5 |
| 1687 | S+ | 4 | 1837 | M+ | 4 | 1969 | M− | 3 |
| 1692 | S | 3 | 1844–E46 | M/S+ | 5 | 1972–E73 | S | 5 |
| 1695 | M | 2 | 1850 | M | 4 | 1976 | M | 5 |
| 1697 | M+ | 3 | 1852 | M | 3 | L1982–M83 | VS | 5 |
| 1701 | S+ | 5 | 1854 | M | 3 | 1987 | M | 4 |

occurrence of a particular type of weather patterns in one area does not guarantee an expected value in another area, the statements of Griffiths (1972), indications from the bottom panel in Figure 1.8, p. 16 of Lockwood (1985), and Figure 6.1 would tend to support the *Gazetteer* statement.

As explained by Griffiths (1972), the variations of climate over the Ethiopian highlands during the year are largely associated with the macroscale pressure changes and the monsoon flows related to these changes. In summer the large low pressure system over India and the Arabian Sea dominates the airflow and brings strong, persistent southwesterlies to the southern part of the zone. At the same time, the intertropical convergence zone (ITCZ) lies just to the north of Eritrea so that the resulting convergence can bring much rain to the highlands of Ethiopia. It is the summer monsoon rainfall that is so important to East Africa and India, and it determines the strength of the Nile River flood through input of the Ethiopian rivers. It is primarily the Blue Nile and secondarily the Atbara rivers, originating in Ethiopia, that supply the waters of the annual flood; the White Nile plays very little part in the flood [based on the data in Shahin (1985)]. When the SOI is high, it indicates that the large low pressure system extending over India and the Arabian Sea is well developed and the summer monsoon rainfall is likely to be heavy; when the SOI is low, it indicates the large low pressure system is not well developed and/or is displaced to the east and the summer monsoon rainfall is likely to be deficient.

Parthasarathy and Pant (1984) established an excellent association between the SO and the Indian summer monsoon for the period 1871–1974. Earlier Pant and Parthasarathy (1981) showed the relationship between the SOI and Indian summer monsoon rainfall. They concluded that good monsoon years over India are characterized by a high SOI, with low pressure and relatively warm temperatures over the Indonesian region and increased pressure and cool temperatures over the eastern Pacific Ocean; in poor monsoon years, the pressure and temperature changes are reversed. The works of Rasmusson and Carpenter (1983) and Khandekar and Neralla (1984) also substantiated the association between the two ocean areas by showing the relationship between SSTs in the equatorial Pacific and the summer monsoon rainfall over India. Parthasarathy et al. (1988) show some interesting correlation values between Indian monsoon rainfall and the SO and western components of the SO.

From studies of changes taking place on the large-scale, it is apparent, in general, that the low index phase of the SO relates, on the east side of the 'seesaw,' to the El Niño, anomalously heavy subtropical Chilean rainfall, anomalously heavy rainfall and above normal SSTs over the equatorial Pacific. In contrast, on the west side, it relates to the eastern and northern Australian drought, an east monsoon drought over Indonesia, deficient summer monsoon rainfall over India, and a weak Nile flood (deficient summer monsoon rainfall over the highlands of Ethiopia). The high index phase of the SO relates, in general, on the east side of the 'seesaw,' to cool anti-El Niño conditions over the northwestern South American coastal region and its adjacent upwelling ocean waters, anomalously low subtropical Chilean rainfall, an equatorial Pacific dry zone extending far to the west and underlain by cool upwelled equatorial Pacific sea water. In contrast, on the west side, it relates to anomalously heavy rainfall over eastern and northern Australia, anomalously heavy east monsoon rainfall over Indonesia, anomalously heavy summer monsoon rainfall over India, and an anomalously large supply of water entering the Nile River system from its Ethiopian source region during the summer monsoon season.

To justify further the use of the Nile River record and show its interrelationships with the large-scale ENSO and other low SOI-related regional features, Table 6.3 was constructed, using the El Niño to represent the more frequently discussed activity on the east side of the ocean-atmosphere 'seesaw' [as reported in e.g., Quinn and Burt (1970, 1972), Namias (1973), Wyrtki (1973), Quinn (1974), Wyrtki (1975), Wyrtki et al. (1976), Quinn et al. (1978), Quinn and Neal (1983), and many others]; regional features included on the west side of the 'seesaw' are as listed in the previous paragraph. The selection of the 1824–1973 record was based on the availability of an unbroken record of Nile River data over this period. The strengths of the large-scale ENSO events are based on the number and respective intensities of the various regional events activated by them, and are as indicated in Table 6.1. El Niño occurrences and strengths are as discussed in Quinn et al. (1987), Quinn and Neal (1992), and as further modified here in Table 6.2. East monsoon droughts over Indonesia are reported by Van Bemmelen (1916), Berlage (1957), and Flohn (1986). Australian droughts are as reported by Nicholls (1988, 1992 this volume). Records of deficient summer monsoon rainfall over India between 1824 and 1870 are based on information in Martin (1858–61), Elliot (1867), Walford (1879), Hunter (1882), Hurst (1891), *The Imperial Gazetteer of India* (1908), Bhatia (1967), Mooley and Parthasarathy (1979), and Mooley and Pant (1981). Here we were interested in the deficient summer monsoon rainfall periods (June–September) which generally relate to a low SOI. However, one must remember that many of the drought periods reported in the various publications include the following dry season's conditions which can persist through May of the following year. From 1871 on, deficient summer monsoon rainfall years were as noted in Mooley and Parthasarathy (1984). However, I have also listed in Table 6.3 those years which fell below their mean summer monsoon rainfall level, but which did not meet their deficient rainfall criteria, and designated them as SBM (slightly below mean) since they were at least in the right direction for the developments we were evaluating. For the maximum levels of the annual Nile River flood and/or monthly and annual river discharge values for various locations, I had the following data records available for consideration over the period 1824–1973:

(1)  Lyons's (1906) data and graphic plots for 1825–1905;
(2)  Toussoun's (1925) tabular data and some anecdotal information for 1824–1921;

Table 6.3 Years (Yr) with large-scale ENSO events and their related regional manifestations for 1824–1973. The regional El Niño is used to represent related activity on the east side of the ocean-atmosphere 'seesaw'; on the west side, related regional features include Indonesian east monsoon drought, Australian drought, deficient India summer monsoon, and weak Nile flood (representing deficient summer monsoon rainfall over Ethiopia). Strengths (Str) are included for ENSOs and El Niños; at times years are modified by E (early), M (mid), or L (late). Weak Nile floods are rated by degree (Deg) of deficiency or reduction (1–5) as noted in Table 6.4, with confidence (Conf) ratings (1–5) based on number of confirmation sources. Notation N.D. indicates no data available. (See text for details.)

| ENSO Yr | Str | El Niño Yr | Str | E. Monsoon drought | Australia drought | Deficient India summer monsoon | Nile Yr | Deg | Conf |
|---|---|---|---|---|---|---|---|---|---|
| 1824–25 | S | 1824 | M+ | N.D. | 1824 | 1824–25 | 1824 | 4 | 2 |
|  |  |  |  |  |  |  | 1825 | 4 | 3 |
| 1827–28 | S+ | 1828 | VS | N.D. | 1828 | 1827–28 | 1828 | 2 | 2 |
| 1830 | M | 1830 | M | N.D. | – | – | 1830 | 2 | 3 |
| 1832–33 | S+ | 1832 | M+ | 1833 | 1832 | 1832–33 | 1832 | 1 | 3 |
|  |  |  |  |  |  |  | 1833 | 4 | 3 |
| 1835–36 | M | – |  | 1835 | – | – | 1835 | 4 | 2 |
|  |  |  |  |  |  |  | 1836 | 3 | 2 |
| 1837–39 | S | 1837 | M+ | 1838 | 1837 | 1837–38 | 1837 | 4 | 3 |
|  |  |  |  |  |  |  | 1838 | 2 | 2 |
|  |  |  |  |  |  |  | 1839 | 3 | 2 |
| 1844–E46 | VS | 1844–46 | M/S+ | 1844–45 | 1845 | 1844 | 1844 | 1 | 2 |
|  |  |  |  |  |  |  | 1845 | 3 | 2 |
| 1850 | S | 1850 | M | 1850 | 1850 | 1850 | 1850 | 2 | 2 |
| 1852–53 | M | 1852 | M | 1853 | – | 1852 SBM, 1853 | 1852 | 2 | 2 |
| 1854–55 | S | 1854 | M | 1855 | 1854 | 1855 | 1855 | 3 | 3 |

Table 6.3 (cont.)

| ENSO | | El Niño | | E. Monsoon drought | Australia drought | Deficient India summer monsoon | Weak Nile flood | | |
|---|---|---|---|---|---|---|---|---|---|
| Yr | Str | Yr | Str | | | | Yr | Deg | Conf |
| 1857–59 | M+ | 1857–58 | M | 1857 | 1857 | 1858–59 SBM | 1857 | 1 | 3 |
| | | | | | | | 1858 | 2 | 3 |
| | | | | | | | 1859 | 2 | 3 |
| 1860 | M | 1860 | M | – | – | 1860 | – | | |
| 1862 | M– | 1862 | M– | – | 1861–63 | – | – | | |
| 1864 | S+ | 1864 | S | 1864 | 1864 | 1864 SBM | 1864 | 4 | 4 |
| 1865–66 | M+ | 1866 | M+ | 1866 | 1866 | 1865–66 | – | | |
| L1867–E69 | S+ | L1867–68 | M+ | 1868 | 1868 | 1868 | 1867 | 1 | 3 |
| | | | | | | | 1868 | 4 | 4 |
| 1871 | M | 1871 | S+ | – | 1871 | – | – | | |
| 1873–74 | M+ | 1874 | M | 1873 | 1874 | 1873 SBM | 1873 | 2 | 3 |
| L1876–78 | VS | 1877–78 | VS | 1877 | 1877 | 1877, 1876 SBM | 1877 | 5 | 4 |
| 1880–81 | M+ | 1880 | M | 1881 | 1880 | 1880 SBM | 1881 | 1 | 2 |
| | | | | | | | 1882 | 2 | 2 |
| 1884–85 | M+ | 1884 | S+ | 1884–85 | 1884 | – | 1884 | 2 | 2 |
| | | | | | | | 1886 | 1 | 2 |
| L1887–E89 | S | L1887–E89 | M–/M+ | 1888 | 1888 | 1888 SBM | 1888 | 4 | 4 |
| 1891 | M | 1891 | VS | 1891 | – | 1891 SBM | – | | |
| 1896–97 | M+ | 1897 | M+ | 1896–97 | 1896 | 1896 SBM | 1897 | 1 | 3 |
| 1899–M1900 | VS | 1899–E1900 | S | 1899 | 1899 | 1899 | 1899 | 5 | 4 |
| L1901–02 | S+ | 1902 | M+ | 1902 | 1902 | 1901, 1902 SBM | 1902 | 2 | 4 |
| 1904–05 | S | 1904–05 | M– | 1905 | 1905 | 1904 SBM, 1905 | 1905 | 1 | 3 |

| | | | | | | | | | |
|---|---|---|---|---|---|---|---|---|---|
| 1907 | M+ | 1907 | M+ | – | 1907 | 1907 SBM | 1907 | 2 | 3 |
| 1911–12 | M+ | 1911–12 | M | 1911 | 1912 | 1911, 1912 SBM | 1912 | 1 | 3 |
| M1913–M15 | S+ | 1914–E15 | M+ | 1913–15 | 1914 | 1913 SBM, 1915 SBM | 1913 | 5 | 4 |
| | | | | | | | 1915 | 2 | 4 |
| 1918–E20 | S+ | 1918–19 | M | 1918–19 | 1918 | 1918, 1920 | 1918 | 1 | 4 |
| 1923 | M | 1923 | M | 1923 | 1923 | 1923 SBM | – | | 4 |
| 1925–26 | S | 1925–26 | VS | 1925–26 | 1925 | 1925 SBM | 1925 | 2 | 3 |
| L1929–E31 | M+ | L1930–E31 | M | 1929–30 | 1930 | 1929 SBM, 1930 SBM | 1930 | 1 | 3 |
| 1932 | M+ | 1932 | S | 1932 | 1932 | 1932 SBM | – | | |
| 1939 | M | 1939 | M+ | – | – | 1939 SBM | 1939 | 1 | 4 |
| 1940–41 | VS | L1940–41 | S | 1940–41 | 1940 | 1941 | 1940 | 2 | 4 |
| | | | | | | | 1941 | 3 | 4 |
| 1943–44 | M | 1943 | M+ | 1944 | 1943 | – | 1943 | 1 | 2 |
| | | | | | | | 1944 | 1 | 2 |
| 1951–E52 | M+ | 1951 | M– | 1951 | 1951 | 1951, 1952 SBM | 1951 | 2 | 3 |
| | | | | | | | 1952 | 1 | 2 |
| 1953 | M | 1953 | M+ | 1953 | 1953 | – | 1953 | | |
| 1957–58 | S | 1957–58 | S | 1957 SBM | 1957 | 1957 SBM | 1957 | 1 | 2 |
| 1965–66 | S | 1965 | M+ | 1965 | 1965 | 1965–66 | 1965 | 2 | 2 |
| | | | | | | | 1966 | 3 | 2 |
| M1968–69 | M– | 1969 | M– | – | – | 1968 SBM | 1968 | 2 | 2 |
| | | | | | | | 1969 | 2 | 2 |
| 1972–73 | S+ | 1972–E73 | S | 1972 | 1972 | 1972 | 1972 | 5 | 2 |
| | | | | | | | 1973 | 3 | 2 |

(3)  *World Weather Records* for 1931–1940 which includes maximum annual
     height readings from the Roda gauge at Cairo for 1918–1943;
(4)  *World Weather Records* for 1941–1950 which includes height data at
     Roda for 1943–1954, and monthly and annual Nile River discharge data
     at Aswan for 1869–1954;
(5)  Popper (1951) for data on flood levels for 1826–1889;
(6)  Shahin (1985) which provides monthly and annual discharges for various
     locations along the White Nile, the Main Nile, Blue Nile, and Atbara
     rivers for 1912–1973.

It was essential to use the height data for 1824–1868. However, Bliss (1926)
considered the discharge data to be superior in quality, so from 1869 on discharge
data were used to evaluate the annual floods. Where the two measurements
overlapped, results were similar, particularly when departures from average were
large. As reported in Shahin (1985), the works on the first Aswan dam and some
of the barrages on the Nile in Egypt were completed in 1902. Shahin also infers
that the history of scientific study for the hydrology of the Nile began with the
introduction of current meters by Lyons (1906) in about 1902; prior to this, flow
measurements were made by floats. The old Aswan dam was heightened twice,
once in 1912 and again in 1934 (Shahin 1985). This dam, along with other storage
works on the Blue Nile and White Nile changed the Nile from Aswan to the sea
into a partially regulated river instead of a naturally flowing one. The High
Aswan Dam has been in operation since 1964; and full regulation has almost been
achieved as a result of the formation of Lake Nasser upstream of the High Aswan
Dam in 1965 (Shahin 1985).

There was a large change in both the discharge data and the height data near
the turn of the century. The very strong ENSO of 1899–early 1900 appeared to
set in between the two flow regimes. Between 1869 and 1898, the average cumulative discharge amount at Aswan for the July–October period was $7428.8 \times
10^7 \, \mathrm{m^3 \, yr^{-1}}$ and for 1900–1929 it was $5775.3 \times 10^7 \, \mathrm{m^3 \, yr^{-1}}$. The average max-
imum annual height of the Nile at Cairo for 1869–1998 was 19.782 m above
Mediterranean Sea level at Alexandria, and for 1900–1929 it was 18.964 m above.
Despite the fact that the average height of the Nile has typically increased with
time due to sedimentary buildup of the river bottom, one notes a distinct fall in
height as was also noted in the discharge amount over this period. Maximum
height data for 1824–1868 were taken from Toussoun (1925) with degree of
reduction in height based on an average maximum height of 19.29 m. The data
used for later computations were taken from Toussoun's maximum heights for
1869–1921; and the remaining height and discharge data were taken from the
*World Weather Records* (to obtain the above computations). *World Weather
Record* data were obtained from the Nile Control Department, Ministry of
Public Works, Cairo, Egypt; maximum level readings are from the Roda Gauge,
Cairo; discharge data refer to the discharge downstream from the Aswan Dam,
gauge at 23°45′N, 32°50′E. The highest discharge months at Aswan are usually

July–October (representing the months of June–September rainfall which are ordinarily the heaviest over the highlands of Ethiopia), with the maximum flow peak usually in September or October. The above break in discharge records was clearly evidenced by Jarvis (1935), Kraus (1956) and Riehl et al. (1979) and was readily apparent from a visual scan of the data records. By using the above findings, there is now a rough relationship available between the annual maximum height values for the Nile River at Cairo and the July–October discharge values at Aswan prior to the installation of the High Aswan Dam and Nasser Lake regulatory facilities. This relationship indicates that an 81.8 cm drop in annual maximum height is roughly equivalent to an average drop in discharge amount for July through October at Aswan of $1653.5 \times 10^7 \, m^3 \, yr^{-1}$. On the basis of these figures, Table 6.4 was constructed for identifying events and determining their intensity, with a drop of 27 cm below the long-term average maximum height of the Nile at Cairo representing a minimal degree 1 intensity event and equivalent to a minimal degree 1 discharge value of $546 \times 10^7 \, m^3$ below the long-term average for July–October at Aswan. Kraus (1956) found that changes in the rainfall regimes about the same time could be observed in the records of most tropical stations at places as far apart as northern Australia and Central America. He considers that the change here probably represents an actual discontinuity of the Nile regime and, although the change in the Nile flow appears to be more clear-cut than that of most rainfall stations, it is due to the integrating

Table 6.4 *Relationship between decrease in annual maximum height of the Nile River flood at Cairo below applicable long-term average maximum height and departure of annual cumulative discharge amount for July–October below applicable annual average amount of July–October at Aswan Dam (23°45′N, 32°50′E), Dongola (19°10′N, 30°29′E), and the combined Blue Nile/Atbara river discharges at, respectively, 15°37′N, 32°30′E/17°42′N, and 33°58′E, with degrees of decrease/ reduction (1–5) indicated. Relationship valid up to about 1964; see text for details.*

| Departure of annual maximum height below a specified average annual maximum height for the Nile River at Cairo in meters above Mediterranean Sea level at Alexandria (m) | Degree of reduction in maximum flood height or discharge amount from long-term averages | Departure of annual cumulative discharge amount for July–October below a specified annual average amount for July–October for the indicated river discharge sites ($\times 10^7 \times m^3$) |
|---|---|---|
| 0.27–0.53 | 1 | 546–1091 |
| 0.54–0.80 | 2 | 1092–1637 |
| 0.81–1.07 | 3 | 1638–2183 |
| 1.08–1.34 | 4 | 2184–2729 |
| 1.35– | 5 | 2730– |

effect of a river catchment. A similar reduction in rainfall near the turn of the century is noted in Diaz et al. (1989).

Bell and Menzel (1972) comment further on the abrupt, large fall in the high water level of the Nile in A.D. 1899, and they refer to Hurst who had written on various occasions that it must be a genuine climatic shift and cannot be explained away as an effect of the first dam on the Nile at Aswan, which was put in opera- tion in 1902. They also believed that this decrease in the high water level of the Nile was undoubtedly related to the findings of Kraus (1954, 1955, 1956) that a change to a drier regime set in around 1900 at a number of tropical and east coast (of Australia and North America) stations.

Table 6.4 was used in conjunction with the various listed data sources to pro- vide the degree of reduction in annual maximum height and/or July–October discharge below applicable average values. These degree of reduction values (1–5) and their related confidence figures were then entered in Table 6.3. The confidence levels (1–5) refer to the number of sources of evidence for the event evaluation, with 5 indicating 5 or more sources.

Table 6.5 summarizes Table 6.3 information with regard to relationships of the regional manifestations to their parent large-scale developmental feature, the ENSO. Although there is widespread agreement in occurrences for S–VS ENSOs, there is considerable event-to-event variability when the moderate inten- sity ENSOs are included; and as indicated in a previous section, weak ENSOs are likely to show a very spotty activity pattern.

To identify further the summer monsoon rainfall contribution over the highlands of Ethiopia (and its resulting effects on the Nile River flood) with the Southern Oscillation-related climatic activity, Figures 6.3 and 6.4 were con- structed. In an attempt to reduce effects of regulatory features to some extent, I used the July–October discharge data for Dongola (19°10′N,30°29′E) on the Main Nile for 1912–1973 [taken from the monthly and annual data of Table. 15 in Shahin (1985)]. I selected this location since it was down-river from the critical inputs of the Blue Nile and Atbara rivers and it was up-river from the High Aswan Dam and Lake Nasser controls. The average discharge amount for

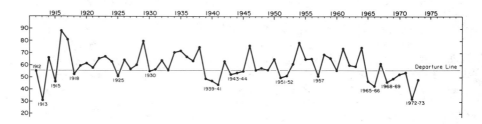

Fig. 6.3   Annual discharge of the Main Nile River for July–October at Dongola (19 °10′N, 30 °29′E) in $10^9\,m^3\,yr^{-1}$, with the horizontal line at $55.92 \times 10^9\,m^3\,yr^{-1}$ used to signify the discharge level for the minimal degree 1 reduction from the long- term average discharge value. These years are idenfied below the corresponding points in the graph. (See text and accompanying Table 6.4 for details.)

Table 6.5 *Summary of Table 6.3 data showing number (No.) of regional events (El Niño, Indonesian east monsoon drought, Australian drought, deficient India summer monsoon rainfall, and weak Nile Flood) activated by ENSOs 1824–1973 and 1832–1973 (since earlier data on Indonesian east monsoon drought are missing) and percentage (%) of activations they represent if all ENSOs moderate (M)–very strong (VS) are considered and if only strong (S)–VS ENSOs are considered.*

| ENSO Events | | | El Niños | | E. Mons. Drt. | | Austral. Drt. | | Def. Ind. Sum. Mons. | | Wk Nile Fld. | |
|---|---|---|---|---|---|---|---|---|---|---|---|---|
| Onset years | Strength | No. | No. | % | No. | % | No. | % | No. | % | No. | % |
| 1824–1972 | M-VS | 44 | 43 | 97.7 | | | 37 | 84.1 | 37 | 84.1 | 36 | 81.8 |
| 1824–1972 | S-VS | 21 | 21 | 100.0 | | | 21 | 100.0 | 21 | 100.0 | 21 | 100.0 |
| 1832–1972 | M-VS | 41 | 40 | 97.6 | 35 | 85.4 | 35 | 85.4 | 35 | 85.4 | 33 | 80.5 |
| 1832–1972 | S-VS | 19 | 19 | 100.0 | 19 | 100.0 | 19 | 100.0 | 19 | 100.0 | 19 | 100.0 |

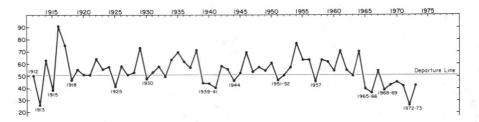

Fig. 6.4   Combined annual discharge of the Blue Nile and Atbara rivers for July–October at respectively 15 °37′N, 32 °E and 17 °42′N, 33 °58′E, in $10^9$ m$^3$ yr$^{-1}$, with the departure line at 51.11 × $10^9$ m$^3$ yr$^{-1}$ to signify the combined minimal degree 1 reduction from the long-term combined average discharge value. See text and accompanying Table 6.4 for details.

July–October for 1912–1964 was 6138 × $10^7$ m$^3$ yr$^{-1}$ and after reducing that by 546 × $10^7$ m$^3$ yr$^{-1}$ (to arrive at the minimal level for a degree 1 event), we have a departure line at 5592 × $10^7$ m$^3$ yr$^{-1}$ (which is about 55.9 × $10^9$ m$^3$ yr$^{-1}$ on Fig. 6.3) to separate out the event years with their lower discharge values. Figures for 1964–1966 are fairly reasonable but they continue to diverge progressively more from reality with time as the reservoir formed by the High Aswan Dam fills. Figure 9.12 of Shahin (1985) shows this year-to-year filling. Shahin reports that the natural supply of water at Aswan in 1973 was quite small and that the reservoir had to be emptied by 30.36 × $10^9$ m$^3$ in the period from 1 January to 8 July to cope with demand. Figure 6.4 shows a graph of the combined annual discharge values for July–October for the Blue Nile and Atbara rivers at 15°37′N, 32°32′E and 17°42′N, 33°58′E, respectively. These data were obtained from Tables 11 (Blue Nile) and 14 (Atbara) in Shahin (1985). The average combined discharge amount for these rivers for July–October of 1912–1964 was 5657 × $10^7$ m$^3$ yr$^{-1}$ and after reducing that by 546 × $10^7$ m$^3$ yr$^{-1}$ (to roughly arrive at the minimal level for a degree 1 event contribution), we have a departure line at 5111 × $10^7$ m$^3$ yr$^{-1}$ (which is about 51.1 × $10^9$ m$^3$ yr$^{-1}$ on Fig. 6.4) to separate out event years with their lower discharge contributions. Again the discharge values are fairly reasonable up to 1964–1966, but values deteriorate with time as controls on the river continue to take place. It must be realized that a summer monsoon contribution from the highlands of Ethiopia is also made through the Sobat River, but it enters the Nile River system over 800 km upriver from Khartoum and its effect which is relatively small, when compared to the Blue Nile, is carried as a part of the White Nile by Shahin (1985). It is interesting that the combined annual average for the Blue Nile and Atbara rivers for July–October (5657 × $10^7$ m$^3$ yr$^{-1}$) when compared to the average discharge for the main Nile at Dongola over the same months and years (1912–964) of 6138 × $10^7$ m$^3$ yr$^{-1}$ is about 92% of the discharge for the average July to October period. Of course, it must be realized that a lot of evaporation takes place between Atbara and Dongola. It is also interesting that the average Blue Nile contribution (of 4454 × $10^7$ m$^3$ yr$^{-1}$) is about 3.7 times the contribution of the Atbara River (1203 × $10^7$ m$^3$ yr$^{-1}$) for the July–October period.

## Historical overview

Known history of the Nile River dates back to just before 5000 B.C. (Shahin 1985). The record of Nile levels dates back to about 3000 to 3500 B.C.; the river gauge is called a Nilometer (Shahin 1985). Three types of Nilometers were used; the first type consisted simply of marking the water levels on cliffs on the banks of the river; the second type consisted of a scale, usually of marble, on which the water level was observed; the standard gauge consisted of a series of steps or pillars built into the river bank to each of which a section of the scale was fixed (Shahin 1985). The third and most accurate Nilometer used, brought water of the Nile to a well and the water level was marked either on the walls of the well or on a central pillar.

The keepers of the early Nilometers were Copts (Popper 1951). Joseph himself is said to have built the first Nilometer at Bedreshen on the west bank of the Nile near the remains of a wall which was said to be part of the 'Granary of Joseph' at the foot of which the Nile used to flow (Popper 1951). Since the Nile appeared to rise in flood at about the same time every year its behavior might be described as regular. Nevertheless, as mentioned by Shahin (1985): The Bible tells us, however, '. . . there came seven years of great plenty throughout the land of Egypt. And there shall rise after them seven years of famine . . .' (*Genesis* 41, 29–30). The interpretation by Joseph to this dream of the Pharaoh of Egypt was probably the first indication of persistence in the hydrologic time series (Shahin 1985). On the basis of information provided by Walford (1879), *Encyclopaedia Judaica* (1971), and Biswas (1970), it appears that the 7-yr famine foreseen by Joseph set in about 1708 B.C.

The beginning of the Moslem era was A.D. 622. There was a Nilometer in the ancient city of Memphis, and according to Arabic tradition the levels of the Nile were still read in Memphis during the first century after the Arab conquest of Egypt in A.D. 641 (Popper 1951). The Arab rulers retained Copts as keepers of the Nilometer for 120 yr after the conquest, until A.D. 861, and apparently it was from them that they received certain recorded statistics of the Nile covering the 20 yr preceding the conquest. The first definitely known Arab Nilometer was built in A.D. 715, at the southern end of Roda Island (Popper 1951). This Nilometer was rebuilt in A.D. 861, and statistics of the Nile subsequent to that date undoubtedly derive from the Roda Nilometer. It is uncertain at which of several other Nilometers then in existence around Cairo the earlier recorded levels of the Nile were read; little is known about any of these Nilometers (Popper 1951).

Popper (1951) found that the earlier data on the Nile River minimum and maximum levels came chiefly from the Arab chroniclers Ibn Taghri Birdi and Ibn al-Hijazi who provided parallel, though not always identical figures, for almost all of the years A.D. 641–1469. Ibn al-Hijazi had data also for A.D. 622–640, before the Arab conquest of Egypt. Ibn Iyas had minima and maxima for the period A.D. 1504–1522; for the missing years A.D. 1470–1503, he had dates of plenitude from which the general behavior of the Nile can be inferred (Popper

1951). Textual information in Toussoun (1925) also covers the missing years in this first part of the record. Ibn Aibak had a third set of data for the first 600 yr of record, and several other Arab chroniclers offer data for shorter portions of this early period or for individual years (Popper 1951). There are practically no data for A.D. 1523–1586 after which there are occasional breaks in the maximum level data up until 1824. (The large break after 1522 was due to the Turkish takeover of Egypt.) However, since information on the eastern side of the tropical ENSO region becomes available from 1524 on, there is the capability of having a roughly complete record on low SOI-related climatic activity available for A.D. 622 to the present. Ibn al-Hijazi's data were used by Toussoun (1925) as the basis for his tables; he supplied missing data from Ibn Taghri Birdi's figures (Popper 1951), and his metric data were based on height above Mediterranean Sea level at Alexandria. Of course, the original measurements were in cubits and fingers and cubits may consist of 24 or 28 fingers, depending on the regime in charge at the time (Popper 1951).

As mentioned earlier, the Nile's headwaters in the highlands of Ethiopia (especially the Blue Nile and Atbara), supply the waters of the annual flood; the White Nile plays very little part in the flood (Brooks 1926). The Nilometers near Cairo have been of particular interest. Although not complete, these records form perhaps the longest written record of any meteorological phenomenon (in that they register the summer monsoon rainfall over Ethiopia). Lyons (1905, 1906, 1908) provides excellent physiographic background information on the Nile River system and its controlling sources, the Nile flood and its variation, and also some excellent data on flood heights and rainfall. Jarvis (1935) provides plots of heights of the Nile maximum annual stages and minimum annual stages as presented by various authors, and some valuable commentaries by many scientists with excellent backgrounds on the Nile River system. Popper (1951) provides a very comprehensive evaluation of the various sources of the Nile River data. His tables, plots, and evaluations are extremely helpful to the construction of the SOI-related climatic records.

## Problematic aspects

Computed rises in flood stages of the Nile ascribable to sedimentation vary considerably from one investigator to another. Bell (1970), Bell and Menzel (1972), and many other investigators accept a rise of 10 cm per century. In Popper's (1951) findings, the increase in the average level of the Nile, and consequently the rise in the level of the Nile bed (through sedimentation), was abrupt in certain years or short periods of years; it did not take place at a uniform rate per century, as generally reckoned. Moreover, if the suggested changes in the scale of measurement are approximately correct, the rise of the bed was greater than the generally accepted rise. Popper's cumulative average plot of the maximum Nile levels (Popper 1951, Graph 1, p. 167) shows the variation of Nile maximum levels up to A.D. 1469. Jarvis's (1935: Table 1, p. 1026) computed rise trend in flood stages ascribable to sedimentation shows variability over various periods ranging generally from 10 to 15 cm per century.

In comparing measurements used by the various authors, one must frequently return to the source data in cubits and fingers and find from what base level the measurements were taken. Popper (1951; Table 2) shows the three scales of measurement used. Each one starts at a different zero level. Popper (1951) uses Scale I up to A.D. 1522 with the zero level at the Nilometer floor, which is 8.15 m above Mediterranean Sea level. Toussoun appears to be using Popper's Scale III with zero level at 0.66 m above the Nilometer floor and 8.81 m above Mediterranean Sea level (Med. S.L.).

Plenitude is the Nile level of 16 cubits (cu) at Cairo. It means the same as the Arabic term 'wafa,' fulfillment. Originally it was the level most favorable for the irrigation of Egypt and was close to the normal annual maximum level at Cairo. When in later centuries this ceased to be the case, it still remained by tradition an eagerly awaited level and was the occasion of formal ceremony. An earthen dam was constructed annually across the Cairo Canal near its mouth to prevent waters of the Nile from entering before they reached the desired level. About the 12th century A.D. it became customary to cut the dam only after plenitude. However, as explained earlier, the rise of the river bed due to sedimentation resulted in higher river levels. In A.D. 1050, the ordinary rise of the Nile was said to be 17 cu (about 17.62 m above Med. S.L.); less was said to be insufficient. It was said that above 17 cu, 'the people are satisfied and celebrate.' Abd al-Latif about A.D. 1200 defines the necessary flood limits as 16 and 18 cu; at 16 cu about half the land begins to be watered and crops are sufficient for 1 yr food supply; between a little over 16 cu and 18 cu the crops are sufficient for 2 yr supply. From A.D. 1342 on, the average annual maximum levels were significantly higher than the 16 cu level and it is expected that a realistic plenitude level was about 17 cu at that time and higher still after about A.D. 1379.

In Muhammadan countries, chroniclers dated all events in terms of the Muhammadan lunar calendar. The Muhammadan year rotates through all months of the solar year once in each cycle of about 33 of its years. Therefore, at regular intervals, the minimum and maximum of one Nile season will fall in different lunar years, and as a result, there is frequently confusion in the records, which is intensified when the skip year is also neglected. The skip year, in the Muhammadan calendar is the 33rd year of a cycle when equated with a cycle of 32 solar years. Problems with regard to time arise, since the chroniclers often recognize different skip years. As a result, records on events (weak or excessive Nile flood years), as reported by different chroniclers, may often differ by a year in the older records; however, in Toussoun's (1925) report there are occasional large discrepancies noted in the textual material dates on the catastrophic occurrences that he obtained from other authors. Fortunately, they do not show up in his tabulated data; however, he does not comment on them in his report. The following empirical formula obtained from Albert Galloway, a professional numismatist, was found to be very useful for converting the A.H. (anno Hegirae) dates of the Mohammedan era to A.D. dates:

A.H. date − (0.0303 × A.H. date) + 622 = A.D. date.

## Poor Nile flood years for A.D. 622–1522

In the construction of Table 6.6, Toussoun's (1925) tabulated data and textual information was considerably modified by tabulated data, graphic plots, and corrections from Popper (1951). Popper's tables included data on the stronger developments from Ibn Taghri Birdi, Ibn al-Hijazi and Ibn Aibak. Table 6.6 lists

Table 6.6 *Years (Yr), over the period A.D. 622–1522, with weak Nile floods (those below plenitude and/or specified average annual maximum flood levels) at Cairo, rated by degree (Deg) of deficiency (1–5) as noted in Table 6.4, with confidence (Conf) ratings based on the number of confirmation sources. See text for details.*

| Yrs | Deg | Conf | Yr | Deg | Conf | Yr | Deg | Conf | Yr | Deg | Conf |
|---|---|---|---|---|---|---|---|---|---|---|---|
| 629 | 4 | 2 | 713 | 4 | 5 | 770 | 2 | 2 | 803 | 4 | 5 |
|     |   |   |     |   |   | 771 | 1 | 2 |     |   |   |
| 632 | 2 | 2 | 721 | 1 | 4 | 772 | 1 | 2 | 811 | 1 | 2 |
|     |   |   |     |   |   | 773 | 1 | 2 | 812 | 4 | 3 |
| 642 | 2 | 2 | 723 | 1 | 3 |     |   |   |     |   |   |
|     |   |   |     |   |   | 776 | 2 | 2 | 817 | 2 | 2 |
| 650 | 5 | 3 | 726 | 1 | 4 |     |   |   | 818 | 1 | 2 |
|     |   |   |     |   |   | 779 | 1 | 2 |     |   |   |
| 662 | 1 | 4 | 733 | 2 | 2 | 780 | 1 | 2 | 828 | 1 | 2 |
|     |   |   |     |   |   | 781 | 1 | 2 |     |   |   |
| 678 | 1 | 2 | 735 | 2 | 3 | 782 | 3 | 5 | 830 | 4 | 5 |
|     |   |   |     |   |   |     |   |   |     |   |   |
| 683 | 1 | 2 | 737 | 1 | 2 | 785 | 1 | 2 | 832 | 3 | 5 |
|     |   |   |     |   |   |     |   |   | 833 | 2 | 2 |
| 687 | 3 | 2 | 740 | 1 | 2 | 788 | 1 | 5 | 834 | 1 | 2 |
| 688 | 3 | 5 | 756 | 2 | 5 | 789 | 1 | 5 |     |   |   |
| 689 | 5 | 5 |     |   |   |     |   |   | 836 | 1 | 2 |
|     |   |   | 759 | 1 | 2 | 791 | 2 | 2 | 837 | 2 | 2 |
| 691 | 1 | 4 |     |   |   | 792 | 1 | 2 |     |   |   |
|     |   |   | 761 | 1 | 5 |     |   |   | 841 | 4 | 5 |
| 693 | 3 | 5 | 762 | 1 | 2 | 794 | 1 | 5 | 842 | 5 | 4 |
| 694 | 5 | 5 | 763 | 1 | 2 |     |   |   |     |   |   |
| 695 | 3 | 3 | 764 | 2 | 2 | 796 | 1 | 2 | 847 | 1 | 2 |
| 696 | 4 | 3 | 765 | 1 | 2 | 797 | 2 | 2 | 848 | 2 | 2 |
|     |   |   |     |   |   |     |   |   |     |   |   |
| 702 | 3 | 5 | 767 | 1 | 2 | 799 | 2 | 2 | 850 | 1 | 2 |
|     |   |   |     |   |   |     |   |   | 851 | 3 | 2 |
| 705 | 4 | 5 | 769 | 2 | 2 | 802 | 3 | 3 | 852 | 1 | 5 |

Table 6.6 (*cont.*)

| Yrs | Deg | Conf | Yr | Deg | Conf | Yr | Deg | Conf | Yr | Deg | Conf |
|---|---|---|---|---|---|---|---|---|---|---|---|
| 860 | 1 | 2 | 981 | 1 | 2 | 1294 | 3 | 4 | 1402 | 1 | 3 |
| | | | 982 | 2 | 5 | | | | 1403 | 4 | 4 |
| 881 | 3 | 4 | | | | 1297 | 1 | 2 | | | |
| | | | 989 | 1 | 2 | 1298 | 2 | 2 | 1408 | 3 | 4 |
| 885 | 1 | 2 | | | | | | | | | |
| | | | 996 | 1 | 2 | 1305 | 1 | 2 | 1418 | 1 | 2 |
| 887 | 1 | 5 | | | | | | | | | |
| 888 | 1 | 5 | 1007 | 3 | 5 | 1309 | 1 | 4 | 1420 | 1 | 3 |
| | | | 1008 | 4 | 5 | | | | | | |
| 894 | 1 | 2 | | | | 1313 | 1 | 3 | 1424 | 2 | 3 |
| 895 | 3 | 5 | 1023 | 3 | 5 | | | | 1427 | 3 | 3 |
| | | | | | | 1321 | 1 | 3 | | | |
| 897 | 1 | 2 | 1036 | 2 | 2 | | | | 1433 | 2 | 3 |
| | | | 1037 | 1 | 2 | 1326 | 1 | 3 | 1449 | 1 | 3 |
| 903 | 5 | 5 | | | | 1334 | 1 | 3 | 1450 | 5 | 4 |
| | | | 1057 | 1 | 2 | | | | 1451 | 1 | 3 |
| 907 | 1 | 2 | | | | 1337 | 1 | 3 | | | |
| | | | 1066 | 2 | 2 | 1338 | 1 | 3 | 1459 | 1 | 3 |
| 917 | 1 | 4 | | | | | | | | | |
| | | | 1072 | 1 | 2 | 1340 | 1 | 3 | 1461 | 2 | 3 |
| 927 | 3 | 4 | 1085 | 3 | 5 | | | | 1462 | 1 | 3 |
| 931 | 2 | 2 | 1096 | 5 | 1 | 1348 | 1 | 2 | | | |
| 939 | 2 | 2 | 1122 | 1 | 2 | 1350 | 1 | 2 | 1466 | 1 | 2 |
| | | | | | | 1351 | 1 | 2 | | | |
| 941 | 1 | 1 | 1144 | 5 | 2 | | | | 1468 | 1 | 2 |
| 942 | 1 | 1 | | | | 1363 | 1 | 2 | | | |
| | | | 1159 | 4 | 5 | | | | 1474 | 2 | 2 |
| 945 | 1 | 5 | | | | 1370 | 1 | 2 | | | |
| 946 | 1 | 5 | 1200 | 5+ | 5 | | | | 1484 | 1 | 2 |
| 947 | 1 | 5 | 1201 | 1 | 2 | 1373 | 2 | 3 | | | |
| 948 | 3 | 5 | 1202 | 1 | 2 | | | | 1490 | 1 | 2 |
| 949 | 3 | 4 | | | | 1380 | 1 | 2 | | | |
| 950 | 1 | 3 | 1210 | 3 | 4 | | | | 1492 | 1 | 2 |
| 951 | 1 | 3 | | | | 1385 | 2 | 3 | | | |
| | | | 1219 | 3 | 4 | | | | 1497 | 3 | 3 |
| 963 | 1 | 2 | | | | 1389 | 3 | 3 | | | |
| 964 | 2 | 2 | 1230 | 5 | 3 | | | | 1504 | 1 | 1 |
| 965 | 1 | 2 | 1231 | 1 | 2 | 1393 | 2 | 3 | | | |
| 966 | 3 | 2 | 1234 | 1 | 2 | 1394 | 2 | 3 | 1510 | 1 | 3 |
| 967 | 5+ | 5 | 1244 | 4 | 4 | | | | | | |
| | | | | | | 1399 | 1 | 3 | 1518 | 2 | 3 |
| 977 | 1 | 2 | 1290 | 1 | 2 | 1401 | 2 | 3 | 1520 | 1 | 3 |

the years of poor Nile floods, their degree of weakness (1–5) and a confidence rating (1–5). For about the first 460 yr, the smallest flood deficiency level (1) was just below 16 cu, the original plenitude level [16 cu = 17.35 m above Med. S.L. at Alexandria in Toussoun's (1925) data, which is a little below normal for the period]. For the other degrees of weakness (2–5), the reduction values, based on 50-yr averages for the annual Nile River maximum levels at Cairo, are as indicated in Table 6.4. After A.D. 1080, the degree of reduction, using the applicable 50-yr average, pertains to all degrees as it did for Table 6.3 data. Also, the confidence level (1–5) is based on the number of sources of evidence [e.g., Toussoun, Popper, Popper's Arab sources (Ibn Taghri Birdi, Ibn al-Hijazi, Ibn Aibak)].

Table 6.6 shows a total of 178 weak Nile flood years with varying degrees of deficiency (1–5); during this 901-yr period the figures on the average would indicate a weak Nile flood occurrence about every 5 yr. However, the distribution of these flood years makes it very difficult to relate them to ENSO events. Considering what we see in Table 6.3 and between applicable parts of Table 6.1 and Table 6.7, it appears possible that those extended weak Nile flood periods of 4–7 yr in a row, that occur in the first 346 yr of Table 6.6, may be associated with two or more ENSOs, with one setting in at the onset of the period and another 2 or 3 yr later. There will be further consideration of these extended periods of activity in the following section. Out of the 901-yr record with 178 yr of weak Nile floods, there were 97 in degree 1, 33 in degree 2, 25 in degree 3, 12 in degree 4, and 11 in degree 5. Table 6.7 also shows some of the longer-term features of this early record. Over the period A.D. 622–999 there were 105 yr of weak Nile floods, occurring in approximately 27.8% of the years. For the period A.D. 1000–1290 there were 23 yr of weak Nile floods, occurring in about 8% of the years. For the period A.D. 1291–1522 there were 50 yr of weak Nile floods, occurring in about 21.6% of the years. There were several cases of extended records during the earliest period but none in the other two periods. It was now essential to tie in these earlier findings with the later records on the Nile; and although there were several breaks in the record between 1523 and 1823, it was possible to use available data and some anecdotal information from Lyons, Toussoun, and Popper to obtain a record for A.D. 1694–1899 as shown in Table 6.7. Over this period weak Nile floods occurred about 35% of the years. Also, several cases of extended weak Nile floods occurred over this period. During the recent period 1900–1973 weak Nile floods occurred about 31% of the years, but there were no periods of extended activity noted. The period A.D. 1694–1899 which would be considered the latter part of the Little Ice Age (LIA), shows frequent occurrences of weak Nile floods and also shows several extended periods of weak Nile flood. In contrast, A.D. 1000–1290, which would be considered to represent the Little Climatic Optimum (LCO), shows a very low percentage of weak Nile floods and no extended periods for such floods. A.D. 1291–1522 was considered to be an interim period between the LCO and LIA. On the basis of the above findings it would appear that activity over A.D. 622–999 would be

Table 6.7 *The years A.D. 622–1973, with the exception of 1523–1693 (when the record showed many breaks), are broken into 5 periods for considering the occurrence of weak Nile River floods (WNRFs): 622–999, a relatively cool period; 1000–1290, representing the Little Climatic Optimum (LCO); 1291–1522, considered to be an interim period between the LCO and Little Ice Age (LIA); 1694–1899, representing the latter part of the LIA; and 1900–1973, covering the recent period. The number of WNRFs by degree (as specified in Table 6.4) and total number are listed for each period, as are cases where extended (4 or more years in a row) of WNRFs occurred.*

| Period | Feature | Number of WNRFs by degree | | | | | Total number | Cases where WNRFs occur 4 or more years in a row |
|---|---|---|---|---|---|---|---|---|
| | | 1 | 2 | 3 | 4 | 5 | | |
| 622–999 | Cool period | 56 | 20 | 15 | 8 | 6 | 105 | 693–696, 761–765, 769–773, 779–782, 945–951, 963–967 |
| 1000–1290 | LCO | 9 | 2 | 5 | 3 | 4 | 23 | None |
| 1291–1522 | Interim period | 32 | 11 | 5 | 1 | 1 | 50 | None |
| 1694–1899 | LIA | 18 | 23 | 14 | 11 | 6 | 72 | 1713–1716, 1782–1785, 1790–1797, 1835–1839 |
| 1900–1973 | Recent period | 9 | 9 | 3 | 0 | 2 | 23 | None |

more representative of a cool period. This would be in agreement with Maejima and Tagami (1986) who noted a cool age during the 7th–9th centuries in Japan.

## Discussion

With regard to the ENSO/El Niño relationship, which is quite close in the moderate through very strong intensities, two enigmatic cases arise, in that I cannot find any evidence for large-scale ENSOs to relate to the El Niños of 1910 and 1917. Both of these regional events are well documented in Quinn and Neal (1991), yet there is no large-scale representation and both occur during high SOIs which is also enigmatic. However, the El Niño is closely allied with what occurs in the eastern core area of the SO (the southeast Pacific subtropical high pressure cell with its control over the southeast trades), and I did note a relaxation in the high SOIs during early 1917 in both the Santiago-Darwin and Juan Fernandez-Darwin SOIs; and, of course, any substantial relaxation of that nature at that time of the year could release an effective equatorial Kelvin wave.

I have often been queried about using rainfall data to verify the river data. One must realize that it would take a couple of thousand rainfall and evaporation stations to attempt to provide the excellently integrated data available through the river discharge systems.

I tend to agree with Popper (1951) on his views concerning the sedimentary buildup of the Nile River bottom. There appeared to be very little buildup over about the first four centuries. It is expected that average rainfall was less over this early cool period, when a cooler atmosphere would contain less precipitable water. It is expected that the rainfall amount was greater and less variable during the Little Climatic Optimum.

The extended periods of low Nile flood appear to be a feature of the cooler periods. The 7-yr periods of plenty and 7-yr periods of famine, as referenced in the Bible, no longer appear unusual when reviewing these long records. In fact, the low Nile flood period of 1790–1797 can be verified by three separate sets of data – Lyons (1906), Toussoun (1925), and Popper (1951). Moreover, the findings here would indicate that Joseph lived during a cool period (prior to and after 1700 B.C.).

The weakest Nile floods occurred in A.D. 650, 689, 694, 842, 903, 967, 1096, 1144, 1200, 1230, 1450, 1553, 1641, 1650, 1694, 1715, 1716, 1783, 1877, 1899, 1913, and 1972. This investigation suggests that the most disastrous Nile flood failure occurred in A.D. 1200, and it is quite likely that its parent large-scale ENSO was similarly unusual and also spawned the exceptionally strong El Niño that led to the cataclysmic 'Chimu flood' in coastal Peru, the evidence of which was reported in Nials et al. (1979).

The deficient summer monsoon rainfall over the highlands of Ethiopia and the El Niño along the coast of southwest Ecuador and northwest Peru are at opposite ends of the 'seesaw' of ocean-atmosphere conditions, yet they are both integral features of the large-scale ENSO. However, as indicated above in reference to

Griffiths (1972), the summer monsoon rainfall is primarily affected by the development and location of the equatorial low core of the SO; and, since Ethiopia lies on the western periphery of this equatorial low, extensive (timewise) drought periods could result from a relatively small eastward displacement of the equatorial low for several summer monsoon seasons. On the other hand, the El Niño is more directly affected by the fluctuating development and strength of the South Pacific subtropical high pressure core of the SO, which causes variations in the strength of the southeast trades, the relaxation of which, during a breakdown phase of this high (indicative of a low SOI) results in the release of an equatorial Kelvin wave and a strengthening of the equatorial countercurrents that cause the El Niño (Wyrtki 1975; Wyrtki et al. 1976; Quinn 1987). The nature of the prior buildup, the Kelvin wave release and the resulting El Niño features would certainly not be conducive to a singular event of 4- to 7-yr duration.

Figure 6.2 shows a long-term change as reflected in the 3-mo running mean plot of the Rapa-Darwin sea-level pressure anomalies. Here, over the period April 1976–March 1988 the SOI anomalies averaged out to 1.5 mb below mean. All other pressure indices showed a similar drop over this 12-yr period. It is interesting that one of the strongest recorded ENSOs (1982–83) occurred in the midst of this significantly below-normal SOI period (Quinn and Zopf 1984). Following that 12-yr period of below-average SOIs, the plot moved up rapidly into the high SOI feature of 1988 (Fig. 6.2), which was accompanied by a period of extremely heavy rainfall and flooding in the west sector of the SO. Le Comte (1989) reported that: 'Heavy rainfall in the Blue Nile's catchment basin in the Ethiopian highlands during late July and early August contributed to major flooding along the Nile in Sudan.' The Egyptian newspaper *al-Ahram*, in its 13 August 1988 edition, stated that the two Nile tributaries the day before spilled across a region south of Khartoum 'until only tree tops remained visible.' This was the first such period of flooding rains over this region in more than a decade. Since mid-1989 it appears that we have slipped back into the below normal index mode.

One of the interesting findings from this study is the fact that the large-scale ENSO developments often show up many months earlier in the regional features on the western side of the 'seesaw' than they do on the eastern side. This was of course very clearly shown in the case of the very strong 1982–83 ENSO.

This study shows that apparently the LCO and LIA may have caused some significant changes in the SO-related activity. Quinn and Neal (1991) noted that the LIA caused an increase in the length and strength of subtropical Chilean droughts during the 17th, 18th, and early 19th centuries. In fact, during the peak drought period, 1770–1814, there was only one ENSO strong enough to bring about an above normal subtropical Chilean rainfall; this was the very strong ENSO of 1782–84 that caused the heavy Chilean rainfall and floods of 1783 (1783 was a very unusual year for atmospheric phenomena, as reported by Wood (1984)). The Laki volcanic eruption occurred in Iceland in 1783 along with other unusual phenomena.

*Acknowledgments* I thank the Chief of the Meteorological Service of French Polynesia; the Director of the Australian Bureau of Meteorology; the Climatic Analysis Center, NWS, NOAA; Forrest R. Miller of the Inter-American Tropical Tuna Commission; and the Interlibrary Loan Service of Oregon State University for continued support to this project. I thank Mrs. Barbara McVicar for processing this information into readable form. Great appreciation is expressed to the editors and reviewers of this item for their comments and suggestions. Support for this research through National Science Foundation Grant ATM-8808185 is gratefully acknowledged and greatly appreciated.

## References

BELL, B., 1970: The oldest records of the Nile floods. *Geographical Journal*, 136: 569–573.

BELL, B. and MENZEL, D. H., 1972: Research directed toward the observation and interpretation of solar phenomena. Final Report, Contract No. F19628-69-C-0077, Air Force Cambridge Research Laboratories, Bedford, Massachusetts. 16 pp.

BERLAGE, H. P., 1957: Fluctuations of the general atmospheric circulation of more than one year, their nature and prognostic value. *Koninklijk Nederlands Meteorlogisch Institut, Mededelingen en Verhandelingen*, 69. 152 pp.

BHATIA, B. M., 1967: *Famines in India*. 2nd ed. New Delhi: Asia Publishing House. 389 pp.

BISWAS, A. K., 1970: *History of Hydrology*. Amsterdam: North-Holland Publishing Company. 336 pp.

BLISS, E. W., 1926: The Nile flood and world weather. *Memoirs of the Royal Meteorological Society*, 1(5): 79–85.

BROOKS, C. E. P., 1926: *Climate through the Ages*. London: Ernest Benn. 439 pp.

DIAZ, H. F., BRADLEY, R. S., and EISCHEID, J. K., 1989: Precipitation fluctuations over global land areas since the late 1800's. *Journal of Geophysical Research*, 94(D1): 1195–1210.

DONGUY, J. R., HENIN, C., MORLIERE, A., and REBERT, J. P., 1982: Appearances in the western Pacific of phenomena induced by El Niño in 1979–80. Tropical Ocean-Atmosphere Newsletter No. 10: 1–2. Seattle: Pacific Marine Environmental Laboratories.

ELLIOT, H. M., 1867: *The History of India, as told by its own historians. The Muhammadan period*. Eight volumes. London: Trubner and Company (Material of interest is in Vols. I, III, V, VI, VII, and VIII.)

*Encyclopaedia Judaica*, 1971: *Joseph*, Volume 10: 202–218. Jerusalem: The Macmillan Company.

FLOHN, H., 1986: Indonesian droughts and their teleconnections. *Berliner geographische Studien*, 20: 251–256.

GRAHAM, N. E. and WHITE, W. B., 1988: The El Niño cycle: A natural oscillator of the Pacific Ocean-Atmosphere system. *Science*, 240: 1293–1302.

GRIFFITHS, J. F., 1972: Ethiopian highlands. *In* Griffiths, J. F. (ed.), Climates of Africa. *World Survey of Climatology, Vol. 10*. New York: Elsevier, 369–381.

HUNTER, W. W., 1882: *The Indian Empire: Its History, People and Products*. London: Trubner and Company. 568 pp.

HURST, J. F., 1891: *Indika – The Countries and People of India and Ceylon*. New York: Harper & Brothers. 794 pp.

*Imperial Gazetteer of India, The*, 1908: The Indian Empire. Oxford: The Clarendon Press, Vol. I, 568 pp. Vol. III. 520 pp.

JARVIS, C. W., 1935: Flood-stage records of the River Nile. *Transactions of the American Society of Civil Engineers*, 101: 1012–1071.

KILADIS, G. N. and DIAZ, H. F., 1989: Global climatic anomalies associated with extremes in the Southern Oscillation. *Journal of Climate*, 2: 1069–1090.

KHANDEKAR, M. L. and NERALLA, V. R., 1984: On the relationship between the sea surface temperatures in the equatorial Pacific and the Indian monsoon rainfall. *Geophysical Research Letters*, 11(11): 1137–1140.

KRAUS, E. B., 1954: Secular changes in the rainfall regine of SE. Australia. *Quarterly Journal of the Royal Meteorological Society*, 80: 591–601.

KRAUS, E. B., 1955: Secular changes of tropical rainfall regimes. *Quarterly Journal of the Royal Meteorological Society*, 81: 198–210.

KRAUS, E. B., 1956: Graphs of cumulative residuals. *Quarterly Journal of the Royal Meteorological Society*, 82: 96–98.

LE COMTE, D., 1989: The rains return to the tropics. *Weatherwise*, 42: 8–12.

LOCKWOOD, J. G., 1985: *World Climatic Systems*. London: Edward Arnold. 292 pp.

LYONS, H. G., 1905: On the Nile flood and its variation. *Geographical Journal*, 26: 249–272, 395–421.

LYONS, H. G., 1906: The physiography of the River Nile and its basin. Survey Department, Egypt. Cairo National Printing Department. 411 pp.

LYONS, H. G., 1908: Some geographical aspects of the Nile. *Geographical Journal*, 32: 449–480.

MAEJIMA, I. and TAGAMI, Y., 1986: Climatic change during historical times in Japan-reconstruction from climatic hazard records. *Geographical Reports of Tokyo Metropolitan University*, 21: 157–171.

MARTIN, R. M., 1858–61: *The Indian Empire*, Vol. 1. London: London Printing and Publishing Company. 582 pp.

MOOLEY, D. A. and PANT, G. B., 1981: Droughts in India over the last 200 years, their socioeconomic impacts and remedial measures for them. *In* Wigley, T. M. L., Ingram, M. J., and Farmer, G. (eds.), *Climate and History*. Cambridge: Cambridge University Press, p. 465–478.

MOOLEY, D. A. and PARTHASARATHY, B., 1979: Poisson distribution and years of bad monsoon over India. *Archiv für Meteorolgie, Geophysik und Bioklimatologie*, Ser. B, 17: 381–388.

MOOLEY, D. A. and PARTHASARATHY, B., 1984: Fluctuations in all-India summer monsoon rainfall during 1871–1978. *Climate Change*, 6: 287–301.

NAMIAS, J., 1973: Response of the Equatorial Countercurrent to the subtropical atmosphere. *Science*, 181: 1245–1247.

NIALS, F. L., DEEDS, E. E., MOSLEY, M. E., POZOROSKI, S. G., POZOROSKI, T. G., and FELDMAN R., 1979: El Niño: The catastrophic flooding of coastal Peru. *Field Museum of Natural History Bulletin*, 50(7): 4–14, 50(8): 4–10.

NICHOLLS, N., 1988: More on early ENSOs: Evidence from Australian documentary sources. *American Meteorological Society Bulletin*, 69: 4–6.

NICHOLLS, N., 1992: Historical El Niño/Southern Oscillation variability in the Australasian region. *In* Diaz, H. F. and Markgraf, V. (eds.), *El Niño: Historical and*

*Paleoclimatic Aspects of the Southern Oscillation*. Cambridge: Cambridge University Press, 151–173.

PANT, G. B. and PARTHASARATHY, B., 1981: Some aspects of an association between the Southern Oscillation and Indian summer monsoon. *Archiv fur Meteorologie, Geophysik und Bioklimatologie*, Ser. B, 29: 245–252.

PARTHASARATHY, B. and PANT, G. B., 1984: The spatial and temporal relationships between the Indian summer monsoon rainfall and the Southern Oscillation. *Tellus*, 36A: 269–277.

PARTHASARATHY, B., DIAZ, H. F., and EISCHEID, J. K., 1988: Prediction of all-India summer monsoon rainfall with regional and large-scale parameters. *Journal of Geophysical Research*, 93(D5): 5341–5350.

PATZERT, W. C., 1975: Results of the 1975 El Niño expedition. *Memorias de la Primera Reunion de los Centros de Investigacion de Baja California y la Institucion Scripps de Oceanografia (CIBCASIO)*. Centro de Investigacion Cientifica y de Educacion Superior de Ensenada (CICESE). Ensenada, Baja California, 31–37.

POPPER, W., 1951: *The Cairo Nilometer*. Berkeley and Los Angeles: University of California Press. 269 pp.

QUINN, W. H., 1974: Monitoring and predicting El Niño invasions. *Journal of Applied Meteorology*, 13: 825–830.

QUINN, W. H., 1979: The false El Niño and recent related climatic changes in the southeast Pacific. *In: Proceedings of the Fourth Annual Climate Diagnostics Workshop, Institute for Environmental Studies, University of Wisconsin, Madison, Wisconsin, Oct. 16–18, 1979*. Washington D.C.: U.S. Department of Commerce, NOAA, 93–110 (available as NTIS-PB80-201130).

QUINN, W. H., 1980: Monitoring and predicting short-term climatic changes in the South Pacific Ocean. *Investigaciones Marinas*, 8(1–2): 77–114.

QUINN, W. H., 1987: El Niño. *In* Oliver, J. E. and Fairbridge, R. W. (eds.), *The Encyclopedia of Climatology*. New York: Van Nostrand Reinhold, 411–414.

QUINN, W. H. and BURT, W. V., 1970: Prediction of abnormally heavy precipitation over the equatorial Pacific dry zone. *Journal of Applied Meteorology*, 9: 20–28.

QUINN, W. H. and BURT, W. V., 1972: Use of the Southern Oscillation in weather prediction. *Journal of Applied Meteorology*, 11: 616–628.

QUINN, W. H. and NEAL, V. T., 1983: Long-term variations in the Southern Oscillation, El Niño, and Chilean subtropical rainfall. *Fishery Bulletin, U.S.*, 81: 363–374.

QUINN, W. H. and NEAL, V. T., 1984: Recent long-term climatic change over the eastern tropical and subtropical Pacific and its ramifications. *In: Proceedings of the Ninth Annual Climate Diagnostics Workshop*. Washington D.C.: U. S. Department of Commerce, NOAA, 101–109 (available as NTIS-PB85-183911).

QUINN, W. H. and NEAL, V. T., 1992: The historical record of El Niño events. *In* Bradley, R. S. and Jones, P. D. (eds.), *Climate Since A.D. 1500*. London: Routledge, Chapman and Hall, 623–648.

QUINN, W. H. and ZOPF, D. O., 1984: The unusual intensity of the 1982–83 ENSO event. *Tropical Ocean-Atmosphere News Letter*, No. 26: 17–20.

QUINN, W. H., NEAL, V. T., and ANTUNEZ de MAYOLO, S. E., 1987: El Niño occurrences over the past four and a half centuries. *Journal of Geophysical Research*, 92(C13): 14,449–14,461.

QUINN, W. H., ZOPF, D. O., SHORT, K. S., and KUO YANG, R. T., 1978: Historical trends and statistics of the Southern Oscillation, El Niño, and Indonesian droughts. *Fishery Bulletin, U. S.*, 76: 663–678.

RASMUSSON, E. M. and CARPENTER, T. H., 1983: The relationship between eastern equatorial Pacific sea surface temperatures and rainfall over India and Sri Lanka. *Monthly Weather Review*, 111: 517–528.

RIEHL, H., EL-BAKRY, M., and MEITIN, J., 1979: Nile River discharge. *Monthly Weather Review*, 107: 1546–1553.

ROPELEWSKI, C. F. and HALPERT, M. S., 1987: Global and regional scale precipitation patterns associated with the El Niño/Southern Oscillation. *Monthly Weather Review*, 115: 1606–1626.

SHAHIN, M., 1985: *Hydrology of the Nile Basin.* Amsterdam: Elsevier. 575 pp.

TOUSSOON, O., 1925: *Mémoire sur l'Histoire du Nil, in Mémoires a L'Institut D'Egypte.* Cairo: Imprimerie de l'Institut Francaise d'Archeologie Orientale. 544 pp.

VAN BEMMELEN, W., 1916: Droote-jaren op Java. *Natuurwetenschappelijk Tijdschrift voor Nederlandsch Indie*, 75: 157.

WALFORD, C., 1879: *Famines of the World, Past and Present.* New York: Burt Franklin. 303 pp.

WALKER, G. T., 1937: World weather VI. Memoirs of the Royal Meteorological Society, Vol. IV(39): 119–139.

WOOD, C. A., 1984: Amazing and portentous summer of 1783. *EOS*, 26 June: 410.

*World Weather Records*, 1931–1940: Smithsonian Institution Miscellaneous Collection, 105. Washington D.C. 646 pp.

*World Weather Records*, 1941–1950: U.S. Department of Commerce. Washington, D.C. 1361 pp.

WYRTKI, K., 1973: Teleconnections in the equatorial Pacific. *Science*, 180: 66–68.

WYRTKI, K., 1975: El Niño – The dynamic response of the Equatorial Pacific Ocean to atmospheric forcing. *Journal of Physical Oceanography*, 5: 572–584.

WYRTKI, K., STROUP, E., PATZERT, W., WILLIAMS, R., and QUINN, W., 1976: Predicting and observing El Niño. *Science*, 191: 343–346.

# 7

# Historical El Niño/Southern Oscillation variability in the Australasian region

NEVILLE NICHOLLS

*Bureau of Meteorology Research Centre, PO Box 1289K, Melbourne 3001, Australia*

## Abstract

The El Niño/Southern Oscillation has a major effect on Australasian climate. The phenomenon amplifies the interannual variability of the climate and imposes temporal patterns, phase-locked to the annual cycle, on droughts and wide-spread, heavy rainfall episodes. The native vegetation and wildlife are clearly adapted to the pattern of climate fluctuations, especially rainfall variations, imposed by the El Niño/Southern Oscillation. This suggests that the El Niño/Southern Oscillation has been affecting the Australasian region for a very long time. The clear adaptation of the fauna and flora to the patterns of climate produced by the El Niño/Southern Oscillation indicates that paleoclimatic studies in the Australasian region may help determine when the phenomenon started to operate.

## Introduction

India suffered a severe drought and famine during 1877 (Kiladis and Diaz 1986). Sir Henry Blandford, the first Director of the Indian Meteorological Service, noted the very high atmospheric pressures over Asia at the time and requested pressure information from other meteorologists around the world. Sir Charles Todd, the South Australian Government Observer, in response to Blandford's request, included pressure observations from various Australian stations in an annual series of publications recording monthly observations made in South Australia and the Northern Territory. Pressures were high during 1877 over Australia too, and much of the country suffered from drought that year.

The coincidence of high pressures and droughts in India and Australia obviously stuck in Todd's mind. In 1888 Australia was again struck by a severe drought. An extensive discussion between the Government Observers from South Australia, New South Wales, and Victoria, on the cause of the drought, was published in *The Australasian* on 29 December 1888. Todd suggested that Indian and Australian droughts usually coincided: 'Comparing our records with those of India, I find a close correspondence or similarity of seasons with regard to the prevalence of drought, and there can be little or no doubt that severe droughts occur as a rule simultaneously over the two countries'.

By 1896, H. C. Russell, the New South Wales Government Observer was also convinced that Indian and Australian droughts often coincided. Russell (1896) attempted to demonstrate the periodicity of droughts but, almost incidentally, indicated the coincidence of droughts in the two countries. Droughts identified in the two countries in Russell (1896) are tabulated in Table 7.1. Not all droughts coincided, but many did. This remarkable observation has since been confirmed (e.g., Williams et al. 1986) and forms part of the suite of climate teleconnections we now call the Southern Oscillation (SO). Recent empirical studies (e.g.

Table 7.1 *Australian and Indian droughts, 1788–1886*

| Australian droughts | Indian droughts |
| --- | --- |
| 1789–1791 | 1790–1792 |
| 1793 | |
| 1797 | |
| 1798–1800 | |
| 1802–1804 | 1802–1804 |
| 1808–1815 | 1812–1813 |
| 1818–1821 | |
| 1824 | 1824–1825 |
| 1827–1829 | 1828 |
| 1833 | 1832–1833 |
| 1837–1839 | 1837–1839 |
| 1842–1843 | |
| 1846–1847 | |
| 1849–1852 | |
| 1855 | |
| 1857–1859 | 1856–1858 |
| 1861–1862 | |
| 1865–1869 | 1865–1866 |
| 1872 | |
| 1875–1877 | 1875–1877 |
| 1880–1881 | |
| 1884–1886 | 1884 |

After Russell (1896).

Ropelewski and Halpert 1987; Kiladis and Diaz 1989) have once again demonstrated that Indian and Australian droughts both tend to occur during El Niño episodes. El Niño – associated droughts in both countries usually start around May. The discovery of the first hints of the Southern Oscillation is usually attributed to Hildebrandsson (1897). Todd deserves some credit, however, for identifying, a decade earlier, the global-scale of these teleconnections.

When Sir Gilbert Walker named and documented the SO in the early decades of the 20th century, its close relationship with Australian rainfall quickly became apparent (e.g., Walker and Bliss 1932). Walker's work indicated that north Australian summer rainfall could be predicted with an index of the SO. Quayle (1929) indicated that spring rainfall farther south could be predicted in the same way. After that, only a trickle of papers discussed the relationship between the SO and Australian climate, until the mid-1970s, when the worldwide attention on El Niño led to a resurgence of interest among Australian meteorologists.

There was rather more interest in the El Niño/Southern Oscillation (ENSO) and its effects on Indonesia after the 1930s than was the case in Australia. Little was published on the relationship of ENSO with the climate of New Guinea, New Zealand, or the rest of the southwest Pacific, until the 1980s.

The next two sections of this chapter discuss the characteristics of Australian climate that are caused by ENSO and some of their impacts on the Australian environment. A further section deals with the effects of ENSO on Indonesia, New Guinea, New Zealand, and the southwest Pacific and a final section discusses how this information could lead to paleoclimatic studies of ENSO. This review is not meant to be an exhaustive review of the effects of ENSO on the climate of Australasia; Allan (1988, 1991) provides such reviews. Rather, the intention here is to provide a guide to the climate response and the possible impacts on the environment and biota. The aim is to stimulate paleoclimatic studies to determine how long ENSO has been affecting this region. Australasia appears to be an ideal, and largely untapped, field for such studies. A secondary aim is to show that ENSO has had a substantial and consistent effect on Australasian climate throughout the period of record.

The term 'Australasian Region' is used here, somewhat unconventionally, to include Indonesia and New Guinea as well as Australia, New Zealand, and the southwest Pacific. The effects of ENSO on Indonesia and New Guinea have not received as much attention, in recent years, as in many other areas. Yet the effects here are strong and result in major impacts.

There is often some confusion about the nature of ENSO and about the terminology used to describe the phenomenon. Many studies have related regional climate fluctuations to 'El Niño' episodes (i.e., the years when the east equatorial Pacific is warmer than normal) and 'La Niño' episodes (when the east Pacific is cold). Recent work (e.g., Meehl 1987; Xu and von Storch 1989; Rasmusson et al. 1990; Barnett 1991) suggests that ENSO is in fact quasi-cyclic, with an underlying time scale of 2 to 3 yr. So El Niño and La Niña 'episodes' are then really the extremes of the quasicyclic and continuous ENSO. Throughout this paper, the terms 'El Niño' and 'La Niña' refer to the extremes of ENSO. The term 'El Niño/Southern Oscillation,' or its acronym 'ENSO,' is used when the complete phenomenon, rather than just one of its extremes, is meant.

## Effects of ENSO on Australian climate

### *Rainfall fluctuations*

#### *Coincidence of El Niño and Drought*

The best-known characteristic of ENSO is the tendency for rainfall anomalies to appear in many areas at about the same time. Thus, droughts in India, north China, Australia, and parts of Africa and the Americas tend to occur during El Niño episodes (e.g., Walker and Bliss 1932; Williams et al. 1986; Ropelewski and Halpert 1987; Whetton et al. 1990), i.e., when the Southern Oscillation Index (SOI) is strongly negative. Figure 7.1 (from Allan, 1985) illustrates the relationship between widespread Australian drought and low values of the SOI. It also indicates that above average rainfalls over much of Australia accompany large positive SOI values, i.e., La Niña episodes.

Most recent studies linking El Niño and Australian drought have used the period of quantitative data, i.e., from about 1875. Documentary evidence of rainfall is available from before that date. New South Wales, later to become a state of the Commonwealth of Australia, was colonized by Britain in 1788. Nicholls (1988a) examined reports of the governors of the colony to the colonial secretary of the British Government in London for references to drought from 1788 to 1841. Droughts were mentioned in the El Niño years of 1791, 1804, 1814, 1817, 1819, 1824, 1828, and 1837. Droughts were also mentioned in 1796, 1798, and 1810, none of which was an El Niño year according to Quinn et al. (1987). The moderate El Niño events of 1821 and 1832 were not mentioned as severe droughts in the governors' correspondence. However, Russell (1877), in a comprehensive study of Australian climate, lists both these years as drought years. On the other hand, he does *not* indicate that severe droughts occurred in 1796 and 1798.

Russell also lists droughts in the El Niño years of 1845, 1850, 1854, 1857, 1864, 1866, and 1868. Droughts are recorded in 1861–1863, which were not El Niño years. El Niño events were also noted by Quinn et al. (1987) for the years 1860, 1871, and 1874. None of these was a drought year according to Russell (1877), although Russell (1896) notes that 1871 and 1874 were dry. He also notes that 1877, a major El Niño year (Kiladis and Diaz 1986), was a severe drought year. It is evident that, between 1788 and 1879, nearly all El Niño events were accompanied by Australian droughts. This association confirms that observed in the more recent quantitative rainfall data (e.g., Ropelewski and Halpert 1987). The coincidence of El Niño events and Australian droughts has existed from, at least, the start of European colonization in 1788 (Nicholls 1988a).

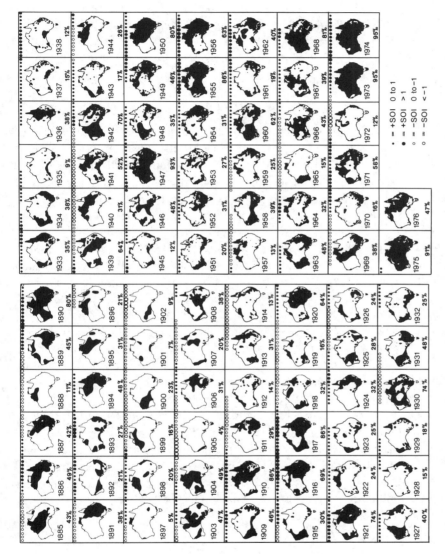

Fig. 7.1  Area of Australia with above average rainfall. The numbers below each map indicate the percentage of Australia with above average rainfall. The circles above the maps indicate the value of the SOI. (Reprinted with permission from Allan 1985.)

Recent studies (e.g., Pittock 1975; McBride and Nicholls 1983; Ropelewski and Halpert 1987, 1989) have shown that the link with ENSO is most consistent with east and north Australian rainfall. Rainfall in the west of the continent is less strongly related to ENSO. Nicholls (1989a) identified a pattern of sea surface temperature variation in the Indian Ocean related to rainfall fluctuations in the central and southern parts of the continent. This pattern is only weakly related to ENSO and appears to be a somewhat independent factor affecting Australian rainfall.

### High variability

Conrad (1941) examined the dependence of interannual rainfall variability on the long-term mean annual rainfall. He defined the relative variability of annual rainfall as the mean of the absolute deviations of annual rainfalls from the long-term mean, expressed as a percentage of the long-term mean. Conrad found that relative variability decreased, in general, as the mean precipitation increased. Over some large areas, however, the relative variability deviated in a consistent way from the global relationship with mean rainfall.

Some of these deviations were due to the influence of ENSO. Nicholls (1988b), using Conrad's data, compared the relationship between relative variability and mean rainfall in areas affected by ENSO with the relationship elsewhere. The relative variability was typically one-third to one half higher for ENSO-affected stations compared with stations with the same mean rainfall in areas not affected by ENSO. Nicholls and Wong (1990) confirmed, on recent data and using the coefficient of variation as a measure of relative variability, that ENSO does amplify rainfall variability in the areas it affects, relative to the variability elsewhere. This was so even after the effect of latitude on variability (tropical rainfall tends to be more variable) was accounted for.

Australia certainly is regarded as a land of highly variable rainfall. It is clear from Nicholls (1988b) and Nicholls and Wong (1990) that this highly variable rainfall is the result of ENSO. The impacts of this high variability are substantial. Some are discussed later in this chapter. Nicholls (1989b) suggested that much of the Australian wildlife was adapted to the patterns of rainfall variability produced by ENSO, implying that the phenomenon has been affecting Australia for much longer than two centuries.

### Continental spatial scales

Australian rainfall fluctuations, as well as being more severe (i.e., more variable) because of ENSO's influence, also operate on very large spatial scales. The 1982/83 drought severely affected over half the continent, an area as large as Western Europe and 25 times the size of England and Wales.

The continental-scale of the 1982/83 drought is typical of many years, although it was more severe than most. Streten (1981) listed the percentage area of Australia receiving above average rainfall for each year from 1950 to 1969. In 3 of the 20 yr more than 80% of the continent received above average rainfall;

in 3 other years less than 20% had above average rain. The likelihood of even 1 yr deviating so far from the expected 50% of the continent receiving above average rainfall is negligible, unless strong spatial correlations exist. Gibbs and Maher (1967) found that in 22% of the years between 1885 and 1965 there was no part of Australia with annual rainfall in the lowest decile. Again, the chance of this happening even once in the 80 yr is negligible, if Australian rainfall fluctuations were random, independent point processes.

The large spatial scale of the ENSO phenomenon is obviously a major reason for the large spatial scale of Australian rainfall fluctuations, but the lack of major mountain ranges inland also contributes. The parts of eastern Australia where the influence of ENSO is weakest lie around the coastal fringe, where the Great Dividing Range can produce local rainfall effects thereby complicating the large-scale influence of ENSO. Inland of the coastal mountains, however, there is little orography to differentiate the reaction to the broad-scale influence of ENSO, leading to widespread, coherent rainfall fluctuations.

### Long time scales

A 'drought' in England is defined as a period of 15 d without rainfall. At the other extreme, the Sahel has been in 'drought' for the past two decades. Many Australian droughts, and the extensive wet periods, tend to last about 12 months. The 1982/83 drought is a good example. The drought started early in 1982 and broke over much of the country in March 1983. El Niño and La Niña episodes both tend to last about 12 mo (e.g., Rasmusson and Carpenter 1982; Ropelewski and Halpert 1987, 1989) and this sets the time scale of Australian rainfall fluctuations.

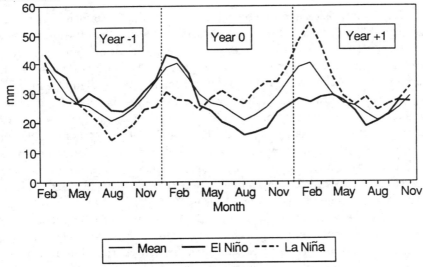

Fig. 7.2 Composite 3-mo mean precipitation for Bourke, Australia for 26 El Niño episodes and 20 La Niña episodes. Years of El Niño and La Niña episodes from Kiladis and Diaz (1989). 'Year 0' indicates year of onset of El Niño or La Niña. The long-term (1876–1982) 3-mo mean rainfall is also shown.

Ropelewski and Halpert (1987) examined the relationship between El Niño episodes and Australian rainfall. They concluded that for much of eastern Australia, rainfall was deficient from about February of an El Niño year through to the following February (e.g., Fig. 7.2). Above-average rainfall is usually observed in this area, over a similar time frame (e.g. Fig. 7.2), during La Niña episodes (Ropelewski and Halpert, 1989). So, the droughts and wet periods associated with the extremes of ENSO both tend to last about a year.

### Phase-locking to the annual cycle

The results of Ropelewski and Halpert (1987, 1989) and many others also demonstrate that these extended periods of drought or extensive rains do *not* occur randomly in time, in relation to the annual cycle. The ENSO phenomenon, and the Australian rainfall fluctuations associated with it, is phase-locked with the annual cycle. The heavy rainfall associated with a La Niña episode tends to start early in the calendar year and finish early in the following year (Fig. 7.2). The dry periods associated with El Niño episodes tend to occupy a similar period (Fig. 7.2). This means that if an extensive drought or wet period is established by the middle of the calendar year it is unlikely to 'break' until at least early the following year. The 1982/83 drought again provides a good example. The drought started about April 1982 and broke over much of the country in March and April 1983. Such phase-locking has been found in most other variables associated with ENSO (e.g., Rasmusson and Carpenter 1982; Allan and Pariwono 1990).

### Biennial cycle

This phase-locking is related to a biennial cycle that is a fundamental element of ENSO variability (e.g., van Loon and Shea 1985; Meehl 1987; Rasmusson et al. 1990; Barnett 1991). There is also a lower frequency variation, but it is the biennial mode that captures the major features associated with ENSO 'episodes.' The biennial cycle is observed over the equatorial Pacific and Indian Oceans and is tightly phase-locked with the annual cycle. It varies in amplitude from cycle to cycle and sometimes changes phase. It is not a strictly 2-year cycle so it may be categorized better as a quasi-biennial cycle. Nicholls (1979, 1984a) and Allan and Pariwono (1990) discussed how ocean-atmosphere interaction around Indonesia and northern Australia, modulated by the seasonal cycle, could result in a biennial cycle phase-locked to the annual cycle. Rainfall over much of Australia displays a quasi-biennial cycle (e.g., Kidson 1925).

The biennial rhythm means that El Niño events will often precede and/or follow La Niña events, and vice-versa. This means that year-to-year changes in rainfall can be extreme. The annual rainfall in the district around Bourke in 1950, a La Niña year, was 841 mm (cf. annual mean of 347 mm). The following year's rainfall was only 217 mm. In 1982 the district received only 162 mm; the following year 552 mm. The biennial cycle in Australian rainfall, and the tendency for wet and dry periods to succeed each other, can be found in the composites of Ropelewski and Halpert (1987, 1989).

The change from El Niño-related drought to La Niña and widespread flooding throughout much of eastern Australia can be rapid, and usually occurs early in the calendar year. An El Niño-related drought in 1888, for instance, was broken on New Year's Day 1889 by heavy rains that caused severe flooding in many parts of eastern Australia (Nicholls 1993). The 1982/83 drought broke over much of Australia during March 1983, again with heavy rains.

The descent into drought can also be rapid. The Bourke district received 418 mm of rain in January and February 1976, at the end of the 1975 La Niña. This was five times the average for these 2 mo. Only 98 mm fell in the next 6 mo, about half the average for this period.

## Winds, temperatures, and tropical cyclones

Rainfall is not the only aspect of the Australian climate affected by ENSO. It seems likely that frosts will tend to be more common in inland Australia during El Niño events, because low rainfall is associated with decreased cloud cover, allowing increased radiative cooling at night. On the other hand, maximum temperatures may be higher during El Niño, also because of the decreased cloud cover.

There are also strong variations in wind between El Niño and La Niña episodes, although these variations are different in different parts of the country. North of about 25°S low-level winds in *winter* during El Niño events tend to be about three to four times stronger than during La Niña events (Drosdowsky 1988). In the southeast the winds in *summer* tend to be two to three times stronger in El Niño. These relationships with ENSO suggest that wind variability may also be high in Australia, as is the case with rainfall. No comparative studies have yet been made to verify this.

Tropical cyclones are more frequent around northern Australia during La Niña events (e.g., Nicholls 1984b). In an El Niño, cyclone activity tends to shift eastwards from the eastern Australian coast to the central South Pacific (e.g., Revell and Goulter 1986; Hastings 1990).

## Predictability of climate fluctuations

The biennial cycle underlying the ENSO phenomenon, and the phase-locking of this cycle to the annual cycle, provides some regularity to the phenomenon and to climate variables associated with it. This regularity provides a degree of predictability. The phase-locking means that ENSO, or at least smoothed indices of the phenomenon (e.g., the SOI), will tend to change phase around March–May and only rarely at other times of the year. Thus if the SOI is strongly positive ('La Niña') during the austral winter, it will probably stay in that phase until early the following year. Climate variations normally associated with this phase of ENSO, and that occur towards the end of the calendar year, may therefore be predictable simply by monitoring the SOI earlier in the year. This observation underpins the *Seasonal Climate Outlook* service provided by the Australian Bureau of Meteorology. These outlooks use observed values of the SOI to predict seasonal rainfall fluctuations over eastern and northern Australia, from early winter until the end of the year. The tendency for the SOI to persist from the southern winter through to the summer also allows the prediction of Australian seasonal tropical cyclone activity (Nicholls, 1985). Nevertheless, these associations are not fool-proof and large intraseasonal swings in the SOI have been observed.

## Australian impacts of ENSO

The last section described the features of the Australian climate that are the result of the influence of ENSO. In this section some of the *impacts* of these ENSO-related climatic characteristics are discussed. This is done for two reasons. Firstly, detailing the impacts will illustrate the pervasiveness and strength of ENSO's influence on Australia. The societal and ecological impact of El Niño in the east Pacific has been well documented, but the only Australian effects that have received much attention have been the close association between drought and El Niño, and the predictability of seasonal rainfall. A more extensive demonstration of how ENSO affects Australia may lead to initiatives to offset its deleterious impacts. The second reason for detailing ENSO's impacts is to show that many opportunities exist for Australian paleoclimatic work to determine the past behavior of ENSO.

Attention is focussed on the apparent impact of ENSO on the native wildlife and vegetation, rather than on crops or the economy or human health, even though ENSO certainly affects these aspects of Australia (e.g., Nicholls 1988c). For instance, the high variability of Australian rainfall was noticed by farmers soon after European settlement. The problems faced by farmers introducing crops was a major indicator of this higher variability. Rolls (1981) notes that 'If a year was dry, the unsuitable English wheats set no seed. If a year was wet the fungus disease, rust, turned crops into a red mush as they headed.'

The focus on the native biota was chosen because of its potential use in paleoclimatic studies to determine prehistoric effects of ENSO on Australia. Some comments are also made on the impact of ENSO on floods, erosion, and fire. The impacts discussed suggest a variety of strategies for paleoclimatic investigations of prehistoric effects of ENSO.

### *Impact of variable rainfall on wildlife*

Nicholls (1989b) discussed several apparent adaptations of Australian wildlife to the highly variable rainfall produced by ENSO. Some examples follow.

## Red kangaroo

Australia's largest herbivore, the red kangaroo (*Macropus rufus*), inhabits the open arid and semiarid plains that cover most of the continent. It shows no seasonal pattern of reproduction but breeds opportunistically in response to good conditions by producing young in rapid succession (Tyndale-Biscoe 1973; Hodson 1979). During good years the female kangaroo has a young or 'joey' at heel, an offspring attached to a teat in the pouch, and a quiescent embryo from which another young develops to replace the pouch offspring as soon as it leaves the pouch. The kangaroo then mates, producing a new embryo that develops to a certain point and then becomes quiescent until the pouch offspring needs to be replaced. This mechanism enables the female red kangaroo to produce an independent offspring every 240 d under good conditions, although it takes 600 d for each offspring to develop from conception to independence.

With the onset of drought and the decline of food there is initially a heavy mortality of the young when they leave the pouch. Matings continue after the birth of each young and there is a continual supply of young kangaroos, most of which die in the transition from pouch life. If conditions are improved by rain then the young will survive. Thus young kangaroos are able to take advantage of the improved conditions almost immediately.

If conditions do not improve, lactation ceases and the pouch young dies. The mother will then give birth to another young that may enter the pouch, attach to the teat and then die. The mother will mate again and so a continual supply of young is produced although in severe drought they die at an early stage.

Under prolonged drought the kangaroos stop breeding. Drought-breaking rains trigger an immediate hormonal response. The females return rapidly to breeding, and may be found with young in the pouch after 60 d. In favorable environmental conditions females become sexually mature when 15 to 20 mo old. Drought delays the onset of sexual maturity and after 2 yr of drought a population may include females aged 3 yr or more that have never produced young. After rain these animals come into breeding condition almost immediately (Strahan 1983).

The life-history strategy of the red kangaroo is clearly adapted to highly variable rainfall. The strategy ensures that the interval between the return of good conditions and the recruitment of young into the population is short. The population can increase rapidly after a drought. Large numbers of adults will face the next drought, increasing the chances that some will survive.

## Long-haired rat

Spasmodic population explosions of long-haired rats (*Rattus villosissimus*) have frequently inflicted misery on settlers in inland Australia. Population explosions follow periods of exceptionally good rains or floods. The long-haired rat has an extremely high reproductive rate and, when favorable conditions lead to an abundance of grass seeds and green plants, this potential is realized. Populations rapidly increase and spread into areas where they are not found at other times. As the country dries, most die and distribution again becomes restricted to small refuge areas where water and plant food is available even during droughts (Strahan 1983).

The rapid population increases and plagues ensure that rats spread, during good conditions, across a very large area. As conditions deteriorate, many potential drought refuges can be occupied. Even if some of these refuges dry out, others are likely to persist and sustain a small population of rats through the drought. When the drought breaks, these rats can rapidly multiply and migrate across the country again. This life-history strategy is well suited to highly variable rainfall and frequent droughts.

### Birds

Adaptations of Australian birds that can be attributed to the unpredictable environment (Keast 1981a) produced by ENSO include the following:

*Nomadism.* Many Australian birds (perhaps 30% of all species) are true nomads, wandering erratically. This nomadism ensures that the population moves to where conditions are good. The area of concentration of many inland species varies from year to year.

*Reproduction and rainfall.* Many inland species can only reproduce successfully when conditions are good. Over most of the continent, nesting is irregular and aseasonal and is initiated by sudden falls of rain.

*Variations in clutch and multiple broods.* Some inland species have smaller clutches in dry years. In very good years various species may have multiple breeds in succession.

*Precocious breeding.* Some species (e.g., the zebra finch) are capable of breeding at a very young age, permitting rapid population buildup during good years.

*Multiple helpers-at-the-nest.* The habit of older offspring assisting the breeding male in raising siblings of subsequent broods is widespread amongst Australian birds. Such behavior allows a rapid recovery under good conditions by freeing the female to lay and incubate the next clutch.

### Other animals

Adaptations to variable rainfall are found in some members of most groups of Australian plants and animals (Keast 1981b). Some respond directly to the climate fluctuations associated with ENSO. For instance, the numbers of green turtles (*Chelonia mydas*) breeding around the Great Barrier Reef seems to be regulated by ENSO (Limpus and Nicholls 1988). Epidemics of Murray Valley encephalitis, a virus spread by mosquitoes, in southeast Australia only occur after La Niña episodes when widespread flooding allows mosquito numbers to increase dramatically (Nicholls 1986).

## Are these adaptations unusual?

The reproductive strategies of many animals and much of the vegetation in other arid and semiarid conditions, not just in Australia, are such that they are able to take advantage of favorable conditions. Ecologists have, however, observed that many Australian adaptations to variable rainfall are unusually marked. Lee and Cockburn (1985) note that 'the strategies employed by the desert kangaroos provide one of the most striking examples of mammalian life histories that are suited to environmental uncertainty.' Hodson (1979), also discussing the kangaroo's adaptations, notes that 'it is a remarkable, and probably unique, example of an adaptation to an opportunistic way of life.'

In the case of rodents, Wagner (1981) suggests that plagues are not a typical feature of arid environments, even though life-history strategies that result in such plagues are a useful adaptation to variable rainfall. He goes on to say that 'as exceptions that may prove the rule, the most persistent reports of plagues in desert rodent populations come from Australia.'

But it is for birds that the clearest differences between Australian life-histories and those in arid areas elsewhere have been noted. Serventy (1971) suggested that nomadism is much more restricted in the proportion of species displaying it and in the extent of movement in other arid parts of the world. The link between precipitation and breeding also seems highest in Australian birds (Wagner 1981; Serventy 1971), as does the aseasonality of breeding. Wagner (1981) concludes that 'the emancipation of reproductive seasons from photoperiodic timing, and its complete aseasonal linkage to precipitation appears to be a significant reality only in Australia.'

There thus appears to be a consensus among ecologists that much of the Australian flora and fauna is adapted to a highly variable rainfall, and that this adaptation is more complete than in other areas. At first these adaptations were attributed to the arid or semiarid environment but they are largely absent in some other arid areas. Wagner (1981) notes, for instance, that evidence of avian reproduction in deserts responding to precipitation irrespective of season 'seems to have come more from the Australian experience than anywhere else, and may have promoted a general impression that the phenomenon is more common in deserts than it is.' Serventy (1971) suggested that the image of nomadism among desert birds has been strongly colored by its prevalence in Australia. The absence of such strong adaptations in some other arid and semiarid areas suggests the higher variability of Australian rainfall may have influenced the Australian adaptations. ENSO is the cause of the modern-day high interannual variability of Australian rainfall. So the adaptations of much of Australia's wildlife to highly variable rainfall may indicate a long association with the ENSO phenomenon. Paleoecological studies might indicate when animals adapted to highly variable rainfall became predominant over much of Australia. This might suggest, in turn, when the ENSO began to affect Australia's rainfall.

## Characteristics of Australian vegetation attributable to ENSO

It is not just the animals that appear to be adapted to the pattern of climate fluctuations imposed by ENSO. The following is a list of just some of the characteristics of Australian vegetation that may be, at least in part, attributable to ENSO's influence on the climate (Nicholls, 1991). The list is not comprehensive and is provided to illustrate that the 'patterning of the climate' caused by ENSO can affect Australia's vegetation. Most of the information that follows was taken from Recher et al. (1979, 1986) and Stafford Smith and Morton (1990).

### Absence of succulents

Cacti, although adapted to arid climates and requiring little moisture, need regular rain. Such plants would be poorly equipped to survive 12-mo droughts and to take advantage of extensive wet periods when they do occur, i.e. they seem unsuited to the high rainfall variability ENSO produces over much of Australia. Succulents are almost totally absent from the Australian arid and semiarid regions. The existence of a few succulents indicates that this is not simply due to isolation from other arid areas.

### Establishment dependent on extended wet periods

Successful establishment of bladder saltbush and mulga (found over much of semiarid Australia) requires extended periods of wet weather. Mulga needs heavy summer rains to produce large numbers of flowers, and subsequent heavy winter rains to set seed. The seeds then germinate in summer and further heavy rains are needed to allow the young seedlings to establish. As noted earlier, La Niña episodes often produce heavy rains starting in late summer and extending well into the next summer, as required for the establishment of mulga seedlings. Most recent cases of widespread establishment of mulga appear to have been during major La Niña episodes.

### Drought tolerance/avoidance

Many Australian plants are remarkably tolerant of severe, extensive droughts. Well-developed tolerance or avoidance strategies are essential because of the frequent severe droughts caused by ENSO. Mulga, once established can withstand all but the most severe droughts and can survive more than 50 yr. Such longevity is a useful characteristic in a country of highly variable rainfall where the conditions for successful establishment occur rarely. Some other species rely on seed reserves to re-establish after the more exceptional drought events (Westoby, 1980). Many Australian plants have large root systems, relative to their above-ground structures. Such plants are more tolerant of drought because they can gather soil moisture long after it has disappeared from near the surface. Spinifex, for example, have very long root systems. Various drought avoidance mechanisms are used by Australian plants. Surviving droughts as seeds is a favorite avoidance technique. Some Australian plants have life cycles tailored to take advantage of the consistent aspects of the climate to set seed, rather than relying on the more variable aspects. For instance some tropical annuals mature at the earliest time at which the rainy season might end, thus ensuring they set seed every year (Andrew and Mott 1983). If they aimed to mature at the 'average' time for the end of the rains, there would be many years in which they would fail to set seed.

### Diverse life-history strategies

In a predictable climate, there are usually only a few optimal strategies in the balance between drought tolerance and establishment vigor. Where conditions vary more from year to year, different establishment strategies are preferred in different years and a wider range of strategies can coexist (e.g., persistence as seed, dormant tuber or growing adult). The 'extreme irregularity of the Australian arid zone climate thus contributes to a relatively high diversity of life history strategies' (Stafford Smith and Morton 1990).

### Vegetation height

Inland Australia has a larger number of trees than most other areas of similar aridity. Milewski (1981) attributes this to the occasional heavy rainfalls which result in deep water penetration. Larger trees, with large root structures able to remove water from far below the surface, would be favored by such rainfalls, relative to an area where rainfall was lighter and more frequent. The heavy rainfall events usually accompany La Niña episodes and persist for about a year or so. This persistence would also favor establishment of larger trees, relative to an area of less variable rainfall.

### Fire resistance/dependence

Much of the Australian flora is fire-resistant or even dependent on fire for successful reproduction. Some plants, e.g., spinifex, actively promote burning by producing flammable oils in their leaves. The fire releases the nutrients and clears away the old plants, allowing young ones to germinate. Fire usually induces substantial seed release from eucalypts. An enhanced 'seed rain' after fire temporarily exceeds losses to seed predation by harvester ants, thus improving the chances of successful establishment (Noble et al. 1986). In the least fertile environments (e.g., spinifex-dominated communities) many species persist only as seeds or dormant root-stock between fires that release scarce nutrients from perennial tissue, and make space amongst perennials. As noted below, ENSO enhances the likelihood of wildfire so fire resistance or dependence can be interpreted, in part, as an adaptation to ENSO's influence on the Australian environment.

## ENSO, floods, erosion, and fire

Many of the climate characteristics caused by ENSO would appear to lead to an increased risk of soil erosion, by both flood and wind. The extreme rains associated with La Niña episodes cause large flows of water that are major creators of landscape structure. McMahon et al. (1987) have shown that streamflow tends to be more variable in Australia than is the case in similar areas in the rest of the world. Because of the biennial nature of ENSO, heavy rains will often follow soon after a major drought, which would have reduced vegetation cover, leaving the soil bare and vulnerable. So the biennial rhythm and the high variability of Australian rainfall, both the result of ENSO, increase the risk of erosion.

The inland lakes of central Australia appear to fill only after La Niña events (Allan 1985). The widespread heavy rainfall throughout eastern inland Australia during these events leads to the filling of streams that drain into these inland lakes. During the 1974–76 flooding and filling of Lake Eyre, in a strong La Niña phase, the lake area of some 10,000 km$^2$ had up to 56 km$^3$ of water in it. Enormous amounts of sediment are shifted during such events.

Wind erosion is also favored by ENSO. Strong winds often accompany El Niño events (section above). Again, with the vegetative cover removed by the drought the strong winds can result in major erosion. The tendency for the strong winds to occur in the hotter and drier seasons (in the winter in the north; in the summer in the south) would aggravate this since the high evaporation in these seasons would also tend to increase the vulnerability of the soil to wind erosion. Southeastern Australia has been subject to immense dust storms towards the end of El Niño episodes, as a result of the high winds and desiccation of the soil during these events.

The long wet periods associated with La Niña, followed by long dry periods (because of the biennial cycle of ENSO), would enhance the likelihood of wildfires. Rapid growth during the wet period would dry quickly during the ensuing drought and then 'dry' thunderstorms could ignite the bush. The stronger winds associated with El Niño-related droughts would further increase the likelihood of wildfire. Any fire ignited by a 'dry' thunderstorm during an El Niño could spread farther and faster because of the stronger winds. The latitudinal variation in the season in which these ENSO-amplified winds occur (winter in the north, summer in the south) means that the strong winds occur at the time of year when fires are most likely anyway.

### Summary

The effects of the Australian climate characteristics such as high variability, extended droughts, etc., on the biota and environment are marked and pervasive. It seems impossible to understand the functioning of the Australian environment and ecology without recourse to the patterning of climate produced by the ENSO phenomenon. This is not surprising, given the strong control that ENSO has on the climate. Many species appear to have adapted to the climate characteristics caused by ENSO. This suggests that the phenomenon has had a long influence on the Australian environment. In turn, this implies that paleoecological studies might provide useful information about prehistorical variations in its influence on Australia, even answering questions such as when ENSO began operating.

## ENSO elsewhere in Australasia

ENSO affects the climate elsewhere in Australasia, so ecological impacts might also be expected. To date, however, little documentation of the ecological impacts of ENSO in these other regions has appeared.

### Indonesia and New Guinea

Braak (1919, 1921–29) and Berlage (1927, 1934) demonstrated that rainfall during the early part of the Indonesian wet season (September–December) was significantly related to the Southern Oscillation. Abnormally low atmospheric pressure at Darwin, Australia (an indication that ENSO is in its 'cold' phase) usually signals an early start to the wet season. Nicholls (1973) showed that droughts in New Guinea often coincided with El Niño episodes. Quinn et al. (1978) confirmed the relationship between ENSO and Indonesian rainfall, pointing out that droughts during the 'dry' season of easterly surface winds (May–November) usually coincided with El Niño events. The effect of El Niño events on Indonesian rainfall is substantial enough to be observable in tree rings (Murphy and Whetton, 1989). Barry (1978) and Wright et al. (1985) showed that cloudiness was reduced during El Niño events. Nicholls (1981) confirmed, on new data, that Indonesian rainfall in the early part of the wet season could be predicted with a simple index of ENSO, in this case pressure at Darwin.

The pattern of Indonesian and New Guinea rainfall fluctuations was considered again, using more extensive data, by Ropelewski and Halpert (1987, 1989). Ropelewski and Halpert (1987) demonstrated that 80% of the El Niño events from 1879 to 1982 were accompanied by below average rainfall between June and November. In their 1989 paper they found that 90% of La Niña episodes were abnormally wet in Indonesia – New Guinea between July and December.

Allen (1989) and Allen et al. (1989) investigated the occurrences of drought and frost in the New Guinea highlands, using both climatological and documentary data. They found that most droughts and frost events occurred during El Niño episodes. They also described the impact of El Niño-related drought and frost in the New Guinea highlands on the indigenous peoples and their crops. Malingreau (1987) showed that Indonesian crops are adversely affected by El Niño-related droughts, such as those of 1972, 1976 and 1982. The major El Niño of 1982/83 led to enormous fires through the equatorial rain forests of Borneo (Malingreau 1987).

## New Zealand

Gordon (1986) describes the relationship between ENSO and New Zealand temperature and rainfall. Temperature spatially averaged over New Zealand is usually below normal during El Niño events, when southerly or southwesterly anomalous flow is usually observed across the country. Anomalous northerly flow during La Niña episodes leads to warmer conditions. This relationship is strongest during autumn and spring, and weakest around February–March.

Interaction between the flow anomalies related to ENSO and the New Zealand orography leads to strong regional variations in the relationship between rainfall and ENSO. In general, an El Niño episode is accompanied by drier conditions in the areas to the north that are sheltered from southerly or southwesterly winds. The southern parts, which are not sheltered, often receive above average rainfall during El Niño events.

## Southwest Pacific

Streten (1973, 1975) noticed that the South Pacific Convergence Zone shifts east during El Niño years. Van Loon and Shea (1985) demonstrated that, during an El Niño, the surface atmospheric trough in the South Pacific extends well into the tropics, leading to southerly anomalies across the southwest Pacific. These southerlies are associated with the shift in the convergence zone that leads to increased rainfall in the south-central Pacific, with a corresponding drought in the islands of the southwest Pacific (Wright et al. 1985). The anomalously low rainfall during an El Niño is very evident in the Fiji–New Caledonia area (Ropelewski and Halpert 1987) where dry conditions extend from autumn for about 12 mo. Abnormally heavy rainfall in this area usually occurs during La Niña episodes (e.g., Ropelewski and Halpert 1989). The largest rainfall anomalies occur around November–February. Dennett et al. (1978) showed that the Southern Oscillation could be used in seasonal rainfall forecasting in Fiji. Also, tropical cyclone activity shifts eastward (Revell and Goulter 1986; Hastings 1990). Tahiti is rarely affected by tropical cyclones; the most notable exceptions have been El Niño years (e.g., 1877, 1982/3). The impact of these rare tropical cyclones in the central south Pacific during El Niño can be severe.

## Concluding remarks

The climatological effects of ENSO are pervasive throughout Australasia. Quantitative evidence has been used to establish the pattern of climate fluctuations associated with El Niño episodes and with La Niña events. These patterns were first noticed around the start of this century and have been confirmed repeatedly since then. Documentary evidence has confirmed that the effects of ENSO have been felt in Australasia at least for 200 yr.

The pattern of rainfall variability imposed by ENSO distinguishes regions affected by the phenomenon from other places. The ENSO affected regions have highly variable rainfall, droughts and wet periods lasting about 12 mo and phase-locked to the annual cycle, and a quasi-biennial cycle so that droughts often follow immediately after extended wet periods and vice versa. All these characteristics, the results of ENSO, produce a climatic environment very different from that experienced in areas where the phenomenon does not markedly affect the climate, e.g., much of the mid-latitudes of the Northern Hemisphere.

The impacts of this characteristic pattern of climate fluctuations should be wide-ranging and substantial. The biota in areas where ENSO affects the climate should have life-strategies adapted to the marked climate fluctuations. Australia is the only part of the Australasian region where any attempt has been made to determine how the native wildlife and vegetation have been affected by ENSO. Some of the apparent impacts on the Australian biota and physical environment have been described. The apparent response of the wildlife and vegetation to ENSO-related climate fluctuations should provide a wide-range of possibilities for paleoclimatic investigations of ENSO. For instance, if the appearance of wildlife clearly adapted to ENSO-related climate fluctuations can be dated, this might indicate when ENSO started affecting Australasia. Other indicators of ENSO events affecting Australasia, e.g., dust storms over eastern Australia during El Niño or floods during La Niña, could provide other avenues for paleoclimatic ENSO studies. Finally, the possibility exists for similar studies elsewhere in Australasia.

## References

ANDREW, M. H. and MOTT, J. J., 1983: Annuals with transient seed banks: the population biology of indigenous Sorghum species of tropical north-west Australia. *Australian Journal of Ecology*, 8: 265–276.

ALLEN, B., 1989: Frost and drought through time and space, Part I: The climatological record. *Mountain Research and Development*, 9: 252–278.

ALLEN, B., BROOKFIELD, H., and BYRON, Y., 1989: Frost and drought through time and space. Part II: The written, oral, and proxy records and their meaning. *Mountain Research and Development*, 9: 279–305.

ALLAN, R. J., 1985: The Australasian summer monsoon, teleconnections, and flooding in the Lake Eyre Basin. Royal Geographical Society of Australasia. *South Australian Geographical Review Papers*, 2.47 pp.

ALLAN, R. J., 1988: El Niño – Southern Oscillation influences in the Australasian region. *Progress in Physical Geography*, 12: 4–40.

ALLAN, R. J., 1991: Australasia. *In* Glantz, M., Katz, R., and Nicholls, N. (eds.), *Teleconnections Linking Worldwide Climate Anomalies*. Cambridge: Cambridge University Press, 73–120.

ALLAN, R. J. and PARIWONO, J. I., 1990: Ocean-atmosphere interactions in low-latitude Australasia. *International Journal of Climatology*, 10: 145–178.

BARNETT, T. P., 1991: The interaction of multiple time scales in the tropical climate system. *Journal of Climate*, 4: 269–285.

BARRY, R. G., 1978: Aspects of the precipitation characteristics of the New Guinea mountains. *Journal of Tropical Geography*, 47: 13–30.

BERLAGE, H. P., 1927: East-monsoon forecasting in Java. *Verhandelingen Koninlijk Magnetisch en Meteorologisch Observatorium te Batavia*, 20.42 pp.

BERLAGE, H. P., 1934: Further research into the possibility of long range forecasting in Netherlands-India. *Verhandelingen Koninlijk Magnetisch en Meteorologisch Observatorium te Batavia*, 26.31 pp.

BRAAK, C., 1919: Atmospheric Variations of Short and Long Duration in the Malay Archipelago. *Verhandelingen Koninlijk Magnetisch en Meteorologisch Observatorium te Batavia*, 5.57 pp.

BRAAK, C., 1921–29: The climate of Netherlands Indies, *Verhandelingen Koninlijk Magnetisch en Meteorologisch Observatorium te Batavia*, 8, vol. 1.257 pp.

CONRAD, V. 1941: The variability of precipitation. *Monthly Weather Review*, 69: 5–11.

DENNETT, M. D., ELSTON, J., and PRASAD, P. C., 1978: Seasonal rainfall forecasting in Fiji and the Southern Oscillation. *Agricultural and Forest Meteorology*, 19: 11–22.

DROSDOWSKY, W., 1988: Lag relations between the Southern Oscillation and the troposphere over Australia. Bureau of Meteorology Research Centre, Melbourne, Australia, *BMRC Research Report*, 13.201 pp.

GIBBS, W. J. and MAHER, J. V., 1967: Rainfall deciles as drought indicators. Bureau of Meteorology, Melbourne, Australia, *Bulletin*, 48.33 pp.

GORDON, N. D., 1986: The Southern Oscillation and New Zealand weather. *Monthly Weather Review*, 114: 371–387.

HASTINGS, P. A., 1990: Southern Oscillation influences on tropical cyclone activity in the Australian/Southwest Pacific region. *International Journal of Climatology*, 10: 291–298.

HILDEBRANDSSON, H. H., 1897: Quelques recherches sur les centre d'action de l'atmosphère. *Konglica Svenska Vetenskapsakademiens Handlingar*, 29.33 pp.

HODSON, A., 1979: The one and the many: animal populations. *In* Recher, H. F., Lunney, D., and Dunn, I. (eds.), *A Natural Legacy: Ecology in Australia*. Oxford: Pergamon Press. 276 pp.

KEAST, A., 1981a: The evolutionary biogeography of Australian birds. *In* Keast, A. (ed.), *Ecological Biogeography of Australia*. The Hague: Junk, xxx–xxx.

KEAST, A., 1981b: Distributional patterns, regional biotas, and adaptations in the Australian biota: a synthesis. *In* Keast, A. (ed.), *Ecological Biogeography of Australia*. The Hague: Junk, xxx–xxx.

KIDSON, E., 1925: Some periods in Australian weather. Bureau of Meteorology, Melbourne, Australia, *Bulletin*, 17: 5–33.

KILADIS, G. N. and DIAZ, H. F., 1986: An analysis of the 1877–78 ENSO episode and comparison with 1982–83. *Monthly Weather Review*, 114: 1035–1047.

KILADIS, G. N. and DIAZ, H. F., 1989: Global climatic anomalies associated with extremes in the Southern Oscillation. *Monthly Weather Review*, 2: 1069–1090.

LEE, A. K. and COCKBURN, A., 1985: *Evolutionary Ecology of the Marsupials*. Cambridge: Cambridge University Press. 274 pp.

LIMPUS, C. J. and NICHOLLS, N., 1988: The Southern Oscillation regulates the annual numbers of green turtles (*Chelonia mydas*) breeding around northern Australia. *Australian Journal of Wildlife Research*, 15: 157–161.

MCBRIDE, J. L. and NICHOLLS, N., 1983: Seasonal relationships between Australian rainfall and the Southern Oscillation. *Monthly Weather Review*, 111, 1998–2004.

MCMAHON, T. A., FINLAYSON, B. L., HAINES, A., and SRIKANTHAN, R., 1987: Runoff variability: A global perspective. *In* Solomon, S. I., Beran, M., and Hogg, W. (eds.), *The Influence of Climate Change and Climatic Variability on the Hydrologic Regime and Water Resources*. IAHS Publication No. 168, 3–12.

MALINGREAU, J-P., 1987: The 1982–83 drought in Indonesia: assessment and monitoring. *In* Glantz, M., Katz, R., and Krenz, M. (eds.), *Climate Crisis*. Nairobi, Kenya: United Nations Environment Programme, 11–18.

MEEHL, G. A., 1987: The annual cycle and its relationship to interannual variability in the tropical Pacific and Asian monsoon regions. *Monthly Weather Review*, 115: 27–50.

MILEWSKI, A. V., 1981: A comparison of vegetation height in relation to the effectiveness of rainfall in the mediterranean and adjacent arid parts of Australia and South Africa. *Journal of Biogeography*, 8: 107–116.

MURPHY, J. O. and WHETTON, P. H., 1989: A re-analysis of a tree ring chronology from Java. *Proceedings Koninklijke Nederlandse Akademie van Wetenschappen*, B92: 241–257.

NICHOLLS, N., 1973: The Walker Circulation and Papua New Guinea rainfall. Bureau of Meteorology, Melbourne, Australia, *Technical Report*, 6.13 pp.

NICHOLLS, N., 1979: A simple air-sea interaction model. *Quarterly Journal of the Royal Meteorological Society*, 105: 93–105.

NICHOLLS, N., 1981: Air-sea interaction and the possibility of long-range weather prediction in the Indonesian Archipelago. *Monthly Weather Review*, 109: 2435–2443.

NICHOLLS, N., 1984a: The Southern Oscillation, sea surface temperature, and interannual fluctuations in Australian tropical cyclone activity. *Journal of Climatology*, 4: 661–670.

NICHOLLS, N., 1984b: The Southern Oscillation and Indonesian sea surface temperature. *Monthly Weather Review*, 112: 424–432.

NICHOLLS, N., 1985: Predictability of interannual variations of Australian seasonal tropical cyclone activity. *Monthly Weather Review*, 113: 1144–1149.

NICHOLLS, N., 1986: A method for predicting Murray Valley Encephalitis in southeast Australia using the Southern Oscillation. *Australian Journal of Experimental Biology and Medical Science*, 64: 587–594.

NICHOLLS, N., 1988a: More on early ENSOs: evidence from Australian documentary sources. *Bulletin of the American Meteorological Society*, 69: 4–6.

NICHOLLS, N., 1988b: El Niño-Southern Oscillation and rainfall variability, *Journal of Climate*, 1: 418–421.

NICHOLLS, N., 1988c: El Niño – Southern Oscillation impact prediction. *Bulletin of the American Meteorological Society*, 69: 173–176.

NICHOLLS, N., 1989a: Sea surface temperature and Australian winter rainfall, *Journal of Climate*, 2: 965–973.

NICHOLLS, N., 1989b: How old is ENSO?, *Climatic Change*, 14: 111–115.

NICHOLLS, N., 1991: The El Niño–Southern Oscillation and Australian vegetation. *Vegetatio*, 91: 23–36.

NICHOLLS, N., 1993: The Centennial Drought. *In* Webb, E. (ed.), *Windows on Australian Meteorology*. Melbourne: Australian Meteorological and Oceanographic Society (in press).

NICHOLLS, N. and WONG, K. 1990: Dependence of rainfall variability on mean rainfall, latitude, and the Southern Oscillation. *Journal of Climate*, 3: 163–170.

NOBLE, J. C., HARRINGTON, G. N., and HODGKINSON, K. C., 1986: The ecological significance of irregular fire in Australian rangelands. *In* Joss, P. J., Lynch, P. W., and Williams, O. B. (eds.), *Rangelands: A Resource under Seige*. Canberra: Australian Academy of Science. 634 pp.

PITTOCK, A. B., 1975: Climatic change and the patterns of variation in Australian rainfall. *Search*, 6: 498–504.

QUAYLE, E. T., 1929: Long range rainfall forecasting from tropical (Darwin) air pressures. *Proceedings of the Royal Society of Victoria*, 41: 160–164.

QUINN, W. H., ZOPF, D. O., SHORT, K. S., and KUOYANG, R. T. W., 1978: Historical trends and statistics of the Southern Oscillation. *Fisheries Bulletin*, 76: 663–678.

QUINN, W. H., NEAL, V. T., and ANTUNEZ de MAYOLO, S. E., 1987: El Niño occurrences over the past four and a half centuries. *Journal of Geophysical Research*, 92: 14,449–14,461.

RASMUSSON, E. M. and CARPENTER, T. H., 1982: Variations in tropical sea surface temperature and surface wind fields associated with the Southern Oscillation/El Niño. *Monthly Weather Review*, 110: 354–384.

RASMUSSON, E. M., XUELIANG WANG, and ROPELEWSKI, C. F., 1990: The biennial component of ENSO variability. *Journal of Marine Systems*, 1: 71–96.

RECHER, H. F., LUNNEY, D., and DUNN, I., 1979 & 1986: *A Natural Legacy: Ecology in Australia*. (1st and 2nd editions.) Oxford: Pergamon. 276 pp & 443 pp.

REVELL, C. G. and GOULTER, S. W., 1986: Southern Pacific tropical cyclones and the Southern Oscillation *Monthly Weather Review*, 114: 1138–1145.

ROLLS, E. C., 1981: *A Million Wild Acres*. Melbourne: Nelson. 465 pp.

ROPELEWSKI, C. F. and HALPERT, M. S., 1987: Global and regional scale precipitation patterns associated with the El Niño-Southern Oscillation. *Monthly Weather Review*, 115: 1606–1626.

ROPELEWSKI, C. F. and HALPERT, M. S., 1989: Precipitation patterns associated with the high index phase of the Southern Oscillation. *Journal of Climate*, 2: 268–284.

RUSSELL, H. C., 1877: *Climate of New South Wales*. Sydney: Government Printer. 189 pp.

RUSSELL, H. C., 1896: Notes upon the history of floods in the river Darling, paper read to the Royal Society of NSW, 3 November 1896.

SERVENTY, D. L., 1971: Biology of desert birds. *In* Farner, D. S. and King, J. R., (eds.), *Avian Biology*. London: Academic Press, 287–339.

STAFFORDSMITH, D. M. and MORTON, S. R., 1990: A framework for the ecology of arid Australia. *Journal of Arid Environments*, 18: 255–278.

STRAHAN, R., 1983: *The Australian Museum Complete Book of Australian Mammals*. Sydney: Angus and Robertson. 530 pp.

STRETEN, N. A., 1973: Some characteristics of satellite-observed bands of persistent cloudiness over the Southern Hemisphere. *Monthly Weather Review*, 101: 486–495.

STRETEN, N. A., 1975: Satellite derived inferences to some characteristics of the South Pacific atmospheric circulation associated with the Niño event of 1972–73. *Monthly Weather Review*, 103: 989–995.

STRETEN, N. A., 1981: Southern hemisphere sea surface temperature variability and apparent associations with Australian rainfall. *Journal of Geophysical Research*, 86: 485–497.

TYNDALE-BISCOE, H., 1973: *Life of Marsupials*. London: Edward Arnold. 254 pp.

van LOON, H. and SHEA, D. J., 1985: The Southern Oscillation Part IV: The precursors south of 15°S to the extremes of the Oscillation. *Monthly Weather Review*, 113: 2063–2074.

WAGNER, F. H., 1981: Population dynamics. *In* Goodall, D. W. and Perry, R. A. (eds.), *Arid-land Ecosystems*. vol. 2. Cambridge: Cambridge University Press, 125–168.

WALKER, G. T. and BLISS, E. W., 1932: World Weather V. *Memoirs of the Royal Meteorological Society*: 4: 53–84.

WESTOBY, M., 1980: Elements of a theory of vegetation dynamics in arid rangelands. *Israel Journal of Botany*, 28: 169–194.

WHETTON, P., ADAMSON, D., and WILLIAMS, M., 1990: Rainfall and river flow variability in Africa, Australia and East Asia linked to El Niño – Southern Oscillation events. *Geological Society of Australia Symposium Proceedings*, 1: 71–82.

WILLIAMS, M. A. J., ADAMSON, D. A., and BAXTER, J. T., 1986: Late Quaternary environments in the Nile and Darling basins. *Australian Geographical Studies*, 24: 128–144.

WRIGHT, P. B., MITCHELL, T. P., and WALLACE, J. M., 1985: Relationships between surface observations over the global oceans and the Southern Oscillation. Seattle: National Oceanic and Atmospheric Administration, *Data Report ERL PMEL-12*. 61 pp.

XU, JIN-SONG, and von STORCH, H., 1989: Principal Oscillation Pattern – Prediction of the State of ENSO. *Max Planck Institut für Meteorologie Report* 35.28 pp.

# 8

# A comparison of Southern Oscillation and El Niño signals in the tropics

HENRY F. DIAZ

*NOAA/ERL, 325 Broadway, Boulder, Colorado 80303, U.S.A.*

ROGER S. PULWARTY

*Cooperative Institute for Research in Environmental Sciences, University of Colorado, Boulder, Colorado 80309, U.S.A.*

## Abstract

A contingency table and spectral analysis of the El Niño and Nile River flood event records compiled by W. Quinn and colleagues is performed. The purpose of this study is to assess and compare the long-term variance characteristics of these two measures of the El Niño/Southern Oscillation (ENSO) phenomenon. Several other modern indices of ENSO are also examined, such as sea surface temperature in the upwelling regions of the eastern equatorial Pacific, the Tahiti–Darwin sea-level pressure index, and rainfall in the normally dry areas of the central equatorial Pacific.

Consistent with previous studies of long-term variations in El Niño development in relation to the occurrence of SOI-negative episodes, it was found that the relative timing of these two manifestations of the ENSO system have varied in relation to one another over the past century. Another important feature of changes in the El Niño system is the fact that the spectral signature of El Niño event data shows a concentration of variance within relatively narrow bands at both short and long time scales. Relatively rapid transitions have occurred in the frequency of occurrence of El Niño events since the mid-16th century.

A comparison of the El Niño and Nile River flood event record shows strong similarities, but also some differences. Over the common period 1821 to 1941, one difference noted is that return intervals for Nile flood deficit (all categories) are longer than those of moderate and stronger El Niño events. For the two strongest event categories, the first half of the common record has very nearly the same return period, whereas for the second half, the return interval for strong and very strong El Niños has been about twice as long as that for severe (categories 3 and 4) Nile flood deficits. Nevertheless, as a measure of the long-term variability of the greater ENSO system, the Nile flood intensity record compiled by Quinn (this volume) represents a very useful contribution to ENSO studies.

## Introduction

The distinction between El Niño (EN) and its atmospheric counterpart, the Southern Oscillation (SO), has been somewhat blurred, although, as Deser and Wallace (1987) have shown, the two phenomena are more loosely coupled than most studies have tended to assume. It is the purpose of this study to examine the temporal synchronicity of these events, by examining various indices that specifically characterize EN and SO events, and to assess their relative magnitudes during such occurrences. We utilize spectrum analysis of various EN and SO-sensitive indices and contingency table analysis as the principal methods for comparison.

Through the work of Quinn and colleagues, compilations of EN occurrences (Quinn et al. 1987; Quinn and Neal 1992) and SO-sensitive proxies have been produced which span several centuries (Quinn 1992, this volume). The long-term character of these data allow them to be used in an analysis of the low-frequency changes associated with these two related, but different manifestations of the large-scale system of air-sea interaction we know as the El Niño/Southern Oscillation (ENSO) phenomenon (see Philander 1990).

The main features of ENSO have been described by a number of investigators (see, e.g., Rasmusson and Carpenter 1982; Rasmusson and Wallace 1983; Wright et al. 1988; Enfield 1989; Trenberth 1991). A summary of the principal ENSO features may be given as follows. Large positive sea surface temperature (SST) anomalies develop in the eastern equatorial Pacific and along the South American coast from the equator to 12°S, with a maximum along the Peruvian coast around May. This constitutes the typical El Niño event. Warming of the entire equatorial Pacific eastward of 170°E builds to a peak around December associated with a partial or total collapse of the sea-level pressure (SLP) gradient across the tropical South Pacific. This corresponds to a negative swing of the Southern Oscillation, which, on average, peaks during northern autumn. An eastward shift of the belt of heavy rainfall over the maritime continent toward the central Pacific, beginning in April or May is observed, tied to the development of westerly wind anomalies at the surface west of the region of heavy rainfall, with peak anomalies typically observed late in the year. The reader is referred to Diaz and Kiladis (1992, this volume) for a fuller description of the global climatic patterns associated with ENSO.

Wright (1985) identifies several 'core' SO regions mostly in or near the equatorial Pacific. The key characteristics of a core region is that the field exhibits the SO signal throughout the year and that the SO signal dominates patterns of interannual variability (Wright 1985; see also Nicholls 1988). Hence, mean anomalies over a core region may be used as an index of the SO. This by no means assumes that all events occur in the same manner with the same climatic variations (see Diaz and Kiladis 1992, this volume).

In this paper, the consistency of ENSO relationships over long periods will be tested using an event/no event classification. Thus, only certain kinds of relationships can be investigated. The record of moderate or stronger EN events assembled by Quinn et al. (1987) and updated by Quinn and Neal (1992) for the period 1525 to 1988 will be investigated for variability at high (periods less than ~10 yr) and low frequencies (20–100 yr). The recent record (1882–1988) of El Niño will be compared with SSTs and rainfall in the Pacific, and the Tahiti minus Darwin SLP difference indicative of swings in the SO. Results of these tests will be compared with previous studies. In particular, differences between the SO and EN will be documented. Quinn (1992, this volume) points out that the high index phase of the SO is related to anomalously high rainfall over the Ethiopian Highlands region, resulting in a high Nile River discharge. Conversely, the low index phase of the SO relates to a weak Nile flood. Fairbridge (1984) has described the record of Nile flood levels as 'the world's longest, best quantified climate proxy, providing year by year values spanning an interval of 13 centuries' (see also Riehl et al. 1979; Riehl and Meitin 1979; Hassan 1981). The usefulness of the Nile record as a proxy for EN occurrences will be investigated by comparing a common record of these two features over the period 1824 to 1941. The nature of these data (i.e. annual values representing the occurrence or nonoccurrence of EN events and Nile flood deficit years) requires, for proper comparison with other continuous measurements of ENSO phenomena, the calculation of suitable statistical measures. The analysis procedures are described below.

## Data and methods of analysis

The data of Quinn et al. (1987), updated in Quinn and Neal (1992) identifying the years with EN events since A.D. 1522 (hereafter referred to as NIÑO) are used to distinguish significant variability along the tropical Pacific coast of South America (see, also, Quinn 1992, this volume). The first year of occurrence of each event is assigned a value of one; multiyear events are identified by their first year, all other years are assigned a zero value. There are 29 moderate or stronger EN events in the 1882–1988 period for an average recurrence of ~3.7 yr, with nine strong or very strong events, 38 events from 1824–1941 (roughly every 3.1 yr), and 107 events from 1525–1988 (mean recurrence very near 4 yr).

The Tahiti — Darwin SLP index of the Southern Oscillation (SOI) is calculated for consecutive 3-mo seasons from 1882 to 1988. The index is a standardized measure and we have assigned a value of 1 for that year if the SOI for three consecutive seasons averages under $-0.5$. For three-season periods which overlap two calendar years, the year containing two of the seasons is set as the event year. When event intensities are being considered then a value of 2 is assigned to that year with seasonal SOI values averaging below $-1.0$. This convention yields 23

SO warm events, with 12 strong events. Calculations are similarly carried out using positive Tahiti minus Darwin SLP values. These SO 'cold events' number 15, with six strong events.

Similar categorizations are carried out for SST in the El Niño 1 and 2 regions (see Rasmusson and Carpenter 1982), hereafter referred to as SST2, and for the larger equatorial Pacific Basin (0–10°S, 90°W–170°E, labelled PSST), and for rainfall in the Central Pacific (after Wright et al. 1988).

Two dimensional contingency tables are calculated on the basis of coincidence between any two distributions, where a coincidence means simultaneous (yes/yes or no/no events). Four frequency categories (yes/yes, yes/no, no/yes and, no/no) are tabulated; any year with missing data from one variable is pairwise deleted for comparison with other variables. Measures of association are derived for a number of contingency tables. These measures are based on maximum likelihood, chi-square, and entropy methods. The chi-square and entropy values are calculated using programs given in Press et al. (1989). Further details are given in Whitely (1983). While chi-square based correlation coefficients, such as Cramer's V used in this study ($0 \leq r \leq 1$) are useful for determining the significance of a relationship, entropy gives a better estimate of the strength of an association with previously determined significance. Entropy values range from 0 to 1. A value of 1 indicates that when one variable is known, then no new information can be provided by knowledge of additional variables (i.e. the association is perfect). Details of the maximum likelihood method ($-1 \leq r \leq 1$) are given in Hamdan (1970), Martinson and Hamdan (1971), and IMSL (1989). Martinson and Hamdan (1971) also provide a useful discussion of the relative merits of different measures of association.

For the recent record, singular spectrum analysis (SSA) is carried out on a 9-yr moving average of such event series. This time filtering converts the event/no event series (0s and 1s) into a smoother-valued time series with values ranging from zero (no event in a given 9-yr interval) to one (nine events in a given 9-yr period). Spectra are calculated for the El Niño, the Nile, and the SOI event series. SSA is used for the purpose of highlighting the dominant modes of variability in these data (for reviews and applications of SSA, see Vautard and Ghil 1989, and Rasmusson et al. 1990). The reconstructed series and the eigenvectors are then analysed using the Fast Fourier Transform (FFT), and their spectra compared. Variability of the entire EN record (1525–1988) is investigated using SSA on a 19-yr moving-average version of these data. FFT spectra are calculated on the reconstructed time series (using the first 12 EOFs) within a moving 91-yr window (see Michaelsen 1989; Berger et al. 1990, for examples of this type of analysis). In this manner, the redistribution of variance to different parts of the spectrum can be plotted over time.

Quinn (1992, this volume) has compiled a record of SO-related climatic variability in the western part of the SO-influenced core region (tropical eastern Africa and India) that incorporates Nile River flood data. Gaps in the Nile record occur during the Turkish period (A.D. 1520–1700), and following the Napoleonic occupation (A.D. 1800–1825) (Riehl et al. 1979; Fairbridge 1984). We have used this proxy of SO-activity in this study and have followed the classification scheme used by Quinn (1992, this volume). We refer the reader to Quinn's chapter (this volume) for full details of the development of this data set. Episodes of Nile flood deficit are ranked on a scale of 0 to 4, where 4 indicates severe deficits and 0 indicates normal or above normal discharge. Nile flood deficits will be compared with the occurrence of El Niño events for 1824–1941.

### El Niño: A.D. 1525–1988

Anderson (1989) and Enfield and Cid (1991) have examined some of the long-term changes in EN behavior. In a separate treatment, Anderson (1992, this volume) presents a brief analysis of the combined NILE and NIÑO data. Briefly, Anderson found significant spectral peaks near 90 and 50 yr in this combined record. Below, we analyse the historical El Niño event record of Quinn (1992, this volume).

Figure 8.1 illustrates some of the low-frequency changes that have occurred since the first historical information was gathered on EN-related climatic variations around Ecuador and Peru. The graph shows the mean recurrence period of the EN events in running 19-yr segments. The choice of temporal filter is arbitrary; it was chosen merely to illustrate the low frequency changes in the recurrence interval between events without regard to event intensity. The long-term average recurrence period for moderate or stronger EN events is close to 4.5 yr. The longest recurrence period occurred in the 17th century, while the 19th century saw the shortest recurrence intervals.

Figures 8.2a,b show the results of a sliding 91-yr FFT on the times series of El Niño occurrences. Two realizations, one retaining information within the 2-to

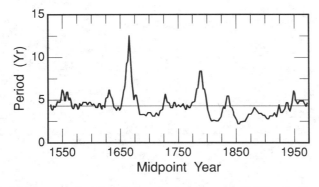

Fig. 8.1   Mean recurrence interval of moderate or stronger El Niño events calculated for overlapping 19-yr segments. Values have been smoothed with a 1-2-1 filter applied twice. (Data from Quinn 1992, this volume.)

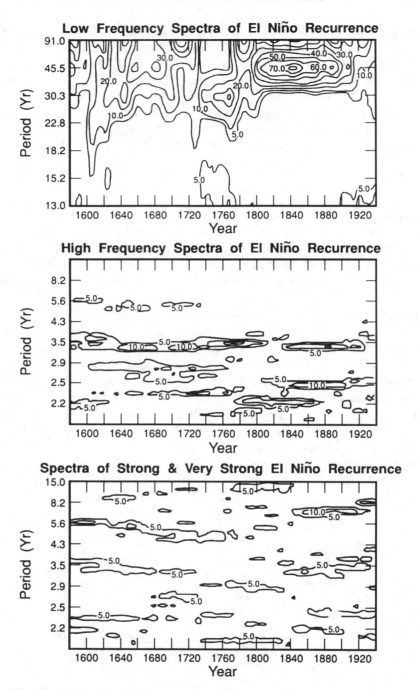

Fig. 8.2 Moving 91-yr FFT spectra of moderate and stronger El Niño recurrence, shown as contours of percentage of total variance explained by different spectral lines. Input values are 19-yr moving averages of event occurrence, 1525–1988, for low-pass, and high-pass (residuals) filtered values (top panel and middle panel, respectively), and high-pass filtered values (bottom panel) calculated from a 31-yr moving average of the recurrence of strong and very strong El Niño events only.

13-yr time scale, and the other for time scales of 13 to 91 yr are reconstructed from 12 significant eigenvectors. Most noticeably in Figure 8.2a are persistent periods of low-frequency oscillation. These center near 18.2 yr (1580–1640), 45.5 (1630–1670 and 1790 – present), 30.3 yr (1680–1730), and 22.8 yr (1730–1775). The ~46-yr periodicity was strongest from 1810 to 1870. We note that a Maximum Entropy Spectral Analysis of these data (not shown) indicates a strong variance peak at about 46.9 yr. Switching of these frequency modes occurs four times during the record. A switch in both low and high frequency modes is recorded at around 1780, and also in the low frequency spectra toward the end of the 1800s.

These results would indicate that EN events were relatively more frequent during 1680 to 1770 and again from about 1820 to 1930 than in the other parts of the record. Our results are in good agreement with those of Michaelsen (1989) who used a set of southwestern United States and northwestern Mexico tree-ring records as a proxy for ENSO (see Lough 1992, this volume). We also repeated the moving spectrum technique using only strong and very strong EN categories. The results, illustrated in Figure 8.2c, also agree quite closely with those of Michaelsen (1989).

Enfield (1988) reviewed evidence from different proxy data and concluded that EN statistics are not significantly different between the Little Ice Age and the present (see also Enfield 1992, this volume). Enfield and Cid (1991) hypothesize, on the basis of an analysis of the Quinn et al. (1987) data, that when solar activity and the quasi-biennial oscillation (QBO) are weak ENSO develops at a shorter periodicity (~3–4 yr), while high Solar activity with a well-developed QBO favors ENSO development on closer to a 4- to 5-yr time-scale. A similar difference in recurrence interval was found for strong and very strong events only. Our analysis shows that large amounts of variance are found at both long and short frequencies (see Fig. 8.2), but the record is also characterized by the presence of sudden transitions in EN recurrence frequency (see also, Michaelsen 1989). The presence of two dominant modes of ENSO variance at about 2.6 and 3.7 year (see Fig. 8.2b) is documented in much greater detail by Rasmusson et al. (1990) for the 1950–1987 period.

## The recent record: A.D. 1882–1988

Strong correlations are observed, as expected, among SOI, PSST, RAIN, and SST2 (Table 8.1). However, while one would also expect NIÑO to have similarly large correlations with other variables (except possibly NILE), and in particular with SOI with which it is sometimes regarded as synonymous, we will show that changes in the association for different parts of the record occur over time (see Diaz and Kiladis 1992, this volume).

The first half of the ENSO record (1882–1938) contains four EN events (1887, 1910, 1917, 1930), with onset years prior to a large-scale negative SO occurrence, while from 1939 to 1988 the onset years of EN and SO occur simultaneously for

Table 8.1 *Contingency table analysis using maximum likelihood* $(-1 \leqslant r \leqslant 1)$, *top entry, chi statistic* $(0 \leqslant r \leqslant 1)$, *middle entry, and entropy methods* $(0 \leqslant r \leqslant 1)$, *bottom entry, for all 'warm event' signals. Correlation coefficients are multiplied by 100. Period of record is 1882–1988 among all variables except for Nile flow (1882–1941)*

|       | NIÑO | NILE | SOI | SST2 | PSST |
|-------|------|------|-----|------|------|
| NILE  | 63   |      |     |      |      |
|       | 40   |      |     |      |      |
|       | 14   |      |     |      |      |
| SOI   | 64   | 51*  |     |      |      |
|       | 41   | 30   |     |      |      |
|       | 14   | 8    |     |      |      |
| SST2  | 78   | 60   | 83  |      |      |
|       | 55   | 44   | 59  |      |      |
|       | 23   | 16   | 29  |      |      |
| PSST  | 75   | 42** | 87  | 86   |      |
|       | 49   | 22   | 63  | 60   |      |
|       | 21   | 4    | 34  | 32   |      |
| RAIN  | 69   | 31†  | 84  | 75   | 73   |
|       | 46   | 16   | 59  | 51   | 45   |
|       | 17   | 2    | 30  | 21   | 18   |

Variables are defined as follows: NILE, Nile River flood event data (from Quinn 1992, this volume); NIÑO, El Niño event data (from Quinn and Neal, 1992); SOI, Southern Oscillation Index warm events; SST2, El Niño events in the Niño 1 + 2 region (data from COADS); PSST, El Niño events in the area encompassed by the Niño 3 and 4 regions (data from COADS, see text). All maximum likelihood estimates and chi-values significant at 1% (with chi-square probability <0.005) except * = 2.5% (0.01) and ** = 10% (0.02), for entry marked†, correlation is not significant.

nine events. The differences between the two halves of the record are illustrated in Tables 8.2 and 8.3. It is noteworthy that while the El Niño of 1917 was considered to be a strong one, the SO was strongly positive for all of that year. The more recognizable conditions of an ENSO event (with associated variation in the central Pacific) returned during the El Niño of 1918. This suggests that in the first half of the record, negative SOI events may have developed to greatest intensity toward the end of year 0 (year of NIÑO onset) continuing into year +1. In the second half of the record the SOI may have reached its strongest phase earlier in year 0.

For the entire record there were three occurrences where large negative swings of the SOI occurred during the year *preceding* El Niño events. Meehl (1987) looked at changes in SO indices occurring from one year to the next. Under this criterion, he identified more instances with SO-type signatures than by looking

Table 8.2 *Contingency table analysis for NIÑO event data and SOI for the first half of the record (1882–1935)*

|  | SOI | SST2 | PSST | RAIN |
|---|---|---|---|---|
| NIÑO | 37** | 67 | 52* | 46* |
|  | 21 | 43 | 31 | 29 |
|  | 4 | 15 | 8 | 7 |
| SOI |  | 88 | 75 | 82 |
|  |  | 66 | 50 | 52 |
|  |  | 36 | 22 | 25 |

Each row entry refers to analysis method defined in Table 8.1. Significance values and symbols as in Table 8.1.

Table 8.3 *Contingency table analysis for NIÑO event data and SOI for the period 1935–1988*

|  | SOI | SST2 | PSST | RAIN |
|---|---|---|---|---|
| NIÑO | 88 | 89 | 91 | 90 |
|  | 64 | 65 | 66 | 65 |
|  | 34 | 38 | 39 | 38 |
| SOI |  | 84 | 98 | 87 |
|  |  | 61 | 77 | 65 |
|  |  | 30 | 57 | 33 |

Each row entry refers to analysis method defined in Table 8.1.

at individual deviations from a reference mean. Identification of an event signal depends on whether the criteria (and information) used to determine the onset year of El Niño varied over the years. Deser and Wallace (1987) in their comparison of Darwin pressure and coastal SST records from 1925 to 1986 also show that EN events (i.e. episodes of above normal SSTs along the coast of Peru) have occurred both in advance of and subsequent to major negative swings of the SO. EN events and negative swings of the SO have also occurred separately (Deser and Wallace, 1987). Their results together with those of Wright et al. (1988), are in agreement with our own findings. Our analysis indicates that while the large-scale structure of SO events was consistent in both halves of the record, its relationship with NIÑO was more variable.

Time series based on 9-yr moving averages of SOI and NIÑO are reconstructed from a singular spectrum analysis using the first 12 eigenvectors and the variance spectrum is computed. Fast Fourier Transforms were deemed suitable for this analysis since some degree of statistical significance concerning the peak

amplitudes is desired, and the comparison of signals is purely exploratory. Figure
8.3 shows the calculated spectra for the length of the record considered. In Figure
8.4 the spectra from the residual (high frequency) series over each half of the
record are shown. In the first half of the record (1882–1935) the largest peaks in
SOI occur at about 7.1 and 2.33 yr, while in NIÑO the only significant peak
occurs at about 2.6 yr. For the recent record (1935–1988) both the SOI and NIÑO
have peaks at 6.25, 3.8, and 2.4 yr.

Linear correlations between SST and SO indices were shown by Elliott and
Angell (1988) to be largest in the earlier and most recent periods of the record,
with lesser values in the middle (see also, Trenberth and Shea 1987). This was also
observed in the present study but is not reflected in Tables 8.2 and 8.3, where
only 'warm' event signals are compared. However a contingency table analysis
between SOI and SST2 'cold' event signals, for the early and later half of the
study do show substantial differences (Table 8.4).

While correlations of SOI with PSST and RAIN for cold event signals remain
relatively consistent at the 1% level, the relationship with SST2 is not significant.
Bivariate normal correlations of SST2 with PSST and RAIN during the first half
of the record (table not shown) are at 0.77 and 0.61, respectively. Trenberth and
Shea (1987) in a study of the evolution of the Southern Oscillation concluded that

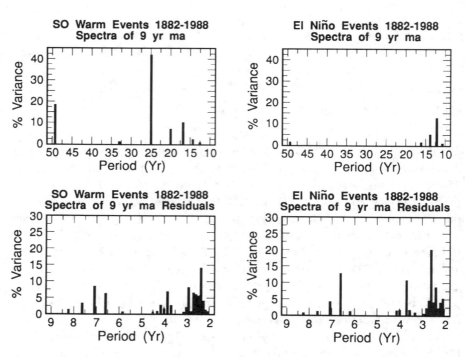

Fig. 8.3   FFT spectra calculated from a singular spectrum reconstruction of a 9-yr (yr)
moving average (ma) of SO-negative and El Niño events for the period 1882–1988. Top
panels: low-pass filtered values, bottom panels: high-pass filtered (residuals) values.

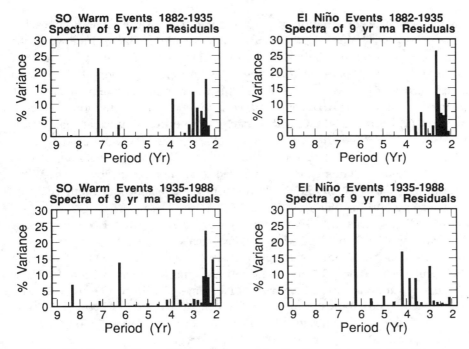

Fig. 8.4 Same as Figure 8.3, except, separately, for the periods 1882–1935 (top panels) and 1935–1988 (bottom panels).

Table 8.4 *Contingency table analysis for 'cold' event signals. Row entries are as described in Table 1*

|  | SST2 | PSST | RAIN |
|---|---|---|---|
| SOI | 17 | 72 | 64 |
|  | 8 | 45 | 37 |
|  | 1 | 19 | 13 |

Associations between SOI and SST2 (SST in the Niño 1 + 2 region) are not statistically significant at the 10% level. Other entries are significant at the 1% level.

the 1925–1935 period was one where large-scale coherent fluctuations like the SO were largely absent.

## Nile flow and El Niño events

The present contributions of the Blue Nile (from Ethiopia) and the White Nile (from Uganda) to the annual Nile flood discharge in Cairo are 68% and 32%, respectively (Fairbridge 1984). Runoff from the Atbara and Blue Nile in the Ethiopian Highlands peaks during the summer monsoon period (July–

September). The contribution of the Ethiopian tributaries to Nile flow reaches 95% in August. Water level then reaches a minimum in March–June, when the Ethiopian contribution diminishes and the White Nile is the dominant source of water with up to 83% of the total flow (Hassan 1981; Fairbridge 1984; Hassan and Stucki 1987). Hassan and Stucki (1987) describe three main periods of Nile flow contribution: (1) A.D. 650–1250, with about equal contribution from both sources, (2) A.D. 1250–1480, with greater contribution from equatorial Africa; and (3) A.D. 1480–1870, with lower discharge from the equatorial sources. Hassan (1981) and Bell (1970) note, however, that during the 1960s an increase in discharge of the main Nile corresponded to an increase in flow of the White Nile by as much as 150 to 250%, relative to a 1911–1960 mean, without a concurrent increase in Blue Nile flow.

According to Hassan (1981) the years 1725 to 1800 (when EN recurrence is ∼5.0 yr) and 1830 to 1870 (when EN recurrence is ∼3.3 yr) were periods of relatively high Nile floods, with relatively low values from 1800 to 1830 (when EN recurrence was near 2.7 yr). These periods roughly coincide with shifts in low frequency variability of the EN record (see Fig. 8.2). From 1730 to 1810 (EN recurrence ∼5.0 yr) most of the low-frequency variance lies between 22- and 35-yr periods. A shift in regime is noted from around 1780 to 1800 (EN recurrence around 3.0 yr), while after 1830 the low-frequency variability of EN event recurrence is dominated by a periodicity of about 45 yr. This change is also documented in the paper by Michaelsen (1989). Periods of changes in EN frequency seem to be associated with low flood episodes in the Nile region, while periods of low frequency in EN occurrence tend to coincide with high floods. Rossignol-Strick (1983) has found a strong correlation between warm periods and heavy rainfall in tropical Africa extending back into the Pleistocene (see also Hassan and Stucki 1987). If this EN-Nile flow relationship is robust, then the missing periods of the Nile flow record may be estimated as follows: 1580–1610 (4.3-yr EN recurrence) normal flow, 1610–1700 (4.4–4.5 yr) high flow, and 1700–1730 (3.3-yr recurrence) low flow.

As noted earlier, the years from A.D. 1520 to 1700 have few direct measurements of Nile flood levels (Riehl et al. 1979; Fairbridge 1984). H. H. Lamb (1977) and Hassan and Stucki (1987) have provided estimates of Nile flood levels during this time. In addition, Faure and Gac (1981) provide estimates of Sahelian rainfall variations over several centuries based on recorded fluctuations of the level of Lake Chad, and more recently, based on annual streamflow variations of the Senegal River. The reconstruction of the Nile flood record indicates a high-flow period beginning near 1590 and ending around 1630. This period is also shown by Faure and Gac (1981) to be associated with high Lake Chad levels. As shown in Figure 8.1, EN occurrences over this period are stable, with recurrence intervals near the long-term mean (∼4 yr). A decrease in reconstructed Nile flow from around 1630 to 1680 roughly coincides with longer recurrence intervals between El Niño events (∼5.5 yr). Likewise, an increase in the heights of Nile floods which begins around 1700 and extends to about 1750 roughly coincides with more frequent EN events (3.8-yr recurrences). These last two periods appear to be exceptions to the general relationship indicated here. The variations in Nile flood levels for the period 1650–1750 are also in rough agreement with the reconstructed variations of Lake Chad levels.

The 'connection' between EN events and Nile flood deficits is now investigated in greater detail using the more recent data of both El Niño and Nile flow. Table 8.5 shows that the bivariate normal correlations between EN events and Nile flood level is high (mostly above 0.5 for several combinations of NIÑO/NILE event intensities). As might be expected, actual predictive power from these associations are small (values in the third row of each set of correlations shown in Table 8.5). While the number of matching pairs of NILE and NIÑO event years is relatively high (48% of the total set of events), the information gained about one variable by knowing information about the other is relatively small. Interestingly, with the Nile record lagging the EN record by 1 year (column entries labelled LAG 1), the correlations are also significant (but negative) as for simultaneous occurences, suggesting that Nile flow may tend to be higher than normal prior to EN events. For comparison, a time series of rainfall anomalies based on all available station data in the Ethiopian Highlands region is shown in Figure 8.5. A decreasing rainfall trend is evident since the 1960s, in agreement with the well-documented Sahel drought (P. J. Lamb 1982; Nicholson 1985), but with no strong indication as to whether a cold or warm event is typically associated with increased or decreased rainfall in a given year. We note, however, that this rainfall index substantially underestimates the likely interannual changes in catchment basin precipitation, since most of the stations are located

Table 8.5 *Contingency table analysis for NIÑO and NILE data for the period 1824–1941*

|  | Nile events | | | |
|---|---|---|---|---|
|  | [1234] | [12] | [123] | [34] |
| NIÑO events |  |  |  |  |
| ALL | 76 | 67 | 74 | 47* |
|  | 50 | 40 | 50 | 24 |
|  | 20 | 15 | 20 | 6 |
| S/VS | 67 |  |  | 83 |
|  | 35 |  |  | 54 |
|  | 12 |  |  | 28 |
| M |  | 72 |  |  |
|  |  | 46 |  |  |
|  |  | 19 |  |  |

Nile flood deficits are ranked from 1 through 4 (where 4 represents the most severe deficit). For NIÑO events: M, moderate; S, strong; VS, very strong.

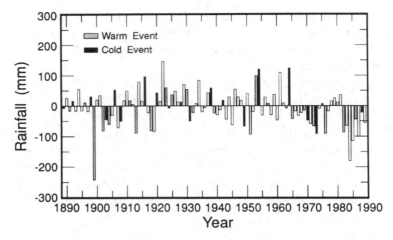

Fig. 8.5  Index of summer (June–August) rainfall anomalies over the Ethiopian Highlands, 1890–1988. Warm and cold ENSO events are highlighted.

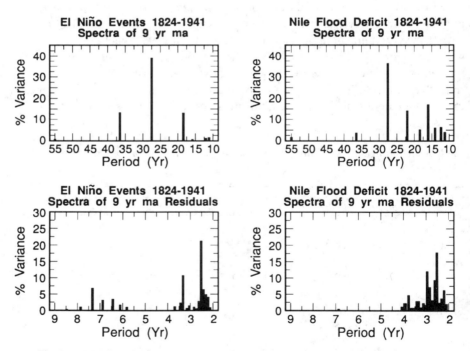

Fig. 8.6  FFT spectra calculated from a singular spectrum reconstruction of 9-yr (yr) moving average (ma) of the NIÑO data and NILE data, 1824–1941. Top panels: low-pass filtered values, bottom panels: high-pass filtered (residuals) values.

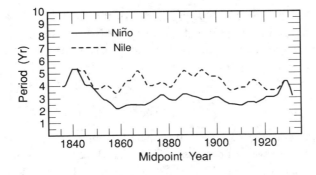

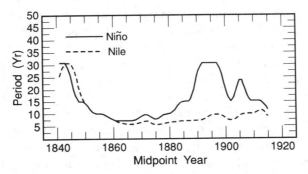

Fig. 8.7 Mean recurrence interval of moderate or stronger NIÑO and categories 1–4 NILE data calculated for overlapping 19-yr segments (top panel), and (bottom panel) overlapping 31-yr segments of strong and very strong EL Niño events and severe Nile streamflow deficiency (categories 3 and 4). Values have been smoothed with a 1-2-1 filter applied twice. (Data from Quinn 1992, this volume.)

at lower elevations in the region. Also, the number of available stations changes through time.

For the NILE flood record (1824–1941), singular spectrum analysis of a 9-yr moving average of events reveals major spectral peaks (>10% of the total variance) at 27.5, 22, and 16 yr (Fig. 8.6). A similar analysis for the NIÑO record shows major peaks at about 37, 27.5, and 18 yr. For the analysis of residuals, two major peaks at around 3–3.5 and 2.5 yr are found for NILE data as well as for the NIÑO data.

The above results point to significant areas of agreement in regards to the characteristic time scales of variability for both Nile flood events and El Niño events. Nevertheless, there is some uncertainty as to the usefulness of the Nile record as a direct proxy for a NIÑO-type series prior to A.D. 1525. While strong events (strong El Niño/severe Nile deficit) have large correlations, the small number of very severe NILE deficits (event category 4 only occurred twice in the recent record) means that a large correlation could have arisen by coincidence. Despite these caveats, we find that strong and very strong EN events are well correlated with the combined Nile categories of three and four. Care must also be taken in determining the physical source of the Nile discharge maxima during different periods (which may be difficult to reconstruct) and in choosing the categories with which to compare variations in El Niño events.

## Conclusions

A simple contingency table analysis of event occurrences is found to be consistent with findings from several previous studies regarding the relationship between the Southern Oscillation and El Niño. As concluded by Deser and Wallace (1987) and others, SST changes in the strong upwelling areas of the eastern equatorial Pacific and coastal Peru are sometimes not in synchrony with variations in other features of the Southern Oscillation occurring farther to the west (see also, Waylen and Caviedes 1986; Wright et al. 1988 and the discussion by Enfield 1992, this volume). Long-term changes in the relationship between the El Niño and its effects on the local climates along the Pacific coast of southern Ecuador and Peru and in the equatorial upwelling zone to the west, and climatic signals due to large-scale circulation changes in the Pacific Basin suggest the need for further study to improve our confidence in the use of distant proxy data as extensions of the ENSO record. The degree of variability of such data independent of the SO over time must also be evaluated. If these connections prove reliable in future studies, then these and other continuous data records which are sensitive to ENSO fluctuations, and which span several centuries in length could be useful to studies of such climatic epochs as the Medieval Warm Period and the Little Ice Age, and hence would be an invaluable addition to Quinn et al.'s (1987) already impressive compilation.

*Acknowledgments* We are grateful to David Enfield and Gerald Meehl for their constructive review of our manuscript, and to William Quinn for furnishing the data upon which this analysis is based. We also thank M. Cecile Penland for providing us a copy of her singular spectrum analysis program, and J. Eischeid for graphics support. Rainfall data over the Central Pacific (RAIN) was kindly provided by Clara Deser.

## References

ANDERSON, R. Y., 1989: Solar-cycle modulations of ENSO: A possible source of climatic change. In Betancourt, J. and MacKay, A. (eds.), *Proceedings, Sixth Annual Pacific Climate (PACLIM)* Workshop, March 5–8, 1989. California Department of Water Resources, Interagency Ecological Studies Program Technical Report 23, 77–81.

ANDERSON, R. Y., 1992: Long-term changes in the frequency of occurrence of El Niño events. *In* Diaz, H. F. and Markgraf, V. (eds.), *El Niño: Historical and Paleoclimatic Aspects of the Southern Oscillation.* Cambridge: Cambridge University Press, 193–200.

BELL, B., 1970: The oldest records of the Nile floods. *Geographical Journal,* 136: 569–573.

BERGER, A., MÉLICE, J. L., and VAN der MERSCH, I., 1990: Evolutive spectral analysis of sunspot data over the past 300 years. *Philosophical Transactions of the Royal Society of London*, Ser. A, 330: 529–541.

DESER, C. and WALLACE, J. M., 1987: El Niño events and their relation to the Southern Oscillation: 1925–1986. *Journal of Geophysical Research*, 92: 14, 189–14,196.

DIAZ, H. F. and KILADIS, G. N., 1992: Atmospheric teleconnections associated with the extreme phases of the Southern Oscillation. *In* Diaz, H. F. and Markgraf, V. (eds.), *El Niño: Historical and Paleoclimatic Aspects of the Southern Oscillation*. Cambridge: Cambridge University Press, 7–28.

ELLIOTT, W. P. and ANGELL, J. K., 1988: Evidence for changes in the Southern Oscillation relationships during the last 100 years. *Journal of Climate*, 1: 729–737.

ENFIELD, D. B., 1988: Is El Niño becoming more common? *Oceanography*, 1: 23–27.

ENFIELD, D. B., 1989: El Niño, past and present. *Reviews of Geophysics*, 27: 159–187.

ENFIELD, D. B., 1992: Historical and prehistorical overview of El Niño/Southern Oscillation. *In* Diaz, H. F. and Markgraf, V. (eds.), *El Niño: Historical and Paleoclimatic Aspects of the Southern Oscillation*. Cambridge: Cambridge University Press, 95–117.

ENFIELD, D. and CID S., L., 1991: Low-frequency changes in El Niño-Southern Oscillation. *Journal of Climate*, 4: 1137–1146.

FAIRBRIDGE, R. W., 1984: The Nile floods as a global climate/solar proxy. *In* Mörner, A. and Karlén, W., (eds.), *Climatic Changes on a Yearly to Millennial Basis*. Dordrecht: Reidel, 181–190.

FAURE, H. and GAC, J.-Y., 1981: Will the Sahelian drought end in 1985? *Nature*, 291: 475–478.

HAMDAN, M. A., 1970: The equivalence of tetrachoric and maximum likelihood estimates of in 2 × 2 tables. *Biometrika*, 57: 212–215.

HASSAN, F., 1981: Historical Nile floods and their implications for climatic change. *Science*, 212: 1142–1145.

HASSAN, F. and STUCKI, B., 1987: Nile floods and climatic change. *In* Rampino, M., Sanders, J. E., Newman, W. S., and Konigsson, L. K., (eds.) *Climate: History, Periodicity, and Predictability*. New York: Van Nostrand Reinhold, 37–46.

IMSL, 1989: Correlation measures for contingency tables. IMSL STAT/LIBRARY Vol. 1, Houston, pp. 268–270

LAMB, H. H., 1977: *Climate: Past, Present and Future*. London: Methuen. 835 pp.

LAMB, P. J., 1982: Persistence of Subsaharan drought. *Nature*, 299: 46–48.

LOUGH, J. M., 1992: An index of the Southern Oscillation reconstructed from western North American tree-ring chronologies. *In* Diaz, H. F. and Markgraf, V. (eds.), *El Niño: Historical and Paleoclimatic Aspects of the Southern Oscillation*. Cambridge: Cambridge University Press, 215–226.

MARTINSON, E. and HAMDAN, M. A., 1971: Maximum likelihood and some other asymptotically efficient estimators of correlation in two-way contingency tables. *Journal of Statistical Computation and Simulation*, 1: 45–54.

MEEHL, G. A., 1987: The annual cycle and interannual variability in the tropical west Pacific and Indian Ocean regions. *Monthly Weather Review*, 115: 27–50.

MICHAELSEN, J., 1989: Long-period fluctuations in El Niño amplitude and frequency reconstructed from tree-rings. *In* Peterson, D. H. (ed.), *Aspects of Climate Variability in the Pacific and the Western Americas*. Geophysical Monograph 55. Washington D.C.: American Geophysical Union, 69–74.

NICHOLLS, N., 1988: El Niño-Southern Oscillation and rainfall variability. *Journal of Climate*, 1: 418–421.

NICHOLSON, S. E., 1985: Sub-Saharan rainfall 1981–84. *Journal of Climate and Applied Meteorology*, 24: 1388–1391.

PHILANDER, S. G., 1990: *El Niño, La Niña, and the Southern Oscillation*. San Diego: Academic Press. 289 pp.

PRESS, W. H., FLANNERY, B. P., TEUKOLSKY, S. A., and VETTERLING, W. T., 1989: *Numerical Recipes: The Art of Scientific Computing*. New York: Cambridge University Press, 476–483.

QUINN, W. H., 1992: A study of Southern Oscillation-related activity for A.D. 622–1990 incorporating Nile River flood data. *In* Diaz, H. F. and Markgraf, V. (eds.), *El Niño: Historical and Paleoclimatic Aspects of the Southern Oscillation*. Cambridge: Cambridge University Press, 119–149.

QUINN, W. H., NEAL, V., and ANTUNEZ de MAYOLO, S., 1987: El Niño occurrences over the past four and a half centuries. *Journal of Geophysical Research*, 92: 14,449–14,461.

QUINN, W. H. and NEAL, V. T., 1992: The historical record of El Niño events. *In* Bradley, R. S. and Jones, P. D. (eds.), *Climate Since A.D. 1500*. London: Routledge, 623–648.

RASMUSSON, E. M. and CARPENTER, T. H., 1982: Variations in tropical sea surface temperature and surface wind fields associated with the Southern Oscillation/El Niño. *Monthly Weather Review*, 110: 354–384.

RASMUSSON, E. M. and WALLACE, J. M., 1983: Meteorological aspects of El Niño/Southern Oscillation. *Science*, 222: 1195–1202.

RASMUSSON, E. M., WANG, X., and ROPELEWSKI, C. F., 1990: The biennial component ENSO variability. *Journal of Marine Systems*, 1: 71–96.

RIEHL, H., EL-BAKRY, M., and MEITÍN, J., 1979: Nile River discharge. *Monthly Weather Review*, 107: 1546–1553.

RIEHL, H. and MEITÍN, J., 1979: Discharge of the Nile River: A barometer of short-period climate variations. *Science*, 206: 1178–1179.

ROSSIGNOL-STRICK, M., 1983: African monsoons and immediate climatic response to orbit insolation. *Nature*, 304: 46–49.

TRENBERTH, K. E., 1991: General characteristics of El Niño-Southern Oscillation. *In* Glantz, M. H., Katz, R. W., and Nicholls S. N. (eds.), *Teleconnections Linking Worldwide Climate Anomalies*. Cambridge: Cambridge University Press, 13–42.

TRENBERTH, K. and SHEA, D., 1987: On the evolution of the Southern Oscillation. *Monthly Weather Review*, 115: 3078–3096.

VAUTARD, R. and GHIL, M., 1989: Singular spectrum analysis in nonlinear dynamics, with applications to paleoclimatic time series. *Physica D*, 35: 395–424.

WAYLEN, P. and CAVIEDES, C. N., 1986: El Niño and annual floods on the north Peruvian littoral. *Journal of Hydrology*, 89: 141–156.

WHITELY, P., 1983: The analysis of contingency tables. *In* McKay, D., Schofield, N., and Whitely, P. (eds.), *Data Analysis and the Social Sciences*. New York: St. Martin's Press, 72–119.

WRIGHT, P. B., 1985: The Southern Oscillation: An ocean-atmosphere feedback system? *Bulletin of the American Meteorological Society*, 66: 398–412.

WRIGHT, P., WALLACE, J. M., MITCHELL, T., and DESER, C., 1988: Correlation structure of the El Niño/Southern Oscillation phenomenon. *Journal of Climate*, 1: 609–625.

# 9

# Long-term changes in the frequency of occurrence of El Niño events

ROGER Y. ANDERSON

*Department of Geology, University of New Mexico Albuquerque, New Mexico 87131, U.S.A.*

## Abstract

A composite record of changes in the frequency of occurrence of El Niño/Southern Oscillation (ENSO) events since the A.D. 622, based on a reconstruction of historical records and Nile flood history, reveals a pattern in which El Niños are more frequent in cycles of ~90, ~50, and 24 to 22 yr. ENSO events are most frequent around A.D. 800 and spectral structure is similar for events based on Nilometer data before A.D. 1522 and historical data from South America after 1522. During the Medieval Warm Period, ENSO-related activity was less common and had a periodicity of ~90 yr.

## Introduction

Quinn et al. (1987) published an historical reconstruction for all El Niño events to A.D. 1800 and for strong El Niño events to A.D. 1522. They and Enfield (1988) noted that El Niños were more common at some times than others. Anderson (1990) applied a simple moving average to the Quinn et al. (1987) historical data in order to illustrate graphically how the incidence of El Niño events changed between 1522 and 1986. Changes in the frequency of occurrence of El Niños since 1700 were recognized as being somewhat systematic, appearing to mimic changes in the frequency of sunspot numbers, and correlating with changes in the Southern Oscillation Index (SOI) (Anderson 1990). Enfield (1992, this volume) shows that an association between El Niño events and sunspot numbers is not ruled out in the historical data set of Quinn et al. (1987).

Quinn (1992, this volume), by including streamflow records from the River
Nile since A.D. 622, extended his reconstruction of ENSO events to the year
A.D. 629. He suggested that the effects of a low-index phase of the Southern
Oscillation (El Niño) in the western Pacific extend to Ethiopia and include the
effects of drought on Nile floods. The difference between the previous mean high
maximum flood and the low maximum of a weak flood year for the Nile is
believed to simulate the shift from high SOI conditions (La Niña) to an El Niño.
The Nile record contains gaps, but Quinn (1992, this volume) has interpreted a
reasonably complete sequence of El Niño events to the year 1522, making it
possible to combine the two kinds of historical information.

In order to compare the Quinn et al. (1987) reconstruction from historical
records back to 1522 with the longer reconstruction based on Nile data, I applied
the moving-average method used in Anderson (1990) to several data sets,
including: (A, all events from 1800 to 1984; B, strong events from 1525 to 1800;
A + B, events from 1525 to 1984; C, (Nile) events from 622 to 1525; and
A + B + C, all events from 622 to 1984. (see Figure 9.1.) This chapter is not,
however, intended as a comprehensive analysis of the Quinn et al. (1987) and
Quinn (1992, this volume) reconstructions of El Niño events. Diaz and Pulwarty
(1992, this volume) provide a statistical comparison of the Quinn's Nile events
and El Niño and SOI records since 1525 and have taken the moving-average
method a step farther by applying singular spectrum reconstruction and spectral
analysis to the record since 1525. They find that strong Nile events are correlated

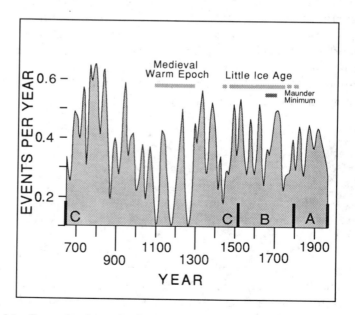

Fig. 9.1  Composite time-series for recurrence of El Niño events since A.D. 622.
Linearly weighted 19-yr running mean. Data from Quinn et al., 1987; Quinn 1992,
this volume). Segment C (Nile), 622–1525; segment B, 1525–1800; segment A,
1800–1984.

with El Niño records but caution that further studies are needed before the entire Nile record can be evaluated. The purpose of this chapter is simply to illustrate a composite plot of the frequency of events since 629 and to suggest that there appears to be some consistency between the older and younger records.

## Long-term changes in recurrence of El Niño events

Changes in the incidence of El Niño events over time can be visualized by transforming unranked events, by means of a simple, weighted moving-average, into a plot in which the frequency of occurrence, in addition to being plotted along the time axis also is expressed as the number of events per year. In this type of plot, amplitude on the $y$ axis represents the frequency of events only, and not the true magnitude (amplitude) of individual El Niño events. No attempt has been made to illustrate the amplitude or ranking of events as assigned by Quinn et al. (1987) and Quinn (1992, this volume) because a connection between rank and frequency of occurrence is not well established (Anderson 1990). However, frequency and amplitude of El Niño events, as determined in tree-ring data (Michaelsen 1989), are related.

In order to examine only those changes related to frequency of occurrence, each Nile event was assigned a unit value (1) and historical data were transformed to numerical data by means of a moving-average. Events in two adjacent years were considered to be a single event and assigned to the first of the two consecutive years. For the purpose of illustration (Fig. 9.1), a 19-yr linearly weighted moving-average was used and number of events per unit time is plotted along the $y$ axis. Different methods were used by Quinn et al. (1987) and Quinn (1992, this volume) for reconstructing El Niño events between A.D. 622 and 1522 (Nilometer data) and after 1522 (historical records). The average time interval between events differs in segments where different methods were applied. To obtain a composite plot in which the $y$ axis represents frequency of events at the same scale, the average frequency of events in segments B and C was adjusted (factors: B = frequency value × 2.50; C = frequency value × 1.95) to correspond to the average frequency for all events in the segment based on the most complete historical record (Segment A).

For spectral analysis, a 9-yr linearly weighted moving average was used to transform El Niño events into a numerical time series. Power spectra were calculated using methods employed by Anderson and Koopmans (1963) and Jenkins and Watts (1968) and using a computer program adapted at Brown University for paleoclimatic data. Significance is not implied for spectral density and differences in the method of data collection and adjustments in the average frequency of events in the different segments imposes limitations on the reliability of spectra for composite time series. Even so, a consistent spectral structure is evident within and between the different segments, suggesting that the combined record (Fig. 9.1) and associated spectra (Fig. 9.2) represent an approximation of long-term changes in the frequency of warm ENSO events.

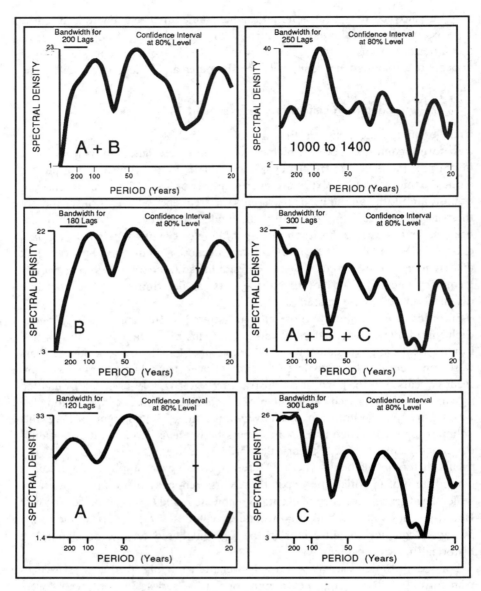

Fig. 9.2  Log power spectral density for segments of the reconstructed El Niño time series. Data converted to events per year by applying a linear nine-term running filter to unranked event series. Segment A, all events 1800–1984; segment B, strong events 1525–1800; segment C, Nile flood SO reconstruction A.D. 622–1525.

## Discussion

### *Patterns of recurrence*

The longest and strongest trend is recognized in the Nile segment (Segment C) where the long-term minimum (fewest El Niño events) occurs between ~A.D. 1000 and ~1400, and corresponds approximately to the Medieval Warm Period. The long-term maximum is near the A.D. 800 and frequency of recurrence between 700 and 900 is higher than in Segments B or A. This could be an artifact of using a different method for the two segments. The recurrence of events during the Maunder Minimum and Little Ice Age is greater than in Segment A and the middle part of segment C (Nile segment).

The recurrence of ENSO events appears to have a similar spectral structure in different segments of the time-series (Fig. 9.2). Segment A, for example expresses a period between 40 and 50 y that continues in Segment B, and also continues in the long segment C (Nile record). Similarly, a period of slightly less than 100 yr is displayed in segments B, combined segments A and B, and again in segment C where it shifts to a slightly higher frequency (~90-yr period) and is strongest during the Medieval Warm Period between 1000 and 1400 (Figs. 9.1, 9.2). A weak increase in spectral density near 33 yr appears only in the Nile segment. Expression of a period slightly longer than 20 yr occurs in spectra for both segment B and segment C. A recurrence cycle near 20+ yr has been removed from Fig. 9.1 by the 19-yr filter but both 40−, 50−, and ~90-yr cycles can be identified in the plot of a 19-yr moving average. A period near 20+ yr is poorly developed in the spectrum for events since A.D. 1525 illustrated by Diaz and Pulwarty (1992, Fig. 8.2a, this volume) but a recurrence cycle between 40 and 50 yr is the strongest one expressed in their analysis and appears in illustrated spectra of the same time segments (Fig. 9.2).

The variance in characteristic spectra (Fig. 9.2) is concentrated in relatively narrow frequency bands, reflecting patterns of recurrence that also are visible in filtered time series (Fig. 9.1, 90-yr and 40-to 50-yr cycles; Fig. 8.1 in Diaz and Pulwarty 1992, this volume, 40-to 50-yr cycles). Considering that segments A, B, and C reflect different physical recording systems and represent data acquired by different methods, the continuity suggests that the Nile data and Quinn's (1992, this volume) reconstruction may accurately portray longer and stronger regional patterns of recurrence associated with El Niño.

### *Origin of long-term cyclicity*

Anderson et al. (1990) reported that the climatological, biological, and geological effects of changes in ocean circulation that are related to El Niño's primary ~4-yr cycle were repeated again at lower-than-ENSO frequencies. That is, features such as productivity, abundance of life forms, geochemistry, and sedimentation that occur during El Niño conditions and during La Niña conditions also are found in cycles of ~20 to ~80 yr. In another example, ~50- and

~ 80-yr cycles in $\delta^{18}$O in *Globigerina bulloides* generally parallel the recurrence of El Niño events over the last ~ 200 yr in the Santa Barbara Basin (Fig. 20.2 in Anderson et al. 1992, this volume). Long-term associations between climatic proxies and El Niño also may occur in East African lakes (Halfman and Johnson 1988) and Fairbridge (1984) cites a number of studies of Nile records that have recognized periods near ~ 80 and ~ 20 yr and that may be linked to El Niño and the quasi-biennial oscillation (QBO).

Comparison of El Niño frequency with the tree-ring record (Michaelsen 1989) and with the SOI (Anderson 1990) suggests a relationship in which El Niños are strongest when they are most frequent. This association is not clear in the historical reconstruction (Quinn et al. 1987) but the large maximum in the number of El Niño events between A.D. 700 and 900 in the Nile record (Fig. 9.1) also occurred at a time when there were many strong events (see Quinn 1992, this volume) and long-term changes in amplitude may accompany changes in frequency of occurrence.

Speculation about associations between ENSO and solar activity may seem premature but there is increasing evidence that the solar cycle modulates wind fields in the stratosphere and troposphere (Tinsley and Deen 1991; Anderson 1992), thereby providing a potential mechanism for a link between solar-geomagnetic effects and ENSO phenomena. Hence, examples of the recurrence of ENSO phenomena at long periods of solar activity are worth noting as possible evidence of a solar-climate connection. The Gleissberg cycle, for example, has a period of 87 to 88 yr and has been confirmed in a 1000-yr record of solar (auroral) activity (Feynman and Fougere 1984) and also is reported in a 9700-yr $^{14}$C record from tree rings (Stuiver et al. 1991). Michaelsen (1989) showed that both the number and strength of El Niño events in a 400-yr tree-ring record expressed a cycle between 80 and 100 yr. The analysis of Nile data by Diaz and Pulwarty (1992, this volume) suggests that Quinn's (1992, this volume) 1300-yr compilation of El Niño events may measure long-term variability in the ENSO system. If so, the recurrence of events in ~ 90-yr cycles seems to confirm Michaelsen's (1989) analysis, and Enfield's (1992, this volume) analysis, and suggests that the recurrence of El Niño events may be another example of an association with a long solar cycle.

Comparison of the recurrence of El Niño events and changes in sunspot number reveals an association in which El Niños are more common when solar activity is weak (Anderson 1990). If the recurrence of El Niño events were modulated by long-term solar changes (e.g. Maunder Minimum, Medieval Grand Maximum), the effect would lengthen the period between El Niño events at times when solar activity is strong. Assuming that the Medieval Warm Period was a time of increased solar activity (Damon 1988), the long-term minimum in the recurrence of El Niño events between A.D. 1000 and 1400 (Fig. 9.1), and the appearance of a strong ~ 90 yr cycle in the same interval (Fig. 9.2), is consistent with having weaker and fewer El Niño events during episodes of strong solar activity (Anderson 1990).

Paleoclimatic studies have not been carried out in enough detail to firmly establish or dismiss a link between changes in climate and long-term cycles in the frequency or intensity of El Niño events. The record reconstructed by Quinn and associates, however, suggests that changes in frequency and intensity occur over time scales of many decades. A continued search of the paleoclimatic record for evidence of such long-term changes may be rewarded by additional examples of associations with ENSO, and perhaps by examples that confirm or deny an association with solar activity (see Anderson et al. 1992, this volume).

*Acknowledgment* This study was supported by National Science Foundation Grant ATM8707462.

## References

ANDERSON, R. Y., 1990: Solar-cycle modulations of ENSO: a mechanism for Pacific and global climate change. *In* Betancourt, J. L. and MacKay, A. M. (eds.), *Proceedings, Sixth Annual Pacific Climate (PACLIM) Workshop, March 5–8, 1989.* California Department of Water Resources, Interagency Ecological Studies Program Technical Report 23, 77–81.

ANDERSON, R. Y., 1992: Possible connection between surface winds, solar activity and the earth's magnetic field. *Nature*, 358: 51–53.

ANDERSON, R. Y. and KOOPMANS, L. H., 1963: Harmonic analysis of varve time series. *Journal of Geophysical Research*, 68: 877–893.

ANDERSON, R. Y., LINSLEY, B. K., and GARDNER, J. V., 1990: Expression of seasonal and ENSO forcing in climatic variability at lower than ENSO frequencies: Evidence from marine varves off California. *In* Meyers, P. A., and Benson, L. V. (eds.), *Paleoclimates: The Record from Lakes, Ocean, and Land. Palaeogeography, Palaeoclimatology, Paleoecology*, 78: 287–300.

ANDERSON, R. Y., SOUTAR, A., and JOHNSON, T. C., 1992: Long-term changes in El Niño/Southern Oscillation. *In* Diaz, H. F. and Markgraf, V. (eds.), *El Niño: Historical and Paleoclimatic Aspects of the Southern Oscillation.* Cambridge: Cambridge University Press, 419–433.

DAMON, P. E., 1988: Production and decay of radiocarbon and its modulation by geomagnetic field-solar activity changes with possible implications for global environment. *In* Stephenson, F. R. and Wolfendale, A. W. (eds.), *Secular, Solar, and Geomagnetic Variations in the Last 10,000 Years.* Dordrecht: Kluwer, 267–285.

DIAZ, H. F. and PULWARTY, R. S., 1992: A comparison of Southern Oscillation and El Niño signals in the tropics. *In* Diaz, H. F. and Markgraf, V. (eds.), *El Niño: Historical and Paleoclimatic Aspects of the Southern Oscillation.* Cambridge: Cambridge University Press, 175–192.

ENFIELD, D. B., 1988: Is El Niño becoming more common? *Oceanography*, 1: 23–27.

ENFIELD, D. B., 1992: Historical and prehistorical overview of El Niño/Southern Oscillation. *In* Diaz, H. F., and Markgraf, V. (eds.), *El Niño: Historical and Paleoclimatic Aspects of the Southern Oscillation.* Cambridge: Cambridge University Press, 95–117.

FAIRBRIDGE, R. W., 1984: The Nile floods as a global climate/solar proxy. *In* Mörner, A. and Karlén, W. (eds.), *Climatic Changes on a Yearly to Millennial Basis.* Dordrecht: Reidel, 181–190.

200 Historical records in ENSO reconstructions

FEYNMAN, J. and FOUGERE, P. F., 1984: Eighty-eight year periodicity in solar-terrestrial phenomena confirmed. *Journal of Geophysical Research*, 89: 3023–3027.

HALFMAN, J. D. and JOHNSON, T. C., 1988: High-resolution record of cyclic climatic change during the past 4ka from Lake Turkana, Kenya. *Geology*, 16: 496–500.

JENKINS, G. M. and WATTS, D. G., 1968: *Spectral Analysis and Its Applications*. San Francisco: Holden-Day, 171–208.

MICHAELSEN, J., 1989: Long-period fluctuations in El Niño amplitude and frequency reconstructed from tree rings. *In* Peterson, D. H. (ed.), *Aspects of Climate Variability in the Pacific and Western Americas*. Geophysical Monograph 55. Washington, D.C.: American Geophysical Union, 69–74.

QUINN, W. H., NEAL, V. T., and ANTUNEZ de MAYOLO, S. E., 1987: El Niño occurrences over the past four and a half centuries. *Journal of Geophysical Research*, 92: 14,449–14,461.

QUINN, W. H., 1992: A study of Southern Oscillation-related climatic activity for A.D. 622–1990 incorporating Nile River flood data. *In* Diaz, H. F., and Markgraf, V., (eds.), *El Niño: Historical and Paleoclimatic Aspects of the Southern Oscillation*. Cambridge: Cambridge University Press, 119–149.

STUIVER, M., BRAZIUNUS, T. F., BECKER, B., and KROMER, B., 1991: Climatic, solar, oceanic, and geomagnetic influences on late-glacial and Holocene atmospheric $^{14}C/^{12}C$ change. *Quaternary Research*, 35: 1–24.

TINSLEY, B. A. and DEEN, G. W., 1991: Apparent tropospheric response to MeV-GeV particle flux variations: a connection via electrofreezing of supercooled water in high-level clouds? *Journal of Geophysical Research*, 96: 22,283–22,296.

# Paleoclimate reconstructions of El Niño/Southern Oscillation from tree-ring records

# 10

# Using tree rings to study past El Niño/Southern Oscillation influences on climate

EDWARD R. COOK

*Tree-Ring Laboratory Lamont-Doherty Geological Observatory Palisades, New York 10964, U.S.A.*

## Abstract

The tree-ring studies in this book describe various applications of dendro-chronology to the identification and reconstruction of the El Niño/Southern Oscillation (ENSO) effects on extratropical climate in the Northern Hemisphere, with particular emphasis on North America. In this contribution, those studies are briefly introduced. Then with reference to those applications, a detailed introduction into the basic principles and practices of statistical dendroclimatology is provided. Special emphasis is placed on the importance of regression model validation as a means of assessing the true accuracy of tree-ring estimates of past climate and ENSO teleconnections.

## Introduction

The tree-ring studies in this book describe various applications of dendrochronology to the identification and reconstruction of El Niño/Southern Oscillation (ENSO) effects on extratropical climate in the Northern Hemisphere, with particular emphasis on North America. The value of annual tree-ring widths as proxy records of past climate is well established and amply reviewed in Fritts (1976), Hughes et al. (1982), and Cook and Kairiukstis (1990). The uniqueness of tree rings as paleoclimatic records lies in their fixed annual resolution and the absolutely dated time series that they produce. Together, these attributes allow for monthly, seasonal, and annual reconstructions of past climate from tree rings, a level of detail rarely achieved by any other paleoclimatic data source. Both high-frequency resolution and absolute dating control are crucial to the study of past ENSO influences on climate because ENSO variability is strongly band-limited (Julian and Chervin 1978), with most of its power either biennial or in the 4- to 5-yr bandwidth (Rasmusson et al. 1990).

Although the quantitative reconstruction of past ENSO effects on climate would seem to be the ultimate goal of dendroclimatology, it is not always necessary or practical to do this. In regions where climate is extremely limiting to tree growth and where ENSO teleconnections have also been identified (e.g. Ropelewski and Halpert 1986; Kiladis and Diaz 1989), it is possible to analyse tree-ring chronologies directly as proxies of past climate for evidence of an ENSO signal. Such is the case in the semiarid regions of the western United States, especially in the southwestern region of Arizona, New Mexico, and northern Mexico.

The direct analysis of tree rings for an ENSO signal has been applied in quite different ways in two of the tree-ring studies in this book. The first by D. M. Meko (1992, this volume) uses spectral analysis techniques to examine and quantify the frequency domain strength and regional coherence of the band-limited ENSO signal in six regional tree-ring chronologies covering most of the western United States. His results indicate that the strongest ENSO signal is probably in the Arizona-New Mexico region, with weaker signals in southern California and Colorado. However, the ENSO signal is relatively weak in all regions compared to lower frequency fluctuations at wavelengths longer than 20 yr in duration. The second study by T. W. Swetnam and J. L. Betancourt (1992, this volume) relates the prevalence of wildfires, precisely dated fire scars on trees, and variations in winter-spring precipitation in Arizona and New Mexico to changes in the Southern Oscillation Index (SOI). They also find an inverse correlation between fire scar frequency and regional tree growth (a proxy for winter-spring precipitation) that indicates a long-term relationship between ENSO variability, precipitation amount, and fire frequency in Arizona and New Mexico.

Three additional tree-ring studies in this book have used a variety of statistical methods to reconstruct the relationships between tree growth, climate, and ENSO in various parts of North America. The first by R. D. D'Arrigo and G. C. Jacoby (1992, this volume) shows strong evidence for an ENSO signal in a 1000-yr-long tree-ring reconstruction of winter precipitation for northwestern New Mexico and in the individual tree-ring chronologies used to produce it. They also demonstrate the feasibility of directly reconstructing the SOI using tree rings from both New Mexico and Java, with the addition of the latter adding significant explained variance to the model. The second study by J. M. Lough (1992, this volume) uses a grid of 65 tree-ring chronologies from the western United States to characterize the effect of ENSO on regional tree growth anomalies. Since 1900, warm (El Niño) event years have been associated with increased tree growth in Colorado, Utah, Arizona, New Mexico, and Mexico. In contrast, cold (La Niña) event years have been associated with decreased growth over the same region. Unfortunately, when similar growth anomaly patterns prior to 1900 were identified in the data, they proved not to be uniquely associated with ENSO extremes. Even so, Lough is able to develop a useful reconstruction of seasonal SOIs back to 1601 using canonical regression analysis. The third by M. K. Cleaveland, E. R. Cook, and D. W. Stahle (1992, this volume) examines long tree-ring reconstructions of past drought in the eastern United States for evidence of ENSO influences using superposed epoch analysis. They find that only the Texas region shows a consistent ENSO signal that relates wet years to El Niños and dry years to La Niñas. They also reconstruct the SOI back to 1699 using tree-ring chronologies from northern Mexico and Texas and use spectral analysis and bandpass filtering to describe the frequency and time domain properties of this record.

Next, I will briefly describe some of the basic principles and practices of statistical dendroclimatology that allow for the accurate reconstruction of past climate and ENSO from tree rings. This section is intended to provide the unacquainted reader with some background information on the subject, which will, I hope, make the tree-ring studies in this book more understandable and easier to evaluate critically. In it, I will emphasize the problem of artificial predictability in statistical model development and the need for model validation in paleoclimatic reconstructions. I emphasize these two points because dendroclimatology is virtually unique in the field of paleoclimatology for the rigor that it uses in statistically validating its climatic reconstruction models. This review will not be exhaustive and the reader should refer to Fritts (1976), Webb and Clark (1977), Gordon (1982), Lofgren and Hunt (1982), and Fritts et al. (1990) for more comprehensive treatments. I will also not review the various aspects of tree-ring chronology development, which include sampling, dating, measuring, and standardizing the ring widths. The reader should refer to Fritts (1976), Graybill (1982), Cook (1987), Cook et al. (1990), and Pilcher et al. (1990) for reviews of those procedures.

## Dendroclimatic reconstruction and verification

The statistical basis for most dendroclimatic reconstructions is the general linear model, with particular emphasis on multiple regression analysis. A time period common to both the tree-ring and climate variables is selected as the calibration period (Fritts 1976) for estimating a regression model that can be used to retrodict or reconstruct climate during times when only the tree-ring data are available. In matrix algebra form, the regression of a dependent climate variable (the predictand) of length $n$ on a suite of $p$ independent tree-ring variables (the predictors) is expressed as

$$\mathbf{Y} = \mathbf{Xb} + \mathbf{e} \tag{1}$$

where $\mathbf{Y}$ is an $n \times 1$ vector of predictands, $\mathbf{X}$ is an $n \times p$ matrix of predictors, $\mathbf{b}$ is a $p \times 1$ vector of regression coefficients estimated as $(\mathbf{X'X})^{-1}\mathbf{X'Y}$, and $\mathbf{e}$ is a $n \times 1$ vector of residuals that represent the variance in $\mathbf{Y}$ not explained by the predictors. The residual vector is assumed to be independent, homoscedastic, and normally distributed. This model is easily extended to the multiple predictand case for producing joint space-time reconstructions by using either canonical regression analysis (Glahn 1968; Fritts et al. 1971) or the orthogonal spatial regression method of Briffa et al. (1986). In fact, the study by Lough (1992, this volume) uses canonical regression analysis to simultaneously reconstruct four seasons of the SOI from tree rings. However, for simplicity I will assume that only one predictand is being reconstructed.

When $p > 1$, it is possible, and indeed sometimes necessary, to transform the original tree-ring predictors into a new set of orthogonal or uncorrelated variables using principal components analysis (PCA) (Cooley and Lohnes 1971; Fritts 1976; Fritts et al. 1990). PCA is commonly used in regression analysis when a suite of predictors is intercorrelated or multicollinear. This is often the case when tree-ring chronologies from the same geographic region are used as predictors because climate exerts more or less the same influence on tree growth. When multicollinearity is high, it can be quite difficult to develop a stable regression model for reconstruction purposes. PCA is also used to reduce the number of predictors because the meaningful variance in the original set of intercorrelated variables is redistributed and concentrated into a reduced subset of orthogonal principal components. These components are ordered by the amount of variance in the original data that each explains, with the first principal component always explaining the most variance in common to all series.

Principal components analysis has become a standard technique of tree-ring analysis. Meko (1992, this volume) uses PCA to develop the six regional tree-ring chronologies used in spectral analysis. D'Arrigo and Jacoby (1992, this volume) and Lough (1992, this volume) also use PCA in their regression models to orthogonalize the tree-ring variables and reduce the dimension of the predictor matrix. For those interested in the more general application of PCA to tree-ring analysis, see the extremely lucid presentation in Fritts (1976).

The classical measure of fit in regression analysis is the coefficient of multiple determination, $R^2$, which is estimated as

$$R^2 = 1 - \frac{\Sigma(Y_i - \hat{Y}_i)^2}{\Sigma(Y_i - \bar{Y})^2} \tag{2}$$

where $\Sigma(Y_i - \hat{Y}_i)^2$ is the sum-of-squares error (SSE) between the actual ($Y_i$) and estimated ($\hat{Y}_i$) predictands, and $\Sigma(Y_i - \bar{Y})^2$ is the sum-of-squares total (SSTO) of $Y_i$ about its mean ($\bar{Y}$). $R^2$ falls exclusively in the range 0–1, with the former being analogous to using the predictand mean (i.e. climatology) as the best predictor of **Y** and the latter being a perfect fit to each observation of **Y**. If the assumptions of regression analysis are met, then $R^2$ can be tested for statistical significance using the $F$-ratio test

$$F_{p,\,n\text{-}p\text{-}1} = \frac{R^2/p}{(1 - R^2)/(n - p - 1)} \tag{3}$$

Although intuitively appealing, $R^2$ and the $F$-ratio test are notoriously unreliable for assessing the true value of a regression model for estimating new values of $Y_i$ from observations of $X_i$ not used in developing the regression model. The predictors are assumed to be known *a priori*, which is virtually never the case. Consequently, data screening and best subset regression schemes are frequently used to reduce a larger pool of $p'$ candidate predictors down to the subset of $p$ 'true' predictors. The potential for artificial predictability (Davis 1976) in best subset models is proportional to $p' : n$ (Rencher and Pun 1980), so that large pools of candidate predictors are virtually guaranteed to inflate the best subset $R^2$. This is best realized by noting that the expected value of $R^2$ in the case of a totally random regression model is

$$E[R^2] = p/(n-1) \qquad (4)$$

(Morrison 1966). Clearly, $R^2$ can only increase as new predictors are added to the model, even if these new predictors have no true relationship with the predictand. The usual probabilities provided by the $F$-ratio test are also incorrect when best subset regression methods are used (Draper and Smith 1981). See Rencher and Pun (1980) for a sobering study of the $R^2$ inflation problem using best subset regression methods. And, for more information on the small sample properties and interpretational difficulties of $R^2$, see Cramer (1987) and Helland (1987).

As a consequence of the inflationary bias noted above, $R^2$ frequently overestimates the ability of a regression model to reconstruct climate outside of the fitted calibration period, resulting in a sometimes marked degradation in the quality of the reconstruction. This shrinkage of $R^2$ (*sensu* Wherry 1931) when a regression model is applied to new data has led to the development of more conservative estimates of explained variance, such as the adjusted $-R^2$

$$R_a^2 = 1 - \left(\frac{n-1}{n-p}\right) \frac{SSE}{SSTO} \qquad (5)$$

(Draper and Smith 1981), which takes into account the loss of degrees of freedom as predictors are entered into the model. Unlike $R^2$, $R_a^2$ can actually decrease if too many predictors are entered in the model. However, $R_a^2$ is still a generally poor measure of true model predictability (Cramer 1987). A related, but less well known estimate of true predictability is the adjusted reduction of error statistic

$$RE' = R_a^2 - \left(\frac{p}{n+1}\right)(1 - R_a^2) \qquad (6)$$

(Kutzbach and Guetter 1980; Fritts et al. 1990), which applies an additional downward correction to $R^2$. $RE'$ is actually the expected value of the reduction of error statistic (formally described below) that is commonly used to verify dendroclimatic reconstructions. While these $R^2$ corrections are undoubtedly useful in determining the true likelihood that a fitted regression model is useful for reconstructing past climate, they are still markedly inferior to verification tests that compare regression estimates of climate to actual data not used in the regression model (Fritts 1976; Kutzbach and Guetter 1980; Fritts et al. 1990). See Snee (1977) for a more general treatment of the regression model validation issue.

There are many possible ways to validate regression models (Gordon 1982). In dendroclimatology, the most commonly used approach is to withhold a portion of the climate data from the calibration exercise and use it to test the validity of regression model estimates. The withheld data typically comprise a contiguous record, which is traditionally called the verification period (Fritts 1976). However, it is perfectly acceptable to withhold randomly selected observations as well. This 'data splitting' procedure is listed by Snee (1977) as one of four general methods of validating regression models, which is distinct from validating on 'new data,' i.e. data totally independent of the climate record used for calibration purposes. Picard and Berk (1990) provide a detailed mathematical treatment of data splitting for model validation purposes. Their results indicate that 'the optimum portion of observations reserved for validation is always less than 1/2 and is usually in the 1/4 to 1/3 range.'

Some of the more commonly used model validation tests are the Pearson product-moment correlation coefficient, the nonparametric sign test, the cross-product mean test, and the reduction of error (*RE*) (Fritts 1976). *RE* is the most rigorous test of those just noted and is estimated as

$$RE = 1 - \left[ \frac{\Sigma (Y_i - \hat{Y}_i)^2}{\Sigma (Y_i - \bar{Y}_c)^2} \right] \tag{7}$$

where $Y_i$ and $\hat{Y}_i$ are the actual and estimated predictands in the verification period and $\bar{Y}_c$ is the calibration period mean. Note that the only difference between $R^2$ and *RE* lies in the denominator term $\bar{Y}_c$. It also differs from its expected value, *RE'*, because it explicitly uses data withheld from the regression model. *RE* has a theoretical range of $-\infty$ to 1.0, with only positive values being acceptable for model verification. A positive *RE* indicates that the regression model has greater predictive skill than the calibration period mean (i.e. climatology). From the viewpoint of model predictability, *RE* has a clear meaning. However, it is not easily interpreted as an estimate of explained variance in the verification period because of its dependence on $\bar{Y}_c$, which can differ markedly from the verification period mean if a trend exists in the data.

For estimating the verification period $R^2$, an extremely stringent statistic is the coefficient of efficiency (*CE*) (Nash and Sutcliffe 1971), which is estimated as

$$CE = 1 - \left[ \frac{\Sigma (Y_i - \hat{Y}_i)^2}{\Sigma (Y_i - \bar{Y}_v)^2} \right] \tag{8}$$

*CE* differs from *RE* only in the denominator term $\bar{Y}_v$, which is the mean of the actual data in the verification period. Obviously, if $\bar{Y}_v = \bar{Y}_c$, then *CE* = *RE*. However, when $\bar{Y}_v \neq \bar{Y}_c$, *CE* < *RE* because the denominator sum-of-squares is fully corrected in estimating $CE$, which is not the case for $RE$. However like $RE$, $CE$ has a theoretical range of $-\infty$ to 1.0, and a $CE > 0$ is usually interpreted as an indication of regression model validity. But since the calibration and verification period means will almost never be identical (i.e. $\bar{Y}_v \neq \bar{Y}_c$), $CE$ will almost always be less then $RE$. Hence, a positive $CE$ is even more difficult to obtain.

The failure of regression models to properly verify can be due to any one of a number of methodological or data problems. A common data problem in the predictor variable set is the lack of time stability between tree growth and climate caused, for example, by purely spurious relationships in the calibration period that break down when applied to the verification period. The same effect can also occur when the relationship between tree growth and climate is truly time dependent, i.e. the relationship actually exists, but only for certain time periods when climatic or environmental thresholds are exceeded. Differentiating between spurious and time-dependent effects is not easy and, perhaps, not relevant to the problem of reconstructing climate from tree rings. In either case, the failure to verify is grounds for rejecting the putative climatic reconstruction. However, the Kalman filter (Harvey 1984; Visser 1986) can explicitly model time-dependent regression relationships and could be used to determine the cause of the model breakdown in more detail.

Data problems can also exist in the predictand variable set when the quality of the instrumental records degrade backwards in time due, for example, to changes in instrumentation, observers, and recording methods. This possibility is relevant to the present volume given questions concerning the reliability of the earliest 19th century SOIs and their suitability for model verification. Unfortunately, the loss of predictand quality back in time can be difficult to identify and quantify. In addition, most climate records used for dendroclimatic calibration and verification are screened and adjusted for inhomogeneities prior to use (e.g. Wright 1989). While not perfect, these adjustments probably do improve the accuracy of the early observations. Therefore, the failure of a reconstruction to verify is usually attributed to the selected predictors.

From a methodological point-of-view, one of the more common causes of spurious regression is autocorrelation in the predictor and/or predictand data. It is well known that climatic times series often possess positive autocorrelation, which imparts 'redness' (i.e. inflated variance in the low frequencies) to the power spectrum (Gilman et al. 1963). Tree-ring series tend to have even more autocorrelation than climate series because of physiological processes that influence the potential for growth in subsequent years (Fritts 1976). Positive autocorrelation reduces the effective sample size or number of independent observations, and this has a direct impact on the assessment of statistical significance between tree rings and climate by producing confidence limits that are too narrow for a given significance level. Consequently, seemingly useful predictors that are completely spurious may be added to a regression model because of autocorrelation in the data.

To guard against spurious relationships between autocorrelated time series, Mitchell et al. (1966) suggested correcting the number of observations downward in proportion to the first-order autocorrelation, $r_1$, in the series as

$$n' = n \left[ \frac{1 - r_1}{1 + r_1} \right] \tag{9}$$

where $n' < n$ when $r_1 > 0$. The degrees-of-freedom for statistical tests are than computed from $n'$ instead of $n$. This correction results in wider confidence intervals and more conservative evaluations of tree-ring and climate relationships, making spurious relationships less likely to be accepted. Fritts et al. (1979) used this correction in developing their dendroclimatic models and in verifying their subsequent reconstructions. In any event, the model residuals should always be checked for autocorrelation using, for example, the Durbin-Watson (DW) statistic (Draper and Smith 1981). In simulation experiments, Granger and Newbold (1977) found that a DW statistic below 1.0 was a strong indicator of spurious regression caused by autocorrelation.

Estimating the effective sample size by using only the first-order autocorrelation coefficient assumes that the persistence is being generated by a first-order Markov or autoregressive process of the form

$$y_t = \phi_1 y_{t-1} + e_t \tag{10}$$

where $y_t$ is a zero-mean process (i.e. $y_t = Y_t - \bar{Y}$), $\phi_1$ is the autoregressive coefficient equal in this case to $r_1$, and $e_t$ is a zero-mean random shock or residual. This model is overly restrictive because it is known that many tree-ring series are better modeled as higher-order autoregressive processes (Meko 1981; Cook 1985). A more general solution to this problem uses Box-Jenkins autoregressive-moving average (ARMA) models (Box and Jenkins 1976) of the form

$$y_t = \sum_{i=1}^{p} \phi_i y_{t-i} + e_t - \sum_{j=1}^{q} \theta_j e_{t-j} \tag{11}$$

to prewhiten the tree rings and climate series as part of developing the regression models for climatic reconstruction. Now, instead of only one autoregressive coefficient for prewhitening, the ARMA($p,q$) model allows for $p$ autoregressive ($\phi_i$) and $q$ moving average ($\theta_j$) coefficients to estimate the persistence. When the appropriate ARMA model is used for prewhitening, the autocorrelation in $y_t$ is removed to yield a new series of serially random residuals (the $e_t$) that do not need any sample size correction of the kind described above. These residuals can than be fed into the regression model as predictors of climate with no concern for the deleterious effects of autocorrelation on the assessment of statistical significance. All that matters is that any persistence due to climate is properly accounted for in the final reconstruction. Meko (1981) investigated the use of ARMA models in developing dendroclimatic reconstructions and found them to be extremely useful. Since that time, ARMA modeling has become a standard method of tree-ring analysis and is used by D'Arrigo and Jacoby (1992, this volume), Lough (1992, this volume), and Cleaveland et al. (1992, this volume) to prewhiten their long tree-ring series prior to correlation and regression analysis with the SOI.

If ARMA modeling is not possible, a very simple transformation of the predictors and predictands to first-differences can be done as

$$\Delta Y_t = Y_t - Y_{t-1} \tag{12}$$

First-differences strictly emphasize the high-frequency changes in a time series at the expense of any long-term trends or multi-year fluctuations. This transformation is recommended if the desire is to emphasize and compare year-to-year changes in time series that also contain trends. Swetnam and Betancourt (1992, this volume) used precisely this transformation for comparing a time series of percentage of fire scarred trees with a regional tree-ring growth record. First-differencing can also be used to check a correlation or regression model based on undifferenced data. If a statistically significant relationship exists between the undifferenced predictors and predictands, then that relationship should also be found in the first-differences as well. Should the relationship breakdown using the first-differences, the original relationship may be spurious. Differencing can also be applied to the verification data (e.g. Gordon 1982; Fritts et al. 1990) to more rigorously test the validity of the regression model.

Autocorrelation also exists in space (Cliff and Ord 1981). Therefore, statistical problems similar to those caused by temporal autocorrelation are likely to occur in developing spatial reconstructions of climate from tree rings or any other field of paleoclimatic proxy records. This problem has received little attention to date (e.g. Livezey and Chen 1983), although it may be partly responsible for some of the extremely high ($> 0.90$) $R^2$s of some published paleoclimatic reconstruction models (e.g. Imbrie and Kipp 1971). Such high $R^2$s probably approach the remeasurement precision and accuracy of the predictand fields themselves, let alone that of the estimated paleoclimatic proxies used for reconstruction.

The fact that paleoclimatic records are estimated is always ignored in reports of $R^2$ because the effect of estimated predictors in regression models is difficult to properly quantify. Standard regression theory assumes that the predictors are measured without error. This allows all of the unexplained variance to be attributable to the predictand. Yet, the paleoclimatic proxy series used as predictors of climate also have error variance due to uncertainty in sampling and measurement. That these effects are not incorporated in the computation of $R^2$, $R^2_a$, or $RE'$ emphasizes even more the need to properly verify paleoclimatic reconstructions of all kinds.

### Concluding remarks

Methods of dendroclimatic reconstruction and model validation rest on a sound statistical foundation. Given a suite of properly sampled and developed climatically sensitive tree-ring chronologies, it is possible now to reconstruct a wide variety of climatic variables, including something as remote as the Tahiti-Darwin SOI. Reconstruction of this index from tree rings in North America is a remarkable achievement considering the influence of so many other environmental influences on tree growth in that region. It attests both to the strength of the ENSO teleconnection with climate over North America and the importance of climate in affecting tree growth.

The following tree-ring papers are all excellent examples of dendroclimatology as it should be applied. A number of techniques have been used to relate tree rings to ENSO; some relatively simple, others more complex; some direct, others indirect. I hope that this introduction into the principles and practices of dendroclimatic reconstruction will provide some useful background information for better understanding these papers.

*Acknowledgement* Lamont-Doherty Geological Observatory Contribution No. 4975.

## References

BOX, G. E. P. and JENKINS, G. M., 1976: *Time Series Analysis: Forecasting and Control.* San Francisco: Holden-Day. 553 pp.

BRIFFA, K. R., JONES, P. D., WIGEY, T. M. L., PILCHER J. R., and BAILLIE, M. G. L., 1986: Climate reconstructions from tree rings: Part 2, Spatial reconstruction of summer mean SLP patterns over Great Britain. *Journal of Climatology*, 6: 1–15.

CLIFF, A. D. and ORD, J. K., 1981: *Spatial Processes: Models and Applications.* London: Pion Press. 266 pp.

CLEAVELAND, M. K., COOK, E. R., and STAHLE, D. W., 1992: Secular variability of the Southern Oscillation detected in tree-ring data from Mexico and the southern United States. *In* Diaz, H. F. and Markgraf, V. (eds.), *El Niño: Historical and Paleoclimatic Aspects of the Southern Oscillation.* Cambridge: Cambridge University Press, 271–291.

COOK, E. R., 1985: *A Time Series Analysis Approach to Tree-Ring Standardization.* PhD dissertation, University of Arizona, Tucson. 171 pp.

COOK, E. R., 1987: The decomposition of tree-ring series for environmental studies. *Tree-Ring Bulletin*, 47: 137–59.

COOK, E. R. and KAIRIUKSTIS, L. A. (eds.)., 1990: *Methods of Dendrochronology: Applications in the Environmental Sciences.* Dordrecht: Kluwer Academic Publishers. 394 pp.

COOK, E. R., BRIFFA, K. R., SHIYATOV, S., MAZEPA, V., and JONES, P. D., 1990: Data analysis. *In* Cook, E. R. and Kairiukstis, L. A. (eds.), *Methods of Dendrochronology: Applications in the Environmental Sciences.* Dordrecht: Kluwer Academic Publishers, 97–162.

COOLEY, W. W. and LOHNES, P. R., 1971: *Multivariate Data Analysis.* New York: Wiley. 363 pp.

CRAMER, J. S., 1987: Mean and variance of $R^2$ in small and moderate samples. *Journal of Econometrics*, 35: 253–266.

D'ARRIGO, R. D. and JACOBY, G. C., 1992: A tree-ring reconstruction of New Mexico precipitation and its relation to El Niño/Southern Oscillation events. *In* Diaz, H. F. and Markgraf, V. (eds.), *El Niño: Historical and Paleoclimatic Aspects of the Southern Oscillation.* Cambridge: Cambridge University Press, 243–257.

DAVIS, R. E., 1976: Predictability of sea surface temperature and sea level pressure anomalies over the North Pacific Ocean. *Journal of Physical Oceanography*, 6: 249–266.

DRAPER, N. R. and SMITH, H., 1981: *Applied Regression Analysis*. New York: Wiley. 709 pp.

FRITTS, H. C., 1976: *Tree Rings and Climate*. London: Academic Press. 567 pp.

FRITTS, H. C., LOFGREN, G. R., and GORDON, G. A., 1979: Variations in climate since 1602 as reconstructed from tree rings. *Quaternary Research*, 12: 18–46.

FRITTS, H. C., BLASING, T. J., HAYDEN, B. P., and KUTZBACH, J. E., 1971: Multivariate techniques for specifying tree-growth and climate relationships and for reconstructing anomalies in paleoclimate. *Journal of Applied Meteorology*, 10: 845–864.

FRITTS, H. C., GUIOT, J., GORDON, G. A., and SCHWEINGRUBER, F., 1990: Methods of calibration, verification, and reconstruction. *In* Cook, E. R. and Kairiukstis, L. A. (eds.), *Methods of Dendrochronology: Applications in the Environmental Sciences*. Dordrecht: Kluwer Academic Publishers, 163–217.

GILMAN, D. L., FUGLISTER, F. J., and MITCHELL, J. M., Jr., 1963: On the power spectrum of red noise. *The Journal of the Atmospheric Sciences*, 20: 182–184.

GLAHN, H. R., 1968: Canonical correlation and its relationship to discriminant analysis and multiple regression. *Journal of the Atmospheric Sciences*, 25: 23–31.

GORDON, G. A., 1982: Verification of dendroclimatic reconstructions. *In* Hughes, M. K., Kelly, P. M., Pilcher, J. R., and LaMarche, V. C. Jr. (eds.), *Climate from Tree Rings*. Cambridge: Cambridge University Press, 58–61.

GRANGER, C. W. J. and NEWBOLD, P., 1977: *Forecasting Economic Time Series*. New York: Academic Press. 333 pp.

GRAYBILL, D. A., 1982: Chronology development and analysis. *In* Hughes, M. K., Kelly, P. M., Pilcher, J. R., and LaMarche, V. C. Jr. (eds.), *Climate from Tree Rings*. Cambridge: Cambridge University Press, 21–28.

HARVEY, A. C., 1984: A unified view of statistical forecasting procedures. *Journal of Forecasting*, 3: 245–275.

HELLAND, I. S., 1987: On the interpretation and use of $R^2$ in regression analysis. *Biometrics*, 43: 61–69.

HUGHES, M. K., KELLY, P. M., PILCHER, J. R., and LAMARCHE, V. C., Jr. (eds.), 1982: *Climate from Tree Rings*. Cambridge: Cambridge University Press. 223 pp.

IMBRIE, J. and KIPP, N. G., 1971: A new micropaleontological method for quantitative paleoclimatology: an application to a last Pleistocene Carribean core. *In* Turekian, K. (ed.), *The Late Cenozoic Glacial Ages*. New Haven: Yale University Press, 71–182.

JULIAN, P. R. and CHERVIN, R. M., 1978: A study of the southern oscillation and Walker circulation phenomenon. *Monthly Weather Review*, 106: 1433–1451.

KILADIS, G. N. and DIAZ, H. F., 1989: Global climatic anomalies associated with extremes in the Southern Oscillation. *Journal of Climate*, 2: 1069–1090.

KUTZBACH, J. E. and GUETTER, P. J., 1980: On the design of paleoenvironmental data networks for estimating large-scale patterns of climate. *Quaternary Research*, 14: 169–187.

LIVEZEY, R. E. and CHEN, W. Y., 1983: Statistical field significance and its determination by Monte Carlo techniques. *Monthly Weather Review* 111: 46–59.

LOFGREN, G. R. and HUNT, J. H., 1982: Transfer functions. *In* Hughes, M. K. Kelly, P. M., Pilcher, J. R., and LaMarche, V. C., Jr. (eds.), *Climate from Tree Rings*. Cambridge: Cambridge University Press, 50–56.

LOUGH, J. M., 1992: An index of the Southern Oscillation reconstructed from western

North American tree-ring chronologies. *In* Diaz, H. F. and Markgraf, V. (eds.), *El Niño: Historical and Paleoclimatic Aspects of the Southern Oscillation*. Cambridge: Cambridge University Press, 215–226.

MEKO, D. M., 1981: Applications of Box-Jenkins methods of time series analysis to the reconstruction of drought from tree rings. PhD dissertation, University of Arizona, Tucson. 149 pp.

MEKO, D. M., 1992: Spectral properties of tree-ring data in the U.S. Southwest as related to El Niño/Southern Oscillation. *In* Diaz, H. F. and Markgraf, V. (eds.), *El Niño: Historical and Paleoclimatic Aspects of the Southern Oscillation*. Cambridge: Cambridge University Press, 227–241.

MITCHELL, J. M., Jr., DZERDZEEVSKII, B., FLOHN, H., HOFMEYR, W. L., LAMB, H. H., RAO, N. K., and WALLEN, C. C., 1966: *Climate Change*. Technical Note No. 79, World Meteorological Organization, Geneva. 79 pp.

MORRISON, D. F., 1966: *Multivariate Statistical Methods*. 2nd ed. New York: McGraw-Hill. 338 pp.

NASH, J. E. and SUTCLIFFE, J. V., 1971: Riverflow forecasting through conceptual models. 1, A discussion of principles. *Journal of Hydrology*, 10: 282–290.

PICARD, R. R. and BERK, K. N., 1990: Data splitting. *The American Statistician*, 44: 140–147.

PILCHER, J. R., SCHWEINGRUBER, F. H., KAIRIUKSTIS, L., SHIYATOV, S., WORBES, M., KOLISHCHUK, V. G., VAGANOV, E. A., JAGELS, R., and TELEWSKI, F. W., 1990: Primary Data. *In* Cook, E. R. and Kairiukstis, L. A. (eds.), *Methods of Dendrochronology: Applications in the Environmental Sciences*. Dordrecht: Kluwer Academic Publishers, 23–96.

RASMUSSON, E. M., WANG, X., and ROPELEWSKI, C. F., 1990: The biennial component of ENSO variability. *Journal of Marine Systems*, 1: 71–96.

RENCHER, A. C. and PUN, F. C., 1980: Inflation of $R^2$ in best subset regression. *Technometrics*, 22: 49–53.

ROPELEWSKI, C. F. and HALPERT, M. S., 1986: North American precipitation and temperature patterns associated with the El Niño/Southern Oscillation (ENSO). *Monthly Weather Review*, 114: 2352–2362.

SNEE, R. D. 1977: Validation of regression models: methods and examples. *Technometrics*, 19: 415–428.

SWETNAM, T. W. and BETANCOURT, J. L., 1992: Temporal patterns of El Niño/southern Oscillation–Wildfire teleconnections in the Southwestern United States. *In* Diaz, H. F. and Markgraf, V. (eds.), *El Niño: Historical and Paleoclimatic Aspects of the Southern Oscillation*. Cambridge: Cambridge University Press, 259–270.

VISSER, H., 1986: Analysis of tree rings data using the Kalman filter technique. *IAWA Bulletin*, 7: 289–297.

WEBB, T., III and CLARK, D. R., 1977: Calibrating micropaleontological data in climatic terms: a critical review. *Annals of the New York Academy of Sciences*, 288: 93–118.

WHERRY, R. J., 1931: A new formula for predicting the shrinkage of the multiple correlation coefficient. *Annals of Mathematical Statistics*, 2: 440–457.

WRIGHT, P. B., 1989: Homogenized long-period Southern Oscillation indices. *International Journal of Climatology*, 9: 33–54.

# 11

# An index of the Southern Oscillation reconstructed from western North American tree-ring chronologies

J. M. LOUGH

*Australian Institute of Marine Science P.M.B. 3, Townsville M.C., Queensland 4810, Australia*

## Abstract

A network of tree-ring chronologies from arid-sites in western North America shows distinctive patterns of growth anomalies in association with recent extremes of the El Niño/Southern Oscillation (ENSO). These anomalies are most marked in the southwestern United States and northern Mexico. Tree-growth tends to be higher in this region in years following an ENSO event and lower in years following an anti-ENSO event. These patterns of tree-growth anomalies associated with recent ENSO extremes are not, however, consistently associated with earlier ENSO extremes. Despite this, it is possible to obtain a statistically significant reconstruction of an index of the Southern Oscillation (SOI) from these tree-ring chronologies, back to A.D. 1601. This reconstruction shows evidence of changes in the character of the SOI over the past few centuries, or at least, a modulation of the North American teleconnection pattern. Climatic information from tree-ring chronologies in the southwestern United States and northern Mexico should complement other sources of proxy climate information, to arrive at the most reliable history of past ENSO events and their teleconnections.

## Introduction

Information about past variations of El Niño/Southern Oscillation (ENSO) can be obtained from trees. Our ability to interpret this information in the tree-ring record depends upon the trees being located at sites where climate is strongly influenced by ENSO events and the ability of the trees to record this influence.

The surface climate in parts of North America experiences one of the strongest and most consistent of all extratropical teleconnections associated with ENSO events (Wright 1977; Ropelewski and Halpert 1986, 1987, 1989; Kiladis and Diaz 1989). H. C. Fritts and coworkers at the Laboratory of Tree-Ring Research, University of Arizona, Tucson, have developed spatially and temporally detailed reconstructions of surface climate variables for North America and the North Pacific back to A.D. 1602 (Fritts et al. 1979; Fritts 1991). These reconstructions were developed from tree-ring chronologies located at sites extending from southwestern Canada to northern Mexico (Fritts and Shatz 1975). The reconstructions of winter temperature, precipitation and, to a lesser extent, sea-level pressure, contain teleconnection patterns associated with ENSO events during the period 1851 to 1900 similar (though with reduced significance and variance) to those observed in instrumental records (Lough and Fritts 1985). Thus, climatic reconstructions from western North American tree-ring chronologies appear capable of resolving high-frequency climatic signals associated with ENSO events.

In this chapter I look in more detail at the strength of the ENSO signal in the network of tree-ring chronologies from arid sites in western North America. I also examine a reconstruction of a Southern Oscillation Index (SOI) from these tree-ring chronologies (Lough and Fritts 1990) for evidence of changes in the character of the SOI over the past few centuries.

## Data and methods

The tree-ring data consist of 65 tree-ring chronologies from arid sites in western North America. The chronologies cover the period A.D. 1600 to 1963 but are best replicated after 1700 (Fritts and Shatz 1975). Tree growth at the 65 sites is related to precipitation as it affects water storage and temperature as it affects evapotranspiration (Fritts 1974, 1976). This is the same network of tree-ring chronologies used to develop the large-scale climatic reconstructions of surface climate variables over North America and the North Pacific which have been described in a number of earlier studies (Fritts et al. 1979; Fritts and Lough 1985; Lough and Fritts 1987; Fritts 1991). Rose (1983) removed the significant amount of autocorrelation in the chronologies using all significant autoregressive-moving-average terms.

Years of Pacific warm (low SOI or ENSO) and cold (high SOI or anti-ENSO) events for the period from 1877 were identified from the work of Kiladis and van Loon (1988). Years of El Niño events since 1600 were identified from the work of Quinn et al. (1987). The reconstructed SOI was developed from the Tahiti-Darwin standardized sea-level pressure series for the period from 1866 (Ropelewski and Jones 1987; data back to 1866 provided by P. D. Jones, pers. comm. 1986).

To examine the teleconnection patterns in the tree-ring chronologies each chronology was converted to standardized anomalies with respect to the mean and standard deviation of the chronology for the period 1600 to 1963. Within the period 1900 to 1963 the anomalies were then averaged for the year that followed each of 13 warm event years and for the year that followed each of 12 cold event years. The difference between the composite warm and cold year anomalies was then obtained.

The composite patterns of tree-growth anomalies associated with the warm and cold events were then correlated with the annual patterns of tree-growth anomalies for each year from 1600 to 1963. Years were identified when the pattern correlation between the composite anomaly pattern for a warm or cold event and the actual anomaly pattern exceeded 0.5 (an arbitrary threshold).

Seasonal values of the Tahiti-Darwin sea-level pressure index were estimated from the principal components of the 65 tree-ring chronologies using canonical regression techniques (Lough and Fritts 1990; Cook 1992, this volume). The model was calibrated over the period 1916 to 1963 and verified over the period 1867 to 1915. The calibration and verification periods were then reversed to test the stability of the regression model.

## Results

### *Teleconnection patterns*

When instrumental data are averaged for years of extreme low and high SOI values, the composite differences reveal distinctive patterns of sea-level pressure, temperature, and precipitation during Northern Hemisphere winter over North

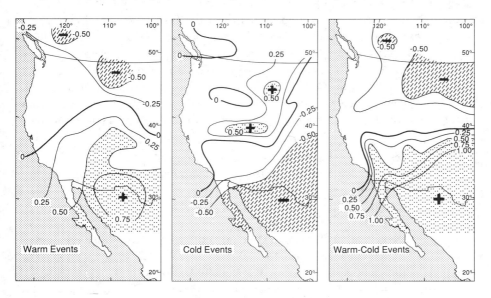

Fig.11.1 Average standardized tree-ring anomalies for 13 warm event years, 12 cold event years, and the difference between the warm and cold event year averages in western North America.

America and the North Pacific. These patterns also existed in reconstructions of the same climatic variables obtained from the western North American tree-ring network for the same time period as the instrumental data and for a time period independent of the instrumental data (Lough and Fritts 1985). As is to be expected, the levels of significance and the magnitude of the anomalies were lower in the reconstructions than the instrumental records.

To examine how strong the ENSO signal was in the tree-ring series, a similar procedure was applied to the original 65 tree-ring chronologies. Warm event years were associated with increased tree growth in Colorado, Utah, Arizona, New Mexico, and Mexico, and slightly reduced tree growth in Montana and southwestern Canada (Fig. 11.1). Average anomalies in these regions exceeded 0.5 standard deviations. The pattern of anomalies following cold event years was nearly opposite to that found for warm event years. The greatest composite differences in tree-growth anomalies were found in the southwestern United States and northern Mexico (see Michaelsen 1989) with opposite sign anomalies (of smaller magnitude) in Montana and southwestern Canada. Sites along the western seaboard of North America in this particular set of data showed no marked anomalies or differences in tree growth associated with the ENSO extremes.

The composite patterns of tree-growth anomalies (Fig. 11.1) were then correlated with the annual patterns of anomalies in the 65 tree-ring chronology data for each year between 1600 and 1963 (Table 11.1). These years of strong pattern representation in the data (pattern correlation equalled or exceeded 0.5) were compared to warm and cold event years of the 20th century (Kiladis and van Loon 1988) and years of El Niño events prior to 1900 (Quinn et al. 1987). As would be expected from their derivation, warm and cold event years of the 20th century were reasonably well associated with warm or cold events. Correspondence between the warm event tree-ring composite and prior El Niño years was, however, very poor. Only 4 of the 22 yr prior to 1900 were associated with an El Niño event (of at least moderate strength) in the preceding year. Some years during the 20th century with a strong representation of the warm composite pat-

Table 11.1 *Years in which the pattern correlation between the 65 tree-ring chronologies and the composite pattern from the 20th century exceeds 0.5*[a]

| | |
|---|---|
| Warm | 1618, 1651, 1701, 1710, 1717, 1718, 1720, 1721(S), 1744, 1746, 1747, 1759, 1783, 1784, 1793, 1800, 1815(S), 1816, 1848, 1858(M), 1869(M), 1889(W/M), 1903(Wm), 1905(Wm), 1919(Wm), 1920, 1924(Wm), 1926(Wm), 1933(Wm), 1941, 1949. |
| Cold | 1811, 1909(Cd), 1921(Cd), 1925(Cd), 1950(Cd). |

[a]Wm or Cd indicates warm or cold event (Kiladis and van Loon 1988) during preceding year (1900–1963) and S, M, and W indicate a strong, moderate, or weak El Niño event (Quinn et al. 1987) during preceding year (1600–1899).

tern were not associated with a prior warm event, such as 1920 and 1949. Similarly, years of known strong ENSO events were not associated with a strong representation of the warm composite tree-ring pattern; for example, for the major ENSO event of 1877/78, the pattern correlation with the warm tree-ring composite was $-0.33$. Strong representation of the cold event composite pattern was also curiously rare, being found only in 1811 prior to the 20th century. Although distinctive and different large-scale tree-ring anomaly patterns seemed to be associated with the extreme phases of ENSO in the 20th century (Fig. 11.1), these patterns of growth anomalies were not uniquely associated with ENSO extremes.

### Reconstruction of an SOI

The 65 tree-ring chronologies were used to estimate seasonal values of the SOI (see Lough and Fritts 1990). Such a reconstruction is limited by (1) the strength and variability between events of the ENSO signal in extratropical regions (e.g. Yarnal and Diaz 1986; Hamilton 1988;); (2) the nonuniqueness of the ENSO signal in North American climate (Lau 1981; Douglas et al. 1982; Simmons et al. 1983; Namias and Cayan 1984) and tree-ring data (shown above); and (3) both climatic and nonclimatic variability in the tree-ring chronologies.

The SOI reconstruction models explained between 18 and 59% of the seasonal SOI variance over the calibration periods with between 10 and 33% of the variance verified over the independent periods (Table 11.2). The majority of the

Table 11.2 *Summary of calibration (CAL) and verification (VER) statistics for the estimates of the SOI from the 65 tree-ring chronologies[a]*

| Season | $r^2$ | $N_p$ | $r^2$ | RE | $N_p$ |
|---|---|---|---|---|---|
| | CAL 1916–1963 | | VER 1867–1915 | | |
| $JJA_{t-1}$ | 0.52 | 5 | 0.04 | $-0.23$ | 1 |
| $SON_{t-1}$ | 0.44 | 5 | 0.10 | 0.05 | 2 |
| $DJF_t$ | 0.57 | 5 | 0.33 | 0.32 | 5 |
| $MAM_t$ | 0.25 | 4 | 0.30 | 0.29 | 5 |
| | CAL 1867–1915 | | VER 1916–1963 | | |
| $JJA_{t-1}$ | 0.36 | 5 | 0.11 | 0.10 | 5 |
| $SON_{t-1}$ | 0.18 | 4 | 0.14 | 0.15 | 4 |
| $DJF_t$ | 0.59 | 5 | 0.29 | 0.30 | 4 |
| $MAM_t$ | 0.45 | 5 | 0.16 | $-0.13$ | 3 |

[a] $r^2$ is correlation coefficient squared; RE is reduction of error statistic; $N_p$ is total number of statistical tests passed at 5% significance level out of a total of 5. The tests are the correlation coefficient, the correlation coefficient of the first differences, the sign test, the sign test of the first differences and the cross-product means test (see Fritts 1991; Cook 1992, this volume).

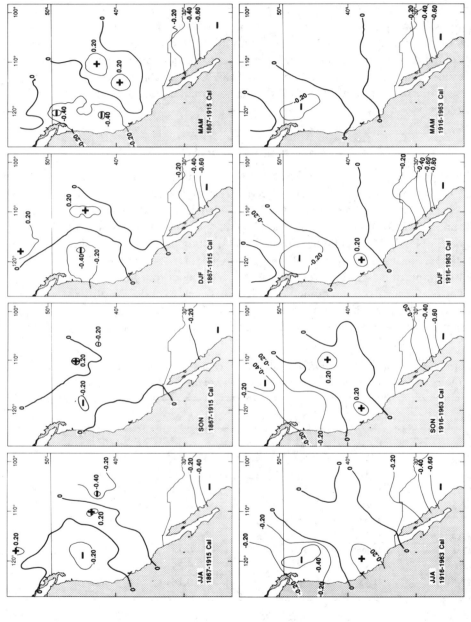

Fig. 11.2  Regression coefficients for the 65 tree-ring chronologies for the SOI reconstruction models based on the 1916–1963 and 1867–1915 calibration periods in western North America.

additional statistical tests ($N_p$) were passed and, with two exceptions, the reduction of error statistic ($RE$) was positive over the verification periods. The most reliable reconstruction over both calibration periods appeared to be the SOI for Northern Hemisphere winter (DJF).

The regression coefficients of the principal components of the predictors were multiplied by the respective eigenvectors to obtain the regression coefficients in terms of the original tree-ring chronologies for the four best seasonal SOI estimates (Fig. 11.2). The pattern of regression coefficients was similar both between seasons and for the two different calibration periods. In each case, the major contribution (i.e. highest regression weights) to the SOI estimates came from the tree-ring chronologies located in northern Mexico (cf. Fig. 11.1).

The reconstruction of the SOI, based on the 1916 to 1963 calibration period, for all four seasons is shown in Figure 11.3. Spectral analysis of the reconstruction showed significant spectral peaks in each of the four seasons at about 10, 5, and 3 yr in the period 1601 to 1962 (with 90 lags). Over three 120-yr subperiods (1601–1720, 1721–1840, and 1841–1960) all four seasons showed significant spectral peaks at about 10 yr in the earlier two periods and at the higher frequency of 3 to 4 yr in the 1841–1960 period. These results suggested that ENSO events may have been more frequent in the past 120 yr than previously.

The number of reconstructed 'low' and 'high' events for each 50-yr period was counted. An event was defined when at least two of the four seasonal estimates had a value 1 standard deviation above or below the mean. The resulting estimated number of events was 17 (1601–1650), 21 (1651–1700), 18 (1701–1750), 17 (1751–1800), 21 (1801–1850), 10 (1851–1900), and 21 (1901–1950). The number of estimated extremes appears to be fairly constant over the time period considered, with the exception of the period 1851 to 1900, which was relatively inactive when only 10 such extreme events were identified. The spectral analyses (see above), however, suggested that 1841 to 1960 was a period of more frequent

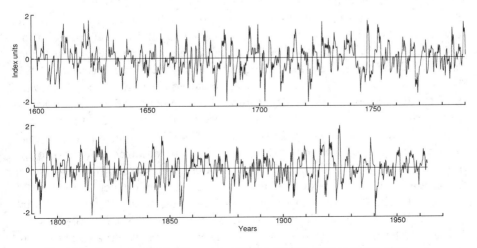

Fig. 11.3   Reconstruction of the seasonal SOI from JJA 1600 to MAM 1963.

ENSO events. This apparent contradiction may be due to the events over this period being more frequent but of smaller magnitude than for earlier time periods (see also Michaelsen 1989).

Years in which all four seasons had negative (or positive) values, with at least two of the seasons having estimated values at least 1 standard deviation below (or above) the mean were defined as the most extreme years (Table 11.3). The 20th century low-index event 1940/41 was estimated to be the third in magnitude in the 360-yr record, with stronger events reconstructed for 1855/56 and 1815/16. The 20th century high-index event 1924/25 was the most extreme in the 360-yr reconstructed record.

Comparison of the estimated SOI values associated with the independently dated strong El Niño events (Quinn et al. 1987) showed negative SOI values were well estimated in 1607, 1761, 1791, 1828, 1844, 1884, 1925, and 1940 (Table 11.4). Moderate negative values were estimated for 1877 (despite the apparent weak tree-ring signal for this year discussed above). Several years did not appear to be well represented in the SOI estimates (e.g. 1614, 1728, 1747, 1891, and 1917). Other sources of data would need to be examined to determine if these were years of El Niño occurrence with weak or unusual mid-latitude teleconnection patterns (cf. Deser and Wallace 1987). The average over all 20 strong El Niño events was not significantly different from the long-term mean.

## Discussion and conclusions

Surface climate reconstructions obtained from a network of tree-ring chronologies from arid sites in western North America (see Lough and Fritts 1985, 1990), and the chronologies themselves contain some information associated with extremes of the SOI (see also Michaelsen 1989; Michaelsen and Thompson 1992, this volume). Parts of North America experience the most consistent extratropical climate teleconnections associated with ENSO extremes. These teleconnection patterns are, however, variable between ENSO events and between ENSO and anti-ENSO extremes (e.g., Yarnal and Diaz 1986; Hamilton 1988; Ropelewski and Halpert 1990) and can occur without prior nonseasonal tropical forcing (e.g., Lau 1981; Douglas et al. 1982; Simmons et al. 1983; Namias and Cayan 1984). The teleconnection patterns with the tree-ring chronologies are also variable. Evidence that some of the ENSO signal is retained within the tree-ring chronologies and associated reconstructions, despite these limitations, is therefore encouraging. Examination of the ENSO signal in the original tree-ring chronology data shows, however, that the major contribution to the appearance of this signal is made by trees located at sites in southwestern North America and northern Mexico. Thus, a comparable reconstruction of the SOI can be obtained from trees at these sites alone (Michaelsen 1989). For future studies of ENSO relationships with tree rings, the number of chronologies from this region should be increased. More recent collections would also allow consideration of the impact of the extreme 1982/83 ENSO event on tree growth.

Table 11.3 *Years of estimated most extreme low and high SOI within the period 1601 to 1962.*

| Rank | Low Index | High Index |
|------|-----------|------------|
| 1 | 1855/56 | 1924/25 |
| 2 | 1815/16 | 1747/48 |
| 3 | 1940/41 | 1622/23 |
| 4 | 1831/32 | 1789/90 |
| 5 | 1854/55 | 1846/47 |
| 6 | 1914/15 | 1695/96 |
| 7 | 1792/93 | 1632/33 |
| 8 | 1680/81 | 1715/16 |
| 9 | 1721/22 | 1920/21 |
| 10 | 1876/77 | 1675/76 |

Table 11.4 *Reconstructed SOI values for 20 yr of strong El Niño events (Quinn et al. 1987) for MAM, JJA and SON for the year of El Niño and $DJF_{t+1}$ and $MAM_{t+1}$ for the following year.*

| Year of El Niño | MAM | JJA | SON | $DJF_{t+1}$ | $MAM_{t+1}$ | Mean |
|-----------------|------|------|------|------|------|------|
| 1607 | −0.3 | −1.1 | −0.9 | −0.6 | −0.4 | −0.7 |
| 1614 | 0.3 | 1.0 | 0.7 | 1.0 | 0.8 | 0.8 |
| 1728 | 0.7 | 0.0 | 0.1 | 0.6 | 0.4 | 0.4 |
| 1747 | −0.4 | 1.7 | 1.1 | 1.5 | 1.1 | 1.0 |
| 1761 | 0.2 | −1.0 | −0.8 | −0.4 | −0.1 | −0.4 |
| 1791 | −0.3 | −0.9 | −0.5 | −0.8 | −0.5 | −0.6 |
| 1803 | 0.4 | 0.0 | −0.6 | −0.2 | −0.1 | −0.1 |
| 1828 | 0.2 | −1.4 | −1.0 | −0.5 | −0.1 | −0.6 |
| 1844 | −0.4 | −1.8 | −1.0 | −0.6 | −0.3 | −0.8 |
| 1864 | 0.4 | 0.1 | −0.2 | −0.3 | −0.1 | 0.0 |
| 1871 | 0.2 | 0.7 | −0.1 | 0.5 | 0.4 | 0.3 |
| 1877 | −0.5 | −0.4 | −0.5 | −0.2 | −0.2 | −0.4 |
| 1884 | −0.2 | −0.8 | −0.7 | −0.3 | −0.1 | −0.4 |
| 1891 | −0.2 | 0.9 | 0.2 | 0.5 | 0.3 | 0.3 |
| 1899 | 0.2 | −0.3 | −0.7 | −0.2 | 0.0 | −0.2 |
| 1917 | 0.7 | 1.0 | 1.0 | 1.1 | 0.7 | 0.9 |
| 1925 | 1.2 | −0.7 | −0.6 | −1.0 | −0.5 | −0.3 |
| 1932 | 0.0 | −0.3 | −0.1 | −0.2 | 0.0 | −0.1 |
| 1940 | 0.1 | −2.2 | −1.1 | −1.4 | −0.8 | −1.1 |
| 1957 | 0.4 | 0.1 | −0.3 | −0.2 | 0.0 | 0.0 |
| Mean | 0.02 | −0.23 | −0.32 | 0.00 | 0.07 | |
| s.d. | 0.37 | 0.97 | 0.62 | 0.67 | 0.43 | |

The practical significance of the estimated SOI obtained from the tree-ring chronologies probably does not match its statistical significance, because we are capturing only part of the regional expression of a global signal. The SOI estimates suggest, however, some variations in the frequency of SOI extremes over the past 360 yr, or at least a modulation of the teleconnection structure over North America. Such variations need to be verified using independent sources of proxy climate information related to ENSO events.

The most reliable estimates of past ENSO events and their climatic impact will only be obtained by combining independent sources of proxy climate information that encompass as many regions of strong ENSO signal as possible. The reliability of the global ENSO signal is obviously more consistent than its regional expressions (see Diaz and Kiladis 1992, this volume). As more high-resolution proxy climate records become available (see Cleaveland et al. 1992, this volume; Cole et al. 1992, this volume) it will be possible to search for this global signal by incorporating the proxy data from several regions of strong response.

*Acknowledgments* This work was largely undertaken while I was working with Professor H. C. Fritts at the Laboratory of Tree-Ring Research, University of Arizona, Tucson. His support and contributions are gratefully acknowledged. Paul Kay and an anonymous reviewer provided helpful comments on an earlier version of this chapter.

## References

CLEAVELAND, M. K., and COOK, E. R., STAHLE, D. W., 1992: Secular variability of the Southern Oscillation detected in tree-ring data from Mexico and the Southern United States. *In* Diaz, H. F. and Markgraf, V. (eds.), *El Niño: Historical and Paleoclimatic Aspects of the Southern Oscillation*. Cambridge: Cambridge University Press, xxx–xxx.

COLE, J. E., SHEN, G. T., FAIRBANKS, R. G., and MOORE, M., 1992: Coral monitors of El Niño/Southern Oscillation dynamics across the equatorial Pacific. *In* Diaz, H. F. and Markgraf, V. (eds.), *El Niño: Historical and Paleoclimatic Aspects of the Southern Oscillation*. Cambridge: Cambridge University Press, xxx–xxx.

COOK, E. R., 1992: Using tree rings to study past El Niño/Southern Oscillation influences on climate. *In* Diaz, H. F. and Markgraf, V. (eds.), *El Niño: Historical and Paleoclimatic Aspects of the Southern Oscillation*. Cambridge: Cambridge University Press, 203–214.

DESER, C. and WALLACE, J. M., 1987: El Niño events and their relation to the Southern Oscillation: 1925–1986. *Journal of Geophysical Research*, 92: 14,189–14,196.

DIAZ, H. F. and KILADIS, G. N., 1992: Atmospheric teleconnections associated with extreme phases of the Southern Oscillation. *In* Diaz, H. F. and Markgraf, V. (eds.), *El Niño: Historical and Paleoclimatic Aspects of the Southern Oscillation*. Cambridge: Cambridge University Press, 7–28.

DOUGLAS, A. V., CAYAN, D. R., and NAMIAS, J., 1982: Large-scale changes in North Pacific and North American weather patterns in recent decades. *Monthly Weather Review*, 110: 1851–1862.

FRITTS, H. C., 1974: Relationships of ring widths in arid-site conifers to variations in monthly temperature and precipitation. *Ecological Monographs*, 44: 411–440.

FRITTS, H. C., 1976: *Tree Rings and Climate*. New York: Academic Press. 567 pp.

FRITTS, H. C., 1991: *Reconstructing Large-Scale Climatic Patterns from Tree-Ring Data: A Diagnostic Analysis*. Tucson: University of Arizona Press. 286pp.

FRITTS, H. C. and SHATZ, D. J., 1975: Selecting and characterizing tree-ring chronologies for dendroclimatic analysis. *Tree-Ring Bulletin*, 35: 31–40.

FRITTS, H. C. and LOUGH, J. M., 1985: An estimate of average annual temperature variations for North America, 1602 to 1961. *Climatic Change*, 7: 203–224.

FRITTS, H. C., LOFGREN, G. R., and GORDON, G. A., 1979: Variations in climate since 1602 as reconstructed from tree rings. *Quaternary Research*, 12: 18–46.

HAMILTON, K., 1988: A detailed examination of the extratropical response to tropical El Niño/Southern Oscillation events. *Journal of Climatology*, 8: 67–86.

KILADIS, G. N. and DIAZ, H. F., 1989: Global climatic anomalies associated with extremes of the Southern oscillation. *Journal of Climate*, 2: 1069–1090.

KILADIS, G. N. and van LOON, H., 1988: The Southern Oscillation. Part VII: Meteorological anomalies over the Indian and Pacific sectors associated with the extremes of the oscillation. *Monthly Weather Review*, 116: 120–136.

LAU, N.-C., 1981: A diagnostic study of recurrent meteorological anomalies appearing in a 15-year simulation with a GFDL general circulation model. *Monthly Weather Review*, 109: 2287–2311.

LOUGH, J. M. and FRITTS, H. C., 1985: The Southern Oscillation and tree rings: 1600–1961. *Journal of Climate and Applied Meteorology*, 24: 952–966.

LOUGH, J. M. and FRITTS, H. C., 1987: An assessment of the possible effects of volcanic eruptions on North American climate using tree-ring data, 1602 to 1900 A.D. *Climatic Change*, 10: 219–239.

LOUGH, J. M. and FRITTS, H. C., 1990: Historical aspects of El Niño/Southern Oscillation – information from tree rings. *In* Glynn, P. W. (ed.), *Global Ecological Consequences of the 1982–83 El Niño-Southern Oscillation*. Oceanography Series 52. Amsterdam: Elsevier, 285–321.

MICHAELSEN, J., 1989. Long-period fluctuations in El Niño amplitude and frequency reconstructed from tree rings. *In* Peterson, D. H. (ed.), *Aspects of Climate Variability in the Pacific and Western Americas*. Geophysical Monograph 55. Washington D.C.: American Geophysical Union, 69–74.

MICHAELSEN, J. and THOMPSON, L. G., 1992: A comparison of proxy records of El Niño/Southern Oscillation. *In* Diaz, H. F. and Markgraf, V. (eds.), *El Niño: Historical and Paleoclimatic Aspects of the Southern Oscillation*. Cambridge: Cambridge University Press, 323–348.

NAMIAS, J. and CAYAN, D. R., 1984: El Niño: the implications for forecasting. *Oceanus*, 27: 41–47.

QUINN, W. H., NEAL, W. T., and ANTUNEZ de MAYOLO, S. E., 1987: El Niño occurrences over the past four and a half centuries. *Journal of Geophysical Research*, 92: 14,449–14,461.

ROPELEWSKI, C. F. and HALPERT, M. S., 1986: North American precipitation and temperature patterns associated with the El Niño/Southern Oscillation (ENSO). *Monthly Weather Review*, 114: 2352–2362.

ROPELEWSKI, C. F. and HALPERT, M. S., 1987: Global and regional scale precipitation patterns associated with the El Niño/Southern Oscillation. *Monthly Weather Review*, 115: 1606–1626.

ROPELEWSKI, C. F. and HALPERT, M. S., 1989: Precipitation patterns associated with the high index phase of the Southern Oscillation. *Journal of Climate*, 2: 268–284.

ROPELEWSKI, C. F. and HALPERT, M. S., 1990: Uncovering North American temperature and precipitation patterns associated with the Southern Oscillation. *In* Betancourt, J. L. and MacKay, A. M. (eds.), *Proceedings of the Sixth Annual Pacific Climate (PACLIM) Workshop, March 5–8, 1989*, California Department of Water Resources, Interagency Ecological Studies Program Technical Report 23, 47–48.

ROPELEWSKI, C. F. and JONES, P. D., 1987: An extension of the Tahiti-Darwin Southern Oscillation index. *Monthly Weather Review*, 115: 1261–1265.

ROSE, M., 1983: Time domain characteristics of tree-ring chronologies and eigenvector amplitude series from western North America. Laboratory of Tree-Ring Research, University of Arizona, Tucson, *Technical Note No. 25*. 42 pp.

SIMMONS, A. J., WALLACE, J. M., and BRANSTATOR, G. W., 1983: Baratropic wave propagation and instability, and atmospheric teleconnections. *Journal of the Atmospheric Sciences*, 40: 1363–1392.

WRIGHT, P. B., 1977: The Southern Oscillation – patterns and mechanisms of the teleconnections and the persistence. Hawaii Institute of Geophysics, MIG-77–13, University of Hawaii. 107 pp.

YARNAL, B. and DIAZ, H. F., 1986: Relationships between extremes of the Southern Oscillation and the winter climate of the anglo-American Pacific coast. *Journal of Climatology*, 6: 192–219.

# 12

# Spectral properties of tree-ring data in the United States Southwest as related to El Niño/Southern Oscillation

DAVID M. MEKO

*Laboratory of Tree-Ring Research University of Arizona Tucson, Arizona 85721, U.S.A.*

## Abstract

Regional tree-ring series, 1700–1964, from the western United States were were studied by spectral and cross-spectral methods for properties relevant to reconstruction of El Niño/Southern Oscillation (ENSO). Regions studied are centered in central Oregon, northern and southern California, northwestern Wyoming, western Colorado, and Arizona-New Mexico. The analysis focuses on the Arizona-New Mexico area, designated as the Southwest. Chronologies were first weighted into regional growth series by principal components analysis to emphasize variance in common. Results indicate that regional tree-ring spectra are generally dominated by low-frequency components, with 32 to 46% of the variance in a 2.8- to 10.2-yr ENSO band. The spatial extent and geographic orientation of significant inter-regional tree-ring correlation in the ENSO band is similar to that in the entire 0 to $\pi$ frequency range. Spatial correlations with the Southwest region are highly significant (99.9% level) and in-phase in a band oriented southwest–northeast from Southern California to Colorado, and weakly significant (95% level) and in-phase with northwestern Wyoming. Spatial correlation of Southwest tree rings with other regional tree-ring series is strongest outside the ENSO band, at wavelengths 20 yr and longer. An important component of low-frequency variance in Southwest trees is a pronounced downward growth trend from the early 1900s to the 1950s. Results support the importance of a previously identified north-south opposition of moisture anomalies in the interior West as a diagnostic ENSO signal, and emphasize the importance of gaining a better understanding of multidecadal climate fluctuations and their relationship to El Niño.

## Introduction

The winter Southern Oscillation Index (SOI) is negatively correlated with December-to-August streamflow anomalies in the United States Southwest in a band extending from southern California through Arizona to southern Colorado (Cayan and Peterson 1989). Although spatial anomaly patterns of winter precipitation in the Southwest have varied greatly from one El Niño episode to another (Namias and Cayan 1984), El Niño is associated with increased fall and spring precipitation in the Arizona–New Mexico region, through mechanisms described by Andrade and Sellers (1988). Southwest tree rings contain a proven signal for interannual moisture variations (Stockton 1975; Rose et al. 1981; Smith and Stockton 1981). To the extent that these moisture variations are related to El Niño/Southern Oscillation (ENSO), Southwest tree rings are a natural source for extending statistical information on ENSO back beyond the period covered by instrumental climate data. Tree-ring data from the Southwest and the larger surrounding area of the western United States have indeed been found to yield a sufficiently strong ENSO signal to be of value in climatological studies (Lough and Fritts 1985; Michaelsen 1989; Lough 1992, this volume; and others).

Since ENSO is apparently band-limited (Julian and Chervin 1978), the spectral properties of tree-ring data are particularly relevant to ENSO reconstruction. Both Michaelsen (1989) and Lough (1992, this volume) studied spectral properties of their tree-ring reconstructions to infer features of pre-instrumental ENSO variation. Michaelsen (1989) band-pass filtered tree-ring chronologies to enhance the ENSO signal before developing a reconstruction model, while Lough (1992, this volume) filtered chronologies beforehand by prewhitening with autoregressive moving-average models.

Filtering operations necessarily impose certain limitations on the spectral properties of resulting tree-ring reconstructions. In attempting to understand the possible ENSO signal in Southwest tree rings, it is useful to examine how the spectral properties of minimally filtered series vary geographically over the Southwest and surrounding areas. The term 'minimally filtered' is used here instead of 'unfiltered' because the conversion from ring widths of individual tree cores into site chronologies generally includes removal of trend (Fritts 1976). This paper examines some spectral and cross-spectral properties of regionally averaged tree-ring data to address several questions relevant to ENSO studies, with special reference to the Southwest. First, how does the fraction of regional tree-ring variance in the ENSO frequency band vary geographically? Second, what is the spatial scale and orientation of significant inter-regional tree-ring correlation in the ENSO band? Third, is spatial correlation between regional tree-growth series in the ENSO band amplified relative to correlation at other frequencies?

## Data

The tree-ring data set consists of 63 ring-width chronologies, each covering at least the period 1700–1964 (Table 12.1). The objective in sample selection was to capture the regional drought signal by obtaining as dense a site coverage as possible over six widely separate regions of the western United States (Fig. 12.1). Regions are centered in central Oregon, northern and southern California, northwestern Wyoming, western Colorado, and Arizona-New Mexico. The region of focus is the Southwest, defined in this paper as Arizona-New Mexico.

Table 12.1 *Tree-ring chronologies used in the study.*

| No. | ID[a] | Region[b] | Species[c] | Elev.(m) | State |
|-----|-------|-----------|------------|----------|-------|
| | | | Site | | |
| 1 | 037640 | OR | PIPO | 1311 | OR |
| 2 | 754540 | OR | PSME | 1067 | OR |
| 3 | TBL640 | OR | PIPO | 1402 | OR |
| 4 | SPR661 | OR | JUOC | 1366 | OR |
| 5 | CAL661 | OR | JUOC | 1463 | OR |
| 6 | COM661 | OR | JUOC | 1486 | OR |
| 7 | FRE661 | OR | JUOC | 1494 | OR |
| 8 | HOR661 | OR | JUOC | 1146 | OR |
| 9 | LIT661 | OR | JUOC | 1650 | OR |
| 10 | LOS641 | OR | PIPO | 1374 | OR |
| 11 | 394649 | NCA | PIJE | 1905 | CA |
| 12 | LEM571 | NCA | PIJE | 1859 | CA |
| 13 | ANT571 | NCA | PIJE | 1480 | CA |
| 14 | ANT641 | NCA | PIPO | 1480 | CA |
| 15 | DON571 | NCA | PIJE | 2265 | CA |
| 16 | SNO641 | NCA | PIPO | 1731 | CA |
| 17 | SJM641 | NCA | PIPO | 1555 | CA |
| 18 | BRY639 | NCA | PIMO | 2073 | CA |
| 19 | 392649 | NCA | PIPO | 1524 | NV |
| 20 | 003529 | SCA | PSMA | 1524 | CA |
| 21 | 005520 | SCA | PSMA | 1463 | CA |
| 22 | 006649 | SCA | PIPO | 2073 | CA |
| 23 | 085651 | SCA | ABCO | 2195 | CA |
| 24 | 323599 | SCA | PIFL | 3048 | CA |
| 25 | 247529 | SCA | PSMA | 1128 | CA |
| 26 | 363429 | SCA | LIDE | 2195 | CA |
| 27 | 054549 | WY | PSME | 1433 | MT |
| 28 | GRDDF9 | WY | MIX | 1768 | MT |
| 29 | 282540 | WY | MIX | 2652 | WY |
| 30 | 283590 | WY | PIFL | 2499 | WY |
| 31 | 051549 | WY | PSME | 1981 | WY |
| 32 | 315549 | WY | PSME | 2133 | WY |

Table 12.1 (*cont.*)

| | | Site | | | |
|---|---|---|---|---|---|
| No. | ID[a] | Region[b] | Species[c] | Elev.(m) | State |
| 33 | 552590 | WY | PIFL | 2195 | WY |
| 34 | 319547 | WY | PSME | 2134 | WY |
| 35 | 421009 | CO | PIPO | 1890 | CO |
| 36 | 452009 | CO | PSME | 1981 | CO |
| 37 | 119620 | CO | PISP | 1859 | CO |
| 38 | 112549 | CO | PSME | 1951 | CO |
| 39 | 118629 | CO | PISP | 2530 | CO |
| 40 | 121549 | CO | PSME | 2164 | CO |
| 41 | 115549 | CO | PSME | 2835 | CO |
| 42 | 116549 | CO | PSME | 2530 | CO |
| 43 | 113629 | CO | PIED | 2164 | CO |
| 44 | 200000 | AZ | MIX | 1830 | AZ |
| 45 | 138549 | AZ | PSME | 2284 | AZ |
| 46 | 163620 | AZ | PISP | 1859 | AZ |
| 47 | 164640 | AZ | PIPO | 1890 | AZ |
| 48 | 177640 | AZ | PIED | 2385 | AZ |
| 49 | 033000 | AZ | PIED | 1700 | AZ |
| 50 | 042000 | AZ | PIPO | 1817 | AZ |
| 51 | 232000 | AZ | PIPO | 2073 | AZ |
| 52 | 393000 | AZ | PIED | 1780 | AZ |
| 53 | 433549 | AZ | PSME | 3200 | AZ |
| 54 | 633000 | AZ | PIED | 1661 | AZ |
| 55 | 643000 | AZ | PIED | 2621 | AZ |
| 56 | 176640 | AZ | PIPO | 2332 | NM |
| 57 | 192000 | AZ | PSME | 2697 | NM |
| 58 | 193540 | AZ | PSME | 2713 | NM |
| 59 | 484640 | AZ | PIPO | 2500 | NM |
| 60 | WCK549 | AZ | PSME | 2134 | NM |
| 61 | SGPEPA | AZ | PIPO | 2070 | AZ |
| 62 | RPPEPA | AZ | PIPO | 2320 | AZ |
| 63 | AFNEPA | AZ | PIED | 2230 | NM |

[a]Site indentification code for access from University of Arizona Tree-Ring Laboratory files.
[b]Region code as used in text and in Figure 12.1.
[c]Species code: PIPO, ponderosa pine (*Pinus ponderosa*); PIED, pinyon pine (*Pinus edulis*); PIMO, singleleaf pinyon pine (*Pinus monophylla*); PIJE, Jeffrey pine (*Pinus jeffreyi*); PIFL, limber pine (*Pinus flexilis*); JUOC, western juniper (*Juniperus occidentalis*); PSMA, bigcone spruce (*Pseudotsuga macrocarpa*); PSME, Douglas-fir (*Pseudotsuga menziesii*); ABCO, white fir (*Abies concolor*); LIDE, incense cedar (*Libocedrus decurrens*); PISP, pine, species unidentified or mixed; MIX, mix of two or more of above species.

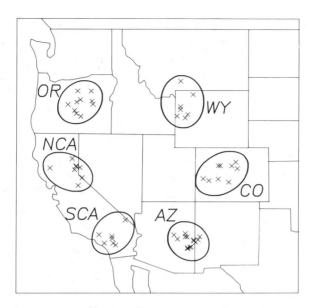

Fig. 12.1   Map of the western United States showing locations of tree-ring chronologies and grouping into regions. Region codes stand for Oregon, northern California, southern California, Wyoming, Colorado, and Arizona. One "X" may represent multiple chronlogies where sites are clustered together.

The only screening done in selection of sites was to exclude high-elevation bristlecone pine (*Pinus aristata*) and other species that may at times respond to climate variables (e.g., summer temperature anomalies) in the opposite sense of drought-sensitive chronologies (LaMarche 1974; Graumlich and Brubaker 1986). The final network includes 10 coniferous species, with greatest representation from ponderosa pine (*Pinus ponderosa*) and Douglas-fir (*Pseudotsuga menziesii*).

Individual chronologies were weighted into regional growth series by principal components analysis (Cooley and Lohnes 1971) on the correlation matrix of tree-ring indices, 1700–1964. The scores of the first component of growth in a region comprise a weighted tree-ring series in which the weights have been selected to maximize the variance in common at all sites. This weighting scheme emphasizes variance due to large-scale environmental factors, such as fluctuations in climate. The percentage of variance explained by the first component in each region is listed below:

| | | |
|---|---|---|
| OR | Oregon | 50 |
| NCA | Northern California | 52 |
| SCA | Southern California | 62 |
| WY | Wyoming | 51 |
| CO | Colorado | 38 |
| AZ | Arizona | 58 |

With the exception of the Colorado region, the first component explains more than half the variance in regional growth. The time series of first-component scores, standardized to zero mean and unit variance, are the subjects of subsequent analysis in this paper. For brevity, these series will be referred to as OR, NCA, etc.

## Methods

Spectra of tree-ring series were estimated by smoothing the periodogram with a succession of Daniell filters (Bloomfield 1976). For a given bandwidth of spectral window, the Daniell filter is preferred over alternative filters for the stability of the spectral estimate (Bloomfield 1976). Each regional series was prepared for spectral analysis by tapering 5% of the data on each end with a raised-cosine filter (Hamming 1983: 102), and padding with zeros to length 512 yr. The periodogram was computed by a fast Fourier transform algorithm, and then smoothed successively by 7-weight and 13-weight Daniell filters to produce the estimated spectrum. Approximate 95% confidence limits for spectral estimates were calculated based on the assumption of a chi-squared distribution (Bloomfield 1976: 196). A null continuum was computed by smoothing the periodogram successively with much broader filters (length 33 and 77) than used for the spectral estimates. The filter lengths given above were arrived at by trial and error – varying the lengths until the spectrum and null continuum captured the main features of the periodogram.

Cross-spectra were also estimated by periodogram smoothing, although with rectangular rather than Daniell filters. Series were not padded or tapered beforehand. Sample length is therefore 265 yr, and Fourier frequencies are $0/265, 1/265, 2/265, \ldots, 132/265$ cycles $yr^{-1}$. A single rectangular filter of length 69 was selected for the significance test of inter-regional squared coherency in the ENSO band. This filter centered on frequency 0.226 $yr^{-1}$ brackets the wavelengths 2.8–10.2 yr. A narrower filter, length 15, was used to smooth the cross periodogram to display the variation in estimated coherency over the 0 to $\pi$ frequency range

The null hypothesis of zero squared coherency was tested by a method described by Bloomfield (1976: 227). Let $s_{X,Y}(\omega)^2$ be the sample squared-coherency and $r_{X,Y}(\omega)$ be the theoretical coherency. If $r_{X,Y}(\omega) = 0$, the probability of a sample squared-coherency less than a given level $\sigma(p)^2$ is $p$, where

$$\sigma(p)^2 = 1 - (1 - p)^{g^2/(1 - g^2)} \tag{1}$$

If the time series has not been tapered or padded, the quantity $g^2$ in the above equation is simply equal to the sum of the squares of the weights of the function used to smooth the periodogram. For example, for a 5-weight rectangular filter, each weight is 0.2, the sum of squares of weights equals 0.2, and the 95% and 99.9999% confidence limits for squared-coherency are

$$\sigma(0.95)^2 = -(1 - 0.95)^{0.2/(1 - 0.2)} = 0.53 \qquad (2)$$

$$\sigma(0.999999)^2 = 1 - (1 - 0.999999)^{0.2/(1 - 0.2)} = 0.97 \qquad (3)$$

These confidence limits apply where the significance of a single squared-coherency value is to be tested, and where the frequency band of interest has been specified before the cross-spectral analysis.

## Results

The Southwest tree-ring series, AZ (Fig. 12.1), is characterized by interannual variations imbedded within 'trends' lasting several decades (Fig. 12.2). The most pronounced of these trends is a gradual decline in growth from the early 1900s to the 1950s. The change in mean level due to this trend is about as large as the interannual variation from a very low growth to high growth value at a given time within the trend. In contrast to this steady but gradual transition is the sharp turnabout from low growth to high growth about 1905.

As shown by the 'Xs' in Figure 12.2, ENSO episodes have occurred in high and low parts of the modern tree-ring 'trend,' in the extreme low-growth period around the turn of the century, and in the period of relatively moderate growth fluctuations in the late 19th century. The low-frequency variations in Southwest tree growth might be expected to hinder detection of an ENSO signal confined to higher frequencies. Because AZ is an average over many sites, it is unlikely that the observed low-frequency variations are artifacts of the biological system or method of data processing.

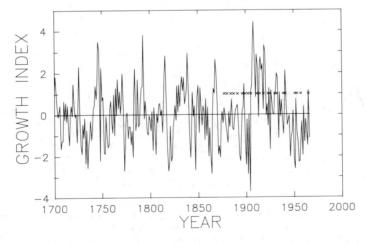

Fig. 12.2  Time series plot of tree-ring series AZ, and occurrences of moderate or strong ENSO events since 1728. Plotted series is an eigenvector score standardized to zero mean and unit variance. (ENSO events (first year only for multiyear events) after Rasmusson 1984.)

### Estimated spectra

The spectrum for AZ is clearly low-frequency (Fig. 12.3). The underlying null
continuum based on a greatly smoothed periodogram slopes gradually upward
toward lower frequencies beginning at a wavelength of about 3 yr. The slope
steepens considerably at about 7 yr. The estimated spectrum itself is relatively
flat from wavelength 2 to 5 yr, except for a pronounced peak at 4.2 yr. The major
peak in the spectrum is at 22.3 yr, a period previously associated with relatively
high variance in the waxing and waning of drought area in the western United
States (Mitchell et al. 1979).

The occurrence of a major peak at long wavelengths (period greater than 21 yr)
is a common feature of all six regional spectra (Fig. 12.4). The period 57 yr
features prominently in 4 of the 6 spectra – as the major peak in OR, NCA, and
WY, and as a minor peak in AZ. A peak near 57 yr has previously been reported
in regionally grouped tree-ring data from the western edge of the northern Great
Plains (Meko 1982). The regional spectra in Figure 12.4 typically change from
relatively flat to upward sloping at a wavelength near 5 yr. The exceptions are
regions SCA and CO, where deep spectral troughs occur at wavelengths 10 yr or
longer.

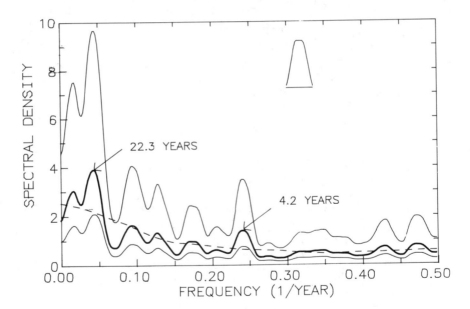

Fig. 12.3   Estimated spectral density of tree-ring series AZ. Spectrum was estimated
by successively smoothing the raw periodogram with Daniell filters of length 7 and 13.
Also shown are (1) 95% confidence bands based on the chisquared distribution, (2) a
null continuum, estimated by smoothing the raw periodogram with Daniell filters of
length 33 and 77, (3) an inset showing the shape of the smoothing filter used to estimate
the spectrum, and (4) periods of notable peaks in the estimated spectrum. Sample
length was 265 yr. The series was tapered (5% at each end) and padded with zeros to
length 512 before computing the periodogram.

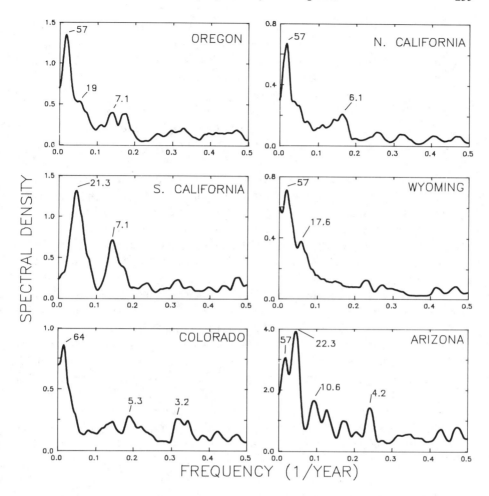

Fig. 12.4 Estimated spectra of six regional tree-ring series. Computation methods were identical to those for the spectrum in Figure 12.3. Bottom right curve (for AZ) is identical to spectrum plotted in Figure 12.3.

The ENSO band as defined here (2.8–10.2 yr) is so broad that it contains a substantial part of the tree-ring variance despite the dominance of the spectrum by lower frequencies. The percentages of total variance in the ENSO band by region are as follows:

OR    38
NCA   40
SCA   41
WY    32
CO    46
AZ    37

The corresponding percentage expected from a white spectrum is the ratio of the number of periodogram ordinates in ENSO band to the total number of periodogram estimates. For the 512-yr padded series, this percentage is 133/256, or 52%. Not surprisingly, all six regional series are deficient in variance in the ENSO band compared to white noise.

The interpretation of spectra is incomplete without some mention of the subject of aliasing. Aliasing refers to the folding of higher frequencies across the Nyquist frequency onto the interval $[0, \pi]$ (Bloomfield 1976: 27). For the series in Figures 12.3 and 12.4, with a Nyquist frequency corresponding to a period of 2 yr, the practical concern is that a spectral feature on $[0, \pi]$ may arise from tree-ring variations with a period less than 2 yr. Aliasing is usually a problem when a time series contains important variability at frequencies higher than those represented by the sampling interval. Aliasing of tree-ring variations due to choice of sampling interval is minimal, since the sample interval (1 yr) is the same as the time increment defining the ring-width variable. The minimum-length period observable in such a time series is 2 yr.

Usually, however, tree-ring spectra are studied not for their information on tree-growth variations per se, but for their information on the environmental variables governing growth. An indirect form of aliasing could plausibly arise from the particular way in which the tree ring 'samples' an environmental variable. For example, consider a hypothetical example of a tree whose cambial growth is governed by spring precipitation, and a climate characterized by an 18-mo periodicity in departures of monthly precipitation from the long-term mean. The spring growth period would repeat its phase relationship to the 18-mo precipitation cycle every 36 mo, possibly resulting in a tree-ring spectral peak at 3 yr. The peak would be real in that it represents a 3-yr periodicity in tree growth, but misleading in that the causative phenomenon has a periodicity of 18 mo. The relevance of such indirect aliasing to the spectra of Figures 3 and 4 is impossible to judge from analysis of annual tree-ring indices alone. It should be noted, however, that variance in monthly sea surface temperatures and SOI is relatively small at wavelengths shorter than 2 yr (Rasmusson and Carpenter 1982; Quiroz 1983).

### Squared coherency

Results of a cross-spectral analysis between the series AZ and other regional tree-ring series are summarized in the form of a map of phase-coherency diagrams in Figure 12.5. Squared coherency (*C*) is proportional to the length of the plotted bar. Circles are plotted at radii corresponding to the 95% and 99.9999% significance levels of squared coherency. For the particular combination of length of time series and width of frequency band, these confidence levels correspond to squared coherency values of 0.043 and 0.183, respectively. The phase between series AZ and other series is proportional to the angle of the bar from the vertical. The phase difference describes the lag of variations at the specified

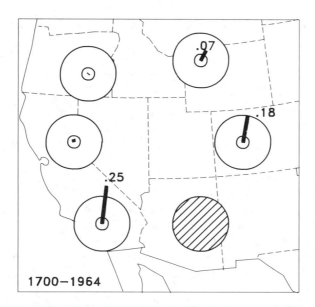

Fig. 12.5   Map of the western United States graphically showing squared coherency and phase for the ENSO band between AZ and other regional tree-ring series. Inner circle corresponds to 95% significance level on squared coherency, outer circle to 99.9999% significance level. Series were not padded or tapered in the cross-spectral analysis, and a rectangular window with 138 degrees of freedom (69 weights) was used in smoothing the raw cross-periodogram. Angle of bar from vertical give phase difference, as described in text.

frequency band in AZ from variations in the paired series. The central frequency of the 2.8- to 10.2-yr ENSO band corresponds to a period of 4.4 yr. Half a cycle corresponds to a phase difference of 2.2 yr. If the arrow on the diagram points north, the variations in the ENSO band are exactly in phase; if the arrow points east variations in AZ lag those in the paired series by a quarter of a cycle, or 1.1 yr.

Squared coherency ($C$) clearly diminishes with distance from AZ, and the high values have a southwest/northeast axis of orientation. Confidence bands (not shown) around the estimated phase widen as squared-coherency decreases, such that the slight departures from the vertical of the bars for SCA, CO, and WY are not significant at the 95% level. AZ is highly coherent and in-phase with SCA and CO, and weakly coherent and in-phase with WY. Coherency is essential zero, however, with the northwestern regions (OR and NCA) of the study area.

Squared-coherency ($C$) is analogous to a (squared) correlation coefficient for a specific frequency band. The squared-coherency values in Figure 12.5 were compared with product moment correlation coefficients (Table 12.2) for evidence of frequency-dependence in the strength of the relationship between regional tree-ring series. The same three regions reach 95% significance with AZ in both squared coherency and correlation, indicating that correlation is not amplified in the ENSO band.

Table 12.2 *Product-moment correlation coefficient between AZ and other regional tree-ring series.*

| Region[a] | r[b] | $r_1^c$ | N$_*^d$ | 95% CL |
|---|---|---|---|---|
| OR | 0.01 | 0.40 | 114 | 0.19 |
| NCA | 0.15 | 0.47 | 96 | 0.20 |
| SCA | 0.57* | 0.41 | 111 | 0.19 |
| WY | 0.30* | 0.54 | 79 | 0.22 |
| CO | 0.52* | 0.27 | 152 | 0.19 |

[a]Follows naming convention in text.
[b]Correlation coefficient between AZ and other regional series. Adjusted for effective sample size, N$_*$. Asterisk marks 95% significance.
[c]First-order autocorrelation coefficient.
[d]Sample size adjusted for autocorrelation, after Mitchell et al. (1966). Unadjusted sample size is 265 yr for all series.

Frequency-dependence in inter-regional tree-ring correlation can be more generally evaluated by cross-spectral analysis using a much narrower smoothing window than in the preceding analysis. Squared coherency between AZ and regional series SCA, CO, and OR over the entire 0 to $\pi$ frequency range is plotted in Figure 12.6. The high correlation of AZ with SCA and CO comes mainly from wavelengths shorter than 3.3 yr and longer than 15 yr. AZ is more highly correlated within the ENSO band (2.8–10.2 yr) with SCA than with CO. Wavelengths outside the ENSO band – shorter than 2.8 yr – contribute greatly to the high correlation between AZ and CO. Consistent with the extremely low correlation coefficient ($r = 0.01$), coherency between AZ and OR is low over the entire frequency range, with the exception of one ordinate that reaches 95% significance between 4 and 5 yr.

Only at longer wavelengths (greater than 20 yr) is AZ highly coherent ($C > 0.5$) with *both* CO and SCA. The high $C$ values are practically significant because the variance of the individual series is also concentrated at these longer wavelengths (Fig. 12.4). The consistency across regions lends support to the idea of a climatic source to low-frequency tree-ring variations.

## Discussion and conclusions

'ENSO-band' variance as defined in this chapter does not distinguish between variance associated with ENSO events and with other climatic (or nonclimatic) influences in the wavelength band 2.8 to 10.2 yr. How consistent are the coherency-phase plots in Figure 12.5 with previous findings of contrasts in regional moisture anomalies associated with ENSO? Two useful maps for this evaluation are the correlations of December-August streamflow with SOI (Fig. 8 in Cayan and Peterson 1989), and the difference between warm and cold event

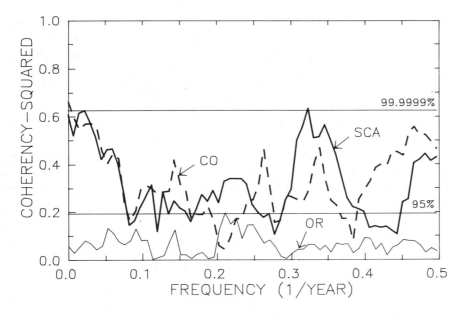

Fig. 12.6  Plots of squared coherency over the full frequency range between tree-ring series AZ and series OR, SCA and CO. Cross-spectral procedure as described in Figure 12.5, except that the window used to smooth the cross periodogram has 15 weights (30 degrees of freedom).

averages of standardized tree-ring anomalies (Fig. 11.1 in Lough 1992, this volume). Both maps indicate that ENSO is characterized by opposing moisture anomalies in Wyoming-Montana and Arizona-New Mexico. The ENSO-band tree-ring correlation between WY and AZ, on the other hand, is significant and positive (Fig. 12.5). Opposing tree-growth anomalies in the two areas might therefore be a distinctive signature for ENSO. On the other hand, the positive ENSO-related correlation between moisture anomalies in southern California and Arizona-New Mexico (Lough and Fritts 1985; Cayan and Peterson 1989) may not be a particularly useful ENSO signature, since high correlation between AZ and SCA is characteristic of the 2.8- to 10.2-yr wavelength band without classification as to ENSO (Fig. 12.5).

The results presented here indicate that Southwest tree rings are highly correlated in the ENSO band with tree rings from southern California to Colorado. The correlation decreases rapidly to the northwest, and is essentially zero with trees from Oregon. High correlation is not restricted to the ENSO band, however, but is also a feature of shorter and longer wavelengths. The high interregional coherency at wavelengths longer than 20 yr is especially interesting, as it implies that multidecadal trends in regional tree-growth anomalies reflect climate variation rather than some artifact such as inadequate detrending in standardizing ring widths. This conclusion is supported by previous comparisons of climatic and tree-ring series in individual regions. For example, the modern trend in Southwest tree-ring series (Fig. 12.2) is a strong feature of precipitation and streamflow data in Arizona (Sellers 1960; Meko and Stockton 1984).

Low-frequency variance obviously complicates the detection of an ENSO signal in Southwest tree-ring data. High-pass or band-pass filtering of tree-ring data will emphasize the proper frequency band for ENSO (Michaelsen 1989), but at the cost of losing considerable climatic variance. Filtering out the low frequencies before reconstruction also eliminates the chance of utilizing the full range of variability of tree-ring data to question the assumption that ENSO is band-limited. Recent studies suggest that ENSO may vary appreciably at wavelengths considerably longer than a decade (Enfield 1988; Anderson 1990; Diaz and Pulwarty 1992, this volume). Effectively extracting the low-frequency climate information in tree rings is essential to testing hypotheses of low-frequency ENSO behavior.

*Acknowledgment* This work was supported by the National Science Foundation, grant ATM-88-14675.

## References

ANDERSON, R. Y., 1990: Solar cycle modulations of ENSO: A possible source of climatic change. *In* Betancourt J. L. and MacKay, A. M. (eds.), *Proceedings, Sixth Annual Pacific Climate (PACLIM)* Workshop, March 5-8, 1989. California Department of Water Resources Interagency Ecological Studies Program Technical Report 23, 77-82 .

ANDRADE, E.R., Jr., and SELLERS, W.D., 1988: El Niño and its effect on precipitation in Arizona and western New Mexico. *Journal of Climatology*, 8: 403-410. Bloomfield, P., 1976: *Fourier Analysis of Time Series: An Introduction*. New York: Wiley. 258 pp.

BLOOMFIELD, P., 1976: *Fourier Analysis of Time Series: An Introduction*. New York: Wiley. 258 pp.

CAYAN, D. R. and PETERSON, D. H., 1989: The influence of North Pacific atmospheric circulation on streamflow in the West. *In* Peterson, D. H. (ed.), *Aspects of Climate Variability in the Pacific and Western Americas*. Geophysical Monograph 55. Washington, D.C.: American Geophysical Union, 375-397.

COOLEY, W. W. and LOHNES, P. R., 1971: *Multivariate Data Analysis*. New York: Wiley. 364 pp.

DIAZ, H.F. and PULWARTY, R. S., 1992: A comparison of Southern Oscillation and El Niño signals in the Tropics. *In* Diaz, H. F. and Markgraf, V. (eds.), *El Niño: Historical and Paleoclimatic Aspects of the Southern Oscillation*. Cambridge: Cambridge University Press, 175-192.

ENFIELD, D. B., 1988: Is El Niño becoming more common? *Oceanography*, 1: 23-27.

FRITTS, H. C., 1976: *Tree Rings and Climate*. London and New York: Academic Press. 567 pp.

GRAUMLICH, L. J. and BRUBAKER, L.B., 1986: Reconstruction of annual temperature (1590-1979) for Longmire, Washington, derived from tree rings. *Quaternary Research*, 25: 223-234.

HAMMING, R. W., 1983: *Digital Filters*. 2nd ed. Englewood Cliffs, N. J.: Prentice-Hall. 257 pp.

JULIAN, P. R. and CHERVIN, R. M., 1978: A study of the Southern Oscillation and Walker circulation phenomenon. *Monthly Weather Review*, 106: 1433–1451.

LAMARCHE, V. C., 1974: Frequency-dependent relationships between tree-ring series along an ecological gradient and some dendroclimatic implications. *Tree-Ring Bulletin*, 34: 1–20.

LOUGH, J. M. and FRITTS H. C., 1985: The Southern Oscillation and tree rings: 1600–1961. *Journal of Climate and Applied Meteorology*, 24: 952–966.

LOUGH, J. M., 1992: An index of the Southern Oscillation reconstructed from western North American tree-ring chronologies. *In* Diaz, H. F., and Markgraf, V. (eds.). *El Niño: Historical and Paleoclimatic Aspects of the Southern Oscillation*. Cambridge: Cambridge University Press, 215–226.

MEKO, D. M., 1982: Drought history in the western Great Plains from tree rings. *In* Johnson, A. I. and Clark, R. A. (eds.), *Proceedings of the International Symposium on Hydrometeorology*. Bethesda, Md.: American Water Resources Association, 321–326.

MEKO, D. M. and STOCKTON, C. W., 1984: Secular variations in streamflow in the western United States. *Journal of Climate and Applied Meteorology*, 23: 889–897.

MICHAELSEN, J., 1989: Long-period fluctuations in El Niño amplitude and frequency reconstructed from tree-rings. *In* Peterson, D. H. (ed.), *Aspects of Climate Variability in the Pacific and Western Americas*. Geophysical Monograph 55. Washington, D.C.: American Geophysical Union, 69–74.

MITCHELL, J. M., Jr., DZERDZEEVSKII, B., FLOHN, H., HOFMEYR, W. R., LAMB, H. H., RAO, K. N., and WALLEN, C. C., 1966: *Climatic Change*. World Meteorological Organisation Technical Note 79. Geneva: World Meteorological Organisation. 88 pp.

MITCHELL, J. M., Jr., STOCKTON, C. W., and MEKO, D. M., 1979: Evidence of a 22-year rhythm of drought in the western United States related to the Hale Solar Cycle since the 17th Century. *In* McCormac, B. M. and Seliga T. A. (eds.); *Solarterrestrial Influences on Weather and Climate*. Dordrecht: Reidel, 125–143.

NAMIAS, J. and CAYAN, D. R., 1984: El Niño: the implications for forecasting. *Oceanus*, 27: 41–47.

QUIROZ, R. S., 1983: Relationships among the stratospheric and tropospheric zonal flows and the Southern Oscillation. *Monthly Weather Review*, 111: 143–154.

RASMUSSON, E. M., 1984: El Niño: The ocean/atmosphere connection. *Oceanus*, 27: 5–12.

RASMUSSON, E. M. and CARPENTER, T. H., 1982: Variations in tropical sea surface temperature and surface wind fields associated with the Southern Oscillation/El Niño. *Monthly Weather Review*, 110: 354–384.

ROSE, M. R., DEAN, J. S., and ROBINSON, W. B., 1981: *The Past Climate of Arroyo Hondo, New Mexico, Reconstructed from Tree Rings*. Arroyo Hondo Archaeological Series, Vol. 4. Santa Fe, New Mexico: School of American Research Press. 253 pp.

SELLERS, W. D., 1960: Precipitation trends in Arizona and western New Mexico. *In Proceedings 28th Annual Western Snow Conference*, Santa Fe, New Mexico: U.S. Soil Conservation Service.

SMITH, L. P. and STOCKTON, C. W., 1981: Reconstructed streamflow for the Salt and Verde Rivers from tree-ring data. *Water Resources Bulletin*, 16(6): 939–947.

STOCKTON, C. W., 1975: *Long-term Streamflow Records Reconstructed from Tree Rings*. Papers of the Laboratory of Tree-Ring Research, Number 5. Tucson, Ariz.: The University of Arizona Press. 111 pp.

# 13

# A tree-ring reconstruction of New Mexico winter precipitation and its relation to El Niño/Southern Oscillation events

ROSANNE D. D'ARRIGO AND GORDON C. JACOBY

*Tree-Ring Laboratory Lamont-Doherty Geological Observatory Palisades, New York 10964, U.S.A.*

## Abstract

A 1000-yr reconstruction of November-May precipitation for northern New Mexico is developed for A.D. 985–1970 based on six millennium-long tree-ring records from moisture-sensitive coniferous sites. Two of the most prolonged and severe droughts occurred during this century: in the 1890s to 1904 and the 1950s to early 1960s. These were exceeded in length and severity only once in the 1000-yr reconstructed record, in the late 1500s, and almost equaled in the early 1200s. The five decadal intervals of greatest drought severity are, in descending order, 1577–1598, 1955–1964, 1895–1904, 1217–1226, and 1778–1787. The wettest periods are 1835–1849, 1905–1928, 1429–1440, 1609–1623, and 1487–1498. The most abrupt shift from severe sustained drought to extremely high precipitation occurred in the 1890s to early 1900s. Dry conditions are indicated for part of the 1400–1600s period around the onset of the Little Ice Age, with near-average conditions reconstructed for the Medieval Warm Period.

The chronologies and reconstruction show correspondence with El Niño/Southern Oscillation (ENSO)-related indices, including Wright's Southern Oscillation Index (SOI). There appears to be the potential for modeling the amplitude and frequency of ENSO in the southwestern United States over the past 1000 yr or more, particularly by integrating tree-ring data from this region with other sources of high-resolution paleoclimatic information. As an example of this potential we show an estimate of an ENSO-related sea surface temperature (SST) index based on combined tree-ring data from the southwestern United States and Java, Indonesia.

## Introduction

The demand for water supplies in the semiarid southwestern United States is of increasing concern (U.S. Water Resources Council 1978; Rind et al. 1990). Climatic variations (such as drought and floods) have great economic and social impact in such regions. There is thus a need to understand the past natural variability of precipitation if we are to predict the future effects of perturbations such as increasing greenhouse gases on climate. Available meteorological records, which are only about a century in length, are not adequate to assess the history of past precipitation variability. For this purpose, proxy and extended historical records are needed.

Very long tree-ring records have been developed from moisture-sensitive conifers in the southwestern United States. Tree-ring data have proven useful in studies of past variations in precipitation-related variables in this general area (e.g. Fritts 1976; Stockton and Jacoby 1976; Meko et al. 1980; Rose et al. 1981, 1982; Swetnam and Betancourt 1990). The tree-ring chronologies published by Dean and Robinson (1978) for New Mexico, Arizona, Colorado, and Utah, derived by merging archaeological (from nonliving wood samples) tree-ring data with data from living trees in the same region, provide continuous time series extending over 1000 yr. The trees from these semiarid sites grow wider rings in response to greater precipitation (primarily during the winter or cooler season) and narrow rings in response to drought.

In this chapter, six of the Dean and Robinson (1978) chronologies were used to reconstruct an approximate 1000-yr-long record (A.D. 985–1970) of November through May precipitation for the Northwestern Plateau Climatic Division in New Mexico. The tree-ring data statistically explain 62% of the variation in recorded precipitation during the common period of tree-ring and meteorological records. The climatic estimates, based on regression analysis, were validated through statistical and graphical comparisons between the estimated and recorded climatic data, and the regression equations were found to be stable in time. This reconstruction is thus considered to be a reliable representation of longer-term precipitation and drought for the region. Detailed procedures for the reconstruction are described in the following sections. There also appears to be a relationship between these reconstructed precipitation variations and ENSO.

## Climate of the study area

The southwestern United States is a semiarid region with spatial diversity in precipitation (e.g. Klein and Bloom 1987; Dean 1988). Annual precipitation is strongly influenced by elevation, with the driest areas being the deserts of southwestern Arizona and southern California and the plateaus of northeastern Arizona and northwestern New Mexico. The climatic regime of some areas of the southwest United States generally has a bimodal precipitation pattern (Hsu and Wallace 1976; Dean 1988), with seasonal distribution of precipitation determined mainly by the region's location between two major sources of atmospheric moisture, the Pacific Ocean and the Gulf of Mexico. Rain that falls during convective storms in summer comes largely from the Gulf of Mexico, but also from the Pacific. By contrast winter precipitation derives primarily from large-scale frontal storms originating from the Pacific and moving eastward across the region. Precipitation maxima occur in summer and winter contrasting with intervals of reduced rainfall in fall and spring. The winter component, which accounts for most of the annual variability, is strongest towards the west and north, and the summer component towards the southeast. For the Northwestern Plateau Climatic Division of New Mexico the greatest amount of precipitation per month falls during the summer months of July and August, with June and October being transitional months between the two seasonal climatic regimes. However the longer winter season (which lasts from about November to May) accounts for about an equal fraction of the annual precipitation.

## Data

The six tree-ring chronologies (site information in Table 13.1 and site locations in Fig. 13.1) are based on annual ring width measurements from two species of pine (*Pinus edulis* Engelm. [pinyon pine] and *Pinus ponderosa* Dougl. [ponderosa pine]) and Douglas-fir (*Pseudotsuga menziesii* [Mirb.] Franco) growing at elevations of about 2000–2100 m. The chronologies were standardized using negative exponential or straight-line curve fits (Rose, M. R., pers. comm. 1985) in

Table 13.1 *Location and species of tree-ring series used in this paper with species and locations (from Dean and Robinson, 1978). Chronologies comprise both living and archaeological data.*

| Site | Species | Lat (North) | Long (West) | Elev(m) |
|---|---|---|---|---|
| 1. Mesa Verde | Douglas-fir *Pseudotsuga menziesii* | 37° 10′ | 108° 30′ | 2075 |
| 2. Cibola | Ponderosa pine *Pinus ponderosa* | 35° 20′ | 108° 30′ | 2075 |
| 3. Cebolleta Mesa | Pinyon pine *Pinus edulis* | 34° 50′ | 107° 45′ | 2075 |
| 4. Gobernador | Pinyon pine *Pinus edulis* | 36° 45′ | 107° 30′ | 2010 |
| 5. Jemez Mtns. | Pinyon pine *Pinus edulis* | 35° 40′ | 106° 30′ | 2010 |
| 6. Chama Valley | Douglas-fir *Pseudotsuga menziesii* | 36° 17′ | 106° 32′ | 2045 |

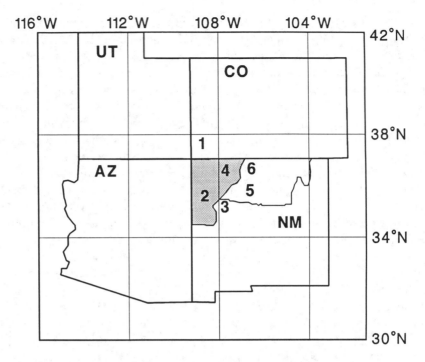

Fig. 13.1   Location map of tree-ring sites used in this study (after Dean and Robinson, 1978). Stippled area is the Northwestern Plateau Climatic Division, outlined area is Northern Mountains Climatic Division. Numbers are for sites in Table 13.1.

order to remove growth-trend or age effects while preserving climatic variation (Fritts 1976).

The chronologies may be considered to be essentially homogenous in time. In some cases the archaeological data were derived from a species different from the living tree data. However, as discussed in Dean and Robinson (1978), the ring width variations of the different records cross-date for the overlapping period and have similar chronology statistics for the earlier (archaeological) and later (living tree-based) time intervals. It was decided for this study to limit the chronology data set to the past 1000 yr in order to avoid any inhomogeneities due, for example, to inadequate sample size in the earlier data.

Divisional climatic records were obtained from the National Climatic Data Center in Asheville, North Carolina. The climate divisions are generally areas of climatic homogeneity and are useful for integrating large-scale climatic variability. The tree-ring sites are located within or adjacent to the Northwestern Plateau Climatic Division of New Mexico (stippled area, Fig. 13.1). The adjacent division of the Northern Mountains (outlined area, Fig. 13.1) extends too far to the east of the sites to justify its inclusion in this analysis. The extended divisional Northwestern Plateau climatic record is composed of two distinct segments. From 1931 to the present there are actual observations based on meteorological recording stations. The 1895 to 1930 record is based on regression estimates derived from state averages since there were few recording stations in many divisions (Karl et al. 1983). Therefore the record for 1931 to present is likely to be a more precise time series of divisional climatic variation than the 1895–1930 record. However, as indicated below the statistical analyses appear to be stable over the two time segments.

## Methods and results

Simple correlation analysis between the chronologies and meteorological records was used to determine the strongest and most consistent relationships with precipitation and the most appropriate season to be reconstructed in regression analysis (e.g. Cook and Jacoby 1983). Correlations were calculated for a 24-mo dendroclimatic year including the prior calendar year and the current year of radial growth. The strongest correlations (significant at the 95% level, using a one-tailed test) were found during the months of November through May indicating that an average of these months was the most appropriate predictand for reconstruction with this data set.

Since bud formation, storage of photosynthates, formation of growth hormones, and other growth processes take place in the year prior to the season of radial growth, the preceding climatic variations can influence ring width of the growth year. Winter precipitation, often occurring as snow, provides recharge to soil moisture that can benefit the trees during the next growing season. Also, since all of the tree species included here are conifers and the temperatures are high enough, photosynthesis can occur in the evergreen needles over a much longer period than the radial growth period (Fritts 1976). It thus seems likely, that the trees can respond to variations in precipitation occurring in the months just prior to the season of actual cambial cell division and hence radial growth.

Before developing a regression equation to reconstruct precipitation, two data transformations were performed. The data were first prewhitened (autoregressive or AR order 2–5) to reduce tree-ring variation not due to climate. This process attempts to remove biological persistence, considered not to be due to climate, from the tree-ring data (Meko 1981), and as discussed by Cook (1992, this volume). The second transformation involved processing the six chronologies through principal components analysis (PCA) to produce uncorrelated, predictor time series which enhance the common variance in the original chronologies (e.g. Fritts 1976; Kutzbach and Guetter 1980).

The first two principal components from the PCA (based on the common interval from A.D. 985–1971) were used as predictors. The first principal component time series contained 71% of the total variance of the six tree-ring chronologies and the second 10%, for a total of 81%. Only one year and the following year's values were used in the equation. The prewhitening should have removed most biological persistence, but precipitation can directly affect the following

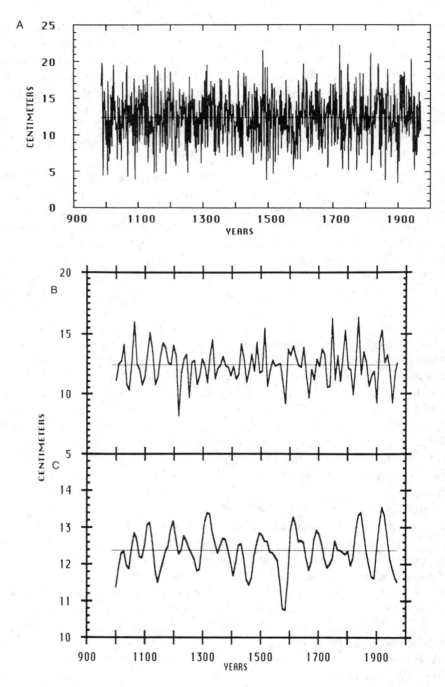

Fig. 13.2  Reconstruction of November through May precipitation for the Northwestern Plateau Climatic Division of New Mexico for A.D. 985–1970. **A.** Year-to-year values. **B.** With 10-yr weighting obtained using a robust smoothing algorithm to emphasize lower-frequency variation. **C.** With 50-yr weighting as in B. The mean lines are also shown.

year's growth (as noted above) so it is reasonable to include the next year's value as a predictor.

The first two principal components and forward lags (representing tree growth in year $t$, the growth period immediately following the November–May season; and year $t + 1$ representing tree growth in the following year) were used to predict November to May precipitation. The multiple linear regression analysis was based on the interval from 1896–1970. The reconstruction extends from A.D. 985–1970 (Figs. 13.2A–C). Figure 13.3 presents the recorded and estimated precipitation values for the 1896–1970 calibration period, and Table 13.2 the calibration and verification statistics (see Cook 1992, this volume). For the full calibration period from 1896–1970 the variance explained or $ar^2$ (adjusted for degrees of freedom) is 62% (Draper and Smith 1981). Calibration and verification tests of the reconstruction model were then made on the 1896–1930 and 1931–1970 time intervals (Table 13.2). For the 1896–1930 calibration period, the $ar^2$ is 71% and the $RE$ (reduction of error) statistic is 0.55. Any positive RE value indicates significant predictive skill (Gordon and LeDuc 1981). When the intervals are reversed the $ar^2$ for the 1931–1970 calibration period is 0.58 and the $RE$ is 0.50. As further validation, the Spearman rank correlation coefficients, sign test, and cross-product means tests were also performed (Fritts 1976). The results of these tests are all highly significant (Table 13.2). The regression weights for the three models are stable in time (Table 13.3). The results therefore indicate that these models are consistent in structure and have high levels of explained variance.

Variance spectral analysis was performed on the reconstructed time series to test for possible periodicities (Jenkins and Watts 1968). Significant spectral

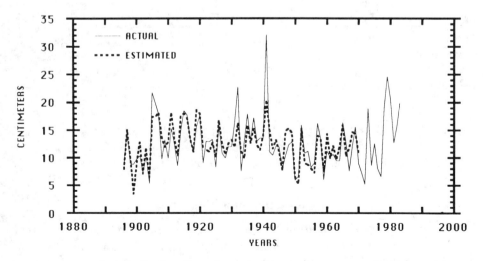

Fig. 13.3 Actual and estimated November through May precipitation for the Northwestern Plateau Division of New Mexico from 1896 to 1970 based on six tree-ring chronologies and station data.

Table 13.2 *Calibration and verification statistics for regression models*[a]

| Calibration | $r$ | $r^2$ | $ar^2$ | Verification | RE | S |
|---|---|---|---|---|---|---|
| 1896–1970 | 0.80 | 0.64 | 0.62 | | | |
| 1896–9130 | 0.86 | 0.74 | 0.71 | 1931–1970 | 0.55 | 0.75 |
| 1931–1970 | 0.79 | 0.62 | 0.58 | 1896–1930 | 0.50 | 0.86 |

| Additional Statistics | | |
|---|---|---|
| Calibration | Sign Test | Cross Products Mean Test |
| 1896–1930 | 30+/9- | 28+/12- |
| | Z-score 3.203 | Z-score 2.3717 |
| | Prob. 0.0007 | Prob. 0.0089 |
| 1931–1970 | 28+/6- | 28+/7- |
| | Z-score 3.6015 | Z-score 3.3806 |
| | Prob. 0.0002 | Prob. 0.0004 |

[a] $r$, multiple $r$; $r^2$, variance explained; $ar^2$, variance adjusted for degrees of freedom; RE, reduction of error; S, Spearman coefficient. Both Spearman coefficients (S) are significant above the 0.0001 level.

Table 13.3 *Regression coefficients for reconstruction models for three calibration periods. Predictors used in the regression analyses are the first and second eigenvectors (EV) at lags $t$ and $t + 1$.*

| Model | Regression coefficients | | | |
|---|---|---|---|---|
| | EV 1 ($t$) | EV 1 ($t + 1$) | EV 2 ($t$) | EV 2 ($t + 1$) |
| 1896–1970 | 0.7406 | 0.2447 | 0.1472 | −0.1544 |
| 1896–1930 | 0.8164 | 0.2612 | 0.1719 | −0.1295 |
| 1931–1970 | 0.7320 | 0.2708 | 0.1838 | −0.0755 |

peaks (at the 95% level) were found at frequencies of 2 to 7 yr and at about 22 yr. Variance spectra of the individual chronologies (without prewhitening or PCA) revealed significant peaks at periods of about 80 to 100 yr.

## Extreme precipitation events

The long record of precipitation presented here can be used to examine the frequency and severity of past extreme precipitation events for northwestern New Mexico. There is considerable interannual (and some longer-term) variation in the reconstruction (Fig. 13.2A–C). We have employed a ranking program to evaluate the five most extreme low and high decadal-scale precipitation events over the past 1000 yr of record. The five intervals of greatest drought are, in descending order, 1577–1598, 1955–1964, 1895–1904, 1217–1226, and 1778–1787. The five wettest periods are 1835–1849, 1905–1928, 1429–1440, 1609–1623, and 1487–1498. Some of these intervals exceed 10 yr in length. The extreme dry interval from 1895–1904 is immediately followed by one of the wettest periods, from 1905–1928. In the 1000-yr record, this is the most abrupt shift from severe sustained drought to extreme high precipitation. However, the shift from dry to wet conditions between the late 1500s and early 1600s is also noteworthy (Fig. 13.2C).

Several dendroclimatological studies conducted near northwestern New Mexico show agreement with these results. Rose et al. (1981) reconstructed spring (March–June) as well as 12-mo (August–July) precipitation at Santa Fe, New Mexico, which is in the Northern Mountains Climatic Division. Rose et al. (1981) used different chronologies from those in our study, including modern and archaeological data. Their record, like ours, extends from A.D. 985–1970. Plots of tree growth indices and growth departures for the Colorado Plateau (Dean, 1988: 156) show the last century and the 1890–1930 period in particular to be the most extreme for almost 2000 yr. A reconstruction of long-term streamflow for the upper San Francisco River in western New Mexico, extending back to 1753, shows low flow in the late 1800s and a prolonged high-flow period in the early 1900s (Stockton 1971). Tree-ring reconstructions of flow for several major streams in the Upper Colorado River Basin and the Upper Basin as a whole show that the interval from 1906-1930 was one of anomalously persistent high runoff, apparently the greatest and longest high-flow period within the last 450 yr (Stockton and Jacoby 1976). This wet period was preceded by a long, persistent low-flow interval (1870–1894). Only one other low-flow period of comparable duration occurred (1566–1595).

The reconstruction of Northwestern Plateau winter precipitation overlaps in time with two major climatic episodes: the Little Climatic Optimum or Medieval Warm Period (from about A.D. 1000–1300) and the Little Ice Age (from about A.D. 1450–1850, although estimates vary [Lamb 1977]). The 10- and 50-yr weighted plots (Figs. 2B and 2C) show that the Little Climatic Optimum was a period of near average reconstructed precipitation in comparison to the entire record for this area. Two prolonged dry periods were found to occur just prior to and during the early part of the Little Ice Age (from about 1400 to the early 1600s). A similar reconstruction was performed without prewhitening in order to evaluate low frequency trends. However the results (for which the calibration-verification statistics were weaker than the prewhitened series) did not reveal any trends which differed from the reconstruction based on prewhitened data.

### Relation of southwestern tree growth to ENSO events

The southwestern United States is one of several regions of atmospheric teleconnection associated with ENSO events (e.g. Ropelewski and Halpert 1986;

Andrade and Sellers 1988; Kiladis and Diaz 1989). Studies of instrumental precipitation records (e.g. Ropelewski and Halpert 1986) indicate that precipitation in the Colorado Plateau and Great Basin regions tends to be higher during ENSO years (warm events). Thus it seems probable that the warm ENSO phase can be positively correlated with growth variations from trees sensitive to moisture and related factors in this region (Michaelsen 1989; Swetnam and Betancourt 1990). Reconstructed indices of ENSO, based largely on tree-ring chronologies from semiarid sites in the southwestern United States, have been developed by Lough and Fritts (1985, 1990) and Michaelsen (1989); see also Lough (1992, this volume).

We have evaluated our reconstruction to test for an ENSO signal over the past 1000 yr of record. The four most positive (wettest) values in the 1000 yr reconstruction are ENSO-related years according to the compilations of Quinn et al. (1987), Quinn and Neal (1990) and Quinn (1992, this volume). In descending order, they are as follows: 1720 (S + in Quinn et al. 1987, VS in Quinn and Neal 1990); 1484 (degree 2 in Quinn 1992, this volume); 1816 (1817, M + in Quinn et al. 1987, Quinn and Neal 1990); and 1941 (S in Quinn et al. 1987, Quinn and Neal 1992; VS in Quinn 1992, this volume). The extreme year 1816 might also be influenced by the 1815 eruption of Tambora (Lamb 1977). The year 1746, the eighth most positive value, was also an ENSO event (1747, S in Quinn et al. 1987, S + in Quinn and Neal 1990).

Comparison of the tree-ring data with Wright's (1989) sea surface temperature index (DT-cap), hereafter referred to as SOI, was made for the period from 1865–1970 using correlation analysis. Correlation coefficients were determined between the seasonalized SOI and the first eigenvector tree-ring scores based on a PCA of five of the (prewhitened) six chronologies used in the reconstruction (Fig. 13.4). Thus the correlations shown in Figure 13.4 are between the four seasonalized SOI values for a given year (DJF-SON) and the tree-ring scores of the prior, current and following years. The Cebolleta Mesa chronology was excluded as it showed the weakest relation with precipitation and the SOI. In Wright's (1989) SOI the highest positive values are associated with warm ENSO events. Starred values in Figure 13.4 are significant at the 95% level using a two-tailed test. As might be expected, the strongest correlations between tree growth and the SOI are during the winter season prior to growth, when the SOI peaks in intensity and the strongest relationship between tree growth and precipitation is found. The negative correlations between tree growth and the SOI in the following year reflect the tendency for relatively drier conditions in the southwestern United States to be associated with colder events which often follow or precede warm ENSO events. Recorded December–May SOI (Wright's [1989] DT-cap index) were estimated based on linear regression analysis with the first eigenvector scores (at lags $t - 1$, $t$ and $t + 1$) as predictors of the recorded data (Fig. 13.5). These predictors explain about 30% ($ar^2$) of the variance in the SOI data over this interval, which is a result comparable to those found in other

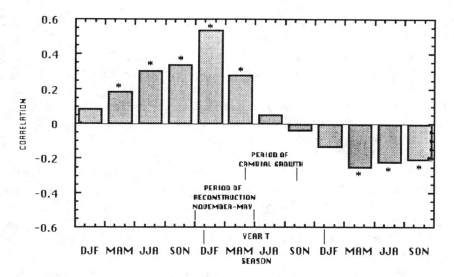

Fig. 13.4 Correlation plots of Wright's (1989) seasonalized SOI and the first eigenvector scores based on five of the chronologies used in the precipitation reconstruction for the years prior to, during (year *t*) and following the year of tree growth. In the Wright SOI series, positive SOI values are associated with warmer or El Niño events. The DJF season is included in the year's value following the month of December. Starred values are significant at the 95% level using a two-tailed test.

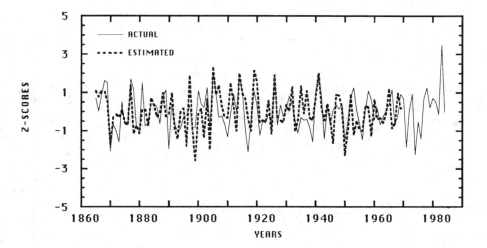

Fig. 13.5 Actual and estimated (from southwestern tree-ring data) December–May SOI based on Wright's (1989) DT-cap SOI series. Z-scores are dimensionless indices (with the mean values subtracted, divided by the standard deviation).

ENSO-related tree-ring studies in the southwestern United States (e.g. Michaelsen 1989; Swetnam and Betancourt 1990).

The ENSO signal in these long chronologies from the southwestern United States could eventually be integrated with additional tree-ring and other high resolution records (e.g. corals, varves, and ice cores) from key sites influenced by ENSO around the globe (Baumgartner et al. 1989; Overpeck and Cole 1990). A preliminary estimate of an ENSO-related SST index (Wright's S-cap, 1989) was made using combined tree-ring records from the southwestern United States and from Java, Indonesia (first developed by Berlage 1931; also discussed in detail by Murphy and Whetton 1989; Jacoby and D'Arrigo 1990). The reader is referred to these articles for more thorough discussions of the Java (teak or *Tectona grandis* L.F.) chronology. The regression estimates (obtained using the southwestern United States and teak time series each as predictors of the SOI-related index in multiple linear regression) for the interval from 1890–1928 are improved over using either data set alone (Fig. 13.6).

Although the ENSO signal in areas of teleconnection may be weaker than in the equatorial Pacific, different spatial modes of ENSO variation are believed to exist (e.g. Deser and Wallace 1987). Proxy data from different areas are therefore needed in order to distinguish local from large-scale variability and to determine the nature of past teleconnections of ENSO. Paleoclimatic records from different areas as well as modeling studies are needed to determine whether ENSO exhibits low frequency variation during different background climates or forcings such as the Little Ice Age, elevated trace gases, or variations in solar activity (Enfield 1989, 1992, this volume).

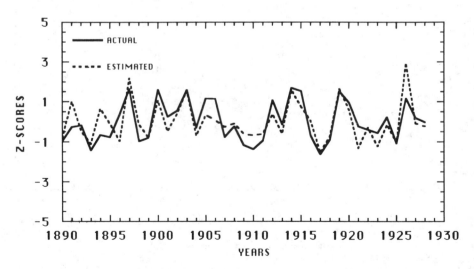

Fig. 13.6   Actual and estimated January–March tropical Pacific SST (S-cap) ENSO index developed by Wright (1989). Estimated values obtained using the southwestern chronologies and a tree-ring time series of teak from Indonesia as predictors in regression analysis. Z-scores as in Fig. 13.5. Calibration period is 1890–1928.

## Summary

A long history of precipitation variability has been described for northwestern New Mexico. The reconstruction is validated through model verification of the regression estimates, and the regression weights are stable in time. The reconstruction compares favorably with other tree-ring studies from the region. Two extreme droughts are revealed within the past century which are exceeded in magnitude only twice in the past 1000 yr, with an unprecedented shift from dry to wet conditions around the turn of the century.

The chronology data show a relationship with ENSO events which has previously been detected in instrumental and other data from the southwestern United States. More studies are needed to investigate in more detail the nature of the ENSO signal in these long records and to integrate them with other instrumental, historical, and proxy data such as the teak data from Indonesia.

*Acknowledgments* This research was supported by the National Science Foundation under Grant Numbers 89–15353 and 87–16630. We thank E. R. Cook and L. D. Ulan for technical assistance. We thank I. S. Ivanciu for performing some of the modeling studies. We thank M. Rose for providing a tape version of the published tree-ring data. Lamont-Doherty Geological Observatory Contribution No. 4964.

## References

ANDRADE, E. R., Jr. and SELLERS, W. D., 1988: El Niño and its effect on precipitation in Arizona and western New Mexico. *Journal of Climatology*, 8: 403–410.

BAUMGARTNER, T. R., MICHAELSON, J., THOMPSON, L. G., SHEN, G. T., SOUTAR, A., and CASEY, R. E., 1989: The recording of interannual climatic change by high-resolution natural systems: tree-rings, coral bands, glacial ice layers, and marine varves. *In* Peterson, D. H. (ed.), *Aspects of Climate Variability in the Pacific and Western Americas*. Geophysical Monographs 55. Washington, D.C., American Geophysical Union, 1–14.

BERLAGE, H. P., 1931: On the relationship between thickness of tree-rings of Djati trees and rainfall on Java. *Tectona*, 24: 935–953.

COOK, E. R., 1992: Using tree rings to study past El Niño/Southern Oscillation influences on climate. *In* Diaz, H. F. and Markgraf, V. (eds.), *El Niño: Historical and Paleoclimatic Aspects of the Southern Oscillation*. Cambridge: Cambridge University Press, 203–214.

COOK, E. R. and JACOBY, G. C., 1983: Potomac River streamflow since 1730 as reconstructed by tree-rings. *Journal of Climate and Applied Meteorology*, 22: 1659–1672.

DEAN, J. S., 1988: Dendrochronology and paleoenvironmental reconstruction on the Colorado Plateaus. *In* Gumerman, G. J. (ed.), *The Anasazi in a Changing Environment*. New York: Cambridge University Press, 119–167.

DEAN, J. S. and ROBINSON, W. J., 1978: *Expanded Tree-Ring Chronologies for the Southwestern United States*. Chronology Series III, Laboratory of Tree-Ring Research, The University of Arizona, Tucson, Arizona. 58 pp.

DESER, C. and WALLACE, J. M., 1987: El Niño events and their relation to the Southern Oscillation. *Journal of Geophysical Research*, 92: 14,189–14,196.

DRAPER, N. and SMITH, H., 1981: *Applied Regression Analysis*. 2nd ed., New York: John Wiley. 709 pp.

ENFIELD, D. B., 1989: El Niño, past and present. *Reviews of Geophysics*, 27: 160–187.

ENFIELD, D. B., 1992: Historical and prehistorical overview of El Niño/Southern Oscillation. *In* Diaz, H. F. and Markgraf, V. (eds.), *El Niño: Historical and Paleoclimatic Aspects of the Southern Oscillation*. Cambridge: Cambridge University Press, 95–117.

FRITTS, H. C., 1976: *Tree-Rings and Climate*. London and New York: Academic Press. 567 pp.

GORDON, G. A. and Le DUC, S. K., 1981: Verification statistics for regression models. Paper presented at *Conference on Probability and Statistics in Atmospheric Science*. American Meteorological Society, California.

HSU, C. P. F. and WALLACE, J. M., 1976: The global distribution of the annual and semiannual cycles in precipitation. *Monthly Weather Review*, 104: 1093–1101.

JACOBY, G. C. and D'ARRIGO, R. D., 1991: Teak (*Tectona grandis* L.F.), a tropical species of large-scale dendroclimatic potential. *Dendrocronologia*, 8: 83–98.

JENKINS, G. M. and WATTS, D. G., 1968: *Spectral Analysis and its Applications*. San Francisco: Holden Day Press. 525 pp.

KARL, T. R., METCALF, L. K., NICODEMUS, M. L., and QUAYLE, R. G., 1983: *Historical Climatology Series 6.1*. National Climatic Data Center, Asheville, North Carolina.

KILADIS, G. N. and DIAZ, H. F., 1989: Global climatic anomalies associated with extremes of the Southern Oscillation. *Journal of Climate*, 2: 1069–1090.

KLEIN, W. H. and BLOOM, H. J., 1987: Specification of monthly precipitation over the United States from the surrounding 700 mb height field. *Monthly Weather Review*, 115: 2118–2132.

KUTZBACH, J. E. and GUETTER, P. J., 1980: On the design of paleoenvironmental data networks for estimating large-scale patterns of climate. *Quaternary Research*, 14: 169–187.

LAMB, H. H., 1977: *Climate: Past, Present and Future. Vol. 2. Climate History and the Future*. London: Methuen. 835 pp.

LOUGH, J. M. 1992: An Index of Southern Oscillation reconstructed from western North American tree-ring chronologies. *In* Diaz, H. F. and Markgraf, V. (eds.), *El Niño: Historical and Paleoclimatic Aspects of the Southern Oscillation*. Cambridge: Cambridge University Press, 215–226.

LOUGH, J. M. and FRITTS, H. C., 1985: The Southern Oscillation and Tree-Rings: 1600–1961. *Journal of Climate and Applied Meteorology*, 24: 952–66.

LOUGH, J. M. and FRITTS, H. C., 1990: Historical aspects of El Niño-Southern Oscillation-information from tree-rings. *In* Glynn, P. W. (ed.), *Global Ecological Consequences of the 1982–83 El Niño-Southern Oscillation*. Oceanography Series 52. Amsterdam: Elsevier, 285–321.

MEKO, D. M., 1981: Applications of Box-Jenkins Methods of Time Series Analysis to the Reconstruction of Drought from Tree-Rings. Ph.D. thesis, University of Arizona. 149 pp.

MEKO, D. M., STOCKTON, C. W., and BOGGESS, W. R., 1980: A tree-ring reconstruction of drought in southern California. *Water Resources Bulletin*, 16: 594–600.

MICHAELSEN, J., 1989: Long-period fluctuations in El Niño amplitude and frequency reconstructed from tree-rings. *In* Peterson, D. H. (ed.), *Aspects of Climate Variability in the Pacific and Western Americas.* Geophysical Monographs 55. Washington, D.C.: American Geophysical Union, 69–74.

MURPHY, J. O. and WHETTON, P. H., 1989: A re-analysis of a tree-ring chronology from Java. *Proceedings, Koninklijke Nederlandse Akademie van Wetenschappen.* Ser. B, 92(3): 241–257.

OVERPECK, J. and COLE, J., 1990: Role of corals, varved sediments and models in understanding global environmental change. *EOS*, 71(29): 983.

QUINN, W. H. 1992: A study of Southern Oscillation-related climatic activity for A.D. 622–1990 incorporating Nile River flood data. *In* Diaz, H. F. and Markgraf, V. (eds.), *El Niño: Historical and Paleoclimatic Aspects of the Southern Oscillation.* Cambridge: Cambridge University Press, 119–149.

QUINN, W. H. and NEAL, V. T., 1992: The ENSO as a climatic forcing factor. *In* Bradley, R. S. and Jones, P. D. (eds.), *Climate Since AD 1500.* London: Routledge, 623–648.

QUINN, W. H., NEAL, V. T., and ANTUNEZ de MAYOLO, S. E., 1987: El Niño occurrences over the past four and a half centuries. *Journal of Geophysical Research*, 92: 14,449–14,461.

RIND, D., GOLDBERG, R., HANSEN, J., ROSENZWEIG, C., and RUEDY, R., 1990: Potential evapotranspiration and the likelihood of future drought. *Journal of Geophysical Research*, 95: 9983–10,004.

ROPELEWSKI, C. F. and HALPERT, M. S., 1986: North American precipitation and temperature patterns associated with the El Niño/Southern Oscillation (ENSO). *Monthly Weather Review*, 114: 2352–2362.

ROSE, M. R., DEAN, J. S., and ROBINSON, W. J., 1981: The Past Climate of Arroyo Hondo New Mexico reconstructed from Tree Rings. *Arroyo Hondo Archaeological Series* Vol. 4. Santa Fe: School of American Research Press. 253 pp.

ROSE, M. R., ROBINSON, W. J., and DEAN, J. S., 1982: Dendroclimatic reconstruction for the southeastern Colorado Plateau. Unpublished manuscript, Laboratory of Tree-Ring Research.

STOCKTON, C. W., 1971: The feasibility of augmenting hydrologic records using tree-ring data. Ph.D. dissertation, University of Arizona.

STOCKTON, C. W. and JACOBY, G. C., 1976: Long-term surface-water supply and streamflow trends in the Upper Colorado River Basin, *Lake Powell Research Project Bulletin* 18. Institute of Geophysics and Planetary Physics, University of California, Los Angeles. 70 pp.

SWETNAM, T. W. and BETANCOURT, J. L., 1990: Fire-Southern Oscillation relations in the southwestern United States. *Science*, 249: 1017–1020.

U.S. Water Resources Council, 1978: *The Nation's Water Resources 1975–2000.* Second National Water Assessment by the U.S. Water Resources Council, Vol. 1, Summary. Washington, D.C.: U.S. Government Printing Office.

WRIGHT, P. B., 1989: Homogenized long-period Southern Oscillation indices. *Journal of Climatology*, 9: 33–54.

# 14

# Temporal patterns of El Niño/Southern Oscillation – wildfire teleconnections in the southwestern United States

THOMAS W. SWETNAM

*Laboratory of Tree-Ring Research, University of Arizona, Tucson, Arizona 85721, U.S.A.*

JULIO L. BETANCOURT

*U.S. Geological Survey, 1675 W. Anklam Rd., Tucson, Arizona 85705, U.S.A.*

## Abstract

Wildland fire occurrence in the southwestern United States is correlated with winter-spring precipitation, tree-ring growth, and the Southern Oscillation. Twentieth century (1905–1985) records from all National Forest lands in Arizona and New Mexico show reduced/increased annual area burned during low/high (respectively) mean December through February Southern Oscillation indices. A 206-yr record of regional fire activity (1700–1905) derived from fire scars in conifer trees shows high correlation with precipitation-responsive tree-ring width chronologies during the 1740s to 1760s and 1840s to 1860s, and lower correlations during the 1790s to 1830s and 1870s to 1900s. Synchronized biennial patterns of high/low tree-ring growth and low/high fire activity, respectively, is evident in the plotted time series (especially during the high correlation periods) and in a superposed epoch analysis. Temporal changes in southwestern fire regimes over the past three centuries may be indicative of ecological changes in these forests (e.g., other natural or human disturbances), or changes in the ENSO phenomenon. Additional study of the temporal modes of ENSO and fire regimes is needed.

## Introduction

Environmental history is recorded by tree rings in a variety of ways. The most familiar type of record is the ring-width chronology, from which centuries-long reconstructions of precipitation, temperature, streamflow, and the Southern Oscillation (SO) can be obtained using regression techniques (Fritts 1976, 1991;

Meko and Stockton 1984; Lough and Fritts 1985; Cleaveland et al. 1992, this volume). Discrete events such as floods, earthquakes, frosts, and forest fires are also recorded by tree rings as injuries or structural changes within annual rings (Fritts and Swetnam 1989). We have found that a tree ring-based forest fire record from the southwestern United States contains a regional signal that is associated with seasonal variations in precipitation and the SO, especially the extreme phases of El Niño and La Niña. Examination of this event record and its changing associations with tree-ring width and precipitation time series offers new insights on the response of terrestrial ecosystems to El Niño/Southern Oscillation (ENSO).

Linkages between ENSO and regional-scale fire activity have been noted before. Catastrophic fires occurred in Australia and Indonesia during the severe 1982–83 ENSO event, while total area burned and numbers of fires reported in the United States in these years were among the lowest on record. Time series studies of Australian bushfire-ENSO relations, however, have yielded mixed results. Gill (1983) found no correlation between a fire activity index for eastern Australia and the SO (1938–1980), but using nonparametric means tests Skidmore (1987) found that greater area was burned during ENSO years than non-ENSO years (1930–1980) in the states of Victoria, New South Wales, and Tasmania. Skidmore found no significant relations between ENSO and area burned in the states of Western Australia or South Australia.

Simard et al. (1985a, 1985b) compiled wildland fire data (annual numbers of fires and area burned) for the United States (1926–1982) and showed that the strongest association between fire activity and El Niño events was in the southeastern United States. The most significant results ($p < 0.01$) from means tests were for the states of Florida, Alabama, Georgia, Tennessee, Louisiana, North Carolina, Missouri, and Iowa. However, their analysis focused entirely on relations between fire occurrence and warm episodes (El Niños) in the tropical Pacific.

We began our investigation with the knowledge that winter-spring precipitation and tree-ring growth in the southwestern United States were teleconnected to ENSO (Douglas and Engelhardt 1984; Lough and Fritts 1985; Andrade and Sellers 1988). Thus, it seemed reasonable that wildland fire occurrence in this region, which is partly a function of seasonal fuel moisture (Deeming et al. 1977), might also be teleconnected to ENSO. We evaluated the effects of both warm (El Niño) and cold (La Niña) episodes on southwestern fire occurrence using 286 yr of fire occurrence data (1700–1985). In this paper we briefly review the results of our initial work (Swetnam and Betancourt 1990), and we report new observations of temporal variations in the ENSO-fire teleconnection.

## Methods

Relations between fire and climate in the 20th century were investigated using fire statistics from all National Forests in Arizona and New Mexico. A time series of annual area burned from 1905 to 1985 was compared to monthly and seasonal precipitation from 14 long-term weather stations (Fig. 14.1) used by Andrade and Sellers (1988) and time series of the SO index (SOI) (Ropelewski and Jones 1987) and Line Island rainfall index (LIRI) (Wright 1989). Some months of the precipitation data are not normally distributed (Andrade and Sellers 1988), so we used the nonparametric Spearman's rank correlation test for combinations that included precipitation data.

To examine long-term relations between climate and fire we compiled tree-ring growth (1700–1960) and fire-scar chronologies (1700–1905) from the southwestern United States. A regional tree-ring chronology was computed as the mean tree-ring width index of 28 chronologies (sites) in Arizona and New Mexico (Fig. 14.1). These chronologies were originally developed by dendrochronologists at the Laboratory of Tree-Ring Research, University of Arizona, for

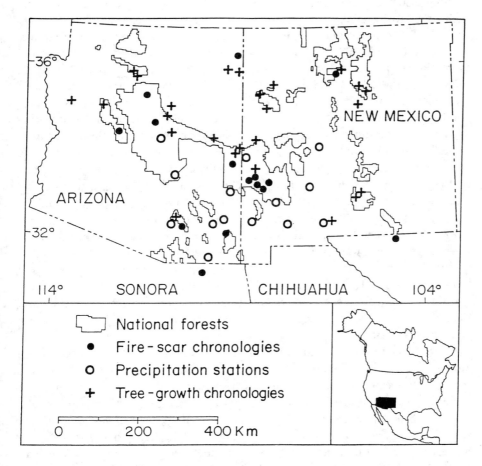

Fig. 14.1   Map of southwestern United States. Locations of National Forests and networks of fire-scar, precipitation and tree-ring data sets are shown.

dendroclimatic reconstructions (this particular network is described by Stockton et al. 1991).

The regional fire-scar chronology was developed from 315 fire-scarred trees growing in 15 sites throughout Arizona, New Mexico, west Texas, and northern Mexico (Swetnam 1990) (Fig. 14.1). A fire-scar index was computed as the mean percentage of trees scarred per year in the 15 sites. The fire-scar record is terminated in 1905 because of the lack of fire scars in the 20th century. The frequent low-intensity surface burns represented by the fire-scar record have dwindled with cessation of aboriginal fires, removal of fine fuels by livestock grazing, and a vigorous program of fire suppression.

Since the fire-scar record did not overlap the 20th century National Forest fire record we could not calibrate the former in terms of an absolute measure of annual area burned. Nevertheless, it is evident that the fire-scar record is a relative measure of regional fire activity. The 15 fire-scar chronologies were broadly dispersed in the two states (Fig. 14.1). The sites were too distant from each other to have recorded contiguous burns, yet higher than expected synchrony of fire dates among the dispersed sites was observed. For example, between 1700 and 1900 more than 50% of all sites (at least 8 sites) recorded fire events in the same year on 14 occasions (1709, 1748, 1752, 1763, 1765, 1772, 1773, 1785, 1806, 1847, 1851, 1859, 1876, 1879). The maximum number of sites (10) recording fires occurred in 1748. If fire occurrence within sites was random and independent of the other sites the expected number of fires jointly occurring in 8 or more sites during the same year was less than one event in this two century period. (This was estimated from the observed fire frequencies and the binomial probability of joint occurrences of fires in 8 or more sites.) The high synchrony of fire dates among the scattered fire-scar sites strongly suggests that climate (e.g., patterns of precipitation, wind, and lightning) is the operative mechanism, since no other known fire-related factor operates on such a large scale.

Pre-20th century fire and climate relations were examined by comparing the fire-scar index chronology with the tree-ring width chronology, which served as a proxy of seasonal precipitation. First differences of the regional fire-scar and tree-ring width index time series were computed (value [year t] − value [year t − 1]) to emphasize year-to-year changes and de-emphasize trends in the data. Pearson correlations were computed for combinations of the SOI, ring-width and fire-scar data.

In addition to the correlation analysis we conducted a superposed epoch analyses to investigate the relations between regional tree-ring growth and the regional fire-scar record. The untransformed tree-ring width and fire-scar chronologies (not the first difference series) were used in this analysis. We followed procedures described by Lough and Fritts (1987) in carrying out the superposed epoch analysis. The largest and smallest fire years, sorted as the highest and lowest quartile (25%) fire-scar index years, were used as 'key-year' dates. The highest and lowest quartile were used because this provided a sufficiently large number of dates (50) for testing, while still emphasizing the regional patterns in the data. The mean tree-ring width departure during the key-years (lag 0), and for lagged years before and after the key-years ( − 5 to + 5 lags), were computed. A random resampling of the same number of key-year dates (50) and the corresponding lagged values was carried out 1000 times. The resampled data were used to estimate confidence intervals for the actual key-year dates, i.e., the probability of the average tree-ring departures associated with the fire years occurring by chance.

Changes in the ENSO/fire association through time were investigated in both the regional fire-scar index versus the regional tree-ring growth comparison (1700–1905) and in the annual area burned versus SOI comparison (1905–1986) by computing and plotting Pearson correlation coefficients for sequential 21 yr periods, overlapping by 10 yr. These 'moving' correlation coefficients were plotted on the central year of the 21-yr period. Temporal changes in the correlation between these series were compared with the frequency of severe and very severe events reconstructed from archival sources (Quinn et al. 1987).

## Results

The modern fire and climate records show significant correlations. Annual area burned in National Forests of Arizona and New Mexico since 1905 was compared with regional precipitation (not shown) and seasonal SOI (Fig. 14.2). A weak signal was anticipated because the fire statistics include both lightning and person-caused fires across a wide spectrum of vegetation types from grassland to boreal forests, each subject to different land use and management practices. However, total area burned closely tracks DJF-SOI (Fig. 14.2) and LIRI until the 1960s (Fig. 14.3), when area burned increased and became less variable, possibly because there was an increased number of person-caused fires or because fire suppression resulted in unusual accumulation of fuels. The earliest part of the fire record (1905–1915) was dominated by very large fires from 1909 to 1912 and

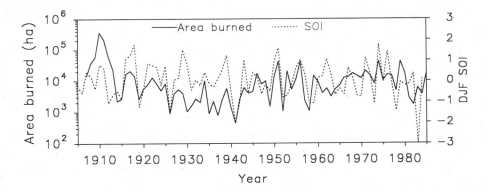

Fig. 14.2   Time series of annual area burned (logarithmic scale) in Arizona and New Mexico, and mean December through February SOI, 1905–1985.

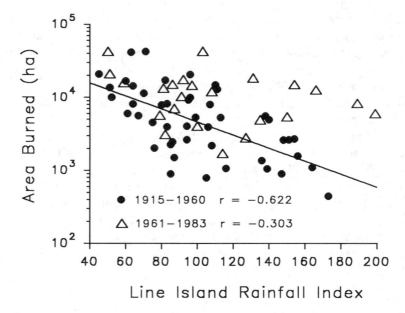

Fig. 14.3   Scatter plot of Line Island Rainfall Index versus area burned in the Arizona and New Mexico National Forests (1915–1983). The pre-1915 and post 1960 periods were omitted to eliminate the overriding effect of exceptional fires that occurred in 1910 to 1912 after several fire free decades, and increasing numbers of person-caused fires after 1960. The regression line was fitted to the 1915–1960 period.

correlation with SOI and precipitation was also lower during this period. The high magnitude of these fire years relative to the rest of the record could be due to woody fuel accumulation since the end of the episodic fire regimes of the 19th century. Our subsequent correlation analysis included comparisons of the full period (1905–1986) and a shortened period (1915–1960). Spring (March–May) precipitation (1915–1960) yielded the highest correlations of any season against both area burned (Spearman's $r = -0.407, p < 0.001$) and the Southern Oscillation (Spearman's $r = -0.402, p < 0.001$).

Figure 14.4 shows a correspondence between first differences of standardized regional tree growth and percentage of trees scarred from 1700 to 1905. The Pearson correlation of this combination was $-0.606, p < 0.001$. The Pearson correlation of the untransformed series (without taking first differences) was $-0.409, p \leq 0.001$. Significant first order autocorrelation was present in both of the first differenced series ($-0.60, p < 0.01$, in the fire-scar series and $-0.38$, $p < 0.01$, in the ring-width series). The autocorrelation was removed from each series using simple AR1 models, resulting in 'residual' series (Box and Jenkins 1976; Biondi and Swetnam 1987). The two residual series had a Pearson correlation of $-0.397, p < 0.001$.

Climatic conditions that favor tree growth suppress fires, whereas reduced growth coincides with extensive fires. An ENSO signal should be expected in the

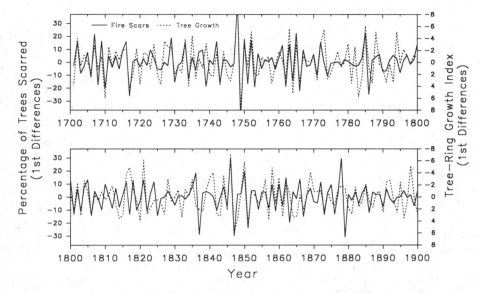

Fig. 14.4   Time series of percentage of trees scarred and tree-ring growth in Arizona and New Mexico.

tree-growth chronologies because precipitation in the fall to spring before the growing season typically exerts the strongest influence on cambial growth in Douglas-fir (*Pseudotsuga menziesii*) and ponderosa pine (*Pinus ponderosa*) (Fritts 1976). Negative correlations have been reported between tree growth in the southwestern United States and the SOI, that is, tree-growth is enhanced during El Niño conditions (Lough and Fritts 1985; Michaelsen 1989; D'Arrigo and Jacoby 1991; Cleaveland et al. 1992, this volume). Thus, the correlation of tree growth and fire occurrence shown in Figure 14.4 is most likely a reflection of an ENSO teleconnection operating through the effects of seasonal precipitation on both tree growth and fire occurrence.

Although the data were correlated across the full length of their common periods, there were clearly periods of higher and lower correlation. These changing temporal relationships are illustrated in Figure 14.5 using moving 21-yr correlation coefficients. Generally, the highest correlations between series were observed during the mid-century periods (ca. 1740–1780 and 1840–1860), whereas the lowest correlations were during the turn-of-the-century periods (ca. 1790–1830 and 1870–1900). This pattern could be explained by at least two general hypotheses: (1) The ENSO-fire signal was weakened during these particular turn-of-the-century periods because other local or regional effects (e.g., other forest disturbances, such as insect outbreaks, or livestock grazing) became stronger and overrode the ENSO signal; (2) the ENSO-southwest United States teleconnection was weakened because of changes in the operative mechanisms (e.g., position of storm tracks as influenced by ENSO) or because of changes in the frequency or amplitudes (or both) of ENSO events.

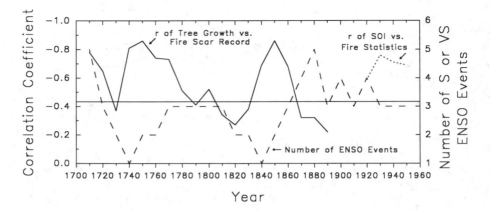

Fig. 14.5  Moving Pearson correlation coefficients, computed for 21-yr periods, are plotted on central years (decades) of the periods for combinations of tree growth chronologies versus fire-scar records (1700–1905) and DJF SOI versus area burned records from forestry documents (1905–1985). Numbers of 'strong' or 'very strong' ENSO events (Quinn et al. 1987) during the same 21-yr periods are also shown (corresponding to right *y*-axis). The solid horizontal line represents the approximate 0.05 significance level for the 21-yr correlations.

Increased livestock grazing in mountain areas of the Southwest (Carlson 1969) during the late 1800s may have interfered with the ENSO-fire signal because grazing reduces fine fuels important for the low-intensity surface fires that scar but do not kill mature trees. There is insufficient historical information to determine if increased livestock grazing, or natural disturbances such as insect outbreaks, may have been a factor during the earlier low correlation period (ca. 1790–1830).

Evidence that might support the second hypothesis (change in frequency and/or amplitude) comes from Quinn et al.'s (1987) reconstruction of ENSO events from South American archival sources. Lower frequencies of 'strong' or 'very strong' ENSO events were observed during the mid-18th and mid-19th centuries and slightly higher frequencies during the late and early parts of these centuries (Fig. 14.5) (also see Enfield 1988). Changes in magnitudes (amplitude) associated with the frequency changes are unclear, especially before 1800 when only severe events were confidently identified from archives (Quinn et al. 1987). Better understanding of these changing temporal patterns will also require historical reconstructions of La Niña events, which figure heavily in ENSO-fire relations (Figs. 14.2, 14.4).

A particularly interesting observation from the tree growth and fire comparison is that periods with highest correlations display a biennial tendency (Fig. 14.4). High tree growth and low fire occurrence years were often followed in the next year by low tree growth and high fire occurrence years, and vice versa (e.g., 1740s to 1760s and 1840s to 1860s). The significant first order autocorrelation in these series reported earlier supports this observation. The results of the

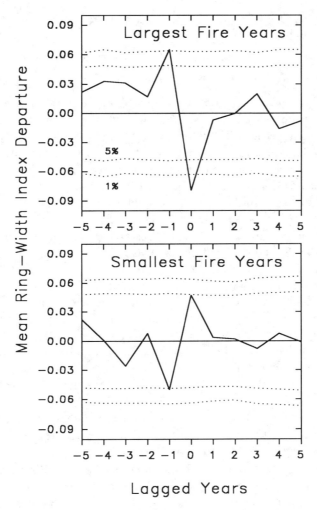

Fig. 14.6   Average regional tree-ring width departures from five years before to five years after fire years determined from the regional fire-scar chronology. The departures for the largest quartile fire years (50 yr) are shown at the top and departures for the smallest quartile years (50 yr) are shown at the bottom. The dotted lines show the 5% and 1% probability levels occurring by chance as determined from 1000 random simulations.

superposed epoch analyses, applied to the entire 1700–1905 period, also revealed a significant inverse pattern of tree-ring growth in the year prior to high and low fire occurrence years (Fig. 14.6). In the case of the largest fire years this probably reflects the importance of prior years of favorable (wet) conditions, when tree-ring growth and production of fine fuels (e.g., grasses and tree leaves) were high, and less favorable (dry) conditions during the fire year when tree-ring growth was lower, fuels were dry, and fire spread was more likely. The reverse pattern is evident for the smallest fire years (Fig. 14.6), and may be explained by less fuels

produced during dry prior years, and subsequent reduced fire activity in the
following wet years.

## Discussion and summary

Regional climatic effects are implicit in the extreme variability of fire occurrence
measured by both the fire-scar record and fire statistics. In general, area burned
was greatest during years with highly positive values of SOI, reduced rainfall in
the Line Islands, and severe winter-spring droughts. Area burned was reduced
after exceptionally wet springs of low-SO phases or El Niño years.

Temporal changes in the relation between ENSO and fires in the Southwest is
suggested by tree growth and climate comparisons with the fire record. Somewhat reduced correlations between these data in the late 20th century may be due
to increased numbers of person-caused fires and/or changes in fuels related to
fire suppression. Reduced correlations are also observed during the late 1700s to
early 1800s, and again in the late 1800s. Interference with the ENSO-fire signal
by other factors, or a change in ENSO itself may explain these observations.

High synchrony of the tree-ring width and fire-scar records was generally
characterized by a biennial pattern of high/low tree-ring growth associated with
low/high (respectively) fire activity. Is it possible that an alternating pattern of
wet and dry years corresponding to low and high fire activity in the southwestern
United States was related in some way to a biennial component of ENSO (Meehl
1987; Rasmusson et al. 1990)? If so, we might speculate that during climatic
periods with a dominant ENSO cycle of about 2 yr the high-frequency southwestern fire regime synchronizes with the tropical Pacific. The observed inverse
relation between frequency of severe or very severe ENSO events and the strength
of the ENSO-fire teleconnection (Fig. 14.5) could be a reflection of changes in
the phasing and amplitude of a biennial component relative to lower frequency
components of the SO (see also Diaz and Pulwarty 1992, this volume). Clearly,
testing of this idea will require further study of the temporal modes of both
ENSO and teleconnected fire regimes.

Synchronous large fires in the Southwest over three centuries, and their
association with the high-SO phase, deficient winter-spring precipitation, and
reduced tree-growth, imply that seasonal climate, and not just fire weather is
important to southwestern fire regimes. ENSO-fire teleconnections may be
expected in other regions where fire occurrence is related to seasonal precipitation influenced by ENSO. Study of such relationships is important because
climate-driven changes in fire regimes are potentially a primary mechanism of
vegetation change on regional and larger spatial scales (Davis 1989; Overpeck
et al. 1990). Reconstructions of fire histories in other regions might utilize combinations of records from tree-rings, charcoal in layered records (e.g., ice cores
and sediments), and historical documents. These reconstructions may also prove
useful in identifying the significant climatologies associated with catastrophic fires, such as those at Yellowstone National Park ($\sim$570,000 ha) in

September 1988 and across Siberia and Mongolia ( ~ 7 million ha) in May 1987 (Salisbury 1989).

## References

ANDRADE, E. R. and SELLERS W. D., 1988: El Niño and its effects on precipitation in Arizona and western New Mexico. *Journal of Climatology*, 8: 403–410.

BOX, G. E. P. and JENKINS, J. M., 1976: *Time Series Analysis*: Forecasting and control. Revised edition. Oakland, Calif.: Holden-Day. 575 pp.

BIONDI, F. and SWETNAM, T. W., 1987: Box-Jenkins models of forest interior tree-ring chronologies. *Tree-Ring Bulletin*, 47: 71–96.

CARLSON, A. W., 1969: New Mexico's sheep industry, 1850–1900: Its role in the history of the territory. *New Mexico Historical Review*, 44: 25–49.

CLEAVELAND, M. K., COOK, E. R., and STAHLE, D. W., 1992: Secular variability of the Southern Oscillation detected in tree-ring data from Mexico and the southern United States. *In* Diaz, H. F. and Markgraf, V. M. (eds.), *El Niño: Historical and Paleoclimatic Aspects of the Southern Oscillation*. Cambridge: Cambridge University Press, 271–291.

D'ARRIGO, R. D. and JACOBY, G. C., 1991: A 1000-year record of winter precipitation from northwestern New Mexico, USA: A reconstruction from tree-rings and its relation to El Niño and the Southern Oscillation. *The Holocene*, 1: 95–101.

DAVIS, M. B., 1889: Insights from paleoecology on global change. *Bulletin of the Ecological Society of America*, 70: 222–228.

DEEMING, J. W., BURGAN, R. E., and COHEN, J. D., 1977: The national fire danger rating system, *U.S. Forest Service General Technical Report* INT-39.

DIAZ, H. F. and PULWARTY, R., 1992: A comparison of Southern Oscillation and El Niño signals in the tropics. *In* Diaz, H. F., and Markgraf, V. (eds.), *El Niño: Historical and Paleoclimatic Aspects of the Southern Oscillation*. Cambridge: Cambridge University Press, 175–192.

DOUGLAS, A. V. and ENGLEHART, P. J., 1984: Factors leading to the heavy precipitation regimes of 1982–1983 in the United States. *In: Proceedings of the Eighth Annual Climate Diagnostics Workshop, Downsview, Ontario*. U.S. Government Printing Office, NTIS PB84-192418, Washington, D.C., pp. 42–54.

ENFIELD, D. B., 1988: Is El Niño becoming more common? *Oceanography Magazine*, 1: 23–027, 59.

FRITTS, H. C., 1976: *Tree Rings and Climate*. New York: Academic Press. 567 pp.

FRITTS, H. C., 1991: *Reconstructing Large-scale Climatic Patterns from Tree-ring Data: A Diagnostic Analysis*. Tucson: University of Arizona Press. 286 pp.

FRITTS, H. C. and SWETNAM, T. W., 1989: Dendroecology: A tool for evaluating variations in past and present forest environments. *Advances in Ecological Research*, 19: 111–188.

GILL, A. M., 1983: Forest fire and drought in eastern Australia. *In: Colloquium on the Significance of the Southern Oscillation – El Niño Phenomena and the Need for a Comprehensive Ocean Monitoring System in Australia*. Australian Marine Sciences and Technologies Advisory Committee (AMSTAC), 161–185.

LOUGH, J. M. and FRITTS, H. C., 1985: The Southern Oscillation and tree rings: 1600–1961. *Journal of Climate and Applied Meteorology*, 10: 952–966.

LOUGH, J. M. and FRITTS, H. C., 1987: An assessment of the possible effects of volcanic eruptions on North American climate using tree-ring data, 1602 to 1900 A.D. *Climatic Change*, 10: 219–239.

MEEHL, G. A., 1987: The annual cycle and interannual variability in the tropical Pacific and Indian Ocean regions. *Monthly Weather Review*, 115: 27–50.

MEKO, D. M. and STOCKTON, C. W., 1984: Secular variations in streamflow in the western United States. *Journal of Climate and Applied Meteorology*, 23: 889–897.

OVERPECK, J. T., RIND, D., and GOLDBERG, R., 1990: Climate induced changes in forest disturbance and vegetation. *Nature*, 343: 51–53.

QUINN, W. H., NEAL, V. T., and ANTUNEZ de MAYOLO, S. E., 1987: El Niño occurrences over the past four and a half centuries. *Journal of Geophysical Research*, 92: 14, 449–14,461.

RASMUSSON, E. M., WANG, X., and ROPELEWSKI, C. F., 1990: The biennial component of ENSO variability. *Journal of Marine Systems*, 1: 71–96.

ROPELEWSKI, C. F. and JONES, P. D., 1987: An extension of the Tahiti-Darwin Southern Oscillation index. *Monthly Weather Review*, 115: 2161–2165.

SALISBURY, H. E., 1989: *The Great Black Dragon Fire: A Chinese Inferno*. Boston: Little, Brown. 180 pp.

SIMARD, A. J., HAINES, D. A., and MAIN, W. A., 1985a: El Niño and wildland fire: An exploratory study. *In* Donoghue, L. R., and Martin, R. E. (eds.), *Weather – The Drive Train Connecting the Solar Engine to Forest Ecosystems. Eighth Conference on Fire and Forest Meteorology, Detroit Michigan, April 29-May 2, 1985*. Bethesda Md.: Society of American Foresters, 88–95.

SIMARD, A. J., HAINES, D. A., and MAIN, W. A., 1985b: Relations between El Niño/ Southern Oscillation anomalies and wildland fire activity in the United States. *Agricultural and Forest Meteorology*, 36: 93–104.

SKIDMORE, A. K., 1987: Predicting bushfire activity in Australia from El Niño/Southern Oscillation events. *Australian Forestry*, 50: 231–235.

STOCKTON, C. W., MEKO, D. M., and BOGGESS, W. R., 1991: Chapter 1, Drought history and reconstructions from tree rings. *In* Gregg, F. and Getches, D. H., Principal Investigators, Severe and Sustained Drought in the Southwestern United States, Phase I Report, U.S. Department of State and Man and Biosphere Program, Grant No. 1753-800554.

SWETNAM, T. W., 1990: Fire history and climate in the Southwestern United States. *In* Krammes, S. J., Technical Coordinator, *Proceedings of Symposium on Effects of Fire in Management of Southwestern Natural Resources. U.S. Forest Service General Technical Report*, RM-191: 6–17.

SWETNAM, T. W. and BETANCOURT, J. L., 1990: Fire-Southern Oscillation relations in the Southwestern United States. *Science*, 249: 1017–1020.

WRIGHT, P. B., 1989: Homogenized long-period southern-oscillation indices. *International Journal of Climatology*, 9: 33–54.

# 15

# Secular variability of the Southern Oscillation detected in tree-ring data from Mexico and the southern United States

MALCOLM K. CLEAVELAND,[1] EDWARD R. COOK,[2] and
DAVID W. STAHLE[1]

1. *Department of Geography University of Arkansas Fayetteville, Arkansas 72701, U.S.A.*
2. *Lamont-Doherty Geological Observatory Palisades, New York 10964, U.S.A.*

## Abstract

Regional drought reconstructions for the eastern and southern United States have been used for superposed epoch analyses (SEA) of the climate response to ENSO forcing. In Texas, El Niño (warm) events are associated ($P < 0.05$) with wet anomalies, and La Niña (cold) events with dry anomalies in the year following onset (yr $+1$). An El Niño response in yr $+1$ was also detected in the Great Lakes Region. Tree-ring chronologies from Oklahoma, Texas, and Mexico were also used to reconstruct a Southern Oscillation Index (SOI) for the winter season (DJF) from 1699 to 1971 with two methods – multiple regression and discriminant analysis. Spectral analysis showed significant ($P < 0.05$) variance peaks around 0.25 cycles/year (period $=$ 4 yr) in both observed and reconstructed SOI. Band-pass filters centered at 4 yr applied to both the observed SOI and the tree-ring reconstruction show synchronous long-term changes in the strength of that frequency component from 1880 to 1973. The filtered reconstruction reveals large changes in the amplitude of the 4-yr frequency component of the SOI from 1699 to 1970. These natural changes in the strength of the ENSO phenomenon need further confirmation with instrumental and proxy evidence, but they have major implications for long-range climate forecasting and the detection of anthropogenic climate change.

## Introduction

The El Niño/Southern Oscillation (ENSO) quasi-periodic behavior is now understood to be a product of coupled oceanic and atmospheric processes with several underlying periodicities (Graham and White 1988; Rasmusson et al. 1990; Diaz and Pulwarty 1992, this volume). ENSO extremes have a significant influence on climate over portions of North America, particularly over the southeastern United States and northern Mexico (e.g., Ropelewski and Halpert 1986, 1987, 1989; Kiladis and Diaz 1989; Cavazos and Hastenrath 1990). This ENSO teleconnection is strongest during winter and will probably play an increasingly important role in improved long-range 'forecasts of opportunity' for cool-season temperature and precipitation over the southeastern United States, where winter temperature forecasts already exhibit the highest skill in the continental United States (Livezey 1990). Because winter climate conditions influence the soil moisture balance and can strongly precondition subsequent tree growth, any ENSO influences on winter and early spring climate may also be registered in the growth of centuries-old trees. Identification of ENSO-sensitive tree-ring chronologies worldwide is an essential first step in the paleoclimatic investigation of ENSO and has been a primary goal of the research surrounding the ENSO workshop this volume is based on.

In this chapter, tree-ring reconstructions of summer moisture amounts extending back to A.D. 1700 in six climate regions of the eastern United States and southeastern Canada are examined for an ENSO response. The geographic distribution of the ENSO signal detected in these summer climate reconstructions are evaluated in light of the ENSO teleconnection evident in meteorological data from eastern North America. Tree-ring data from the region with the strongest ENSO signal were used to reconstruct a Southern Oscillation Index (SOI) extending from 1699 to 1971 (Stahle and Cleaveland, in press). Other investigators (e.g., Trenberth 1984; Rasmusson et al. 1990) have investigated the frequency characteristics of ENSO and have identified two important components in the quasi-biennial and 4- to 5-yr frequency ranges. In this paper, we test our reconstruction for significant concentrations of variance at ENSO frequencies with spectral and cross-spectral analyses. The observed and reconstructed SOI series are then filtered with band-pass filters designed to emphasize the strong 4-yr ENSO frequency component. The filtered reconstruction suggests considerable variation in strength at this ENSO frequency band over the southern United States and northern Mexico during the past 293 yr.

## Drought reconstructions for the eastern United States

A well-replicated network of some 150 tree-ring chronologies spanning most of the deciduous and coniferous forest biomes of the eastern United States was used to reconstruct summer Palmer Drought Severity Indices (PDSI; Palmer 1965) back to 1700 (Cook et al. 1992). Most of these chronologies are sensitive to climatic factors controlling growing season moisture supply, but there are differences in the response between some species and site types. Consequently, the climatic response of each chronology was carefully screened, and only those series sensitive at least in part to summer moisture variation were included in this analysis (i.e., over 100 chronologies, see Fig. 15.1).

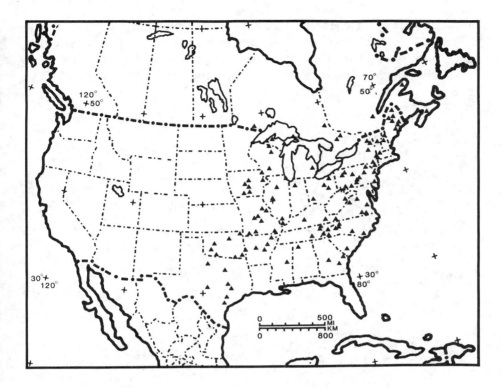

Fig. 15.1  The network of tree-ring sites used to reconstruct eastern United States climate 1700–1972.

Local tree-ring data were used to reconstruct summer PDSI in 24 separate states or regions within the eastern United States and extreme southeastern Canada (exclusive of the Great Plains states north of Texas, and excluding Louisiana, Mississippi, Alabama, Florida, Iowa, Minnesota, and Maine). In the northern states, June–July–August average PDSI was reconstructed with multiple regression using principal component (PC) scores from the tree-ring data as potential predictors of summer PDSI. The PC scores were derived from local tree-ring chronologies in or near each state. Individual tree-ring chronologies were used as potential predictors in stepwise multiple regression analyses with June PDSI in the southern states. These differences in the summer period actually reconstructed were dictated by large phenological differences in the tree growth season from the southern to northern United States, but should not seriously affect the following analyses of drought regions or ENSO influence in the tree-ring reconstructions for the eastern United States (Cook et al. 1992).

The autoregressive properties of the state-average summer drought data were reproduced in the reconstructions, and the tree-ring estimates of summer PDSI were validated by comparisons with independent PDSI data not used for calibration. Only those 23 state or regional drought reconstructions which passed the validation tests on independent data were used in subsequent analyses (Cook et al. 1992). These calibration and verification results indicate that the tree-ring data explain some 40 to 60% of the state-average summer drought indices, and are valid proxies of temporal changes in soil moisture supplies over most of the eastern United States.

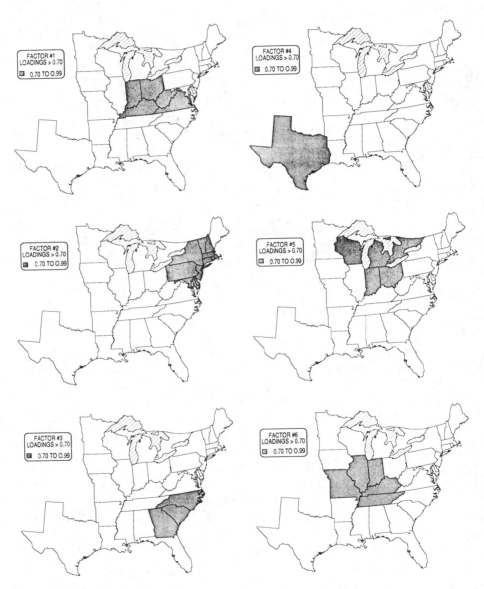

Fig. 15.2  The six obliquely rotated summer season drought factors. Areas with factor loadings ⩾ 0.70 are shaded (from Cook et al. 1992). Factors and area names are: 1. Mideastern, 2. Northeastern, 3. Southeastern, 4. Texas, 5. Great Lakes, and 6. Midwestern.

In order to identify homogeneous drought regions within the eastern United States from 1700 to 1972, the 23 verified summer drought reconstructions were entered into principal component analysis involving oblique rotation (RPCA; Richman 1986). The six drought regions identified (using criteria of an eigenvalue > 1.0, and rotated principal component or factor loadings > 0.70 by the individual state reconstructions) are illustrated in Figure 15.2 (Cook et al. 1992) The corresponding factor scores are shown in Figure 15.3. These drought regions generally agree with the regional drought patterns identified by Karl and Koscielny (1982) in an analysis of gridded PDSI data available from 1895 to 1981 for the entire United States. The agreement between the reconstructed and actual drought regions indicates that these tree-ring data are useful proxies of spatial as well as temporal variability in summer drought over the eastern United States.

## The ENSO signal in summer drought reconstructions for the eastern United States

Superposed epoch analysis (SEA; Haurwitz and Brier 1981) was used to search for an ENSO influence on the six regionalized reconstructions of summer drought for the eastern United States. SEA and related methods have been effectively used to search for the hypothesized impact on noisy climate data from episodic events such as ENSO extremes (Ropelewski and Halpert 1986, 1987; Bradley et al. 1987; Kiladis and Diaz 1989) and explosive volcanic eruptions (Kelly and Sear 1984; Bradley 1988; Mass and Portman 1989).

In this analysis, a homogenized Darwin-Tahiti SOI extending from 1851 to 1984 (Wright 1989) was used to identify the timing of warm and cold events. These El Niño and La Niña 'key years' were identified as departures ± 1.0 standard deviation from the 1851–1984 mean SOI. Note that El Niño events are indicated by positive indices in Wright's SOI, and that the best 'annual' SO indices are believed to represent an average of June through February monthly values, identified by the year of each June value (Wright 1989). For comparison with the results based on Wright's data, and for analyses prior to 1851, key years for El Niño events only were taken from the analysis of archival data by Quinn et al. (1987).

The SEA composites for warm and cold events were separately computed by averaging the reconstructed drought RPCA scores in each region among all key years (year 0), and among the 2 yr before and the 3 yr after each key year (yr −2, −1, +1, +2, and +3). These composites were computed separately for the key years derived from Wright (1989) and Quinn et al. (1987), and were calculated with a biweight robust mean to minimize the undesirable effect of statistical outliers on estimation of the mean (Mosteller and Tukey 1977). The inclusion of 3 yr following an ENSO event more than allows for the identification of any reasonable lag in the teleconnection between ENSO extremes and climate anomalies in eastern North America. There is no plausible physical basis for summer moisture anomalies in eastern North America to lead ENSO extremes in the

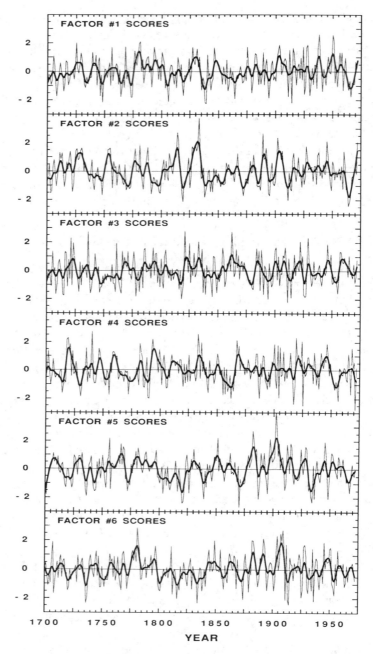

Fig. 15.3  The factor (obliquely rotated principal component; PC) scores of eastern United States summer drought 1700–1972 for regions shown in Figure 15.2 (from Cook et al. 1992). A smoothing filter (heavy line) depicts low-frequency variation.

tropical Pacific, but the 2 yr prior to the ENSO key years were included in the composites to help assess the random variability inherent in the SEA composite.

The key year randomization procedure of Haurwitz and Brier (1981) was used to determine the statistical significance of the reconstructed drought departures in the SEA composites. A pseudo-random number generator was used to select a new set of key years 1000 times, and 1000 random SEA composites were computed. Monte Carlo confidence intervals were determined by ranking the random composites and using the 2.5 and 97.5 percentiles as the lower and upper two-tailed 95% confidence limits.

Results of the superposed epoch analyses are illustrated for each regional drought reconstruction in Figure 15.4. These analyses indicate that El Niño and La Niña events have little influence on the summer PDSI reconstructions in most sections of the eastern United States. Only the Texas June PDSI reconstruction appears to have a statistically significant response to warm and cold ENSO extremes (Fig. 15.4d), which is similar to the ENSO signal in regional climate data. Warm events in yr 0 are followed by above average tree growth and reconstructed June PDSI in yr +1, while cold events are followed by below average tree growth and reconstructed June PDSI in the Texas region (Fig. 15.4d). This response is consistent with the ENSO signal detected in Southern Plains tree-ring data (see below) and with the sign of the ENSO teleconnection to winter and early spring climate variables in Texas (e.g., Douglas and Englehart 1981; Ropelewski and Halpert 1986; Kiladis and Diaz 1989).

Because seasonal tree growth and the Palmer Index for June both integrate the effects of precipitation and temperature anomalies during preceding months, a reconstructed June PDSI response to ENSO extremes during the preceding winter and spring remains plausible. This can be substantiated by correlation analyses between winter (DJF) SOI and the state-average June PDSI for Texas from 1888 to 1982 (Karl et al. 1983), which is statistically significant ($r = -0.39$, $P < .05$), particularly when the correlation is restricted to years of warm or cold ENSO events ($r = -0.58$, $P < .01$).

The Great Lakes summer drought reconstruction also exhibits a strong response to cold events in yr +1 (Fig. 15.4e). This Great Lakes response suggests drier than average conditions in the summer following La Niña episodes, but because this apparent response does not correspond to any consistent, well-documented ENSO teleconnection to Great Lakes meteorological data, its true significance is not clear. The 5-yr SEA means for warm and cold events in this region are quite different, which might partially explain the statistical significance of the yr +1 value for cold events (Fig. 15.4e). We do note, however, that the intense midwestern drought of 1988 developed in the context of a rapid and sweeping reversal of tropical Pacific sea surface temperature anomalies, shifting from El Niño conditions during 1986–87 to La Niña conditions in the spring and summer of 1988 (Trenberth et al. 1988). In addition, significant correlation of SOI extremes with the Palmer (1965) hydrologic drought index for Iowa (Karl et al. 1983) have recently been noted (Cleaveland and Duvick, 1992).

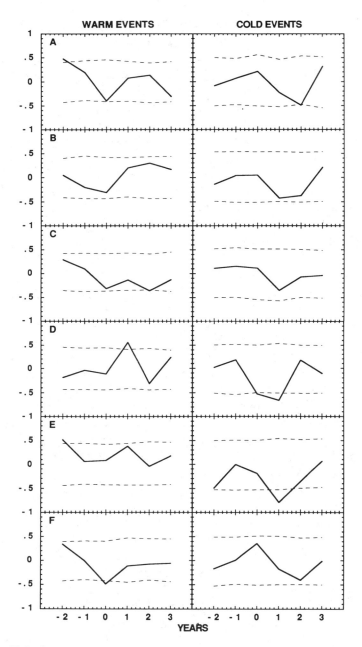

Fig. 15.4   Superposed epoch analysis of drought (obliquely rotated principal component) scores: A = Factor 1, the mideastern Factor; B = factor 2, the northeastern Factor; C = Factor 3, the southeastern Factor; D = Factor 4, the Texas Factor; E = Factor 5, the Great Lakes Factor; and F = Factor 6, the midwestern Factor. Dashed lines are two-tailed 95% confidence limits based on key year randomization. The solid lines are mean normalized drought scores for the years associated with warm (El Niño) and cold (La Niña) events based on stratification of Wright's (1989) SOI, 1851–1984.

To test further the possible El Niño influence on the drought reconstructions, the moderate to very strong El Niño years compiled from historical sources by Quinn et al. (1987) were used as key years in SEA for the period 1700 to 1983. These results were compared to the SEA results based on Wright's (1989) data, and again only the Texas region exhibits a statistically significant response associated with an increase in summer moisture amounts in yr +1 (not shown). However, the significance of this signal from 1699 to 1983 was slightly less than observed with Wright's data from 1851 to 1983 (see also Michaelsen and Thompson, 1992, this volume).

These analyses indicate only a modest ENSO influence on the development of reconstructed summer moisture anomalies over the Southern Great Plains (Texas), and little or no detectable influence elsewhere in the eastern United States and southeastern Canada. Although some strong warm or cold events do evidently influence summer climate over the eastern United States (e.g., the midwestern drought of 1988, linked to anomalies in the tropical Pacific [Trenberth et al. 1988]), the average effect of ENSO events appears to be weak in most of the eastern United States. Texas is the one region studied with a significant signal in both phases of the ENSO, and this is consistent with studies of instrumental climate data which find the strongest ENSO effect to be in Texas and other states immediately adjacent to the Gulf of Mexico (Douglas and Englehart 1981; Ropelewski and Halpert 1986, 1989; Kiladis and Diaz 1989). A strong ENSO influence is not evident in the summer drought reconstructions for the southeastern United States where a teleconnection to cool-season climate data has been detected, but this region is rather poorly represented by climate sensitive tree-ring chronologies.

The ENSO signal in the tree-ring reconstructions for Texas probably reflects in part the early initiation of tree growth, which can begin as early as late February and early March during warm winters (Stahle 1990). Therefore, some of the annual growth of trees in Texas appears to take place during the season when the direct ENSO influence on regional climate often occurs. ENSO induced changes in the soil moisture content during the winter and early spring may then continue to precondition tree growth later in the growing season when the direct climate teleconnection has diminished. Most other available tree-ring chronologies in the eastern United States are derived from trees which initiate radial growth much later in the spring (e.g., April or May), so that the carry-over effects of winter ENSO extremes on soil moisture levels may become more tenuous.

## A tree-ring reconstruction of the winter SOI

Given the distribution of the ENSO signal in climate data for eastern North America, and in the summer drought reconstructions, tree-ring data from the Southern Plains, Texas, and recently acquired chronologies from northern Mexico were used in attempts to reconstruct an ENSO index (Stahle and Cleaveland in press). Ten selected conifer chronologies collected in Mexico by M. A. Stokes,

T. H. Naylor, and others from the University of Arizona were included with several deciduous oak chronologies to compile a regional network extending from the Southern Plains into northern Mexico. This network provides the best available tree-ring chronology coverage of the region adjacent to the Gulf of Mexico where a reasonably strong and consistent ENSO teleconnection has been detected in climatological data compiled for the cool-season (e.g., Douglas and Englehart 1981; Ropelewski and Halpert 1986, 1989; Kiladis and Diaz 1989; Cavazos and Hastenrath 1990; Lough 1992, this volume).

Correlation analyses between a winter (DJF) SOI and each individual tree-ring choronology available from northern Mexico and the Southern Plains reveal the nature of the ENSO influence on climate-sensitive tree growth in this region (Fig. 15.5). This SOI is based on the normalized difference between monthly sea-level pressure at Tahiti and Darwin (Ropelewski and Jones 1987; Jones 1988; Allan et al. 1991). The index extends from 1866 to 1990, and missing values for Tahiti or Darwin during the 19th and early 20th centuries have been estimated from other stations in the tropical Pacific (Jones 1988). El Niño or warm events are indicated by strongly negative SO indices.

The Mexican chronologies tend to be most highly correlated with a cool-season SOI average extending from October to February, while the chronologies from

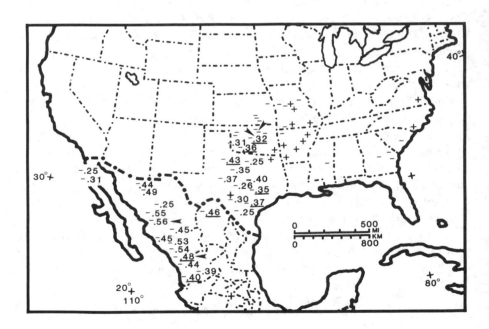

Fig. 15.5   Correlation coefficients between the winter SOI (Jones 1988; Jones 1989 pers. comm.) and tree-ring chronologies in Mexico and the southern United States from 1866 to 1965. Those correlations not significant ($P > 0.05$) show only the sign of the coefficient. Correlations of chronologies used in the multiple regression reconstruction are marked by arrows, those used in the discriminant analysis reconstruction are underlined.

the Southern Plains tend to be most highly correlated with a December to May SOI average. Consequently, the winter (DJF) SOI signal common to most chronologies in this region (Fig. 15.5) was selected for reconstruction. The negative correlation between tree growth and the winter SOI is consistent with the sign and seasonality of the ENSO teleconnection to climate data from the Gulf of Mexico region, and the SOI signal in tree growth appears to weaken from Mexico northward into the Great Plains in a fashion similar to the ENSO signal in rainfall data (Douglas and Englehart 1981; Lough 1992, this volume).

Several multiple regression models incorporating various tree-ring predictors from Mexico and the Southern Plains were considered in attempting to reconstruct winter SOI. All competing models were calibrated with winter SOI from 1900 to 1971 and were tested for accuracy and stability against the independent winter SOI data from 1866 to 1899.

A calibration model involving multiple regression (Draper and Smith 1981) between two regional tree-ring chronology averages from Mexico and Oklahoma and winter SOI provides both plausible and stable estimates of winter SOI (Figs. 15.5, 15.6). This calibration model accounts for 41% of the winter SOI variance from 1900 to 1971 (adjusted for loss of degrees of freedom), and is validated by comparison of the reconstruction with the independent SOI data from 1866 to 1899 ($r = 0.51$, reduction of error statistic [RE; Fritts 1976] = 0.24; Fig. 15.6). The Mexican tree-ring average is based on the Guadalupe and Creel Douglas fir (*Pseudotsuga menziesii*) chronologies from Durango and Chihuahua, respectively. The Oklahoma tree-ring average represents the mean of two post oak (*Quercus stellata*) chronologies from northcentral and southcentral Oklahoma (i.e., the Lake Keystone Reservoir and Lake Arbuckle series). The Mexican average explains the largest fraction of winter SOI variance, but the Oklahoma average contributes significantly to the regression model and

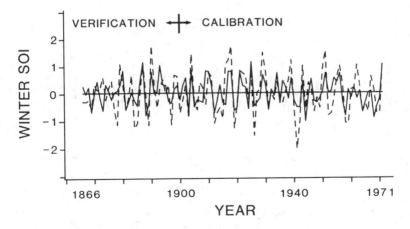

Fig. 15.6   Observed (dashed lined) and reconstructed (solid line) winter SOI (from Stahle and Cleaveland, in press). The multiple regression calibration period is 1900–1971 and the validation period is 1866–1899.

accounts for an additional 6% of the winter SOI variance. This relative contribution of data from Mexico and Oklahoma is reasonable given the geographic distribution of the ENSO signal in this region. A comparison of the regression estimates of winter SOI from 1900 to 1971 based on just the Mexican chronology average ($R^2_{adj.} = 0.35$) with the multiple regression model using both the Mexican and Oklahoma chronology averages ($R^2_{adj.} = 0.41$) indicates that the Oklahoma data contribute primarily during years with large positive or negative SO indices, presumably reflecting an expansion of the area or intensity of teleconnected climate effects during the largest warm and cold ENSO episodes (Stahle and Cleaveland, in press). The Texas chronologies apparently enter stepwise regression after the Oklahoma chronologies (which have equal or lower correlations with the SOI, Fig. 15.5) probably because multicollinearity with Mexican chronologies excludes the Texas series. Although Lough's (1992, this volume) calibration and verification statistics are higher, her tree-ring network coverage has little overlap with ours and may represent a somewhat different extratropical ENSO signal.

The winter SOI reconstruction based on the multiple regression with the Mexican and Oklahoma averages extends from 1699 to 1971 (Fig. 15.7). Figure 15.7 also identifies a number of winter SOI extremes estimated from a selected network of eight SOI-sensitive tree-ring chronologies in Mexico, Texas, and Oklahoma using a classification scheme based on a discriminant function (Cooley and Lohnes 1971) analysis (Stahle and Cleaveland, in press). This eight chronology network includes two of the series in the Mexican and Oklahoma regression analysis averages, but is more evenly distributed across northern Mexico, Texas, and Oklahoma where the strongest ENSO teleconnection has been detected (Fig. 15.5). To classify past SOI extremes, the winter SOI data were first subdivided into one of three categories: cold event, near normal, or warm event

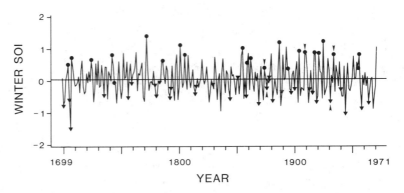

Fig. 15.7  Winter SOI series 1699–1971 reconstructed with multiple regression model. Dots are years classified by discriminant analysis (posterior probability of group membership $\geqslant 0.65$) as El Niño events and triangles are La Niña classifications to 1965. Events marked with arrows 1866–1965 are misclassified. (From Stahle and Cleaveland in press.)

(i.e., SOI $\geq$ 0.5; SOI < 0.5 and > $-0.5$; or SOI $\leq$ $-0.5$, respectively). Three linear discriminant functions were then derived from the growth anomalies recorded at each of the eight sites from 1866 to 1965 and used to classify each year into one of the three SOI categories (some of the Mexican chronologies end in 1965). No prewhitening was performed on either the winter SOI or the tree-ring data. Using a stringent posterior probability for group membership of $\geq$ 0.65 in order to minimize false classifications, the discriminant function was able to correctly classify 39% of all warm or cold event 'extremes' from 1866 to 1965. Of the 27 classified warm or cold extremes, 22 were correct, and none of the 5 misclassifications actually belonged in the opposite extreme (i.e., they were 'near normal' years). These classification results were verified using a subperiod analysis (not shown), which suggests that the classifications of past winter SOI extremes illustrated in Figure 15.7 correctly classify some 39% of past extremes with approximately 80% accuracy, and with a very low incidence of outright misclassifications (i.e., the classification of an El Niño event which was actually a La Niña event or vice versa).

### Spectral analysis of winter SOI

Previous analyses of ENSO indices have shown that most SO variance is concentrated in the 2- to 10-yr frequency range (e.g., Trenberth 1976; Chen 1982; Trenberth and Shea 1987; Schneider and Schönwiese 1989). Recent investigation of sea surface temperature and zonal wind data from the equatorial Pacific shows two dominant frequency modes at the 2- and 4- to 5-yr time scales, and translation of these two frequency components into the time domain suggests that they may cancel or reinforce each other to alter the amplitude of warm and cold events (Rasmusson et al. 1990). On the basis of meteorological and proxy data, however, it also appears that the dominant mode of ENSO variance at low frequencies fluctuates with time (e.g., Rasmusson and Carpenter 1982; Michaelsen 1989; Rasmusson et al. 1990). Proxy data including tree rings may provide valuable insight into the amplitude and frequency of the ENSO phenomenon over longer periods than is possible with instrumental records (see Michaelsen and Thompson 1992, this volume).

Cross-spectral analysis (Jenkins and Watts 1968; Marple 1987) was used to examine the accuracy with which our regression-generated reconstruction reproduces the low frequency characteristics of the winter SOI in the common period from 1866 to 1971. The full reconstruction (1699–1971) was then analysed to examine the strength of its high and low frequency components. The auto-covariance, power spectrum, cross-covariance, and cross-spectrum were computed with the Fourier transform, and spectral estimates were then smoothed with a Hamming window (IMSL Inc. 1982). Spectral densities of observed and reconstructed data are plotted (Fig. 15.8a) along with the squared coherence function for the two series (Fig. 15.8b). Two strong spectral peaks appear in the observation-based winter SOI from 1866 to 1971 at 3.7 to 4.0 and 5.2 to 6.5 yr

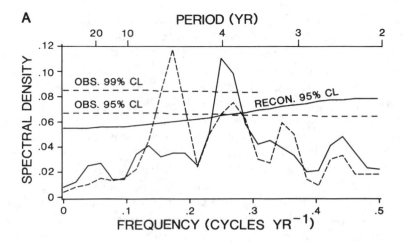

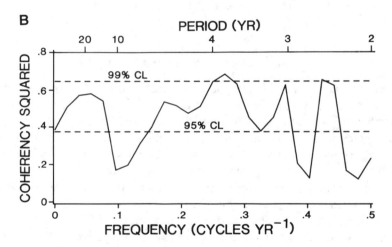

Fig. 15.8 Spectral analysis of reconstructed winter SOI 1866–1971. (A) Power spectrum of observed (dashed line) and reconstructed (solid line) winter SOI. The confidence limits (CL) were computed from the white noise null continuum for the observed SOI data (Jenkins and Watts 1968). The CL were computed from a slightly blue noise null continuum for the reconstructed data during this subperiod (1866–1971), although the null continuum is essentially a white noise process over the full time period of the reconstruction (1699–1971, Fig. 15.9). Bandwidth = 0.0484 cycles/yr, degrees of freedom = 10.26. (B) Coherency squared function for observed and reconstructed winter SOI, measuring the strength of the association of the two series (analogous to the correlation coefficient).

(Fig. 15.8a). Reconstructed winter SOI also contains a strong peak at 3.7 to 4.0 yr, but shows no significant spectral power at the second frequency band (Fig. 15.8a, Fig. 15.9). The difference is not attributable to lack of coherence because the squared coherence function is significant ($P < 0.05$) between 2.7 and 6.5 yr (Fig. 15.8b) and the series are in phase at all frequencies ($P < 0.05$).

The observed winter SOI spectra provide scant evidence for any important amount of variance concentrated around biennial frequencies, as was recently detected in zonal wind and SST data (Rasmusson et al. 1990). The Tahiti-Darwin surface pressure indices do not completely represent the Southern Oscillation phenomenon, however, and our use of only winter average SOI further restricts the degree to which this index measures the SO.

The low frequency (4- to 5-yr) component of ENSO discussed by Rasmusson et al. (1990) does appear to be present in both the observed and reconstructed winter SOI from 1866 to 1971 (Fig. 15.8) and may represent the predominant frequency at which the ENSO influences climate over the Gulf of Mexico region. The tendency for a change in phase of the biennial and low frequency components on every other biennial peak illustrated by Rasmusson et al. (1990) might reinforce the 4- to 5-yr component, with more significant large-scale climate influences in the extratropics, including northern Mexico and the Southern Plains.

Spectral analysis of the full 273-yr reconstruction also shows a concentration of variance around a period of 4 yr (Fig. 15.9). This highly significant spectral peak accounts for over 27% of the variance of the reconstructed winter SOI series. This may be the strongest signature of the ENSO phenomenon in this region, and this particular frequency band may alone account for roughly one quarter of the variance in tree growth and the climate conditions which drive tree growth over northern Mexico and the Southern Great Plains. In fact, the strength of the spectral peak at 4 yr and the lack of evidence for a biennial component (Fig. 15.9) suggests that the 4 yr component alone, or perhaps its interaction with the biennial component of ENSO identified by Rasmusson et al. (1990), may be the frequency at which ENSO has the largest influence on climate in the

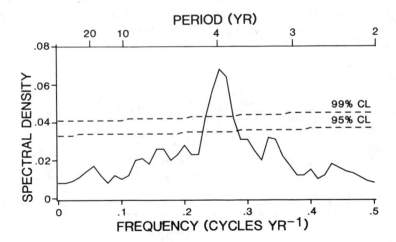

Fig. 15.9   Power spectrum of reconstructed winter SOI, 1699–1971. Confidence limits (CL) are computed from the null continuum. Bandwidth = 0.028 cycles/yr, degrees of freedom = 15.26.

extratropics of North America, if not worldwide. This possibility could be further evaluated with spectral analyses of instrumental and proxy climate data from ENSO-teleconnected regions of the extratropics. We have not analysed meteorological data from Mexico, but instrumental June PDSI from an average of two climatic divisions in southern Texas have a significant spectral peak at 4 yr ($P < 0.05$, Stahle and Cleaveland 1988). June PDSI is a good measure of the primary climate signal in tree-ring data from the Southern Plains, and is probably a reasonable measure in northern Mexico as well. Nevertheless, in spite of the documented ENSO teleconnections to climate in the Gulf of Mexico region, and the presence in the tree-ring reconstruction of a highly coherent low frequency component in a known ENSO frequency band, the similar spectral peaks around 4 yr in the actual and reconstructed SOI may not be entirely causally linked. Other mechanisms are certainly involved in climate variability in this region, and may account for some of the variance in the 4-yr frequency range.

To investigate secular variability in the 4-yr frequency range, we applied 5-, 11-, and 31-weight band-pass filters designed to pass variance centered at a period of 0.25 cycles/year (Rabiner and Gold 1975). A symmetrical Chebyshev filter with 31 weights ($-0.0006$, $0.0113$, $0.0016$, $-0.0084$, $-0.0028$, $-0.0054$, $0.0042$, $0.0389$, $-0.0051$, $-0.0899$, $0.0050$, $0.1466$, $-0.0037$, $-0.1914$, $0.0014$, $0.2084$, $0.0014$, . . . , $-0.0006$) designed by the second author was selected to minimize 'leakage' around the target frequency. Output from this filter is shown in Figures 15.10 and 15.11. The band-pass filtered series (Fig. 15.10) have large changes in amplitude which are synchronized between the observed and reconstructed SOI. The peaks are most phase-locked in the periods when the amplitudes are greatest, and are out of phase only when the amplitude of both series is lowest (Fig. 15.10). This change in the degree of phase-locking from periods of high to low amplitude variations is physically reasonable because ENSO forcing of climate and tree growth in the extratropics should be strongest during periods with strong ENSO oscillations. Because both reconstructed and

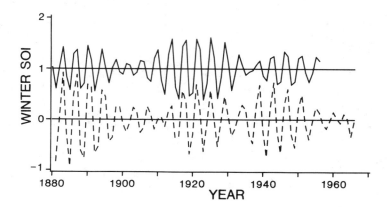

Fig. 15.10 Band-pass filter of the observed (dashed line) and reconstructed (solid line) winter SOI (filter frequency response centered at 4 yr).

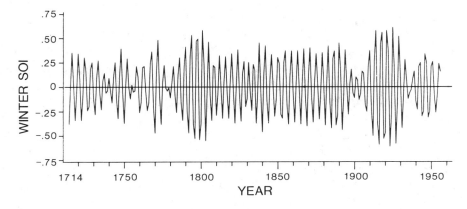

Fig. 15.11   Band-pass filter of the reconstructed winter SOI, 1699–1971 (filter frequency response centered at 4 yr).

observed SOI series show the same response in this frequency band, the variation is not an artifact of reconstruction methodology or tree growth processes unrelated to climate. Other SOI reconstructions appear to show similar changes in variance through time (Michaelsen 1989; Lough and Fritts 1985, 1990), although they have not been band-pass filtered. These ENSO amplitude variations may be caused by changes in tropical oceanic-atmospheric conditions similar to those documented by Elliott and Angell (1988) and Trenberth (1990).

The large-magnitude variance changes seen in the modern period (Fig. 15.10) continue into the past, with other high amplitude periods as shown in Figure, 15.11: 1715–1735, 1745–1755, 1765–1775, 1785–1805, a nearly continuous high-amplitude period 1835–1890, and the period of highest amplitude 1910–1930 (Fig. 15.11). The irregular spacing of periods of low and high variance makes it unlikely that the properties of the filtered series are harmonic artifacts of the filters themselves. These tree-ring estimates, therefore, suggest that winter SOI, or at least the extratropical influence of ENSO, may be subject to simultaneous changes in the frequency and amplitude of both warm and cold events on decadal to multidecadal timescales (see Diaz and Pulwarty 1992, this volume).

## Conclusions

The strongest influence of ENSO on summer drought reconstructions available for the eastern United States and southeastern Canada is located in Texas. This was hardly unexpected because the strongest, most consistent ENSO signal in meteorological data for the eastern United States is located in close proximity to the Gulf of Mexico (Douglas and Englehart 1981; Ropelewski and Halpert 1986; Kiladis and Diaz 1989). This is a cool season teleconnection for the most part, and typically does not persist through the boreal spring. Tree growth in Texas usually begins in early spring when ENSO influences may persist, but tree growth begins later farther north. Tree-ring data in the Midwest and northeastern United

States therefore appear to be less sensitive proxies of ENSO teleconnections. The summer drought reconstruction for the Southeast region does not exhibit a strong ENSO signal, and this may also reflect late spring initiation of radial growth characteristic of the baldcypress trees used in the reconstruction. However, most of these baldcypress (*Taxodium distichum*) chronologies are north of the Gulf coast region, out of the area where the strongest ENSO teleconnection has been detected in meteorological data (e.g., Rogers 1988). If a network of climate sensitive tree-ring chronologies can be developed for a variety of species in the Gulf coastal states, it may be possible to use these data to estimate past ENSO activity.

This study confirms several previous reports and suggests that the strongest ENSO signal in North American tree-ring data may actually be found in conifer chronologies from northern Mexico (e.g., this study; Lough and Fritts 1990; Lough 1992, this volume; Stahle and Cleaveland in press). Nonetheless, statistically significant ENSO influences have been detected in tree-ring data from the Pacific Northwest (Lough and Fritts 1990), the Southwest (Michaelsen 1989), the Southern Plains, and may be detected along the Gulf of Mexico when suitable tree-ring chronologies are available. Analyses of these additional ENSO proxies, and of the instrumental climate record will be necessary to confirm or refute the multidecadal-scale changes in the strength and frequency of warm and cold ENSO events detected by the time series analyses of observed and reconstructed winter SOI. This work can be easily justified because the apparent secular changes in ENSO amplitude have important implications for long range weather forecasting based in part on conditions in the tropical Pacific, and may considerably complicate the detection of anthropogenic climate change.

*Acknowledgments* This research has been supported by the National Science Foundation, Climate Dynamics Program, through grants ATM-8914561 (University of Arkansas) and ATM 87–16630 (Lamont-Doherty Geological Observatory), and by National Oceanic and Atmospheric Administration grant NA89AA-D-AC199. We thank Don Graybill and two anonymous reviewers for constructive criticism to improve the paper. Dave Meko and Tom Harlan of the Laboratory of Tree-Ring Research, University of Arizona, supplied the Mexican tree-ring chronologies and site information. Dr. Dwight Mix, University of Arkansas Department of Electrical Engineering, developed several band-pass filters for our use. Dr. James Dunn, University of Arkansas Department of Mathematics, advised us on discriminant analysis. Lamont-Doherty Geological Observatory Contribution No. 4965.

# References

ALLAN, R. J., NICHOLLS, N., JONES, P. D., and BUTTERWORTH, I. J., 1991: A further extension of the Tahiti-Darwin SOI, early ENSO events and Darwin pressure. *Journal of Climate*, 4: 743–749.

BRADLEY, R. S., 1988: The explosive volcanic eruption signal in Northern Hemisphere continental temperature records. *Climatic Change*, 12: 221–243.

BRADLEY, R. S., DIAZ, H. F., KILADIS, G. N., and EISCHEID, J. K., 1987: ENSO signal in continental temperature and precipitation records. *Nature*, 327: 497–501.

CAVAZOS, T. and HASTENRATH, S., 1990: Convection and rainfall over Mexico and their modulation by the Southern Oscillation, *International Journal of Climatology*, 10: 377–386.

CHEN, W. Y., 1982: Assessment of Southern Oscillation sea level pressure indices. *Monthly Weather Review*, 110: 800–807.

CLEAVELAND, M. K. and DUVICK, D. N., 1992: Iowa climate reconstructed from tree rings, A.D. 1640 to 1982. *Water Resources Research*, in press.

COOK, E. R., STAHLE, D. W., and CLEAVELAND, M. K., 1992: Dendroclimatic evidence from eastern North America. *In* Bradley, R. S. and Jones, P. D. (eds.), *Climate Since A.D. 1500*. London: Routledge, 331–348.

COOLEY, W. W. and LOHNES, P. R., 1971: *Multivariate Data Analysis*. New York: Wiley. 364 pp.

DIAZ, H. F. and PULWARTY, R. S., 1992: A comparison of Southern Oscillation and El Niño signals in the tropics. *In* Diaz, H. F. and Markgraf, V. (eds.), *El Niño: Historical and Paleoclimatic Aspects of the Southern Oscillation*. Cambridge: Cambridge University Press, 175–192.

DOUGLAS, A. V. and ENGLEHART, P. J., 1981: On a statistical relationship between autumn rainfall in the central equatorial Pacific and subsequent winter precipitation in Florida. *Monthly Weather Review*, 109: 2377–2382.

DRAPER, N. R. and SMITH, H., 1981: *Applied Regression Analysis*, Second Edition. New York: Wiley. 709 pp.

ELLIOTT, W. P. and ANGELL, J. K., 1988: Evidence for changes in Southern Oscillation relationships during the last 100 years. *Journal of Climate*, 1: 729–737.

FRITTS, H. C., 1976: *Tree Rings and Climate*. London: Academic Press. 567 pp.

GRAHAM, N. W. and WHITE, W. B., 1988: The El Niño cycle: A natural oscillator of the Pacific Ocean – atmosphere system. *Science*, 240: 1293–1302.

HAURWITZ, M. W. and BRIER, G. W., 1981: A critique of the superposed epoch analysis method: its application to solar-weather relations. *Monthly Weather Review*, 109: 2074–2079.

IMSL Inc., 1982: *IMSL Library Reference Manual*, Vol. 2. Houston: IMSL Inc. FTFREQ-1 to FTFREQ-5.

JENKINS, G. M. and WATTS, D. G., 1968: *Spectral Analysis and Its Applications*. San Francisco: Holden-Day. 525 pp.

JONES, P. D., 1988: The influence of ENSO on global temperatures. *Climate Monitor*, 17: 80–89.

KARL, T. R. and KOSCIELNY, A. J., 1982: Drought in the United States: 1895–1981. *Journal of Climatology*, 2: 313–329.

KARL, T. R., METCALF, L. K., NICODEMUS, M. L., and QUAYLE, R. G., 1983: *Statewide Average Climatic History Texas 1888–1982*. Historical Climatology Series 6–1. Asheville, North Carolina: National Climatic Data Center. 39 pp.

KELLY, P. M. and SEAR, C. B. 1984: Climatic impact of explosive volcanic eruptions. *Nature*, 311: 740–743.

KILADIS, G. N. and DIAZ, H. F., 1989: Global climatic anomalies associated with extremes in the Southern Oscillation. *Journal of Climate*, 2: 1069–1090.

LIVEZEY, R. E., 1990: Variability of skill of long-range forecasts and implications for their use and value. *Bulletin of the American Meteorological Society*, 71: 300–309.

LOUGH, J. M., 1992: An index of the Southern Oscillation reconstructed from western North American tree-ring chronologies. *In* Diaz, H. F. and Markgraf, V. (eds.), *El Niño: Historical and Paleoclimatic Aspects of the Southern Oscillation*. Cambridge: Cambridge University Press, 215–226.

LOUGH, J. M. and FRITTS, H. C., 1985: The Southern Oscillation and tree rings: 1600–1961. *Journal of Climate and Applied Meteorology*, 24: 952–966.

LOUGH, J. M. and FRITTS, H. C., 1990: Historical aspects of El Niño/Southern Oscillation – information from tree-rings. *In* Glynn, P. W. (ed.), *Global Ecological Consequences of the 1982–83 El Niño-Southern Oscillation*. Amsterdam: Elsevier, 285–321.

MARPLE, S. L., Jr., 1987: *Digital Spectral Analysis With Applications*. Englewood Cliffs, N.J.: Prentice-Hall. 492 pp.

MASS, C. F. and PORTMAN, D. A., 1989: Major volcanic eruptions and climate: a critical evaluation. *Journal of Climate*, 2: 566–593.

MICHAELSEN, J., 1989: Long-period fluctuations in El Niño amplitude and frequency reconstructed from tree-rings. *In* Peterson, D. H. (ed.), *Aspects of Climate Variability in the Pacific and the Western Americas*. Geophysical Monograph 55. Washington, D.C.: American Geophysical Union, 69–74.

MICHAELSEN, J. and THOMPSON, L. G., 1992: A comparison of proxy records of El Niño/Southern Oscillation: *In* Diaz, H. F. and Markgraf, V. (eds.), *El Niño: Historical and Paleoclimatic Aspects of the Southern Oscillation*. Cambridge: Cambridge University Press, 323–348.

MOSTELLER, F. and TUKEY, J. W., 1977: *Data Analysis and Regression*. Reading, Mass.: Addison-Wesley. 588 pp.

PALMER, W. C., 1965: *Meteorological Drought*. Weather Bureau Research Paper No. 45. Washington, D.C.: U.S. Department of Commerce. 58 pp.

QUINN, W. H., NEAL, V. T., and ANTUNEZ de MAYOLO, S. E., 1987: El Niño occurrences over the past four and a half centuries. *Journal of Geophysical Research*, 92C: 14,449–14,461.

RABINER, L. R. and GOLD, B., 1975: *Theory and Application of Digital Signal Processing*. Englewood Cliffs, N.J.: Prentice-Hall. 762 pp.

RASMUSSON, E. M. and CARPENTER, T. H., 1982: Variations in tropical sea surface temperature and surface wind fields associated with the Southern Oscillation/El Niño. *Monthly Weather Review*, 110: 354–384.

RASMUSSON, E. M. and WALLACE, J. M., 1983: Meteorological aspects of El Niño/ Southern Oscillation. *Science*, 222: 1195–1202.

RASMUSSON, E. M., WANG, X., and ROPELEWSKI, C. F., 1990: The biennial component of ENSO variability. *Journal of Marine Systems*, 1: 71–96.

RICHMAN, M. B., 1986: Rotation of principal components. *Journal of Climatology*, 6: 293–335.

ROGERS, J. C., 1988: Precipitation variability over the Caribbean and tropical Americas associated with the Southern Oscillation. *Journal of Climate*, 1: 172–182.

ROPELEWSKI, C. F. and HALPERT, M. S., 1986: North American precipitation and temperature patterns associated with the El Niño/Southern Oscillation (ENSO). *Monthly Weather Review*, 114: 2352–2362.

ROPELEWSKI, C. F. and HALPERT, M. S., 1987: Global and regional scale precipitation patterns associated with the El Niño/Southern Oscillation. *Monthly Weather Review*, 115: 1606–1626.

ROPELEWSKI, C. F. and HALPERT, M. S., 1989: Precipitation patterns associated with the high phase of the Southern Oscillation. *Journal of Climate*, 1: 172–182.

ROPELEWSKI, C. F. and JONES, P. D., 1987: An extension of the Tahiti-Darwin Southern Oscillation Index. *Monthly Weather Review*, 115: 2161–2165.

SCHNEIDER, U. and SCHÖNWIESE, C. D., 1989: Some statistical characteristics of El Niño/Southern Oscillation and North Atlantic oscillation indices. *Atmósfera*, 2: 167–180.

STAHLE, D. W., 1990: The Tree-Ring Record of False Spring in the Southcentral U.S.A. Ph.D. dissertation, Arizona State University. 272 pp.

STAHLE, D. W. and CLEAVELAND, M. K., 1988: Texas drought history reconstructed and analyzed from 1698 to 1980. *Journal of Climate*, 1: 59–74.

STAHLE, D. W. and CLEAVELAND, M. K., in press: Southern Oscillation extremes reconstructed from tree rings of the Sierra Madre Occidental and southern Great Plains. *Journal of Climate*.

TRENBERTH, K. E., 1976: Spatial and temporal variations of the Southern Oscillation. *Quarterly Journal of the Royal Meteorological Society*, 102: 639–653.

TRENBERTH, K. E., 1984: Signal versus noise in the Southern Oscillation. *Monthly Weather Review*, 112: 326–332.

TRENBERTH, K. E., 1990: Recent observed interdecadal climate changes in the Northern Hemisphere. *Bulletin of the American Meteorological Society*, 71: 988–993.

TRENBERTH, K. E., BRANSTATOR, G. W., and ARKIN, P. A., 1988: Origins of the 1988 North American drought. *Science*, 242: 1640–1645.

TRENBERTH, K. E. and SHEA, D. J., 1987: On the evolution of the Southern Oscillation. *Monthly Weather Review*, 115: 3078–3096.

WRIGHT, P. B., 1989: Homogenized long-period Southern Oscillation Indices. *International Journal of Climatology*, 9: 33–54.

# Records from ice cores and corals

# 16

# Reconstructing interannual climate variability from tropical and subtropical ice-core records

L. G. THOMPSON AND E. MOSLEY-THOMPSON

*Byrd Polar Research Center, The Ohio State University, Columbus, Ohio 43210, U.S.A.*

P. A. THOMPSON

*Department of Decision and Information Sciences, University of Florida, Gainesville, Florida 32611, U.S.A.*

## Abstract

The patterns and sources of interannual, decadal, and century-scale climatic and environmental variability are of greatest relevance to human activities. Ice-core records from tropical and subtropical ice caps provide unique information about the chemical and physical character of the atmosphere. Annual variations in the amount and chemical composition of precipitation accumulating on these ice caps produce annual laminations which allow precise dating of these stratigraphic sequences. The thickness of an annual lamination reflects the net accumulation, while the physical and chemical constituents (e.g., dust, isotopes, ions) record local atmospheric conditions during deposition. In this chapter interannual climate variability is reconstructed from ice cores on the tropical Quelccaya ice cap, Peru, and the subtropical Dunde ice cap, China.

Except for the annual cycle, the El Niño/Southern Oscillation (ENSO) is the dominant signal in the global climate system on time scales of months to several years. Associated with major dislocations of rainfall regimes in the tropics and subtropics, ENSO events leave signatures in the physical and chemical character of the annually accumulated layers. Variations in annual layer thicknesses, microparticle concentrations, conductivities, and oxygen isotopic ratios are examined for the last 150 yr when limited documentary data exist for evaluation and calibration. When the 1500-yr Quelccaya net balance record is integrated with archaeological evidence, it appears that oceanic/atmospheric linkages which produce ENSO-like precipitation responses may dominate for longer periods. A comparison of decadally averaged net balance on Quelccaya with that on the Dunde ice cap, from A.D. 1600 to 1980 indicates the existence of low-frequency teleconnections across the Pacific Basin through the Walker Circulation.

## Introduction

This chapter evaluates the history of interannual climatic and environmental variability in the southern Andes of Peru as recorded in the Quelccaya ice cap. Interannual variability is a major feature of the global climate system. Since human activities may be inadvertently altering the mean state of the climate system, it is essential to determine whether interannual variability is closely linked to changes in the mean state. Urgency arises from the likelihood that human activities are more sensitive to interannual variability than to the slower changes in the system's mean state.

Characterizing the relationship between interannual variability and changes in the longer-term mean requires long records of climatic variability which encompass times when the climate was different from today (e.g., Little Ice Age, Medieval Warm Period, Last Glacial Stage). Ice-core data provide a multifaceted record of past variations in the Earth's climate and environment. Seasonal variations in precipitation amount and chemical composition produce laminations in both polar and alpine (high elevation) glaciers. Such seasonal variations allow these stratigraphic sequences to be dated precisely (Thompson et al. 1984a), a prerequisite for discerning the interannual character of the record.

This chapter examines the interannual climate variability preserved in ice core records from the Quelccaya ice cap (13°56′S, 70°50′W) (Fig. 16.1) located in the southern Andes of Peru. Here two ice cores, one 154.8-m core (henceforth summit core) containing a record of 1350 yr and a 163.6-m core (henceforth core 1) containing a record of 1500 yr, were recovered in 1983. These records are examined over two time intervals. The first focuses upon the last 150 yr when limited documentary data exist for evaluation and calibration of the record. The second interval is the last 1500 yr for which independent evaluation of the record is very limited. Finally, the decadally averaged net accumulation record from Quelccaya is compared with that from the subtropical Dunde ice cap, China. For two such widely separated regions there is a remarkable similarity in the net accumulation trends over the last 400 yr, although Dunde appears to lag Quelccaya by several decades.

## Recent variations: A.D. 1850 to 1984

### The Quelccaya ice cap record

Figure 16.2 presents the annual averages of microparticle concentrations (MPC), liquid conductivities (LC), net accumulation ($A_n$), and oxygen isotopic ratios ($\delta^{18}O$) from A.D. 1850 to 1984. Both LC and MPC (Fig. 16.2) clearly show that elevated concentrations of both soluble and insoluble material characterized the

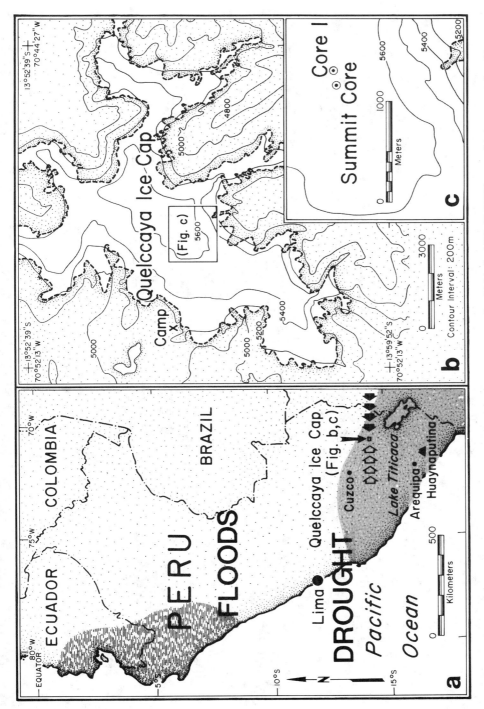

Fig. 16.1  The location of (a) the Quelccaya ice cap relative to precipitation patterns associated with the 1982–83 ENSO event; (b) the topography of ice cap, and (c) the locations of two deep cores drilled in 1983. Droughts occur in the south of Peru while floods occur in the north. The prevailing winds over Quelccaya during the southern hemisphere summer are from the east (shown by the solid arrows) and in the southern hemisphere winter are from the west (shown by open arrows).

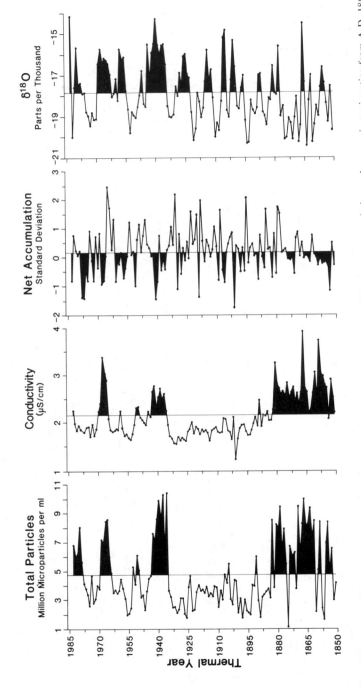

Fig. 16.2  Annual (July–June) averages for particulate concentrations, liquid conductivities, net accumulation and oxygen isotopic ratios from A.D. 1850 to 1984 on the Quelccaya ice cap, Peru. The average (solid line) for that period is included for each parameter.

end of the Little Ice Age (LIA) (Thompson et al. 1986). Most notable is the abrupt concentration decrease from 1882 to 1884 (Thompson and Mosley-Thompson 1987). From 1884 to 1935 the average concentrations of both insoluble dust and soluble materials are below their 1850–1984 averages with the exception of a minor peak in 1891. Greatly elevated concentrations of soluble and insoluble particulates occur from 1935 to 1945, synchronous with the most severe drought recorded in southern Peru prior to 1950 (Newell 1949). Since 1950 there have been four periods of elevated dust concentrations (1951–1954, 1959–1960 [minor], 1965–1970, 1980–1984) with low concentrations during intervening times. Note that from 1980 to 1984 insoluble particles were more highly concentrated than soluble material.

The oxygen isotopic ratios ($\delta^{18}O$) (Fig. 16.2) exhibit a higher frequency of variation than the particulate concentrations. Over much of this period (A.D. 1880–1984) the variability appears somewhat quasi-periodic with an average periodicity of ~ 12 yr. From A.D. 1880 to 1940 the mean $\delta^{18}O$ values gradually increase to a maximum (least negative) in the early 1940s. From the 1940s to late 1970s there has been a gradual decrease (more negative) in the mean $\delta^{18}O$ values, consistent with the northern hemispheric temperature cooling trend from the 1940s to the mid-1970s. These trends are most evident when the annual data are averaged decadally (see Thompson et al. 1986). There is also a gradual change in the frequency and amplitude of $\delta^{18}O$. Prior to A.D. 1882 more negative (depleted in $\delta^{18}O$) values occurred more frequently and with a shorter periodicity. During this interval (1850–1880) there were more years with $\delta^{18}O$ around $-20°/_{oo}$ and decadal averages prior to 1880 were consistently lower.

It has been shown previously (Thompson et al. 1986; Thompson and Mosley-Thompson 1989; Thompson 1992) that more negative oxygen isotopic values characterized the entire LIA period from A.D. 1530 to 1880. Thompson et al. (1986) demonstrated that decadally averaged $\delta^{18}O$ values closely reflect the Northern Hemisphere decadally averaged temperatures (Landsberg 1985) from A.D. 1580 to 1980. Thus, less negative $\delta^{18}O$ values appear to represent warmer temperatures while the more negative oxygen isotopic values represent cooler temperatures. However, the reader is reminded that on an annual basis $\delta^{18}O$ may be altered by other factors such as (1) variations in the $^{18}O$ depletion as water vapor is transported over the Amazon Basin and the air masses rise up the Andes; (2) seasonal and annual variations in snow intensity; and (3) seasonal changes in sublimation of snow on the ice cap surface (Grootes et al. 1989). The longer records (centuries) of $\delta^{18}O$, MPC, and LC reveal that periods of less negative $\delta^{18}O$ values such as during the 1935 to 1945 drought, are consistently correlated with increased concentrations of both insoluble and soluble particulates.

The net accumulation record (Fig. 16.2) reveals a marked increase in the annual variability of both amplitude and frequency beginning abruptly in 1878. In fact, reduced variability and small amplitudes in $A_n$ are characteristic only of the transitional period near the end of the LIA (1850–1878). These striking differences in the frequency and/or amplitude of all four environmental parameters are thought to directly reflect concomitant changes in the circulation regime from the LIA period to the post-LIA period (Thompson et al. 1986). The net accumulation record ($A_n$) shows a general increasing trend from lower values more characteristic of the latter half of the LIA (A.D. 1720 to 1860). This increasing trend in $A_n$ peaked in the 1930s and has decreased since (Thompson et al. 1985). The interpretation of $A_n$ from 1967 to present must be made cautiously as these values are derived from layers above the firn/ice transition where density determinations are less reliable. The relationship between the ice core $A_n$ record and the observational record is discussed further in the next section.

### Complementary observational and historical records

Past observations of the elevation, spatial distribution, and type of cultivation in southern Peru provide a qualitative measure of the prevailing environmental conditions. Many historical records reveal the elevation limits of cultivation at different times during the past few centuries and document the changes in climatic conditions in the high Andes of Peru. Cardich (1985) reports several such observations.

Of particular note for the time span under discussion are the observations made by a German-Swiss naturalist, J.J. von Tschudi, during his travels in Peru from 1838 to 1842. While ascending the narrow Rimac Valley, about two leagues ($\sim$ 10 km) above San Mateo, von Tschudi came to the village of Chicla at 3750 m elevation where he recorded: 'Here, barley is grown in a few protected ravines, but it does not mature and is cut green for forage. This is the last place in the valley where the soil is cultivated' (von Tschudi 1847, in Nuñez 1973, p.50). Cardich (1985) reports that he has visited Chicla twice to observe present conditions. He notes: 'The village remains the same, although the trail that passed through it in von Tschudi's time has become a principal road to the highlands. Today, however, in contrast to 150 years ago, potatoes and other hardy cultigens grow on the slopes extending well above the village. More sensitive grains also mature, including barley, habas, and even wheat, whose thermal requirements are higher. The limit of cultivation in the valley has also been displaced upward several kilometers and is now between Bellavista and Casapalca, where the altitude is 4000 m. This represents an increase of more than 200 m in the elevation at which crops can be grown successfully.'

These historical observations are consistent with and support the inferences drawn from the Quelccaya ice-core records; specifically, that climatic conditions prior to 1880 were characterized by cooler temperatures (more negative $\delta^{18}O$), drought conditions (lower $A_n$), and increased atmospheric dustiness. Undoubtedly both colder temperatures and reduced precipitation would have contributed to the observation of poor crops and a lower elevation limit of cultivation in the Peruvian highlands at this time. However, since roughly 1880 the ice-core data suggest milder temperatures and increased precipitation which

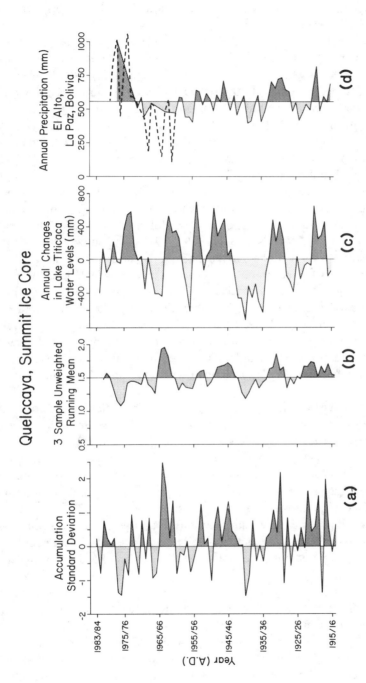

Fig. 16.3   The records from 1915 to 1984 for (a) standardized net annual accumulation ($A_n$) on the Quelccaya ice cap and (b) the three-sample running mean of $A_n$ are compared with (c) annual changes in Lake Titicaca water levels (mm) and (d) annual precipitation (mm) at El Alto, La Paz, Bolivia. The dashed line indicates years with incomplete precipitation records.

again are consistent with Cardich's recent observations of more diverse crops and higher cultivation limits.

Further evidence supports the reliability of the $A_n$ record as a proxy for paleoenvironmental conditions on the Altiplano of southern Peru. Lake Titicaca, the world's highest freshwater lake (3812 m a.s.l.), covers an area of 8446 km$^2$ and is located 300 km south of Quelccaya. Streams originating on the north side of the ice-cap flow into the Amazon, while streams originating on the south side flow into Lake Titicaca. Water level in Lake Titicaca has been monitored since 1914, and its fluctuations should be consistent with changes in the net accumulation on the Quelccaya ice cap, if indeed the ice cap is recording regional precipitation variations.

The Quelccaya $A_n$ record for 1915 to 1984 (Fig. 16.3a,b) has been compared with the annual changes in Lake Titicaca water levels (mm) (Fig. 16.3c) and the annual precipitation (mm) at El Alto, La Paz, Bolivia (Fig. 16.3d). The greatest drought of this century, 1935 to 1945 (Newell 1949), is associated with a 5-m drop in the mean water level in Lake Titicaca. Similarly, at this time the Quelccaya ice cap recorded very low accumulation, elevated dust concentrations (both soluble and insoluble), and less negative $\delta^{18}O$ (Fig. 16.2). Prior to 1970 major increases in Lake Titicaca water levels were associated with above average net accumulation on Quelccaya. Since 1970 the relationship appears to have reversed for some unknown reason. In general decreases in $A_n$, coupled with increases in particulate concentrations (soluble and insoluble), may, however, provide reasonable proxies with which to reconstruct major drought events from the longer ice core records.

From 1915 to 1961 (data sets are complete) the coefficient of determination ($R^2$) between the three-sample running mean of $A_n$ (Fig. 16.3b) and changes in Lake Titicaca water level is 0.304, very close to the $R^2$ (0.325) between annual precipitation at La Paz, Bolivia, and changes in Lake Titicaca water levels (Thompson et al. 1988). Table 16.1 demonstrates that the three-sample running mean of $A_n$ on Quelccaya is most strongly correlated ($R^2 = 0.324$) with annual changes in Lake Titicaca water levels in the subsequent year (T + 1). This probably reflects the time lag between precipitation on the ice cap and runoff into Lake Titicaca. As expected, the annual precipitation (mm) at El Alto near La Paz is equally well correlated with the annual changes in Lake Titicaca water levels in the current year (T).

The relatively low $R^2$, even for the best correlations, between changes in annual water levels in Lake Titicaca and the measured precipitation at La Paz and $A_n$ on Quelccaya arises because lake level is dependent also upon other factors such as evaporation, surface discharge, and groundwater flow into and out of the lake (Street-Perrott and Harrison 1985). Nevertheless, these results suggest that for the period of comparison (1915–1961) the $A_n$ record from Quelccaya provides a reasonable estimate of regional precipitation and is comparable in quality to instrumental observations (La Paz) in the region. Thus, the longer $A_n$

Table 16.1 *Coefficients of determination (R²) for the 1915 to 1961 records of: Quelccaya net accumulation (annual and three-year running mean), changes in water levels in Lake Titicaca, and measured precipitation at El Alto, Bolivia.*

| | Bal Dev | Lake T | Lake T(+1) | Lake (T + 2) | El Alto | Bal 3 s.r.m. |
|---|---|---|---|---|---|---|
| Bal Dev | 1.000 | 0.160 | 0.138 | 0.014 | 0.116 | 0.172 |
| Lake T | 0.160 | 1.000 | – | – | 0.325 | 0.302 |
| Lake T(+1) | 0.138 | – | 1.000 | – | 0.235 | 0.324 |
| Lake T (+2) | 0.014 | – | – | 1.000 | 0.22 | 0.230 |
| El Alto | 0.116 | 0.325 | 0.235 | 0.22 | 1.000 | 0.124 |
| Bal 3 s.r.m | 0.172 | 0.302 | 0.324 | 0.23 | 0.124 | 1.000 |

Bal Dev, Net accumulation (standard deviation);
Lake T, Annual changes in Lake Titicaca water levels (mm);
Lake T(+1), Lake Titicaca record (1 year lag);
Lake T(+2), Lake Titicaca record (2 year lag);
El Alto, Annual precipitation (mm) El Alto, La Paz, Bolivia;
Bal 3 s.r.m., Three sample, unweighted running mean of Quelccaya net accumulation (m ice equivalent).

records from Quelccaya should provide unique and otherwise unavailable information about the hydrological history in this region of southern Peru.

## El Niño/Southern Oscillation events

Except for the annual cycle, El Niño/Southern Oscillation (ENSO) is the dominant signal in the global climate system on time scales of a few months to a few years. It is associated with major dislocations of rainfall regimes in the tropics during which the northern coastal desert regions of Peru and Ecuador experience abnormally high precipitation and the southern highlands of Peru and northern Bolivia experience drought (Fig. 16.1). Quelccaya, situated in southern Peru, experienced a major drought associated during the 1982/83 El Niño. This is apparent in Figure 16.4 which contrasts the ice cap margin in 1978 (a non-ENSO year) with the margin at the same location during the 1983 ENSO. In this section a preliminary evaluation is made of the potential record of ENSO events as recorded in the Quelccaya ice cap by concomitant changes in microparticle concentrations, liquid conductivity levels, oxygen isotopic ratios, and net accumulation over the period where the most complete instrumental and historical documentation exist.

Annual pit studies from 1975 to 1984 on Quelccaya demonstrate that the two major ENSO events which occurred during this time were associated with substantially reduced net accumulation (Fig. 16.5). The annual variations in the

Fig. 16.4  The impact of the 1983 El Niño on the margin of Quelccaya ice cap. The left photograph was taken in a "normal" year (July 1978) while the photograph on the right, which was taken in July 1983, clearly shows the effect of widespread melting.

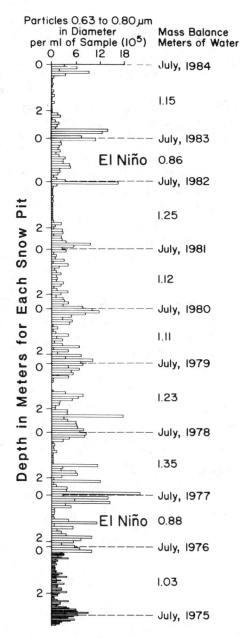

Fig. 16.5   Particle concentrations (0.63 μm ≤ diameter ≤ 0.80 μm) and net annual accumulation were determined from successive pits excavated each year from 1976 to 1984 on the Quelccaya ice cap. The El Niño years of 1976–77 and 1982–83 exhibit marked reductions in net mass accumulation.

particulate concentrations (0.63 $\mu$m $\leq$ diameter $\leq$ 0.80 $\mu$m) are evident with the greatest concentrations associated with the dry season from May to August each year. The dashed line (Fig. 16.5) represents the July snow surface, and thus, the separation between consecutive July surfaces reflects net accumulation between consecutive thermal years (July to subsequent June). Snow layer thicknesses are converted to water equivalent thicknesses using the vertical density profile. Figure 16.5 reveals that the El Niño years of 1976–77 and 1982–83 exhibit marked reductions in net accumulation.

To assess the utility of the Quelccaya record for extracting a longer ENSO chronology, the character of the more recent accumulation can be compared with both observational and historical data. Figure 16.6 illustrates the annual records of the Southern Oscillation index (SOI), sea surface temperatures (SST), and the historical record of major, moderate and minor El Niño phases (Quinn et al. 1987). The 1982–83 El Niño is recorded on Quelccaya by increased particulate deposition, elevated conductivity levels (more soluble material), less negative $\delta^{18}O$, and substantially reduced net accumulation. Figure 16.6 clearly shows that many (although not all) of the historical El Niño phases are associated with less negative $\delta^{18}O$, increased concentrations of dust (soluble and insoluble), and reduced net accumulation. The relationship among the ice-core parameters and the ENSO indicators (SOI and SST) has been explored statistically in an initial effort to develop a transfer function which might be applied to the older part

Table 16.2 *Autocorrelations and cross correlations Thermal Years: A.D. 1936–1983*

a. Correlation of $T_t$ and:

| k | $T_{t+k}$ | $I_{t+k}$ | $D_{t+k}$ | $A_{t+k}$ | $C_{t+k}$ | $P_{t+k}$ |
|---|---|---|---|---|---|---|
| 0 | 1.00 | −0.73 | 0.36 | −0.18 | 0.03 | 0.11 |
| 1 | 0.10 | 0.02 | 0.03 | 0.03 | −0.02 | 0.06 |
| 2 | −0.22 | 0.18 | 0.08 | 0.01 | −0.04 | 0.01 |
| 3 | −0.06 | −0.02 | 0.17 | 0.04 | −0.05 | 0.22 |
| 4 | −0.06 | 0.02 | 0.08 | 0.16 | −0.04 | 0.08 |
| 5 | −0.09 | 0.09 | 0.08 | −0.26 | −0.06 | 0.09 |

b. Correlation of $I_t$ and:

| k | $I_{t+k}$ | $T_{t+k}$ | $D_{t+k}$ | $A_{t+k}$ | $C_{t+k}$ | $P_{t+k}$ |
|---|---|---|---|---|---|---|
| 0 | 1.00 | −0.73 | −0.32 | 0.20 | −0.14 | −0.23 |
| 1 | 0.14 | −0.22 | −0.18 | 0.03 | −0.06 | −0.30 |
| 2 | −0.18 | 0.24 | −0.19 | 0.07 | −0.16 | −0.22 |
| 3 | −0.15 | 0.15 | −0.20 | 0.03 | −0.06 | −0.18 |
| 4 | −0.06 | 0.08 | −0.13 | −0.17 | 0.01 | −0.18 |
| 5 | 0.18 | −0.00 | −0.11 | 0.19 | 0.11 | 0.07 |

The following symbols are used above:

T, sea surface temperature (SST); I, Southern Oscillation index (SOI); D, oxygen isotope ratio ($\delta^{18}O$); A, annual accumulation ($A_n$); C, liquid conductivity (LC); P, concentration of large particles; t, thermal year; k, lead time in thermal years.

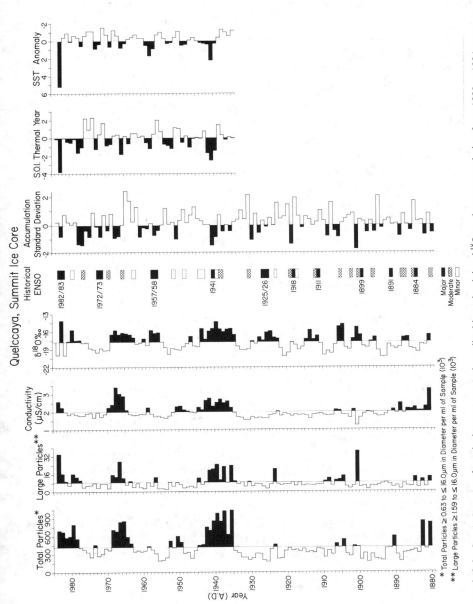

Fig. 16.6 Annual (July–June) averages of particulate concentrations, liquid conductivity, $\delta^{18}O$, and net accumulation from 1880 to 1984 are compared with two common ENSO indicators: the Southern Oscillation Index (SOI) and the sea surface temperature (SST) anomalies from Puerto Chicama, Peru (Quinn et al. 1987). Note SST increases toward the bottom.

of the ice-core record in order to extract a longer ENSO history (Thompson et al. 1987).

Autocorrelations and cross correlations between the ice core parameters, SST, and SOI are given in Table 16.2 (a, b) for lead times ranging from 0 to 5 yr. In general, SOI is more strongly correlated with the ice-core variables than SST; however, the strongest single correlation ($R = 0.36$) is between SST and $\delta^{18}O$. This is logical as the ocean surface waters are ultimately the source of the moisture for the accumulating snowfall. As might be expected, there are strong correlations between SST and SOI ($R = -0.73$), conductivity levels and particle concentrations ($R = 0.69$), particle concentrations and $\delta^{18}O$ ($R = 0.56$), and conductivity levels and $\delta^{18}O$ ($R = 0.44$). All correlations were for 1936 to 1983 during which all six parameters were available. A correlation coefficient ($R$) of at least $\pm 0.29$ is required to consider the relationship as statistically different from zero at a 5% significance level.

In situ observations on the Quelccaya ice cap during both the 1976 and 1982–83 El Niño events reveal a marked change in the general climatic patterns. For example, net accumulation was substantially reduced (Fig. 16.5) and net radiation receipt increased (Thompson et al. 1984b). Figure 16.6 suggests that similar climatic conditions probably characterized most of the earlier ENSO events.

The substantial concentration increases in both soluble and insoluble particles per unit volume of water is a direct result of the reduction in net accumulation. Even under conditions of constant particulate flux the reduced net accumulation would concentrate the material. However, it is certain that there was an associated increase in the flux of particles due to the increased dryness on the Altiplano of southern Peru during most of the ENSO events. Increased surface sublimation due to higher radiation receipt would also tend to concentrate the particles, especially during the dry season.

The reduction in net accumulation is the primary mechanism responsible for the less negative mean oxygen isotope values. Under normal conditions 80% of the snow falls on the ice cap during the wet season, when $\delta^{18}O$ values are most negative. In addition, increased sublimation due to longer periods of surface exposure to radiation between snowfalls leads to an enrichment of $^{18}O$ in the surface snow, producing annual mean $\delta^{18}O$ values which are less negative (Grootes et al. 1989). Thus, the physical mechanisms by which ENSO events are recorded in the Quelccaya ice cap are very clear. Unfortunately, other climatic processes, such as the 1935 to 1945 drought, also may produce similar variations in the ice-core parameters and hence complicate the detection of the short term ENSO events from the record at a single site.

## Longer-term variability: A.D. 450 to 1984

### El Niño/Southern Oscillation events

One of the records potentially available in ice cores taken from tropical and subtropical glaciers is a long-term record of El Niño/Southern Oscillation (ENSO) phases in the equatorial Pacific (Thompson et al. 1984b). Instrumental records of ENSO phases rarely extend more than 100 yr in length. Quinn et al. (1987) used historical documentation to extend the El Niño record back to the arrival of the Spanish in South America. These historical records are based largely on evidence obtained along the west coast of northern South America and from the adjacent Pacific Ocean waters. Unfortunately, the record is much less reliable prior to 1800. In addition, the effects of El Niño phases are not uniform as demonstrated by the large 1982–83 El Niño in which storms moved down the entire coastal area of Peru although the intensity and magnitude of the flooding was not uniform. This is consistent with the large spatial variability in El Niño frequency and intensity reported by Waylen and Caviedes (1986). Studies have shown that El Niño events are oscillatory, but not truly periodic (Wright 1977; Quinn et al. 1978, 1987; Hamilton and Garcia 1986; Quinn and Neal 1992). The ice-core records suggest that ENSO phases fluctuate substantially in both frequency and intensity. The recent major ENSO event (1982–83) focused attention upon this phenomenon and once again spurred a great effort to determine its predictability (Rasmusson and Carpenter 1982; Cane 1983; Rasmusson and Wallace 1983; Cane and Zebiak 1985; Kiladis and Diaz 1986; Yarnal and Diaz 1986). Bradley et al. (1987) demonstrated that the global characteristics of both warm and cold phases of ENSO affect both temperature and precipitation in the Northern Hemisphere.

The Quelccaya records may be compared with the historical El Niño reconstruction by Quinn et al. (1978, 1987) for the period A.D. 1450 to 1984. Figures 16.7 and 16.8 illustrate changes in the concentrations of insoluble (left) and soluble (right) particulates along with occurrences of major, moderate and minor El Niño events. Figures 16.9 and 16.10 illustrate the changes in $\delta^{18}O$ (left) and net annual accumulation (right) along with occurrences of major, moderate and minor El Niño events. As in the 20th century (Fig. 16.6), El Niño events were generally associated with reduced accumulation, higher concentrations in the insoluble and soluble dust, and less negative $\delta^{18}O$. However, as in the 20th century, similar variations may be produced by non-ENSO related events. Thus, reconstructing the ENSO history will require complementary records from more than one Peruvian ice cap. Most preferable is to extract a very high resolution history from several appropriate ice caps situated in northern Peru where the response to major ENSO phases should be in opposition to that on the Quelccaya ice cap. A number of potential ice caps and glaciers may exist in the Cordillera Blanca (Fig. 16.11).

## Linkages with archaeological records

The historical record of man's activities in pre-Spanish Peru is limited, and the process of piecing it together is hampered by the lack of written records. Archaeological sites in Peru may be assigned to either coastal or highland cultures (Fig. 16.12). Civilizations in both areas were largely agrarian and both developed in very climatically sensitive regions (Cardich 1985), where they

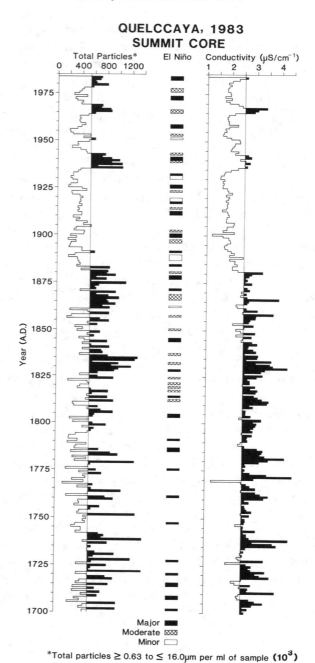

Fig. 16.7  Annual (July–June) averages of particulate concentrations (0.63 μm ≤ diameter ≤ 16.0 μm) and liquid conductivities from A.D. 1700 to 1984 are illustrated along with the historical record of El Niño occurrences (Quinn et al. 1987).

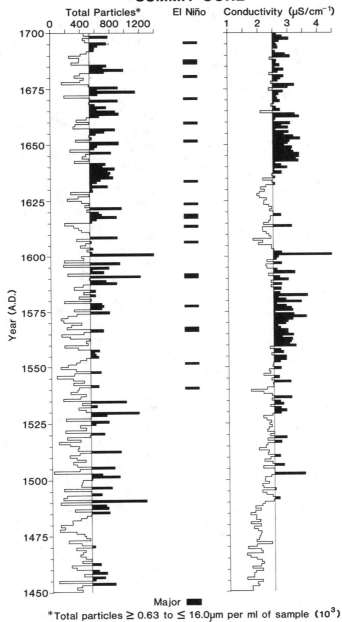

Fig. 16.8 Annual (July–June) averages of particulate concentrations (0.63 μm ≤ diameter ≤ 16.0 μm) and liquid conductivities from A.D. 1450 to 1700 are illustrated along with the historical record of El Niño occurrences (Quinn et al. 1987).

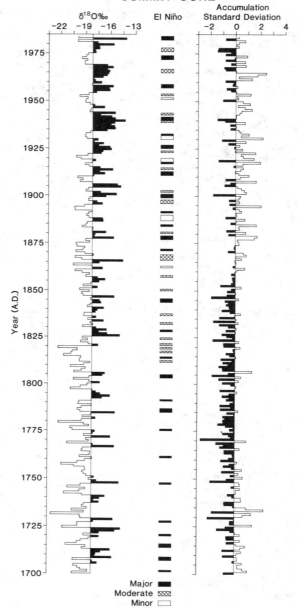

Fig. 16.9   Annual averages of $\delta^{18}$O and standardized net accumulation from A.D. 1700 to 1984 are illustrated along with the historical record of El Niño occurrences (Quinn et al. 1987).

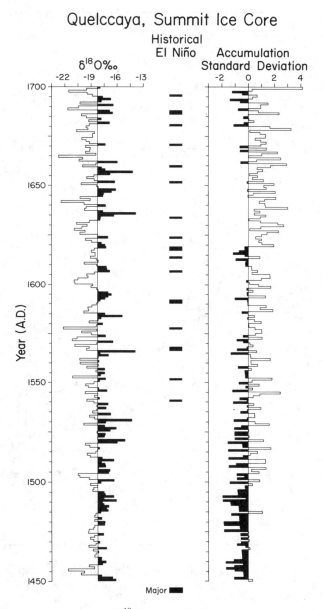

Fig. 16.10   Annual averages of $\delta^{18}O$ and standardized net accumulation from A.D. 1450 to 1700 are illustrated along with the historical record of El Niño occurrences (Quinn et al. 1987).

depended heavily upon water availability. Along the coast agriculture depended upon water from the rivers originating in the Andes. On the Altiplano agriculture extended to the upper limit of both temperature and water availability.

The longer precipitation record for southern Peru derived from the Quelccaya $A_n$ record provides valuable information which can be related to the early

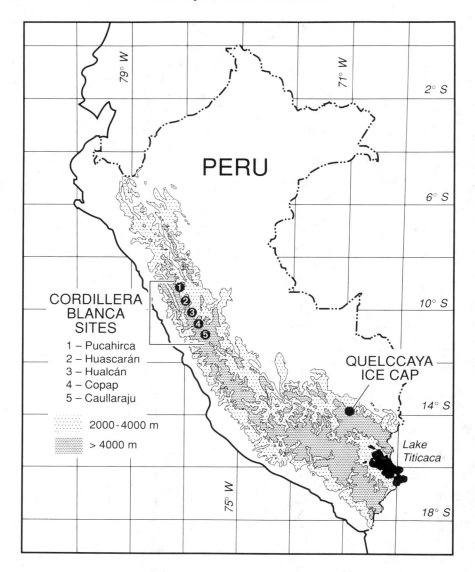

Fig. 16.11   Major archaeological sites in Peru are shown relative to elevation and the Quelccaya ice cap. These sites fall into two main groups, coastal and highland. Potential drill sites for additional ice core records in the Cordillera Blanca are shown.

history of man in this region of South America. Figure 16.12 presents the decadal trends in accumulation from A.D. 470 to 1980. Currently, ENSO events are associated with droughts in the southern Peruvian highlands (Fig. 16.1) and floods in the coastal desert areas of northern and central Peru and southern Ecuador (Thompson et al. 1984b; Lam and Del Carmen 1986). The Quelccaya net balance record for the last 1500 yr has been integrated with archaeological evidence (Shimada et al. 1991). These data strongly suggest that flourishing

# Quelccaya, Peru

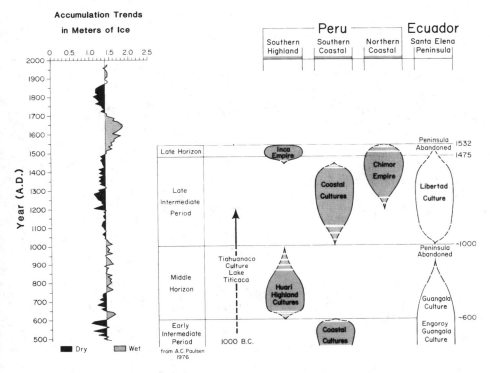

Fig. 16.12   Decadal accumulation (precipitation) trends which are a composite of $A_n$ from core 1 and summit core are shown. Major wet (lighter gray) and dry (darker gray) periods are indicated. On the right, the rise and fall of coastal and highland cultures in Ecuador and Peru (after Paulsen 1976) are shown.

highland cultures appear when moisture conditions in the mountains are above normal and that coastal cultures flourish when the mountains are drier than normal. These data suggest that longer-term ocean/atmosphere linkages similar in character to shorter-term ENSO phases may have persisted over many decades and created conditions similar to those currently associated with brief (1–2 yr) ENSO phases.

Paulsen (1976) reported a similar relationship in the longer-term archaeological records associated with the rise and fall of coastal cultures in Peru and Ecuador. She noted that highland and coastal cultures seemed to flourish out of phase, i.e., highland cultures flourished when coastal cultures declined and vice versa. The cultural record, dated primarily using highly refined ceramic sequences and some [14]C measurements, is included in Figure 16.12 for comparison with the net accumulation record from Quelccaya. Assuming that the same seesaw relationship that currently exists during ENSO events (Thompson et al. 1984b) could have persisted over longer time intervals, wetter coastal conditions would be expected during periods of highland drought. The fact that

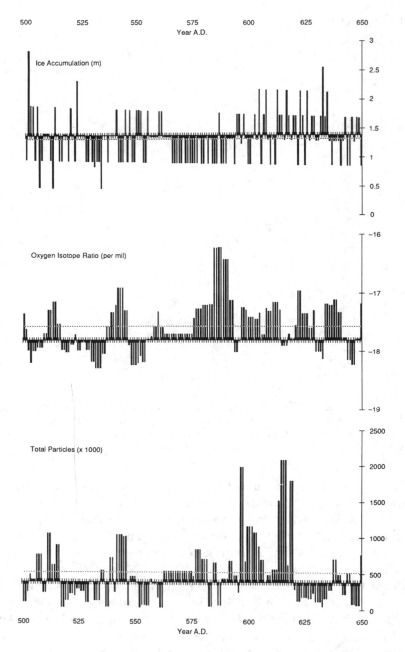

Fig. 16.13   Net annual accumulation, $\delta^{18}O$, and particulate concentrations (0.63 $\mu$m $\leqslant$ diameter $\leqslant$ 16.0 $\mu$m) are displayed around their respective long-term means (A.D. 500–1984). The dashed lines are their respective short-term (A.D. 500–650) means. (After Shimada et al. 1991.)

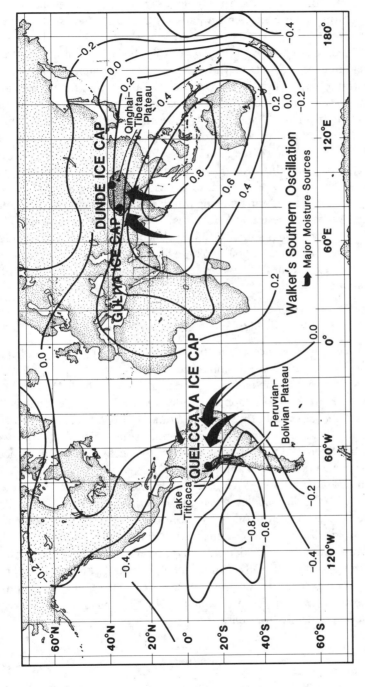

Fig. 16.14   The position of the Quelccaya ice cap (south of the equator) and the Dunde ice cap (north of the equator) relative to Walker's Southern Oscillation.

coastal cultures flourished when the highlands were drier implies that ENSO-like oceanic and atmospheric circulation patterns may have persisted for extended periods of time.

Several other examples of this relationship between climate and cultural development may be drawn from the earlier part of the Quelccaya record. Figure 16.13 illustrates large variations in $A_n$, $\delta^{18}O$, and particulate concentrations from A.D. 500 to 650. Precipitation appears to have decreased 30% during a major drought from A.D. 563 to 594. It seems intuitive that such an abrupt change in water availability could produce major environmental stresses within Andean cultures. Further evidence for this cultural dependence upon climate arise from the fact that this extended drought between A.D. 500 and 600 is contemporaneous with major relocations of the Mochica and other Andean cultures which produced rapid and far-reaching internal transformations (Shimada et al. 1991). It is likely that these relocations resulted from the inhabitants' attempts to control life-giving water sources. Finally, encompassed in this time interval is a large dust event from A.D. 600 to 620, which is believed to reflect the impact of prehistoric agricultural activities on the environment of the Altiplano (Thompson et al. 1988).

### Linkages between South America and Asia

Additional ice-core evidence from the Dunde ice cap in China indicate that decadally averaged net accumulation trends on the Qinghai-Tibetan Plateau are quite similar to those in the southern Andes of Peru from A.D. 1600 to 1980. Although these areas are on opposite sides of the Pacific Ocean basin, they should be related physically through the Walker Circulation (Fig. 16.14).

Figure 16.15 illustrates a preliminary comparison of decadally averaged net accumulation from A.D. 1610 to 1980 for Quelccaya and Dunde ice caps (Thompson et al. 1989, 1990; Thompson 1992). Over the last 400 yr these records show a remarkable similarity in net balance for such widely separated areas. This is consistent with the physical linkage (teleconnection) between these areas which has been recognized since the turn of the century. The ENSO phenomenon perturbs the ocean/atmosphere system of the Pacific Basin episodically and is related to anomalous weather patterns covering much of the globe (Rasmusson and Wallace 1983; Nicholls 1987; Enfield 1989; Kiladis and Diaz 1989). The comparisons in Figure 16.15 support the hypothesis that these teleconnections exist, not only for high frequency events such as ENSO, but also for lower frequency events which may persist for centuries.

### Summary

The Quelccaya (Peru) and Dunde (China) ice caps are situated in regions where the climate exhibits a high degree of interannual variability. Here the preservation of seasonal variations in chemical and physical constituents, make possible

## Mean Decadal Accumulation (m)

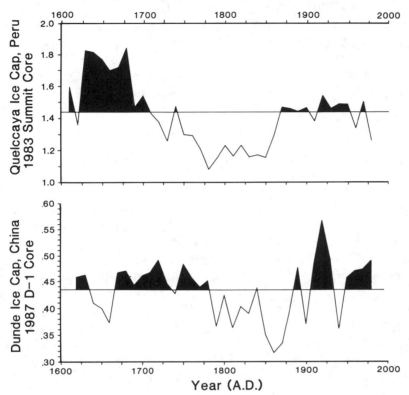

Fig. 16.15  The decadally averaged net accumulation (~A.D. 1600–1980) for the composite of two cores from the Quelccaya ice cap is compared with that from core D-1 on the Dunde ice cap, China.

very precise dating of the annual precipitation layers. Thus, the analyses of cores drilled in these ice caps provide unparalleled opportunities to construct multifaceted paleorecords of both the high- and low-frequency changes of the climate system and the regional environment. The potential for extracting much longer ice core records of the largest ENSO events has been presented and strategies for accomplishing this were discussed briefly. In addition, evidence was presented for the potentially strong impact of high frequency climatic variability upon human activities as expressed in archaeological records.

The Quelccaya ice cap records also reinforce the importance of low frequency oscillations in tropical accumulation over the last 1500 yr in southern Peru. The records from these high-elevation, low- and mid-latitude sites allow the documentation of low frequency oscillations as well as provide a new perspective on tropical teleconnections in space and time. Currently, a new ice core drilling program is planned for glaciers in the Cordillera Blanca at 8°S (Fig. 16.11). These sites are much closer to the zone of maximum ocean/atmosphere response to ENSO events and thus, when coupled with the Quelccaya records, they should provide much clearer proxies for paleo-ENSO events along the coast of South America. Coupling these South American records with those anticipated from the Tibetan Plateau (Dunde and Guliya ice caps) should provide both additional understanding of the physical linkages across the Pacific Basin and a record of the major ENSO events for the last few centuries.

## Acknowledgments

We acknowledge with great appreciation the efforts of the many individuals, Peruvian, Chinese and American, who participated in the numerous field programs. Mary Davis conducted the particulate and conductivity analyses at The Ohio State University and the oxygen isotope analyses were conducted at the University of Washington by P. Grootes and at The Geophysical Laboratory of the University of Copenhagen by N. Gundestrup. The Peruvian program was a cooperative effort with colleagues from ElectroPeru. The Chinese program was a cooperative effort with colleagues from the Lanzhou Institute of Glaciology and Geocryology. This work and the many field programs associated with it were supported by the NSF Office of Climate Dynamics and Division of Polar Programs (ATM75–15513A02; ATM78–1609A01; ATM81–05079A02; ATM82–13601A02; ATM85–19794; ATM89–16635); The National Geographic Society (3323–86); and NOAA (NA89AA-D-AC197). J. Nagy produced the figures and K. Doddroe typed the mansucript. Byrd Polar Research Center Contribution No. 774.

## References

BRADLEY, R. S., DIAZ, H. F., KILADIS, G. N., and EISCHEID, J. K., 1987: ENSO signal in continental temperature and precipitation records. *Nature*, 327: 497–501.

CARDICH, A., 1985: The fluctuating upper limits of cultivation in the central Andes and their impact on Peruvian prehistory. *Advances in World Archaeology*, 4: 293–333.

CANE, M. A., 1983: Oceanographic events during El Niño. *Science*, 222: 1189–1195.

CANE, M. A. and ZEBIAK, S. E., 1985: A theory for El Niño and the Southern Oscillation. *Science*, 228: 1085–1087.

ENFIELD, D. B., 1989: El Niño, past and present. *Reviews of Geophysics*, 27: 159–187.

GROOTES, P. M., STUIVER, M., THOMPSON, L. G., and MOSLEY-THOMPSON, E., 1989: Oxygen isotope changes in tropical ice, Quelccaya, Peru. *Journal of Geophysical Research*, 94(D1): 1187–1194.

HAMILTON, K. and GARCIA, R. R., 1986: El Niño/Southern Oscillation events and their associated midlatitude teleconnections 1531–1841. *Bulletin of the American Meteorological Society*, 67: 1354–1361.

KILADIS, G. W. and DIAZ, H. F., 1986: An analysis of the 1877–78 ENSO episode and comparison with 1982–83. *Monthly Weather Review*, 114: 1035–1047.

KILADIS, G. W. and DIAZ, H. F., 1989: Global climatic anomalies associated with extremes in the Southern Oscillation. *Journal of Climate*, 2: 1069–1090.

LAM, J. A. and CARMEN, DEL, 1986: The evolution of rainfall in northern Peru during the period January 1982 to December 1985, in the coastal, mountain and jungle regions. *Chapman Conference on El Niño*, Guayaquil, Ecuador, April 27–31, 1986.

LANDSBERG, H. E., 1985: Historic weather data and early meteorological observations. *In* Hecht, A. D. (ed.), *Paleoclimate Analysis and Modeling*. New York: Wiley, 27–70.

NEWELL, N. D., 1949: Geology of the Lake Titicaca Region, Peru and Bolivia. *Geological Society of America Memoir, 36*: 11.

NICHOLLS, N., 1987: Prospects for drought prediction in Australia and Indonesia. *In* Wilhite, D. A. and Easterling, W. E. (eds.), *Planning for Drought: Toward a Reduction of Societal Vulnerability*. Boulder, Colo.: Westview, 61–72.

PAULSEN, A. C., 1976: Environment and empire: climatic factors in pre-historic Andean culture change. *World Archaeology*, 8: 121–132.

QUINN, W. H. and NEAL, V. T., 1992: The ENSO as a climatic forcing factor. *In* Bradley, R. S. and Jones, P. D. (eds.), *Climate Since A.D. 1500*. London: Routledge, 623–648.

QUINN, W. H., ZOPF, D. O., SHORT, K. S., and KUO YANG, R. T., 1978: Historical trends and statistics of the Southern Oscillation, El Niño, and Indonesian droughts. *Fishery Bulletin*, 76: 663–678.

QUINN, W. H., NEAL, V. T., and ANTUNEZ de MAYOLO, S. E., 1987: El Niño occurrence over the past four and a half centuries. *Journal of Geophysical Research*, 92: 14, 449–14,461.

RASMUSSON, E. M. and CARPENTER, T. H., 1982: Variations in tropical sea surface temperature and surface wind fields associated with the Southern Oscillation, El Niño. *Science*, 222: 1195–1210.

RASMUSSON, E. M. and WALLACE, J. M., 1983: Meteorological aspects of the El Niño-Southern Oscillation. *Science*, 222: 1195–1202.

SHIMADA, I., SCHAAF, C. B., THOMPSON, L. G., and MOSLEY-THOMPSON, E., 1991: Cultural impacts of severe droughts in the prehistoric Andes: application of a 1,500-year ice core precipitation record. *World Archaeology*, 22: 247–270.

STREET-PERROTT, F. A. and HARRISON, S. P., 1985: Lake levels and climate reconstruction. *In* (A. D. Hecht (ed.), *Paleoclimate Analysis and Modeling*, New York: Wiley, 291–340.

THOMPSON, L. G., 1992: Ice core evidence from Peru and China. *In* Bradley, R. S. and Jones, P. D. (eds.), *Climate Since A.D. 1500*. London: Routledge, 517–548.

THOMPSON, L. G. and MOSLEY-THOMPSON, E., 1987: Evidence of abrupt climate change during the last 1,500 years recorded in ice cores from the tropical Quelccaya ice cap, Peru. *In* Berger, W. H. and Labeyrie, L. R. (eds.), *Abrupt Climate Change. Evidence and Implications*. NATO Advanced Science Institutes Series C: Mathematical and Physical Sciences Vol. 216, Dordrecht: Reidel, 99–110.

THOMPSON, L. G. and MOSLEY-THOMPSON, E., 1989: One-half millennia of tropical climate variability as recorded in the stratigraphy of the Quelccaya ice cap, Peru. *In* D. Peterson (ed.), *Climatic Change in the Eastern Pacific and Western Americas*. Geophysical Monograph 55. Washington D.C.: American Geophysical Union, 15–31.

THOMPSON, L. G., MOSLEY-THOMPSON, E., GROOTES, P. M., POURCHET, M., and HASTENRATH, S., 1984a: Tropical glaciers: potential for ice core paleoclimatic reconstruction. *Journal of Geophysical Research*, 89: 4638–4646.

THOMPSON, L. G., MOSLEY-THOMPSON, E., and ARNAO, B. M., 1984b: El Niño-Southern Oscillation events recorded in the stratigraphy of the tropical Quelccaya ice cap, Peru. *Science*, 234: 361–364.

THOMPSON, L. G., MOSLEY-THOMPSON, E., BOLZAN, J. F., and KOCI, B. R., 1985: A 1500-year record of tropical precipitation in ice cores from the Quelccaya ice cap, Peru. *Science*, 229: 971–973.

THOMPSON, L. G., MOSLEY-THOMPSON, E., DANSGAARD, W., and GROOTES, P. M., 1986: The 'Little Ice Age as recorded in the stratigraphy of the tropical Quelccaya ice cap. *Science*, 234: 361–364.

THOMPSON, L. G., DAVIS, M., MOSLEY-THOMPSON, E., and LIU, K., 1988: Pre-Incan agricultural activity recorded in dust layers in two tropical ice cores. *Nature*, 336: 763–765.

THOMPSON, L. G., MOSLEY-THOMPSON, E., DAVIS, M. E., BOLZAN, J. F., DAI, J., YAO, T., GUNDESTRUP, N., WU, X., KLEIN, L., and XIE, Z., 1989: Pleistocene climate record from Qinghai-Tibetan Plateau ice cores. *Science*, 246: 474–477.

THOMPSON, L. G., MOSLEY-THOMPSON, E., DAVIS, M. E., BOLZAN, J., DAI, J., GUNDESTRUP, N., YAO, T., WU, X., KLEIN, L., and ZICHU, Z., 1990: Glacial stage ice core records from the subtropical Dunde ice cap, China. *Annals of Glaciology*, 14: 288–297.

THOMPSON, P. A., THOMPSON, L. G., and MOSLEY-THOMPSON, E., 1987: Hindcasts of El 'Niño events in the 19th century. Working Paper Series 87–124, College of Business, Columbus, Ohio: The Ohio State University, 31 pp.

VON TSCHUDI, J. J., 1847: *Travels in Peru During the Years 1838–1842 on the Coast, in the Sierra, Across the Cordilleras and the Andes, into Primeval Forest.* London: David Bogue., *reprinted:* Nuñez, Estuardo, 1973, *El Peru Visto por Viajeros.* Lima: Editorial Peisa.

WAYLEN P. R. and CAVIEDES, C. N., 1986: El-Niño and Annual Floods on the North Peruvian Littoral. *Journal of Hydrology*, 89: 141–156.

WRIGHT, P. B., 1977: The Southern Oscillation patterns and mechanisms teleconnections and persistence. Hawaii Institute of Geophysics, HIG-77-13.

YARNAL, B. and DIAZ, H. F., 1986: Relationships between extremes of the Southern Oscillation and the winter climate of the Anglo-American Pacific coast. *Journal of Climatology*, 6: 197–220.

# 17

# A comparison of proxy records of El Niño/Southern Oscillation

JOEL MICHAELSEN

*Department of Geography University of California Santa Barbara, 93106, U.S.A.*

L. G. THOMPSON

*Byrd Polar Research Center The Ohio State University Columbus, 43210, U.S.A.*

## Abstract

Three previously published proxy records of El Niño/Southern Oscillation (ENSO) variability were compared: a documentary record from coastal South America; an ice-core record from the Quelccaya ice cap in Peru; and a tree-ring record record from arid site conifers in the southwestern United States. The records were calibrated with long ENSO instrument records, and all the proxy records showed significant levels of correspondence with the instrument records. In fact, the proxy records appear to capture roughly comparable amounts of ENSO variability as would instrument records from the same regions. The $\delta^{18}O$ ice-core record showed some evidence of missing years, but after adjustment the correlation with the tree-ring record was reasonably consistent back to about A.D. 1630. This result provides additional verification of the two proxy records and also indicates that ENSO variability persisted relatively unchanged through main period of the Little Ice Age. The documentary record does not match the other two proxy records, particularly the tree-ring record, very well prior to the 20th century.

## Introduction

One of the most rapidly developing areas in climatology is the study of climate variability on interannual-to-century time scales. Because instrument records are in general only long enough to provide good information on the shorter end of this scale, much of the work has involved the development and interpretation of high resolution proxy climatic records. High-resolution proxy records are taken here to mean either natural or documentary records with annual or almost annual temporal resolution (Baumgartner et al. 1989). Clearly these time scales are the most relevant for determining the potential impacts of climatic changes over the next century, whether produced by increased carbon dioxide or by other causes.

Both paleoclimatic studies and efforts to model future climate have often focussed primarily on reconstructing or predicting changes in mean climate. For example, most studies of the impact of increased carbon dioxide attempt to predict changes in average temperature and precipitation. Similarly, paleoclimatic studies have established that at least some portions of the Northern Hemisphere experienced significantly colder mean temperatures over the last several centuries during the Little Ice Age (e.g. Lamb 1977; Wigley et al. 1981; Grove 1988). In many cases, however, modest changes in mean conditions could be accompanied by much more significant changes in variability. Determining the nature of changing variability is a much more challenging problem, given the inaccuracies of paleoclimatic data and the shortcomings of modern climate models.

One of the main sources of uncertainty in determining future climate changes is how components of modern climate variability would change in response to large-scale changes in mean climate. There is much to be learned about this question from paleoclimatic studies of the behavior of important aspects of modern climate variability during the most recent period of different mean conditions, the Little Ice Age. The El Niño/Southern Oscillation (ENSO) phenomenon is without doubt one of the most striking sources of large-scale variability in the modern climate, and the degree to which its characteristics as observed during the 20th century are a function of the mean climatic conditions of this period is a key issue. How sensitive is ENSO to overall climate? Did it persist through the Little Ice Age or is it closely tied to modern conditions? If it did persist, are there any indications of changes in its frequency or magnitude which might be related to large-scale climate changes? These and other questions relating to the internal workings of the ENSO phenomenon can best be addressed by studies of high-resolution proxy records, since there are detailed ENSO records for only the last few decades and any evidence from instrument records only for the last century.

It is always important to develop the most accurate proxy records possible and to assess the level of accuracy as carefully as possible, but this is especially true in studies of variability. Simply put, it is much more difficult to estimate variances than means. All proxy sources of climatic information, either natural recording systems or documentary evidence, have unique strengths and weaknesses. Each responds only to certain aspects of environmental variability and contains variability, or noise, caused by nonclimatic factors. None is an ideal climate record, so it is important to compare evidence from different sources to reduce the inaccuracies inherent in individual records and to develop more complete pictures of the climate. In addition, each record reflects environmental variations in a single region. Some of these variations are purely regional, while others are related to variability over larger scales. (This attribute is also shared by instrument records, of course, and has led to the common practice of aggregating information over large geographic areas to study large-scale variability.)

Considering the widespread oceanographic and atmospheric impacts of the ENSO phenomenon, there are many possibilities for developing high-resolution proxy climate indicators which could provide information about ENSO fluctuations over the last several centuries. A preliminary analysis of a number of such indicators over a short period of time was presented in Baumgartner et al. (1989), where it was argued that considerable improvements in accuracy and fidelity could be gained by combining proxy records from different sources and regions.

In this report, three such indicators are compared over the last 400 yr: tree-ring chronologies from southwestern United States and northwestern Mexico (Drew 1976; Michaelsen 1989); ice-cores from Quelccaya ice cap in Peru (Thompson et al. 1984, 1985, 1986, 1988; Thompson and Mosley-Thompson 1989; Thompson et al. 1992, this volume); and documentary records (Quinn et al. 1987). The main objectives of the research are (1) to determine the characteristics of the response of each proxy record to ENSO variability and assess the consistency of the response over time; and (2) to study the variability of ENSO over the last 400 yr in order to shed some light on the questions posed above relating to its robustness in the face of large-scale climate changes.

The analysis involved a calibration stage where the proxy ENSO records were compared to relatively long instrument records published by Wright (1989) and a comparison of the three records, themselves, over the period A.D. 1570–1964. Because the documentary record is ordinal, rather than interval scale, two separate analyses were carried out. The tree-ring and ice-core records were compared with the instrument records and with each other using interval-scale statistical techniques. Then the tree-ring, ice-core, and instrument records were converted to categorical datasets recording the occurrence or nonoccurrence of a warm ENSO event for comparison with the documentary record. Analyses of these data were based on statistical point process techniques. In both analyses emphasis was placed on techniques which can identify temporal variations in quantities of interest.

## Data

The tree-ring reconstruction (Michaelsen 1989) was based on seven chronologies from New Mexico and nothern Mexico selected from the archived records of the Laboratory for Tree-Ring Research at University of Arizona (Drew 1976). They are from a region which generally shows increased precipitation during ENSO years and during the following year (Ropelewski and Halpert 1986; Kiladis and Diaz 1989). The reconstruction extends only through 1964, the ending date of some of the chronologies.

The ice-core record is from the Quelccaya ice cap, situated in the easternmost glaciated mountain chain of the Peruvian Andes at 14°S and an elevation of 5670 m (Thompson et al. 1984, 1985, 1986, 1988; Thompson and

Mosley-Thompson 1989; Thompson et al. 1992, this volume). Precipitation in this region is concentrated during the November–April high-sun season. It is produced primarily by convective activity in moist, unstable air masses moving out the Amazon Basin. The pronounced seasonality in precipitation produces distinct annual layering in the ice cap, with alternating layers of ice and dust accumulation. In general, southern Peru and the Amazon Basin are warmer and slightly drier during warm ENSO years (Kiladis and Diaz 1989). Two cores were extracted, and the the primary data used in this study are from the summit core. The variables measured include concentrations of three different sized particles, electrical conductivity, $\delta^{18}O$, and accumulation rate. The $\delta^{18}O$ record from core 1, drilled 150 m to the east of the summit core, was also examined in an attempt to clarify possible inaccuracies in dating the ice core. Unless otherwise noted, discussion of the ice-core records will refer to measurements from the summit core. All data are totals for the 'thermal year' running from July of one year through June of the following year.

The documentary ENSO record was compiled by Quinn et al. (1987) based on analysis of a wide variety of sources. The information is concentrated in the coastal areas of Peru and the Pacific along the coast of South America. Terrestrial indicators of warm ENSO events included accounts of heavy rains and flooding, and marine indicators included variations in wind as reflected in the time taken for sailing voyages. Quinn et al. (1987) group warm events into several categories based on strength, ranging from weak to very strong. This study will focus on events described as strong or very strong.

The instrument records utilized in the calibration study are three numerical indices of ENSO derived by Wright (1989). They are (1) an index of rainfall at island stations in the tropical Pacific, 1894–1983; (2) sea surface temperature (SST) averages for the central and eastern Equatorial Pacific, 1881–1986; and (3) a Southern Oscillation sea-level pressure (SLP) index based on the Tahiti and Darwin differences for 1935–1984 and reconstructed back to 1852 using a number of other pressure records. All three instrument records were aggregated to annual averages to match the resolution of the proxy records. A uniform sign convention was adopted so that warm ENSO events were positive deviations. This involved reversing the sign of the Tahiti-Darwin pressure index.

## Methods

A detailed description of the tree-ring ENSO reconstruction is given in Michaelsen (1989) and only a summary will be included here. The reconstruction was based on lagged values of the first principal component of the tree-ring records. The component was bandpass filtered to retain variability on time scales from 10 yr to about 3 yr. This approach was taken because ENSO is a band-limited process with variance concentrated between approximately 3 yr and 7 yr (e.g. Doberitz 1968; Julian and Chervin 1978) and because the proxy records are likely to be affected by other phenomena at other frequencies. The explained variance under cross-validation (Michaelsen 1987) was 31%. All the ice-core data used in this study were also bandpass filtered to be comparable with the tree-ring reconstruction.

Since the documentary records are categorical time series, or point processes, two separate analyses were required, both for calibration with instrument records and for intercomparison between the proxy records. The first involved only the continuous tree-ring and ice-core records. In order to include the documentary records in the second phase, all the continuous proxy and instrument records were converted to point processes by establishing a threshold value to define warm events. The thresholds were defined so that all records had the same number of events for the period of overlap between all the proxy and instrument records, 1894–1964. The documentary record listed 12 strong or very strong warm ENSO years during this period, giving an average rate of 0.169 events/yr, or one out of every 5.9 yr. Several other threshold values were tried without appreciably changing the results.

At a more fundamental level some care must be taken in interpreting what is meant by this definition of an warm ENSO event. Ambiguity is introduced by the fact that it is not uncommon for a warm event to span 2 yr. According to the definition of an event based simply on exceeding a threshold, this would constitute two events. A more intuitive approach might be to define a multiyear warm event as a single event. This definition could be particularly useful for identifying changes in the rates of occurrence, or equivalently, in the intervals between events. This idea was implemented by defining a second set of point processes including only the first years of warm ENSO events. Estimates of rates will be presented for the records of all warm ENSO years and the records of warm ENSO onset years.

In an individual record it is almost as easy to define events and analyse rates for warm event onsets as for all warm years, but comparisons between records become much more difficult. Under the straight warm ENSO year definition, a match occurs whenever two records have a warm event during the same year. Attempting to implement this simple matching criterion with the warm event onset definition would present problems in the not uncommon situation when one record has a 2-yr warm event while a second record has only a single-year warm event. If the single-year warm event corresponds to the onset year a match occurs, but if it corresponds to the second year no match is found. It might be possible to come up with a more complicated criterion which would identify a match whenever a single year ENSO in one record occurred in either year of a two year ENSO in the other record, but this would make it difficult to estimate expected numbers of matches required to test the null hypothesis of independence. As a result, comparisons between proxy and instrument records and between different proxy records was only carried out using the original ENSO year event definition.

Standard cross-correlation analyses were utilized to calibrate the continuous records and to compare the tree-ring and ice-core reconstructions. In addition to statistics calculated for the full periods of overlap, time-varying lag correlations were calculated using a sliding window to compare segments of the records. This approach is commonly employed to check the cross-dating of tree-ring records where it is used to identify shifts in one core sample relative to others produced by missing or false growth rings. It can also provide important information about changes in the strength of a relationship between proxy and instrument records or between different proxy records. Consequently, it can be useful for investigating the assumption implicit in most paleoclimatic reconstructions that the characteristics of the recording system do not vary appreciably over time.

The rectangular window,

$$w(s) = \begin{cases} 1/(2h+1) & \text{for } |s| < 1 \\ 0 & \text{otherwise,} \end{cases} \tag{1}$$

is most commonly used. In this study a tapered window, or kernel, based on the biweight function,

$$w(s) = \begin{cases} 15/16(1-s^2)^2/h & \text{for } |s| < 1 \\ 0 & \text{otherwise,} \end{cases} \tag{2}$$

was used. Tapered kernel functions have the advantage over rectangular ones of producing smoother curves since they have no discontinuities. The correlation for a window centered on some time, $t_0$, is calculated by applying the kernel function to both records with

$$s = (t - t_0)/h,$$

where $h$ is a kernel width parameter which gives the width of the portion of the kernel with values above 0.5 and must be specified by the user. A window width of 50 yr was arbitrarily selected for the correlation studies.

The basic statistic of interest for an individual point process is the rate function, $m(t)$, (Cox and Lewis 1966) which gives the probability of an event occurring at time $t$. If the process is stationary, then the rate is constant and can be estimated directly as

$$\hat{m}(t) = \hat{m} = N_T/T, \tag{3}$$

where $N_T$ is the number of events occurring in the time interval of length $T$. In many cases in may be useful to consider an alternative nonstationary process with a variable rate. This is often done by assuming a particular parametric form for the rate, such as an exponential, but Solow (1991) presents an attractive nonparametric alternative using a kernel estimator. This approach is very similar to the sliding window correlation method described above and is directly comparable to the kernel method of density estimation (Silverman 1986). If the point process has N events occurring at times $t_1, t_2, \ldots, t_N$ in the interval $[0,T]$, then the variable rate estimate at time $t$ is

$$\hat{m}(t) = 1/h \sum_{i-1}^{N} w[(t - t_i)/h], \tag{4}$$

where $w(s)$ is the kernel function. Solow (1991) and Silverman (1986) present a technique for estimating the width parameter, $h$, using maximum likelihood cross-validation, along with efficient calculation methods in the density estimation context which can be applied here. Cross-validation estimates are obtained by omitting each event in turn, i.e.

$$\hat{m}_{-i}(t) = 1/h \sum_{j \neq i} w[(t - t_i)/h]. \tag{5}$$

The value of $h$ is selected which maximizes the cross-validated log likelihood function,

$$CV(h) = \sum_{i=1}^{n} \log[\hat{m}_{-i}(t_i)]. \tag{6}$$

Using this criterion, window widths of 40 yr were used for the calibration studies and 80 yr for comparisons of the proxy records.

Solow (1991) also presents a method for obtaining approximate confidence intervals for kernel rate estimates under the assumption that the events are generated by a Poisson process. In a Poisson process the intervals between events are independent, identically distributed exponential variates. An examination of the intervals between warm ENSO events suggests that they do appear to be independent but are not exponentially distributed, particularly in the case of intervals between warm event onset years. A somewhat more general model which appears more appropriate is a renewal process with intervals that are independent, identically distributed gamma variates (Cox and Lewis 1966). The gamma distribution,

$$f(x) = \beta^{-\alpha} x^{\alpha - 1} e^{-x/\beta} / \Gamma(\alpha), \tag{7}$$

is commonly used to model precipitation data (e.g. Ropelewski et al. 1985; Wilks 1990) because it will allow for a wide range of degrees of asymmetry for different values of the shape parameter, $\alpha$. (It includes the exponential distribution for $\alpha = 1$.) Under this assumption, the variance of a constant rate estimate can be approximated as

$$Var(\hat{m}) = IN_T/T^2. \tag{8}$$

The coefficient of dispersion, or squared coefficient of variation (Cox and Lewis 1966) is $I$ and for a gamma distribution is

$$I = Var(x)/E^2 = 1/\alpha. \tag{9}$$

An estimate of the variance of a kernel estimate (under the assumption of a constant rate) is

$$Var(\hat{m}(t)) = \hat{m}/4\hat{\alpha}h_0^2 \tag{10}$$

where

$$h_0 = (3/7)^{1/2}h \tag{11}$$

(Solow 1991). Note that in the case of the Poisson distribution ($\alpha = 1$) the relationship is the same as that given by Solow. Confidence intervals can be obtained by using the normal approximation and the fact that the rate estimates are asymptotically unbiased.

The basic statistic used to measure the degee of association between two point processes will be the closest analog to a correlation for point processes, the cross intensity function (Cox and Lewis 1966). This functon measures the conditional probability of an event occurring in one series, given that an event has occurred in the other series. In can be easily estimated by dividing the number of matches (events occurring in simultaneously in series A and B) by the number of events occurring in series A. A straightforward test of independence can be obtained by noting that under the assumption of independence the conditional probability is equal to the unconditional probability. In this case the number of matches is a binomial variate with a probability equal to the unconditional probability of an event occurring in series B.

## Result

### Calibration

Correlations between the proxy data and the instrument records (Table 17.1) show that the tree-ring index is significantly correlated with all the instrument records. It explains about 30% to 35% of the variance in the ENSO records. The $\delta^{18}O$ record is also significantly correlated with the instrument records at the .01 level, while the total particle correlations are significant at the 0.05 level. Significance levels are based on a full 71 degrees of freedom. A more conservative approach would be to assume that the filtering retains about one-half to two-thirds of the original degrees of freedom. Calculations based on 35 degrees of freedom indicate that the tree-ring and $\delta^{18}O$ correlations are still significant at the 0.01 level while the total particle correlations are no longer significant.

Table 17.1 *Correlations between proxy records and instrument records, 1894–1964*

|  | Rainfall | SST | SLP |
|---|---|---|---|
| Tree-rings | 0.597** | 0.529** | 0.529** |
| *Ice cores* | | | |
| Accumulation | −0.186 | −0.135 | −0.141 |
| All particles | 0.258* | 0.261* | 0.212* |
| Large particles | 0.168 | 0.189 | 0.138 |
| Small particles | 0.192 | 0.211 | 0.144 |
| Oxygen | 0.479** | 0.462** | 0.441** |
| Conductivity | −0.007 | 0.003 | −0.085 |

A single asterisk (*) indicates significance at the 0.05 level, and a double asterisk (**) indicates significance at the 0.01 level.

Due to the intercorrelation between the $\delta^{18}O$ and total particle records, a multiple regression using both does not perform significantly better than one using $\delta^{18}O$ alone. As a result, only the $\delta^{18}O$ data was used in further analyses. It explains about 20 to 25% of the variance in the ENSO records. Reconstructions based on the tree-ring and $\delta^{18}O$ records are shown in Figure 17.1. Combining the tree-ring and $\delta^{18}O$ data produces a regession equation which explains up to 40% of the variance in the ENSO records.

It should be noted that the ice-core records are values based on the July–June period, while the instrument records are calendar year averages. Therefore, correlations are based on 6 mo of overlap (January–June) and 6 mo (July–December) where the instrument records lag the ice-core records. Correlations

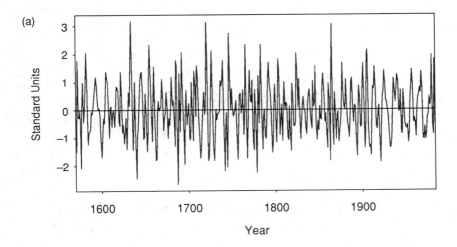

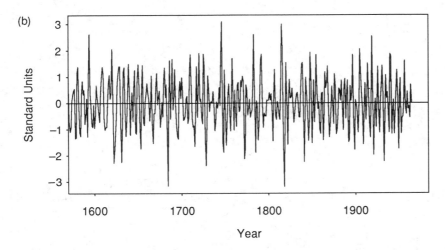

Fig. 17.1   Reconstructions of sea surface temperatures based on (a) the $\delta^{18}O$ record, and (b) the tree-ring record. Units in standard deviations.

between July–June instrument record averages and the ice-core records did not show any appreciable improvement. Comparisons of seasonally averaged instrument records show that the highest correlations with the $\delta^{18}O$ record occur during the January–June overlapping period and remain high through the rest of the year.

The sliding correlation plots (Fig. 17.2) for both proxy records show some evidence of temporal variations in the strengths of the relationships with the ENSO records. Both proxy records have peak correlations with all three instrument records in the first 30 yr of the 20th century, with the tree-ring/instrument record correlations reaching values of 0.65 to 0.70 in the 1920s and the ice-core/instrument record correlations reaching 0.60 around 1910. Both proxy records also show declining correlations with the instrument records in the 1950s, but the ice-core record, which extends into the 1980s, shows some evidence of increasing correlations with three instrument records in the 1970s.

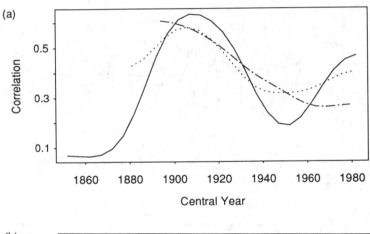

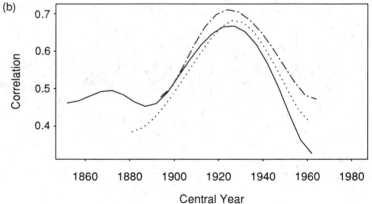

Fig. 17.2  Sliding correlation plots between the pressure record (solid), the SST record (dotted), and the rainfall record (dash-dotted), and (a) the $\delta^{18}O$ record, and (b) the tree-ring record. Window width is 40 yr.

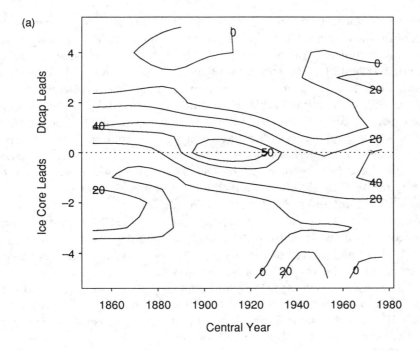

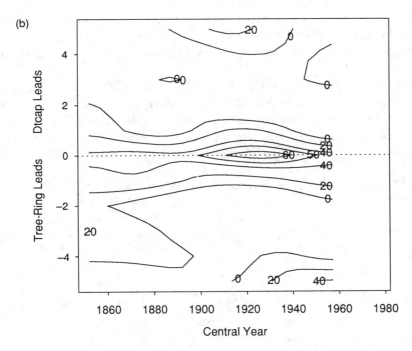

Fig. 17.3   Sliding lag correlations ×100 between the pressure record and (a) the $\delta^{18}O$ record, and (b) the tree-ring record. Leads and lags are given in years, and the window width is 40 yr.

Perhaps the most striking aspect of the correlations is the decline in correlation between the ice-core and SLP records prior to about 1880. This decline reduces the overall correlation for the period 1852–1984 to 0.34, compared to 0.45 for the period 1894–1984. The correlation between the tree-ring and SLP records shows a similar, but less pronounced, pattern. Examination of the sliding lag correlations (Fig. 17.3a) suggests that the decline may be partially caused by a shift in the ice-core record relative to the SLP record so that it lags by one year prior to about 1880. It appears possible that a year was missed in the ice-core record. When a year is added between 1880 and 1881 by inserting the mean, the overall correlation for 1852–1984 increases to 0.46. The sliding lag correlations for the tree-ring record (Fig. 3b) give no indications of any similar shifts.

The $\delta^{18}O$ record for core 1 is highly correlated with the summit core record during this period and shows the same shift relative to the SLP record. Both ice-core records are in phase with the SLP back through the 1880s, showing warm events in 1888 and 1881, but the major 1877 warm event in the SLP record appears in 1878 in the ice-core records. Similarly, the SLP record shows weak-to-moderate warm events in 1855, 1864, and 1868 and cold events in 1863 and 1870, all of which match with events dated 1 yr later in both ice-core records.

Table 17.2 gives the number of matching events for the thresholded instrument records and proxy records. The probability of an event in any year in one record is 0.17 (12/71), so under the independence hypothesis, the probability of events occurring simultaneously in two records is $0.029 = 0.17^2$. Thus, in 71 yr the expected number of matches is two; five or more matches are required to reject the independence hypothesis at the 0.05 level and six or more to reject at the 0.01 level.

All three of the proxy records have enough matches with each of the instrument records to reject the hypothesis of independence at the 0.05 level. The tree-ring record has six matches with each of the instrument records. The $\delta^{18}O$ does slightly better with seven matches with the SST and SLP records, while the documentary record does slightly worse with five matches with the SST and SLP

Table 17.2 *Number of co-occurrences of strong and very strong ENSO events, 1894–1964*

|          | Rainfall | SST    | SLP    | Doc.   | Trees  | Oxygen |
|----------|----------|--------|--------|--------|--------|--------|
| Rainfall | —        | 9      | 9      | 6      | 6      | 6      |
| SST      | 0.0000   | —      | 9      | 5      | 6      | 7      |
| SLP      | 0.0000   | 0.0000 | —      | 5      | 6      | 7      |
| Doc.     | 0.0085   | 0.0384 | 0.0384 | —      | 4      | 6      |
| Trees    | 0.0085   | 0.0085 | 0.0085 | 0.1302 | —      | 8      |
| Oxygen   | 0.0085   | 0.0014 | 0.0014 | 0.0085 | 0.0002 | —      |

Values below the diagonal are probabilities of at least that number of matches for independent records. (12 events).

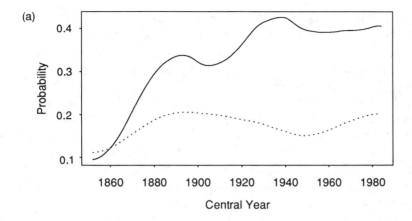

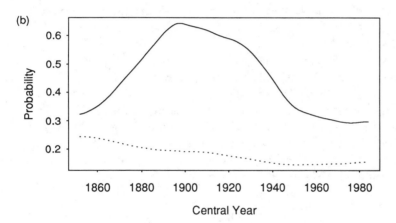

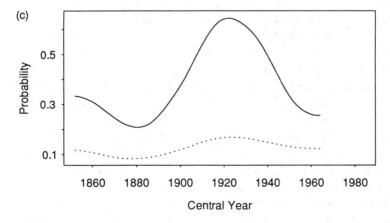

Fig. 17.4  Kernel estimates of the conditional probabilities of a warm ENSO year given an warm ENSO year in the pressure record (solid lines) and unconditional rate estimates (dashed lines) for (a) the documentary record, (b) the $\delta^{18}O$ record, and (c) the tree-ring record. Kernel width is 40 yr.

records. Not surprisingly, the three instrument records are highly related to each other, with nine matches in all cases. The tree-ring and $\delta^{18}O$ records are also highly related by this criterion with eight matches. The $\delta^{18}O$ record also shows a significant number of matches with the documentary record. Only the documentary and tree-ring records are not significantly related, with four matches.

The three proxy records were compared with the longer SLP record to identify any changes in the strength of the correspondence over time. The kernel method described in the previous section was used to estimate time-varying rates of warm ENSO event occurrence for each proxy record and rates of co-occurrence of warm events in the SLP record and the proxy records. The results are displayed in Figure 17.4. In each plot the solid line is the conditional probability of having an ENSO event in the proxy record, given an ENSO event the SLP index, while the dashed line is the unconditional probability of an ENSO event occurring in the proxy record. If the two series were independent the two probabilities would be equal. (Note that the $\delta^{18}O$ record was adjust for a probable missing year as described above.)

In all cases the conditional probabilities vary by at least a factor of two. In general, the conditional probabilities are low prior to 1880 and rise to relative high values by 1900. The documentary record, in particular, has a low correspondence with the SLP index at the beginning of the record, with conditional probabilities as low as the unconditional probabilities. The conditional probabilities rise above 0.30 by 1900 and level off at around 0.40 by 1920. The $\delta^{18}O$ and tree-ring records have higher peaks of about 0.60 around 1900 and 1920 but then decline by 1950 to values around 0.30.

### Proxy record comparisons

The tree-ring and $\delta^{18}O$ records were compared for the full period of overlap, 1570–1964, by calculating cross-correlations for sliding windows (Fig. 17.5). Three more instances where the two series became offset by one year were identified. Correlations between the tree-ring and core 1 $\delta^{18}O$ records (not shown) exhibit a very similar pattern, with one exception. Between about 1760 and 1800 core 1 lags the tree-ring record by only 1 yr while the summit core lags by 2 yr. This discrepancy results from the fact that during this period the two $\delta^{18}O$ records are, themselves, offset by 1 yr, as is evident in Figure 17.6. These results are consistent with those reported by Thompson et al. (1986: 363), indicating that the Quelccaya cores have been dated back to A.D. 1500, with an estimated uncertainty of ± 2 yr and an absolute date of A.D. 1600, defined by the identification of the Huaynaputina ash in both Quelccaya ice cores.

Although is not possible to be certain which series should be shifted, it seems likely that the $\delta^{18}O$ records are responsible, considering the fact that a missing year was located in the comparison with the SLP index and that the tree-ring index is composed of seven different chronologies, each constructed from a number of individual samples. (In a certain sense it is unfair to compare a record

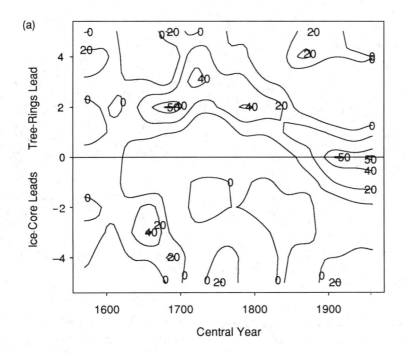

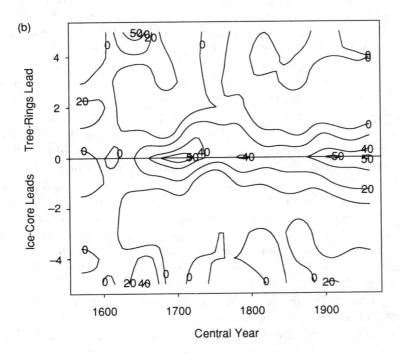

Fig. 17.5   Sliding lag correlations ×100 between tree-ring record and (a) the uncorrected and (b) the corrected $\delta^{18}O$ record. Leads and lags are given in years, and the window width is 50 yr.

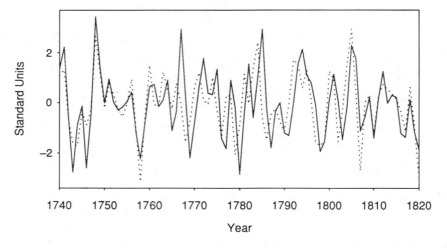

Fig. 17.6   Partial time-series plots of the $\delta^{18}O$ records from the summit core (solid) and core 1 (dotted).

produced from many different trees spread over a region with a single core from a single ice cap.) Consequently, the summit core $\delta^{18}O$ record was adjusted by adding two missing years (1840 and 1731) and deleting one extra year (1692). After adjustment the correlation for the period 1570–1964 with the tree-ring record is 0.351. The cross-correlation analysis also indicates, however, that the relationship between the two records disappears completely prior to 1630. The correlation for the period 1630–1964 increases to 0.426.

The two continuous records were converted to categorical records using the same thresholds as in the calibration study. While this approach fixes the same number of events during the 1894–1964 calibration period, there were substantial differences in the number of events for the full period of overlap, 1570–1964. By this definition, the number of events ranges from 77 (one out of every 5.1 yr is an warm event year) for the ice core record, to 62 (one out of every 6.4 yr) for the tree-ring record, and 55 (one out of every 7.2 yr) for the documentary record. As noted above, however, in many cases warm events persist over more than one year, so there were not as many separate warm events in the records. If the alternative warm event definition is used, and multi-year events are treated as a single event, the discrepancies are not nearly as great. The $\delta^{18}O$ record has 51 events (warm event every 7.7 yr), the tree-ring record has 47 (8.4 yr), and the documentary record still has the least with 40 events (9.9 yr).

The reason for the differences in overall occurrence rates is evident in Figure 17.7. All three records naturally have comparable rates for the 20th century calibration period, but precalibration rates are substantially higher for the $\delta^{18}O$ record, lower for the documentary record, and somewhere in between for the tree-ring record. There are some indications of more frequent ENSOs at around 1650–1700, particularly in the $\delta^{18}O$ and tree-ring records, and frequencies are generally low around 1800. The correspondence between the fluctuations in the three records is not very striking, however, and none of the fluctuations is large enough to reject the hypothesis of a process with a constant rate. The difference in constant rate estimates between the $\delta^{18}O$ and documentary records is significant at the 0.01 level, so combining the decrease in event frequencies in the documentary record with the increase in the $\delta^{18}O$ record does produce significantly different rate estimates, even if the variations within each record do not.

As noted above the disparity in rate estimates is reduced somewhat by treating a multiyear warm event as a single event. Comparison of the warm year rates in Figure 17.7 with the onset occurrence rates in Figure 17.8 suggest that a major reason for this is that the $\delta^{18}O$ warm year rate is much higher than the onset

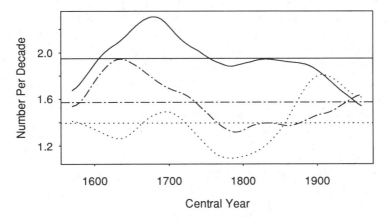

Fig. 17.7   Kernel estimates of warm ENSO year occurrence rates for the $\delta^{18}O$ record (solid), the documentary record (dotted), and the tree ring record (dash-dotted). Kernel width is 80 yr. Horizontal lines give constant rate estimates.

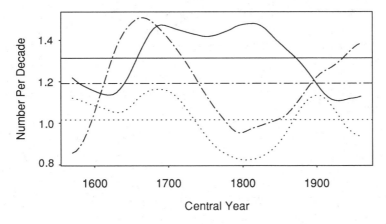

Fig. 17.8   Same as Figure 17.7 for warm ENSO onset even definition.

year rate in the early 1700s (2.3 events/decade compared to 1.5 events/decade). The tree-ring and $\delta^{18}O$ records now show comparable rates during this period. The documentary record onset rate is still substantially lower, but the 20th century peak is not nearly so far out of line with the rest of the record, and its rate fluctuations closely resemble those in the tree-ring record. The main reason the $\delta^{18}O$ rate is still higher than the rates for the other two records is that it does not share in the low onset rates around 1800. As was the case for the ENSO year rate fluctuations, none of the variations in the individual records is large enough to reject the null hypothesis of a constant rate, but the difference between the $\delta^{18}O$ and documentary constant rate estimates is significant.

Tables 17.3 and 17.4 give statistics on the number of matches between the proxy records for the full period, 1570–1964, and for the precalibration period, 1570–1893. The tree-ring and $\delta^{18}O$ records show a high degree of correspondence during both periods, with at least twice the number of matches expected by chance. The tree-ring and documentary records, on the other hand, have no more matches than would be expected by chance. The $\delta^{18}O$ and documentary records have a significant number of matches for the full period, but the correspondence is considerably worse when the modern calibration period is excluded.

The time-varying conditional probabilities are shown in Figure 17.9. The correspondence between all three records is greatest during the calibration period.

Table 17.3 *Number of matches for strong/very strong ENSO events, 1570–1964*

|        | Doc    | Trees   | Oxygen    |
|--------|--------|---------|-----------|
| Doc.   | –      | 8(8.6)  | 19(10.7)  |
| Trees  | 0.5862 | –       | 27(12.1)  |
| Oxygen | 0.0052 | 0.0000  | –         |

Expected number of matches are given in parenthenses, and values below the diagonal are probabilities of getting as many or more matches if the series are independent (based on the normal approximation to the binomial distribution).

Table 17.4 *Number of matches for strong/very strong ENSO events, 1570–1893*

|        | Doc    | Trees   | Oxygen    |
|--------|--------|---------|-----------|
| Doc.   | –      | 4(6.6)  | 12(8.6)   |
| Trees  | 0.8495 | –       | 20(10.0)  |
| Oxygen | 0.1222 | 0.0007  | –         |

Expected number of matches are given in parenthenses, and values below the diagonal are probabilities of getting that number of matches if the series are independent.

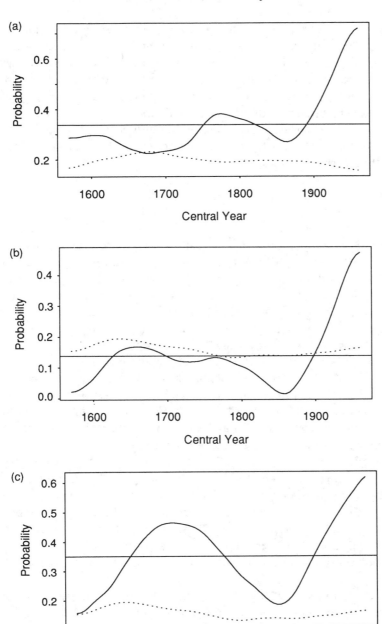

Fig. 17.9 Kernel estimates of conditional probabilities of a warm ENSO event (solid lines) and unconditional probabilities (dashed lines) for (a) $\delta^{18}$O events conditioned on documentary events; (b) tree-ring events conditioned on documentary events; and (c) tree-ring events conditioned on $\delta^{18}$O events. Kernel width is 80 yr. Horizontal line gives constant rate estimates.

The tree-ring and $\delta^{18}O$ records are also closely related through much of the 1600s and 1700s. The documentary record does not match well with either of the other records in the precalibration period, apart from a moderate peak with the ice core record in the late 1700s. It is also noteworthy that none of the records match well during the mid-1800s. The small numbers of events in the matching series and the fact that the conditional probabilities are ratios of two rates make it difficult to determine confidence intervals, but it is interesting that the conditional rates fluctuate by a factor of three or four, a level of variability much greater than for the unconditional rates.

## Discussion

### *Calibration*

The most important point to note is that all three proxy records do appear to contain some information on ENSO variability. All of the skill measures do show a significant level of correspondence. It seems likely that the relationships between the proxy and ENSO instrument records are not a great deal weaker than would be the relationships between actual climatic data and the ENSO records. For example, Kiladis and Diaz (1989) note that precipitation anomalies at Abilene, Texas, were positive during 70% of warm ENSO winters. The tree-ring record has positive values during 85% of the warm ENSO years in the SST record. Similarly, Manaus, Brazil, temperature anomalies were positive during 95% of warm ENSO winters and 76% of warm ENSO springs, while the comparable figure for the $\delta^{18}O$ record was 75%. Recent work has shown that even El Niño variability in the coastal South American regions covered by the documentary records are not as closely linked to basinwide ENSO variability as was previously thought (Deser and Wallace 1987, 1990). Apparently, the proxy records are doing a satisfactory job of capturing the component of regional variability which is related to large-scale ENSO variability, and the main source of error in the ENSO reconstructions is produced by the lack of correspondence between regional conditions and ENSO variability. In other words, the proxy records, while not perfect recorders of regional conditions, do seem to respond well to that portion of regional variability which is related to large-scale ENSO features.

This conclusion strengthens the argument for combining records from different regions to filter out regional variability not related to ENSO. As noted above, combining the $\delta^{18}O$ and tree-ring records does increase the explained variance by around 10%. The same pattern can also been seen with the categorical records. If a warm ENSO year is identified in any one of the proxy records there is at best a 50 to 55% chance it will also be identified as a warm ENSO year in any of the instrument records. For a year classified as a warm ENSO year in at least two of the proxy records, however, the chance of it being in the warm ENSO class in any of the instrument records increases to 70% (seven out of ten).

The identification of an apparent missing year in the sliding correlation between the SLP and $\delta^{18}O$ records presents a clear lesson of the utility of examining time-varying relationships in calibration studies. This gap produces a substantial decrease in the overall correlation, and if left unadjusted, could lead to the conclusion that the relationship is too weak to be of use. The other fluctuations in the strength of the correlations are probably within the expected level of sampling variability, indicating that the relationships are reasonably consistent over the calibration period. They are large enough, however, to potentially produce misleading results if the common technique of splitting the calibration period into training and verification samples was employed. A full cross-validation (e.g. Michaelsen, 1987) would generally produce more stable skill estimates.

Interpretation of variations in the probabilities of proxy record warm events conditioned on instrument record warm events is complicated somewhat by the need to classify events. As can be seen in Table 17.5, there are a number of instrument warm ENSO years when the proxy records show some indications of warm event activity (weak or moderate class for the documentary record or moderate positive values for the tree-ring and $\delta^{18}O$ record). There do appear to be some common features, however, which are worth noting. Most striking is the lack of correspondence between the SLP record and any of the proxy records prior to about 1880. For example, of the five SLP ENSO years before 1877, none are identified in the documentary record and only one in the $\delta^{18}O$ record. There are a couple of possible explanations for this lack of correspondence. First, the SLP record, itself, is likely to be somewhat shaky for the early period. In may not be coincidence that the correspondence starts to increase about 1880, the same time that the Darwin pressure record begins. Second, as noted in the intercomparisons between the proxy records, for much of the 1800s none of the proxy records show good correspondence among themselves. Furthermore, both the documentary and tree-ring records have low ENSO frequencies during the this period. This raises the possibility that the lack of correspondence among the various proxy records and between the proxy records and the SLP record may be in part caused by a period of weak, infrequent ENSOs which did not have a strong impact on regional conditions in the various different areas.

A marked contrast is seen during the first part of the 1900s. For example, the $\delta^{18}O$ record matches five and the tree-ring record six of the seven SLP warm event years between 1905 and 1941. During this period, warm events were relatively frequent (particularly 1900–1920) and strong. As a result, ENSO variability was a significant factor affecting regional conditions. It is also true, of course, that the SLP and the other instrument records become more reliable after 1900. After 1941, however, only the strong warm events of 1957/58 and 1982/83 are clearly identified in the proxy records, suggesting that improved quality of the instrument records is not solely responsible for the improved correspondence during the early 20th century. Instead, it appears that the variations in the correspondence between the proxy and instrument records is more directly

Table 17.5 *Years classified strong/very strong ENSO events in at least one instrument record, 1853–1983. Years classified as ENSOs are show in bold face type*

| Year | Rainfall | SST | SLP | Doc | Trees | Oxygen |
|------|----------|------|------|------|-------|--------|
| 1853 |          |      | 0.88 | –    | –0.72 | –0.36  |
| 1855 |          |      | 1.40 | –    | 1.84  | 1.31   |
| 1856 |          |      | 0.70 | –    | 0.36  | 0.56   |
| 1868 |          |      | 1.92 | W/M  | 1.41  | 0.62   |
| 1869 |          |      | 0.78 | –    | –0.09 | –0.03  |
| 1877 |          |      | 3.14 | VS   | 0.69  | 1.69   |
| 1881 |          | 0.09 | 1.23 | –    | 0.57  | 1.10   |
| 1885 |          | 0.19 | 0.90 | S    | 0.00  | –0.98  |
| 1888 |          | 1.45 | 1.08 | W/M  | 0.94  | 1.15   |
| 1896 | 0.36     | 1.44 | 0.55 | M+   | 0.42  | 0.87   |
| 1900 | 0.55     | 1.69 | 0.88 | S    | 0.12  | 0.93   |
| 1902 | 1.31     | 1.97 | 1.03 | M+   | –0.31 | 0.40   |
| 1905 | 1.91     | 2.16 | 1.65 | W/M  | 2.04  | 2.02   |
| 1911 | 1.03     | 0.61 | 0.53 | S    | 0.77  | –0.04  |
| 1912 | 1.14     | 0.52 | 0.55 | S    | –0.62 | 0.83   |
| 1914 | 2.37     | 2.23 | 2.15 | M+   | 1.52  | 1.55   |
| 1919 | 2.14     | 1.38 | 0.88 | W/M  | 2.53  | 0.15   |
| 1926 | 0.68     | 0.49 | 0.93 | VS   | 1.39  | 1.07   |
| 1930 | 1.54     | 1.23 | 0.85 | W/M  | 0.65  | –0.24  |
| 1940 | 2.14     | 2.01 | 1.77 | S    | 1.87  | 1.27   |
| 1941 | 2.16     | 2.08 | 2.32 | S    | 1.81  | 1.07   |
| 1946 | 0.48     | –0.41 | 0.70 | –   | –0.40 | 0.85   |
| 1951 | 0.58     | 0.86 | 0.50 | W/M  | 0.20  | –0.66  |
| 1953 | 1.10     | 0.81 | 1.35 | M+   | –1.15 | –1.17  |
| 1957 | 1.22     | 1.21 | 0.58 | S    | 1.61  | 1.44   |
| 1958 | 0.99     | 1.07 | 0.75 | S    | –0.12 | 0.92   |
| 1965 | 1.48     | 1.31 | 1.47 | M    |       | 0.56   |
| 1966 | 0.19     | 0.37 | 0.80 | –    |       | 0.79   |
| 1969 | 0.30     | 1.37 | 1.13 | –    |       | 0.86   |
| 1972 | 1.79     | 1.78 | 0.98 | VS   |       | –0.58  |
| 1976 | 1.07     | 0.21 | –0.42 | W/M |       | –0.18  |
| 1977 | 1.27     | 0.70 | 1.55 | –    |       | 0.46   |
| 1978 | –0.32    | –0.12 | 0.80 | –   |       | 0.15   |
| 1979 | 1.00     | 0.68 | 0.15 | –    |       | 0.36   |
| 1980 | 1.12     | 0.52 | 0.73 | –    |       | 1.96   |
| 1982 | 2.00     | 1.86 | 1.72 | VS   |       | –0.86  |
| 1983 |          | 1.81 | 2.17 | VS   |       | 1.81   |

related to the magnitude and frequency of ENSO events. Not surprisingly, the proxy records match well during periods when warm events are strong and frequent and considerably less well during periods when when warm events are less frequent and/or generally weaker.

## Proxy record comparisons

Probably the most striking result of this research is the relatively stable, significant relationship between the $\delta^{18}O$ and tree-ring records extending back to 1630 (after adjusting the $\delta^{18}O$ record). It is very difficult to conceive of any hypothesis to explain this other than that they are both responding to large-scale ENSO variability. This leads to two conclusions. First, there is strong evidence that ENSO variability quite similar to that in the modern period has existed for at least 350 yr. Since this includes the major cold period of the Little Ice Age, which is also evident in the ice core record (Thompson et al. 1986), it appears that the ENSO phenomenon is fairly robust to large-scale climate changes. Second, the two proxy records recorded this variability with reasonable fidelity throughout the period. Clearly, this correspondence provides a level of verification of the proxy records and of the persistence of ENSO variability far beyond what could be obtained from working with a single proxy record.

It should be noted that the correspondence between the two records depends on the validity of the adjustments applied to the $\delta^{18}O$ record. The statistical evidence for the offsets in the $\delta^{18}O$ record is fairly solid, but the physical mechanisms which might introduce errors in dating the ice-core record are not clear. Furthermore, the Huaynaputina ash layer is accurately dated in A.D. 1600 (Thompson et al. 1986), so any missing years would have to be compensated for by an equal number of false years. There are two possible sources of errors in dating. First, cores were extracted in discrete 1- to 1.5-m segments, and some material could be lost between segments. The fact that the $\delta^{18}O$ records from both cores match, with the minor exception of the 1 yr offset between 1760 and 1800, argues against this possibility. Second, the natural record itself might not always faithfully record the annual cycle. While the $\delta^{18}O$ ratios, microparticle concentrations, and conductivities all show clear annual cycles, there are also short-term intra-annual variations which make it difficult to identify annual cycles from any single record. Ambiguities were resolved by integrating the records (Thompson et al. 1986). It is possible that environmental variations could affect all of these measures in a similar way in both cores, either depressing an annual cycle or producing a intra-annual variation which would be indistinguishable from a second annual cycle. If the statistical evidence is to be believed, this is probably the most likely explanation. Further comparisons with other ENSO proxy records will be needed to resolve the issue.

The reason for the lack of correlation prior to 1630 is not clear at present. It is possible that it signals a decline in ENSO variability, but defects in one or both of the proxy records is a more likely explanation. For example, the tree-ring chronologies are based on successively fewer samples going back in time, so their reliability probably decreases accordingly.

A comparison of the ENSO onset rates with the ENSO year rates indicates that some of the more subtle features of the curves may be primarily artifacts of the way an event is defined. A number of alternative event definitions and thresholds were tried, but the major features of the rate curves did not change appreciably.

In particular the $\delta^{18}O$ record has more warm events prior to 1900, and the documentary record has fewer events. All three records show increased event frequencies in the late 1600s and early 1700s. All three records also show reduced frequencies around 1800, although the evidence in the $\delta^{18}O$ record is admittedly less clear. Given the modest degree of correspondence between the rate fluctuations in the different records, along with the fact that none of the fluctuations are statistically significant, it is probably best to conclude that the evidence does not contradict the hypothesis that ENSO variability has been reasonably consistent over the last 350 yr.

One of the more puzzling results of the study is the lack of correspondence prior to the modern period between the documentary record and the other two records, particularly the tree-ring record. This may reflect some combination of changes in the reliability of the documentary record and a lack of correspondence between local El Niño conditions along the coast of Peru and large-scale ENSO variability. The fact that the correspondence with the proximate $\delta^{18}O$ record is somewhat higher could lend support to the latter possibility. In addition, there are a number of instances in the calibration period when ENSOs identified in the other instrument and proxy records are present in the documentary record as weak or moderate events, suggesting part of the problem may again be related to the difficulties in categorizing a continuously varying phenomenon. The reliability of a documentary record over time is very difficult to assess. Clearly decreases in the quality and quantity of documentary evidence will make interpretation more difficult for earlier periods, as will other factors such as the evolution of meaning and usage in languages. As a result, it may be more difficult and challenging to maintain a constant level of reliability in documentary proxy records than in natural proxy records because there is often less change in the structure of natural recording systems than there has been in human recording systems. Clearly, however, the lack of correspondence with the tree-ring record and weak correspondence with the $\delta^{18}O$ record do not convincingly invalidate the documentary record. It is also clear that neither the documentary record nor any other single proxy record can be considered the base against which to test the validity of new proxy records.

## Summary

All three proxy records have significant levels of correspondence with the instrument records during the modern calibration period. After adjusting the $\delta^{18}O$ record for apparent missing years, the correspondence between it and the tree-ring record remains consistent back to about 1630. In addition to providing verification of the reliability of the records, this correspondence suggests strongly that ENSO variability persisted without major changes through a period of significantly different large-scale climatic conditions. This robustness of the ENSO phenomenon to past climatic changes suggests that there is a good possibility that it may continue essentially unchanged through the sorts of climatic changes that are likely to occur during the next century or two. Nevertheless, it should be kept in mind that both the ice cores and the tree-ring record describe variations within the generally colder climate of the so-called Little Ice Age. There is ice-core and archaeological evidence for 'Mega El Niño' events in A.D. 600 and 1100, times that were isotopically warmer than present (see Nials et al. 1979; Shimala et al. 1991).

There is some evidence, albeit less convincing, that there were periods of strong ENSO activity during the early 1700s and the early 1900s and a period of low activity during the early 1800s. Furthermore, there are indications that the level of correspondences between the proxy and instrument records and between the different proxy records are considerably better during periods of high activity. On a methodological note, there are many examples in the study of the utility of examining time-varying statistics in paleoclimatic calibration and reconstruction studies. Finally, the results of this study show very convincingly the importance of comparing different high resolution proxy records. It is hoped that this is just a first step, and as the quality and quantity of ENSO proxy records increases, further studies can remove some of the uncertainties and refine the results presented here.

## References

BAUMGARTNER, T. R., MICHAELSEN, J., THOMPSON, L. G., SHEN, G. T., SOUTAR, A., and CASEY R. E., 1989: The recording of interannual climatic change by high-resolution natural systems: tree-rings, coral bands, glacial ice layers, and marine varves. *In* Peterson, D. H. (ed.), *Aspects of Climate Variability in the Pacific and the Western Americas*. Geophysical Monograph 55. Washington D.C.: American Geophysical Union, 1–14.

COX, D. R. and LEWIS, P. A. W., 1966: *The Statistical Analysis of Series of Events*. London: Chapman and Hall. 285 pp.

DESER, C. and WALLACE, J. M., 1987: El Niño events and their relationship to the Southern Oscillation. *Journal of Geophysical Research*, 92: 14,189–14,196.

DESER, C. and WALLACE, J. M., 1990: Large-scale atmospheric circulation features of warm and cold episodes in the tropical Pacific. *Journal of Climate*, 3: 1254–1281.

DOBERITZ, R., 1968: Cross spectrum analysis of rainfall and sea temperature at the equatorial Pacific Ocean. *Bonner Meteorologische Abhandlungen*, 8: 1–61.

DREW, L. G., 1976: *Tree-Ring Chronologies for Dendroclimatic Analysis: An Expanded North American Grid*. University of Arizona, Tucson.

GROVE, J. M., 1988: *The Little Ice Age*. London: Methuen. 498 pp.

JULIAN, P. R. and CHERVIN, R. M., 1978: A study of the southern oscillation and Walker circulation phenomenon. *Monthly Weather Review*, 106: 1433–1451.

KILADIS, G. N. and DIAZ, H. F., 1989: Global climatic anomalies associated with extremes in the Southern Oscillation. *Journal of Climate*, 2: 1069–1090.

LAMB, H. H., 1977: *Climate: Present, Past and Future*, Vol 2. London: Methuen. 638 pp.

MICHAELSEN, J., 1987: Cross-validation in climate forecasting models. *Journal of Climate and Applied Meteorology*, 26: 1589–1600.

MICHAELSEN, J., 1989: Long-period fluctuations in El Niño amplitude and frequency reconstructed from tree-rings. *In* Peterson, D. H. (ed.), *Aspects of Climate Variability in the Pacific and the Western Americas.* Geophysical Monograph 55. Washington D.C.: American Geophysical Union, 69–74.

NIALS, F. L., DEEDS, E. R., MOSELEY, M. E., POZORSKI, S. C., POZORSKI, T., and FELDMAN, R. A., 1979: El Niño: The catastrophic flooding of coastal Peru. *Field Museum of Natural History Bulletin*, 504 (Part I), 504 (Part II).

QUINN, W. H., NEAL V. T., and ANTUNEZ de MAYOLO, S. E., 1987: El Niño occurrences over the past four and a half centuries. *Journal of Geophysical Research*, 92, 14,449–14,461.

ROPELEWSKI, C. F., JANOWIAK J. E., and HALPERT, M. S., 1985: The analysis and display of real-time surface climate data. *Monthly Weather Review*, 113: 1101–1106.

ROPELEWSKI, C. F. and HALPERT, M. S., 1986: North American precipitation and temperature patterns associated with the El Niño/Southern Oscillation (ENSO). *Monthly Weather Review*, 114: 2352–2362.

SHIMADA, I., SCHAAF, C. B., THOMPSON, L. G., and MOSLEY-THOMPSON, E., 1991: Cultural impacts of severe droughts in the prehistoric Andes: Application of a 1,500-year ice core precipitation record. *World Archaeology: Archaeology and Arid Environment*, 22: 247–270.

SILVERMAN, B. W., 1986: *Density Estimation.* London: Chapman and Hall. 175 pp.

SOLOW, A. R., 1991: The nonparametric analysis of point process data: the freezing history of Lake Konstanz. *Journal of Climate*, 4: 116–119.

THOMPSON, L. G., DAVIS, M., MOSLEY-THOMPSON, E., and LIU, K., 1988: Pre Incan agricultural activity recorded in dust layers in two tropical ice cores. *Nature*, 336: 763–765.

THOMPSON, L. G., MOSLEY-THOMPSON, E., BOLZAN, J. F., and KOCI, B. R., 1985: A 1500 year record of tropical precipitation recorded in ice cores from the Quelccaya Ice Cap, Peru. *Science*, 229: 971–973.

THOMPSON, L. G., MOSLEY-THOMPSON, E., and MORALES AMAO, B., 1984: El Niño-Southern Oscillation events recorded in the stratigraphy of the tropical Quelccaya ice cap, Peru. *Science*, 226: 50–53.

THOMPSON, L. G., MOSLEY-THOMPSON, E., DANSGAARD, W., and GROOTES, P. M., 1986: The 'Little Ice Age' as recorded in the stratigraphy of the tropical Quelccaya ice cap, Peru. *Science*, 234: 361–364.

THOMPSON, L. G. and MOSLEY-THOMPSON, E., 1989: One-half millenia of tropical climate variability as recorded in the stratigraphy of the Quelccaya ice cap, Peru. *In* Peterson, D. H. (ed.), *Aspects of Climate Variability in the Pacific and the Western Americas.* Geophysical Monograph 55. Washington D.C.: American Geophysical Union, 15–31.

WIGLEY, T. M. L., INGRAM, M. J., and FARMER, G., (eds.), 1981: *Climate and History.* Cambridge: Cambridge University Press. 530 pp.

WILKS, D. S., 1990: Maximum likelihood estimation for the gamma distribution using data containing zeros. *Journal of Climate*, 3: 1495–1501.

WRIGHT, P. B., 1989: Homogenized long-period Southern Oscillation indices. *Journal of Climatology*, 9: 33–54.

# 18

# Coral monitors of El Niño/Southern Oscillation dynamics across the equatorial Pacific

JULIA E. COLE*

*Lamont-Doherty Geological Observatory of Columbia University, Palisades, New York 10964, U.S.A., and
Department of Geological Sciences, Columbia University, New York, New York 10025, U.S.A.*

GLEN T. SHEN

*School of Oceanography, WB-10 University of Washington, Seattle, Washington 98195, U.S.A.*

RICHARD G. FAIRBANKS

*Lamont-Doherty Geological Observatory of Columbia University, Palisades, New York 10964, U.S.A., and
Department of Geological Sciences, Columbia University, New York, New York 10025, U.S.A.*

MICHAEL MOORE

*Reef Research Group, Museum of Paleontology, University of California, Berkeley, California 94720, U.S.A*

## Abstract

Variability in the El Niño/Southern Oscillation (ENSO) system generates most
of the interannual variability observed in global climate, yet its long-term history
in the equatorial Pacific remains poorly documented. The fundamental dynamic
components of ENSO variability in the equatorial Pacific include interannual
changes in upwelling, atmospheric convection, and wind speed and direction.
These processes are integral physical components of ENSO variability, and they
produce distinct thermal and chemical signals in the surface ocean. Shallow-
growing corals from sensitive Pacific sites incorporate these anomalies in the
isotopic and trace metal chemistry of their aragonite skeletons. Short ($\sim 20$ yr)
coral records provide independent monitors of the ENSO system at three sites
across the Pacific basin: the Galapagos ($1°S$, $91°W$), Tarawa Atoll ($1°N$, $173°E$),
and Bali ($8°S$, $115°E$). Galapagos Cd/Ca, Ba/Ca, and $\delta^{18}O$ records reflect the
degree of regional upwelling in the eastern Pacific, which is suppressed during
warm ENSO conditions. Oxygen isotopic data from Tarawa Atoll corals record
the intense precipitation that the eastward displacement of the Indonesian Low
brings to this region during warm ENSO periods. An independent record of
Mn/Ca from one of these corals reflects the weakening and reversal of the trade

* Present address Institute of Arctic and Alpine Research, University of Colorado, Boulder,
Colorado 80309-0450, U.S.A.

winds that may trigger the onset of warm ENSO conditions basinwide. Finally, $\delta^{18}O$ from a Bali coral reflects the weakening of the Indonesian monsoon associated with warm ENSO periods. The isotopic and trace metal records from these three sites illustrate how chemical records from coral skeletons yield previously unobtainable information on dynamic aspects of ENSO variability, in many cases at monthly resolution. Individually, each of these records closely monitors an important ENSO component: SST, rainfall, winds, and upwelling. Together they provide information on the variability of climatic and oceano-graphic anomalies throughout the tropical Pacific, including the spatial patterns of evolution and recurrence of both warm-and cool-phase ENSO anomalies. Living corals that reach hundreds of years in age and fossil corals spanning thousands of years are available throughout the Pacific, enabling high-resolution ENSO reconstructions under the altered climate boundary conditions of the late Pleistocene.

## Introduction

Much of the world's interannual climate variability can be traced to the El Niño/Southern Oscillation (ENSO) phenomenon of the equatorial Pacific. Recent advances in clarifying the physical processes responsible for ENSO development have led to viable short-term ENSO forecasts (Zebiak and Cane 1987; Barnett et al. 1988; Zebiak 1989), yet past variations in the behavior of this system remain poorly characterized. Improved paleoclimatic documentation of ENSO variability will advance our understanding of past tropical and global climate changes, for example by detailing the spatial variability of the ENSO phenomenon and its teleconnections to temperate regions. Clarifying the past ENSO response to changing climate boundary conditions will help us to anti-cipate the response of this system to ongoing increases in greenhouse gases and perhaps shed light on tropical climate dynamics during the last glacial maximum. In the region of the primary ENSO signal, the equatorial Pacific, long records of high-frequency climate variability are scarce, and most available information derives from the eastern tropical Pacific (e.g., Quinn et al. 1987). Especially in the central to western oceanic regions, instrumental records are rare and in most cases cover only the period since World War II; a few records extend through this century. To understand the behavior of ENSO over longer periods, and to reconstruct details of spatial variability, we clearly need to explore geological and biological archives sensitive to ENSO-related environmental conditions.

Many of the fundamental dynamical features of ENSO – including changes in upwelling, winds, convection, and sea surface temperatures – produce distinct chemical and thermal changes in the surface waters of the tropical Pacific. Chemical records from the skeletons of shallow-growing equatorial Pacific corals faithfully monitor the full spectrum of variability in these signals (Druffel 1985; Shen et al. 1987; Lea et al. 1989; McConnaughey 1989; Cole and Fairbanks 1990). Corals from sensitive Pacific sites are uniquely suited to provide detailed records of the major features of ENSO from this key region. We present isotopic and trace-metal results from sites spanning the equatorial Pacific that demonstrate excellent possibilities for ENSO reconstructions in this region, which has received little paleoclimatic attention. Comparison of these widely separated sites, each sensitive to distinct features of ENSO, reveals the spatial patterns of ENSO variability over recent decades and yields a rich, dynamic picture of ENSO development and recurrence.

## ENSO

The ENSO phenomenon involves the large-scale oscillation of the tropical Pacific ocean-atmosphere system between two extremes, characterized by warm and cool sea surface temperatures (SSTs). Both warm- and cool-phase ENSO conditions experience coherent suites of associated oceanographic and atmospheric anomalies. Anomalies may propagate to higher latitudes via mechanisms that include the displacement of upper atmospheric pressure patterns and the generation of troughs that penetrate the North American continent (van Loon and Rogers 1981; Rasmusson and Wallace 1983). The influence of ENSO dominates interannual climate variability throughout the equatorial Pacific, and ENSO-related environmental changes can bring catastrophic consequences to ecological and human systems (Glantz 1984; Glynn 1990).

During the warm phase of ENSO, upwelling of cool waters along the South American coast is suppressed. Mean SST anomalies reach nearly 2°C in the eastern tropical Pacific and attenuate to the west, reaching ≤ 0.5°C by the dateline (Rasmusson and Carpenter 1982). West of 180°, the primary signature of warm-phase ENSO conditions is the dramatic rearrangement of precipitation and wind patterns resulting from the northeastward migration of the Indonesian Low pressure system. This development brings drought to the Australasian/Indonesian area and torrential rains to the normally dry islands near the equator and the dateline. Rainfall also increases dramatically over the region of increased sea surface temperatures between the dateline and the South American coast. The normally dominant easterly trades weaken and may reverse to westerly as a consequence of changing SST and sea-level pressure (SLP) fields. Warm-phase ENSO conditions alternate with cooler periods of intensified Walker circulation that experience a coherent set of environmental anomalies similar in nature but opposite in sign to those described above (Philander 1985, 1990; Wright et al. 1988; Deser and Wallace 1990). During these cool periods, eastern Pacific upwelling is intensified, the trans-Pacific SLP gradient is enhanced, and trade winds are stronger than average. Convection over Australasia/Indonesia increases, while the central-western equatorial Pacific islands experience little rain. These cool periods often have teleconnections to temperate climates which are opposite in sign to the warm-phase impacts (Ropelewski and Halpert 1989). The meteorologic and oceanographic features of ENSO are discussed at greater length by several authors (Rasmusson and Carpenter 1982; Rasmusson and

Wallace 1983; Philander 1983, 1990; Cane 1986; Enfield 1989; Deser and Wallace 1990).

The western equatorial Pacific, although neglected by many paleoclimatic studies of ENSO, plays a fundamental role in key aspects of this oscillating ocean-atmosphere system. The weakening and reversal of the trade winds in this region are the dynamical manifestation of the change in SLP that defines the Southern Oscillation, and these wind shifts may trigger the equatorial Kelvin waves that propagate warm sea surface temperatures to the cooler eastern Pacific (Cane 1986). Equatorial rainfall anomalies in the central-western Pacific disrupt atmospheric circulation patterns aloft, generating ENSO-related climate variability in temperate regions (Rasmusson and Wallace 1983), and far western Pacific SST affects the strength of the Northern Hemisphere's climatic response to ENSO (Hamilton 1988). Thus the development of ENSO conditions and the propagation of ENSO anomalies to off-equatorial sites appears to be driven largely by conditions in the western Pacific. In addition, observations indicate that eastern and western equatorial Pacific ENSO components are imperfectly coupled, leading to varying spatial modes of ENSO occurrence (Rasmusson and Wallace 1983; Fu et al. 1986; Deser and Wallace 1987, 1990). For example, the ENSO conditions of the 1980s developed in a fundamentally different spatial pattern from the 'canonical' ENSO, which was based on observations during the previous two decades (Deser and Wallace 1990). These spatial differences may influence the patterns of extratropical climate anomalies associated with ENSO. In the northeastern United States, mild winters in 1982/83 and frigid conditions in 1976/77 resulted in part from different patterns of SST anomaly evolution in the tropical Pacific during those years (Namias and Cayan 1984).

### Coral records

Coral skeletons contain several independent archives of environmental variability that are especially well suited to record certain fundamental features of the ENSO system. The most useful of these include the isotopic content ($\delta^{18}O$, $\delta^{13}C$) of the skeletal aragonite and the concentration of skeletally-bound trace metals, such as cadmium (Cd), barium (Ba), and manganese (Mn). Proxy climate records are only useful, however, in conjunction with well-constrained chronologies. Coral records can be dated in a variety of independent ways which allow the development of precise, subseasonal-resolution chronologies for their paleoclimatic records. With their combination of temporal resolution and sensitive environmental recorders, corals can provide high-resolution information on many of the key dynamical aspects of ENSO variability in the low-latitude Pacific. Living corals in the equatorial Pacific may reach 200 to 350 yr in age, and older live specimens have been recovered from a few locales (Potts et al. 1985). Samples in this age range allow investigation of the tropical ocean conditions during the Little Ice Age climate anomaly (A.D. 1500–1850; Grove 1988). Multiyear 'floating' records from fossil specimens (dated by Th/U mass spectrometry; Edwards et al. 1987; Bard et al. 1990) can provide information on more distant time periods, when climate boundary conditions such as insolation, trace gases, and sea level were drastically different.

## *Isotopic indicators*

The $\delta^{18}O$ of calcium carbonate precipitated in equilibrium with seawater decreases by about 0.22‰ for every 1°C rise in water temperature (Epstein et al. 1953). This relationship was initially obscure in corals, however, because of a variable offset from predicted equilibrium values. However, the offset was shown to be constant within a coral genus (Weber and Woodhead 1972) for rapidly growing portions of a skeleton (Land et al. 1975; McConnaughey 1989). Therefore, in the absence of other influences, coral $\delta^{18}O$ records taken along the axis of maximum growth reflect ambient temperatures at subseasonal resolution (Fairbanks and Dodge 1979; Dunbar and Wellington 1981; Pätzold 1984; McConnaughey 1989). But variations in the $\delta^{18}O$ of the seawater will cause deviations from this simple temperature relationship (Epstein et al. 1953; Fairbanks and Matthews 1979; Swart and Coleman 1980; Dunbar and Wellington 1981; Cole and Fairbanks 1990). In the tropical ocean these variations result from changes in evaporation ($^{18}O$ enrichment), precipitation ($^{18}O$ depletion), or runoff. In regions with fairly constant or well-known temperature histories, the $\delta^{18}O$ therefore reflects these variations in the hydrologic balance (Swart and Coleman 1980; Cole and Fairbanks 1990). In many cases, coral skeletal $\delta^{18}O$ reflects some combination of thermal and hydrographic factors.

The $\delta^{13}C$ signal in coral skeletons is more difficult to decipher in environmental terms, because of the complicated interactions with biological processes which involve strong isotopic fractionation. Environment-related controls on skeletal $\delta^{13}C$ include (1) the isotopic composition of the ambient seawater (Nozaki et al. 1978; Aharon 1985), (2) coral geometry and growth rate (e.g., apex versus side of coral head) (Land et al. 1975; McConnaughey 1989), and (3) photosynthesis of endosymbiotic dinoflagellates (Weber 1974; Goreau 1977; Fairbanks and Dodge 1979; Swart 1983; McConnaughey 1989). This last factor is mediated primarily by the ambient light level, which is determined in a coral's environment by water depth and insolation. Coral skeletal $\delta^{13}C$ has been linked to insolation in various contexts, from depth-dependent variation (Weber and Woodhead 1970; Fairbanks and Dodge 1979; McConnaughey 1989) to annual cycles that mirror rainy (hence cloudy) seasons (Fairbanks and Dodge 1979; Pätzold 1984; McConnaughey 1989; Cole and Fairbanks 1990). McConnaughey (1989) has also shown that certain shallow corals can be light-saturated and experience photoinhibition during brighter periods, while deeper corals may respond to increased light by increasing photosynthesis. These responses produce opposite $\delta^{13}C$ signatures in coral skeletons. Environmental reconstruction from coral $\delta^{13}C$ records therefore requires more information about growth conditions than is usually available in a paleoceanographic context.

### Trace metals

The oceanic distributions of certain metals in trace concentrations are known to reflect specific environmental processes, such as upwelling, advection, aeolian transport, and runoff (Boyle et al. 1976; Martin et al. 1976; Shen and Boyle 1988; Lea et al. 1989; Shen and Sanford 1990; Shen et al. 1991, 1992a). Changes in seawater metal concentrations indicate variations in these governing processes. Several of these trace metals have large divalent ionic radii, similar to $Ca^{2+}$. As a result, these metals appear to substitute readily for Ca in the aragonite ($CaCO_3$) lattice of coral skeletons. Shen and Boyle (1988) demonstrated that an intense series of oxidative and reductive treatments effectively removes from coral samples the contaminant phases of most metals, leaving only the structurally bound phases of interest. Known distribution coefficients between corals and seawater allow the reconstruction of ambient seawater metal concentrations from metal concentrations in the coral skeleton. Trace metal records from corals therefore yield a history of the processes that control trace metal distributions. The most useful metals for coral reconstructions of ENSO include Cd, Ba, and Mn. Shen and Boyle (1987, 1988), Lea et al. (1989), Linn et al. (1990), Shen and Sanford (1990), and Shen et al. (1991) describe these applications in greater detail, including specific techniques for sample cleaning and analysis.

### Cadmium

The modern distribution of cadmium follows that of marine nutrients such as phosphate and silica, with low levels in shallow (euphotic zone) water, reflecting biological removal, and higher levels at depth, reflecting regeneration of organic matter (Boyle et al. 1976; Martin et al. 1976). Cadmium records from foraminifera preserved in deep-sea sediment cores have provided valuable insight into ocean processes (especially deepwater circulation and nutrient concentrations) on glacial-interglacial timescales (Boyle 1990). In more recent coral records, the Cd concentration reflects the presence of Cd-rich upwelled deep water versus Cd-poor oligotrophic waters. Records of Cd from Galapagos corals directly reflect upwelling variations associated both with seasonal cycles (Linn et al. 1990; Shen and Sanford 1990) and interannual ENSO variability (Shen et al. 1987). Preliminary measurements of Cd in a coral from Tarawa Atoll (1°N, 173°E) suggested that intense ENSO-associated rainfall may dilute Cd concentrations in the surface waters, yielding lower skeletal Cd levels during ENSO warm periods (Shen and Sanford 1990); further measurements indicate that this effect does not occur consistently (Shen, unpubl. data).

### Barium

Barium exhibits nutrient-like behavior akin to silica and cadmium (Chan et al. 1977), and because its concentration in both seawater and corals is about 2–3 orders of magnitude greater than Cd, coral Ba records are much less prone to contamination effects. Lea et al. (1989) have shown that seasonal-resolution records of Ba from Galapagos corals also reflect ENSO-mediated upwelling in that region. They found that, compared to Cd, Ba appears to demonstrate greater sensitivity to periods of weak upwelling, possibly due to biological uptake of Cd at rates comparable to the slow rate of supply during these times. Ba is also enriched in continental runoff waters, and Ba records from corals near continental margins may reflect this input (Shen and Sanford 1990). However, Ba records may be complicated by a slight temperature effect upon incorporation into the coral skeleton (Lea et al. 1989), which requires further investigation.

## Manganese

In contrast to Cd and Ba, Mn concentrations typically display high levels in surface waters and decrease with depth. A mid-depth maximum coincides with the local $O_2$ minimum zone, where particulate Mn oxides are reduced and solubilized. Important sources include aeolian and fluvial deposition of Mn from terrigenous sources as well as reducing environments, such as the mid-depth $O_2$ minimum or shelf sediments, where particulate Mn is degraded. In the Galapagos, Mn levels reflect aerosol deposition and long-range advection of Mn-enriched surface water from the South-Central American shelf, where reducing environments generate high Mn levels in the overlying water. This water mass has the greatest influence during the boreal spring and during ENSO warm periods, when low-Mn upwelled waters are absent from the surface. In Galapagos corals, the general Mn signature of reduced upwelling associated with the warm phase of ENSO is an attenuated seasonal cycle (Shen and Sanford 1990; Linn et al. 1990; Shen et al. 1991). In individual reef settings, dissolved levels of Mn in the water column can be augmented by local sediment fluxes. This effect produces very high skeletal Mn concentrations in corals from the Gulf of Panama and certain Caribbean islands (Shen et al., 1991). Diagenetic Mn fluxes from sediments may also prove useful indicators of climate change at Pacific atoll sites, which are far removed from continental sources of Mn (Shen et al. 1992a).

## Chronologies

Paleoclimatic records from corals can be dated by several means. Annual density banding often provides a straightforward calendar, especially in environments with significant seasonal variability. Accuracy of $\pm$ 1 yr is possible with the most clearly banded samples, but corals from low-seasonality sites (such as the western equatorial Pacific) may be poorly banded and the causes of band formation may not be obvious. Density bands may form in response to environmental and/or physiological stimuli (Wellington and Glynn 1983), the timing of which may vary slightly from year to year. Oxygen and carbon isotopes respond to known environmental forcings which often vary annually, and these records can yield accurate chronologies if the seasonal inputs are assumed to occur regularly each year. Radiometric techniques using $^{14}$C and $^{230}$Th have great potential for constructing accurate coral chronologies. New mass spectrometric techniques enable precise Th/U dating of samples ranging in age from a few decades to several hundred thousand years (Edwards et al. 1987). For recent samples, band-counting and Th/U dating of a 180-yr-old coral record yielded the same ages, with similar margins of error (Edwards et al. 1988). Samples up to 30,000 yr in age can also be absolutely dated with radiocarbon, provided that the appropriate corrections are applied (Bard et al. 1990). Bomb-produced radioisotope horizons (e.g., $^{90}$Sr and $^{14}$C; Druffel 1981; Toggweiler and Trumbore 1985) mark the era of atmospheric nuclear testing and can confirm certain age assignments in the 1950s and 1960s.

## Paleoclimatic records of ENSO

Baumgartner et al. (1989) summarize proxy ENSO records from various geological and biological archives, including sediments, ice cores, tree rings, and corals. They conclude that the most accurate recorders of ENSO variability have a direct coupling to a specific fundamental feature of the ENSO phenomenon. The isotopic chemistry and trace metal content of coral skeletons respond directly and predictably to variations in key aspects of tropical Pacific climate. Several studies have begun to develop these climate archives; these studies generally focus on reconstructing variations in eastern Pacific upwelling and the associated SST changes. The Galapagos Islands are particularly sensitive to ENSO-related changes in ocean dynamics, and many studies have used Galapagos corals to reconstruct upwelling-dependent SSTs (Druffel 1985; McConnaughey 1989; Dunbar et al. 1991) and concentrations of trace metals (Shen et al. 1987; Lea et al. 1989; Linn et al. 1990) and radiocarbon (Druffel 1981). The longest record from this sensitive site extends back nearly four centuries at annual resolution and indicates significant decadal-scale variability in surface ocean conditions (Dunbar et al. 1991). Farther north, short oxygen isotope records from five Costa Rican corals demonstrate not only the elevated temperatures associated with the extreme 1982/83 warm event but also the loss and recovery of the thermal gradient in the top 15 m of the water column (Carriquiry et al. 1988). The ENSO signal in Gulf of Panama corals appears less clear; the interaction of changes in hydrologic balance, circulation, and thermal variations produce changes in coral $\delta^{18}O$ that do not reflect ENSO consistently (Druffel et al. 1990) but do provide an integrated record of local variability.

Farther west, Druffel (1985) presented $\delta^{18}O$ records from central Pacific corals that document ENSO-related SST changes at Fanning (4°N, 159°W) and Canton (3°S, 172°W) Islands. Cole and Fairbanks (1990) demonstrated that oxygen isotope records in corals from Tarawa Atoll (1°N, 173°E) reflect changes in the isotopic composition of the surface ocean which result primarily from intense precipitation during ENSO warm conditions. Shen et al. (1992a) suggest that skeletal Mn content of Tarawa corals is controlled by the degree of resuspension of lagoonal sediments during the westerly wind anomalies associated with the onset of strong warm anomalies.

Other proxy records of ENSO variability have also proven useful along the margins of the low-latitude Pacific basin (Baumgartner et al. 1989). Accumulation rates in the Quelccaya ice cap reflect moisture deficits over the Amazon often associated with warm-phase ENSO conditions (Thompson et al. 1984). Rapidly accumulating sediments in the Gulf of California may contain an ENSO signal in their faunal records (Baumgartner et al. 1989) Tree-ring records from New Mexico and Java record rainfall changes that accompany ENSO warm conditions (D'Arrigo and Jacoby 1992, this volume). ENSO proxies also exist at higher latitude sites (Michaelsen 1989; Swetnam and Betancourt 1990; D'Arrigo and Jacoby 1992, this volume), where ENSO teleconnections are strong enough to yield a consistent response to the forcing in the equatorial Pacific. Comparing coral records of equatorial Pacific ENSO dynamics with such records from Pacific margins and temperate regions will provide useful information on how climatic variability in the equatorial Pacific translates to larger-scale impacts, and how these teleconnections may have evolved in response to changing climate boundary conditions.

Over longer time scales, sediments in Galapagos crater lakes record extended times of drought that may reflect the absence of interannual ENSO variability during the last glacial period (Colinvaux 1972, 1982), and reduced vegetation variability in Australasia and South America suggests diminished effect of ENSO in this region during the early Holocene as well (McGlone et al. 1992, this volume). These results imply that large-scale changes in climate boundary conditions during the deglacial may have profoundly affected the ENSO system. Lower sea level at this time would be a likely forcing (Colgan 1990), as the exposure of large areas of land in the Australasian region would impede westward flow of warm surface water (Gordon 1986; Wyrtki 1987) and could have prevented the migration of the Indonesian Low by anchoring it to a more continental heat source. The effects of these alternate boundary conditions on ENSO variability can be tested with corals that predate 8000 yr ago, when sea level remained below the level of the Sunda Shelf ( > 40 m below today's; Fairbanks 1989).

## Coral sites and data

We present isotopic and trace-metal records from corals that grew at three key sites across the Pacific: the Galapagos Islands (1°S, 91°W), Tarawa Atoll (1°N, 173°E), and Bali (8°S, 115°E). These sites were selected for their sensitivity to fundamental dynamical aspects of the ENSO phenomenon, including upwelling, SST, rainfall, and winds, and for the geographic coverage they provide across the equatorial Pacific (Fig. 18.1, Table 18.1). Comparison of records across this range can provide useful information on spatial variability of ENSO conditions. These short records lay the groundwork for longer reconstructions, many now

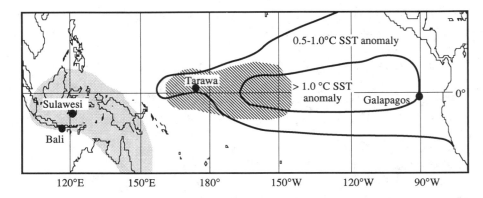

Fig. 18.1 Map of the equatorial Pacific showing the locations of the sites discussed in the text and the spatial extent of typical ENSO-related anomalies. The heavy contours indicate positive SST anomalies during ENSO warm-phase conditions. The shaded region comprises the area of maximum droughts associated with the warm phase of ENSO, and the hatched region designates the area of maximum positive precipitation anomaly during ENSO warm anomalies.

Table 18.1 *Coral sites and available records*

| Site | Location | Major ENSO impacts | Coral records | Reference[a] |
|------|----------|---------------------|---------------|--------------|
| Galapagos | 1°S, 90°W | Upwelling suppressed Warm SST | $\delta^{18}O$ | 1,2 |
| | | Low nutrients | Cd, Ba | 3,4 |
| | | Increased rainfall | $\delta^{13}C(?)$ | 2 |
| Tarawa Atoll | 1°N, 172°E | Intense rain | $\delta^{18}O$ | 5 |
| | | Trade winds weaken/reverse | Mn | 6 |
| Bali | 8°S, 115°E | Weakened monsoon, longer dry season | $\delta^{18}O$ | This chapter |

[a]References: 1, Druffel (1985); 2, McConnaughey (1989); 3, Shen et al. (1987); 4, Lea et al. (1989); 5, Cole and Fairbanks (1990); 6, Shen et al. (1992a).

in progress. However, they provide only the outline of the detailed network of sites needed to address important questions of spatial variability in ENSO and its teleconnections. Table 18.1 lists the available records and the major climate signal for each site, along with the primary references. Age models were developed using oxygen (Galapagos, Bali) or carbon (Tarawa) isotope stratigraphies and were confirmed by density banding (for Galapagos, Bali, and Tarawa coral LDGO 44) and bomb-produced $^{14}C$ and $^{90}Sr$ measurements (for Tarawa coral LDGO 44; Toggweiler 1983).

## *Galapagos*

The Galapagos Islands lie in the core of the normally cool tongue of upwelled waters in the eastern equatorial Pacific. During the warm phase of ENSO, the suppression of regional upwelling replaces these cool, nutrient-rich waters with warm, oligotrophic waters more typical of the tropical surface ocean. Upwelling suppression generates profound thermal and chemical anomalies in Galapagos surface waters: SSTs rise by 1 to 2°C on average, and the levels of nutrients and nutrient tracers (Cd and Ba) drop considerably. Rainfall increases as a consequence of warmer SSTs. Manganese-rich surface waters from the south Central American shelf are advected into the Galapagos region as well, with the southward-flowing Costa Rica Current. The dramatic ecological consequences of this oceanographic reorganization are well documented (Glynn 1990).

Signatures of the thermal and chemical anomalies at Galapagos are incorporated directly into the skeletal chemistry of Galapagos corals. Several studies have shown that isotopic and trace metal records from these corals reflect the history of upwelling and SST changes at this location (Druffel 1981, 1985; Shen et al. 1987, 1991; Lea et al. 1989; McConnaughey 1989; Linn et al. 1990). We show a selection of these records to illustrate the sensitivity of this site to the primary feature of ENSO in the eastern Pacific, the variability of regional upwelling.

Figure 18.2 presents $\delta^{18}O$ (McConnaughey 1989), Ba/Ca (Lea et al. 1989), and composite Cd/Ca (Shen et al. 1987) histories from a Punta Pitt (Isla San Cristobal, Galapagos) coral (at 17 m depth) compared with the SST record from Academy Bay (Isla Santa Cruz, Galapagos). The $\delta^{18}O$ record closely monitors seasonal SST extremes ($R^2 = 0.92$; McConnaughey 1989), capturing the entire range of variability in the SST record. ENSO warm-phase conditions (shaded) as well as cool periods of strong upwelling can be identified in this record. The seasonal-resolution Ba/Ca and Cd/Ca records monitor the contribution of nutrient-like trace metals to the surface ocean by the deeper upwelled waters. These records also reflect the varying intensity of upwelling throughout this period and closely mirror the higher-resolution $\delta^{18}O$ history. The Cd/Ca record is a composite of two datasets from the same Punta Pitt coral, measured at different times and in different labs, corrected for consistency. One of these records was extended back to 1936 to yield a longer proxy upwelling history (Fig. 18.3). A record of SST from Puerto Chicama, Peru, provides a longer record of regional oceanographic change for comparison. All of these records reflect known ENSO variability, and they provide information on the relative intensity of anomalies in the eastern tropical Pacific.

## *Tarawa*

Tarawa Atoll lies in the heart of the positive rainfall anomaly generated by the northeastward migration of the Indonesian Low during warm-phase ENSO conditions. At these times, Tarawa experiences sustained rainfall of $\geq 300\,mm$

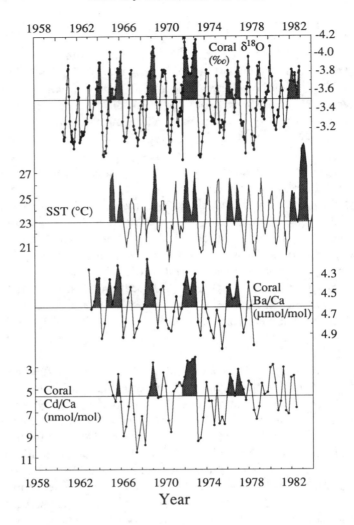

Fig. 18.2 Instrumental and coral records of upwelling variability associated with ENSO warm periods in the Galapagos Islands. From the top: coral $\delta^{18}O$ (McConnaughey 1989), measured SST at Academy Bay (Isla Santa Cruz), coral Ba/Ca record (Lea et al. 1989), and a composite record of Cd/Ca from the Punta Pitt coral. All coral measurements were made on the same coral head, from Punta Pitt, Isla San Cristobal, and all are plotted on reverse vertical axes. Two records run at different times were standardized and averaged to yield the Cd/Ca curve (e.g., Shen et al. 1987). Shaded periods reflect warm-phase ENSO conditions as noted by Quinn et al. (1978, 1987), which result in suppression of regional upwelling and bring warmer, nutrient-poor waters to the Galapagos.

$mo^{-1}$, and 500 to 800 mm of rain may fall during the rainiest month of a warm-phase period. The record of the zonal wind component at Tarawa also shows a strong response to the ENSO system; only during the warmest ENSO anomalies does the wind direction reverse to westerly. The rainfall and wind data

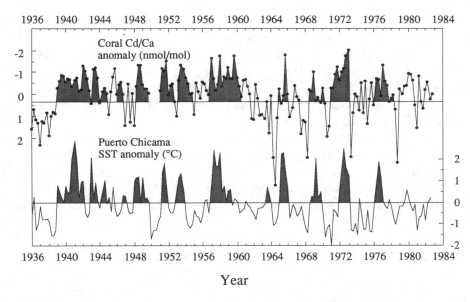

Fig. 18.3 Normalized Galapagos Cd/Ca record (plotted positive downward) and seasonalized anomaly of Puerto Chicama SST record (Shen et al. 1992b). Shaded periods reflect ENSO warm-phase conditions as noted by Quinn et al. (1978, 1987).

demonstrate the sensitivity of this site to the shifts in surface pressure that comprise the Southern Oscillation. Horel and Wallace (1981) found that, of all the single-site climate variables they considered, total annual Tarawa rainfall correlates most strongly with the Southern Oscillation Index (SOI; $R = 0.78$). Wind and precipitation data are directly coupled to this fundamental feature of the ENSO phenomenon, but available SST data suggest only a small thermal response in the nearby surface waters. Monthly SST data are not available from Tarawa, but averaged fields indicate a mean ENSO-related anomaly of $\leq 0.5°C$. Spatially and temporally pooled data suggest seasonality of $\leq 1.5°C$, although the seasonal cycle appears poorly defined (Levitus 1982; Fu et al. 1986).

The rainfall and wind-direction anomalies are sufficiently intense to produce chemical changes in the surface waters near Tarawa, which are incorporated into the skeletal chemistry of shallow-growing corals. The increased rainfall shifts the $\delta^{18}O$ of the surface water to more negative values. The isotopic composition of this rainfall is most likely in the range of $-8$ to $-10‰$ (Cole and Fairbanks 1990). This signal is amplified by the stratification of the water column from the intense precipitation, which thins the surface mixed layer (Lukas and Lindstrom 1991) and thereby confines the isotopically depleted rainfall to a reduced volume of water. These influences should produce ENSO-related changes in surface water $\delta^{18}O$ of between 0.4 and 0.7‰. Coral $\delta^{18}O$ records are also sensitive to changes in SST, but these changes at Tarawa are small and should only produce $\delta^{18}O$ changes of about 0.15‰ during an average warm anomaly, in parallel with precipitation-induced changes. Cole and Fairbanks (1990) discuss these influences in more detail. Their composite $\delta^{18}O$ record from two shallow corals (LDGO 44 and 39, at 2 to 3 m depth) closely match the SOI (Wright 1989), in terms of both timing and amplitude of the full spectrum of ENSO-related anomalies.

Changes in wind speed and direction associated with ENSO variability also generate chemical signals in the surface waters around Tarawa. Tarawa Atoll includes a string of islands on the windward (eastern) side, protecting a shallow lagoon which is exposed to the west. When the prevailing easterly trades weaken and reverse to westerly during strong ENSO warm periods, Mn is apparently remobilized from the mildly reducing environment of lagoonal sediments and redistributed to the surface waters around the atoll. Shen et al. 1992a generated a seasonal Mn record for LDGO 44 that reflects these changes in the flux of Mn, with significantly higher concentrations observed during periods of known westerly winds. A lagoonal source of Mn is supported by preliminary measurements from a coral that grew in a nearby lagoon (Abaiang Atoll, several km. north of Tarawa), showing Mn concentrations 4 to 5 times greater than the open water corals from Tarawa, and by high Mn concentrations measured in lagoonal sediments (Shen et al. 1992a). The Tarawa Mn record provides a clear marker of trade wind reversal (e.g., during 1965 and 1972) and appears to reflect times of weakened easterly trades as well (e.g. 1969, 1963). The Mn and composite $\delta^{18}O$ histories are shown in Figure 18.4, with the SOI, rainfall, and zonal wind records for comparison.

The isotopic and trace metal histories from this site provide independent monitors of distinct climatic phenomena, precipitation and wind direction. Meteorological records indicate varying phase offsets between wind and precipitation anomalies, which can potentially be reconstructed using paired Mn and $\delta^{18}O$ records from the same core. Comparing data from the same core avoids potential chronological errors in addressing the issue of small phase offsets between records. However, the meteorological records for this period indicate offsets of only a few months or less, and the resolution of the existing Mn records is too coarse to capture these differences.

### Bali

At Bali, in the Indonesian archipelago, the Asian monsoon and ENSO interact to produce a climate of strong seasonal as well as interannual variability (Hackert and Hastenrath 1986; Allan and Pariwono 1989). Linkage between these climatic systems has been recognized at sites throughout Asia, including India, Sri Lanka, Tibet, and China, as well as Indonesia (Hastenrath 1988). During the austral summer (December–January–February), the locus of the most intense convection in southern Asia is centered over the Australasian region (the Indonesian Low pressure cell; Hastenrath 1988), bringing intense rains to Indonesia and northern Australia. When ENSO enters a warm phase, this system migrates

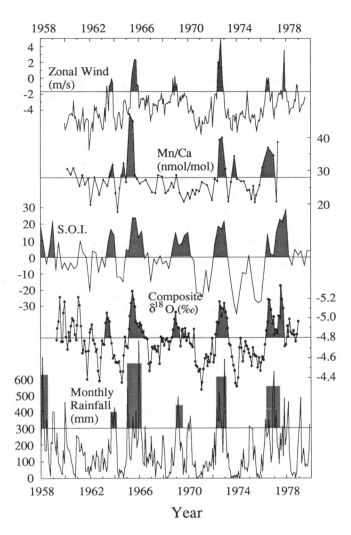

Fig. 18.4 Instrumental and coral records of environmental conditions at Tarawa Atoll. From the top: zonal winds, coral Mn/Ca (Shen et al. 1992a), the Southern Oscillation Index (SOI; Wright 1989); composite coral $\delta^{18}O$ (plotted reverse; Cole and Fairbanks 1990), and monthly rainfall (from *Monthly Climatic Data for the World* and the New Zealand Meteorological Service). Shaded periods reflect ENSO warm-phase conditions as noted by Quinn et al. (1978, 1987), which involve the weakening and reversal of the trade winds and dramatically increased rainfall at this site.

northeastward towards Tarawa, effectively weakening the monsoon in Indo-nesia. Longer dry seasons and drier rainy seasons on Bali result; other sites in Indonesia appear even more sensitive to the influence of ENSO. Recent ENSO-related monsoon failures have had disastrous consequences for the economic and agricultural systems of Indonesia (Glantz 1984). Sea surface temperatures also vary seasonally by 3 to 5°C (Levitus 1982), due to seasonal changes in mean

oceanographic flow patterns, and slight cool anomalies ($\leq 0.5°C$) persist for
several months during ENSO conditions (Nicholls 1984). During the cool phase
of ENSO, and during the rainy season (DJF), the surface ocean around Bali is
warmer and fresher; these conditions act in concert to produce more negative
$\delta^{18}O$ values in coral skeletons. During the dry season and warm-phase ENSO
anomalies, Bali SSTs are cooler and the water is more saline, producing more
positive coral $\delta^{18}O$ values.

A short $\delta^{18}O$ record from a Bali coral shows the influence of both seasonal
and interannual climate variability and suggests moderate sensitivity to the
ENSO system. Figure 18.5 compares the Denpasar (Bali) rainfall record with Bali
coral $\delta^{18}O$. Longer dry seasons occur during ENSO periods and are reflected in
the coral record as longer periods of enriched isotopic values. However, the
absolute magnitude of rainfall variability does not correspond well with the
magnitude of the observed $\delta^{18}O$ changes, suggesting other influences. Although
Indonesian droughts are reliably associated with ENSO recurrence, other factors
(such as changes in mixing of warm, fresh surface waters with cool, salty deeper
waters) may affect the coral isotope record. The relative contributions of thermal
and hydrographic effects cannot be determined by isotopic measurements alone;
analysis of runoff proxies, such as fluorescent banding (Isdale 1984; Scoffin

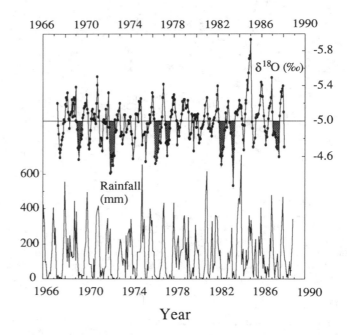

Fig. 18.5   Rainfall (Ho Tong 1975) and coral $\delta^{18}O$ records from Bali, Indonesia.
Shaded periods indicate years of Pacific-wide ENSO warm-phase conditions that
generate drought across Indonesia. The combined effects of annual SST and rainfall
cycles act in parallel to produce the annual cycle observed in the coral record. In the
coral record, droughts are often apparent as longer dry season periods of increased
$\delta^{18}O$ that may be followed by attenuated rainy-season minima in $\delta^{18}O$.

et al. 1989) or barium (Shen and Sanford 1990) might help to deconvolve these signals. Coral records from sites in Indonesia with stronger precipitation anomalies should provide more sensitive monitors of ENSO-related precipitation variability. For example, the southern peninsula of the island of Sulawesi also experiences drought during the warm phase of ENSO, and the higher rainfall in this region should produce enhanced surface ocean sensitivity to changes in hydrologic balance. Cores from this region should track precipitation variability more closely.

## Climatic interpretation of coral records, 1958–1988

Figure 18.6 presents short isotopic and trace metal records from the Galapagos, Tarawa, and Bali. All records have been normalized and are plotted in units of standard deviations. We subtracted the average seasonal cycle from the Bali and Galapagos records, in order to emphasize interannual variability; the lack of strong seasonality at Tarawa made this procedure unnecessary for the Tarawa records. The Tarawa isotope record presented here consists of the previously shown composite record (1959–1979) spliced onto a record from a single core recently collected at the same location (1979–1989). All records reflect known warm-phase ENSO years, as identified by Quinn et al. (1987) and shaded in Figure 18.6. This correspondence demonstrates the sensitivity of these records to the large-scale ENSO forcing. Equally significant, all display subtle differences in timing and magnitude from the generalized ENSO history, and these differences appear in part to reflect known patterns of variability in the specific climatic feature being monitored. We discuss individual years at further length in the following sections on periods of warm and cool ENSO anomalies in the coral records.

### *Warm-phase ENSO anomalies*

Published indices of the state of the ENSO system (Quinn et al. 1978, 1987; Wright 1989) suggest six occurrences of ENSO warm anomalies during the period covered by our coral records: 1963, 1965, 1969, 1972, 1976, 1982–83, and 1986–87. In the Tarawa isotopic record and in all Galapagos coral records, 1963 appears as a period of weak warm-phase anomalies. This interpretation concurs with observed weak anomalies in Tarawa rainfall and eastern Pacific SST. No significant Mn anomaly appears in the Tarawa coral, in agreement with the wind record, which indicates no reversal of the trade winds. By contrast, in 1965 a much stronger anomaly occurs in all records. Intense rains and the reversal of the trade winds at Tarawa generate strong signals in both the $\delta^{18}O$ and Mn records from this coral. In the Galapagos record, the 1965 anomalies in Cd, Ba, and $\delta^{18}O$ are also relatively strong. The relative intensity of this anomaly implied by the coral data is supported by meteorological observations across the Pacific.

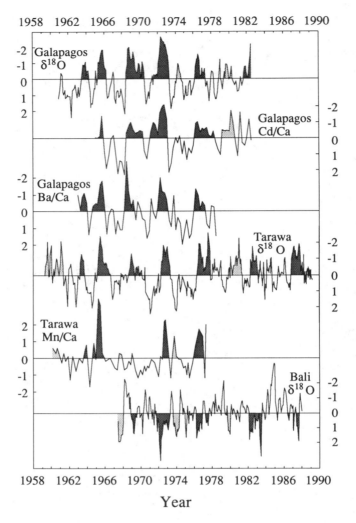

Fig. 18.6 Anomaly representations of short records from Galapagos, Tarawa, and Bali corals. All records were normalized by subtracting the mean and dividing by the standard deviation; for the Galapagos and Bali records, the mean annual cycle was subtracted in order to highlight interannual variability. Darker shaded periods reflect ENSO warm anomalies as noted by Quinn et al. (1978, 1987); light shading denotes conditions similar to the warm phase of ENSO at individual sites. In 1979, weak warm-phase anomalies are recorded by corals at all sites, consistent with observed anomalies in the related climatological variables.

Tarawa isotopic data reflect a weak warm-phase period in 1969, consistent with rainfall data from this region and with the SOI. The Tarawa Mn record does not indicate unusual conditions at this time, consistent with the continued prevalence of easterly winds. The Bali coral record suggests a weak warm-phase anomaly as well. In the Galapagos, this anomaly was somewhat stronger than in the western Pacific, as both coral and instrumental data indicate. The 1972

ENSO anomaly is one of the strongest in the Bali coral record, in agreement with observed drought conditions. In the Tarawa $\delta^{18}O$ record, this anomaly appears somewhat more moderate, in agreement with rainfall data and the SOI. However, persistent westerly winds produce a strong Mn signal in the Tarawa coral. At the Galapagos site, extremely warm SSTs produce a very strong signal in the coral $\delta^{18}O$; Ba and Cd anomalies are also significant. At all but the Bali site, conditions revert to cool-phase levels by mid-1973. The Bali record indicates a brief interruption of the ENSO anomaly at the beginning of 1973 with warm-phase anomalies resuming and persisting throughout this year.

The 1976 ENSO period is clearly recorded as a moderately strong anomaly in all of the coral records we present, consistent with observations. However, major differences between sites appear during 1977 which confirm the climatic sensitivity of these proxy records. During 1977, the Tarawa and Bali records suggest warm-phase ENSO anomalies, but the Galapagos coral indicates a return to normal conditions from the 1976 warming. These data concur with observed strong anomalies in the SOI and in rainfall data from the central-western Pacific, coupled with weak to nonexistent anomalies in eastern Pacific SST records. The 1982–83 ENSO warm period, one of the strongest of this century, appears as a significant drought in the Bali record. The Galapagos record, which ends in the spring of 1982, reflects the very beginning of the intense warming observed at this site. The isotopic record from the Tarawa coral shows only a mild anomaly in 1982–83. These results agree with observed rainfall, which was not extreme, and with satellite-documented patterns of convection in the western Pacific, which show that the Indonesian Low migrated well east of Tarawa during this highly unusual period (Rasmusson and Wallace 1983). During the most recent period of ENSO warm anomalies, 1986–87, the Tarawa coral exhibits a prolonged, moderately intense isotopic anomaly, concurring with rainfall data. The Bali coral record suggests only minor drought anomalies at this time, in line with Denpasar rainfall variations. Instrumental climate records from Pacific sites farther east suggest moderately strong anomalies during this period.

### Cool-phase ENSO anomalies

Most of these coral records capture the full range of variability in tropical Pacific climate, including the cool ENSO phases of strengthened trades, intensified upwelling, and abundant monsoonal rainfall. The Tarawa Mn record might be expected to be less sensitive to this variation, since the record appears to be dominated by strong inputs of Mn during westerly wind bursts and lower variability at other times. The Bali record appears to reflect local anomalies that are not always consistent with basinwide cool-phase conditions. However, the isotopic record from Tarawa and all Galapagos records capture the full range of environmental anomalies observed in the Pacific-wide ENSO indices. Deser and Wallace (1990) have identified periods characterized by basin-wide cool-phase anomalies during 1964, 1967, 1970, 1973, 1975, and 1988. The latter coral records suggest cool-phase anomalies during nearly all of these years, with intensity varying somewhat between sites. In 1964, significant cool-phase anomalies occur at both Tarawa and the Galapagos. Such anomalies are also observed during 1967, although the anomaly is less intense at Tarawa than at Galapagos. During 1970, both sites experience strong cool anomalies. Strong cool-phase anomalies appear in 1973 in all but the Galapagos Ba record. Conditions in 1975 suggest weak cool-phase anomalies. Additionally, both Galapagos records indicate anomalously cool conditions in 1978. The only record that extends through 1988 is the Tarawa isotope history, which shows no clear anomaly at this time. The Bali record shows strongest cool-phase anomalies only during 1968 and 1984. Weaker cool-phase anomalies of only a few months duration are suggested during 1970, early 1973, and 1979.

### Weak anomalies in tropical Pacific climate

Occasionally, weak warm anomalies develop over a large part of the equatorial Pacific but fail to spread across the entire basin or fail to intensify sufficiently to be recognized as an ENSO warm event. Two such periods are evident in these coral records. In 1961, very weak warm anomalies developed over much of the western Pacific, but failed to intensify or spread to the eastern half of the basin. This is evident in the Tarawa isotope record and confirmed by the Tarawa rainfall data as well as other western Pacific records. In 1974, weak warm-phase anomalies are apparent in the records of Bali and Galapagos $\delta^{18}O$ but not in the Tarawa $\delta^{18}O$ record. Consistent with this pattern, supporting anomalies appear in the records of Denpasar rainfall and Galapagos SST, but not in Tarawa rainfall. However, Tarawa winds are normal as well, while a significant anomaly occurs in the Tarawa coral Mn record. In 1979, ENSO-like conditions occurred more consistently across the equatorial Pacific. Weak warm-phase anomalies appear in all of the coral records that cover this period. Instrumental records of climate from across the Pacific support the prevalence of these weak anomalies in 1979 (Donguy and Dessier 1983).

Other years during these records show patterns of variation across the Pacific that do not appear to reflect large-scale variation in the ENSO system; at these times, local factors apparently overwhelmed the large-scale ENSO forcing. In 1967, for example, Bali experienced drier than normal conditions apparently unrelated to Pacific-wide ENSO anomalies. Although the Tarawa Mn record also shows a small response at this time, the winds were not anomalous. During 1959 and 1960, ENSO-like variability in both Tarawa records may also be of largely local origin. In coral records that predate instrumental data, distinguishing this local variability from basin-scale anomalies in the ENSO system will require the comparison of reconstructions from widely separated Pacific sites.

## Climatic interpretation of Galapagos Cd/Ca record, 1936–1982

Seasonal anomalies of Cd/Ca from a Galapagos coral (plotted positive downward) and SST measured at Puerto Chicama, Peru, are shown in Figure 18.3. Overall, periods of positive SST anomaly tend to correspond with negative Cd/Ca anomalies, in agreement with the underlying control of regional upwelling on both records. High Cd/Ca anomalies occur during all of the ENSO warm-phase periods identified by Quinn et al. (1978, 1987; shaded in Fig. 18.3) and during the warm anomalies in SST recorded at Puerto Chicama.

The intensity of regional upwelling as felt in the surface waters of the eastern Pacific is primarily controlled by ENSO variability, but ENSO may not be the only factor modulating the coral Cd/Ca record, especially the relative intensity of observed anomalies. For example, the Cd/Ca record appears more sensitive to periods of strong upwelling than to periods of suppressed upwelling; this is logical if Cd levels fall to near-zero concentrations during ENSO warm conditions. This bias is particularly apparent in the period since 1963, when cool SST anomalies are reflected by very high Cd/Ca anomalies. All of the cool anomalies discussed in the previous section except for 1975 are strongly expressed in this record. Between 1939 and 1963, fewer extreme cool anomalies occur, and high Cd/Ca anomalies are absent. (Note that there are no Cd/Ca data for the coolest years in this interval, 1950–51, due to accidental slab fracturing.) The cool anomaly observed in 1955–56 has no counterpart in the Cd/Ca record. In 1937–38, cool SST's are reflected by high Cd/Ca values. Conversely, although warm anomalies are generally well expressed in the Cd/Ca anomaly record, the maximum anomaly does not vary much between events, especially between 1940 and 1960. This tendency may result from a lower limit of Cd concentration reached in the surface water when upwelling is suppressed during ENSO periods.

Other factors that may affect this Cd/Ca record include interannual to decadal variations in vertical and horizontal transports. The steep vertical gradient of Cd in the water column makes it a sensitive tracer of the depth of origin of the upwelled water. In addition, the surface water concentration varies slightly, and variations in lateral advection can produce small changes through time in concentrations observed at any single site. Shen et al. (1992b) provide an expanded discussion of this record, including comparison with Ba and Mn, and stable isotope histories developed from the same coral head.

## Conclusions

The isotopic and trace metal records from these three sites illustrate how chemical records from coral skeletons can yield important documentation of various dynamic aspects of ENSO variability. Individually, each provides a sensitive monitor of a key component of the ENSO phenomenon: SST, rainfall, winds, and upwelling. Taken together, these coral records provide data on the evolution and spatial variability of climatic and oceanographic anomalies throughout the tropical Pacific, covering the full spectrum of interannual variability in tropical Pacific climate.

The coral records presented here capture much of the interannual variability associated with the ENSO system of the tropical Pacific, including both warm and cool anomalies already identified by instrumental and historical data (Quinn et al. 1978, 1987; Wright 1989; Kiladis and Diaz 1989; Deser and Wallace 1990). The discrepancies between such compilations from historical data or regionalized indices and proxy climate records from individual sites may provide information on the spatial variability of ENSO anomalies. Deser and Wallace (1987, 1990) have shown that ENSO conditions across the central and western Pacific are not perfectly coupled to eastern Pacific SST anomalies, and Fu et al. (1986) demonstrate the existence of different spatial modes of SST anomaly evolution during ENSO conditions. As sensitive monitors of Pacific climate, coral records from carefully chosen sites can enable the reconstruction of these patterns of variability over the past several centuries. The short records that we present span much of the existing historical record at these sites and demonstrate the potential for longer reconstructions from these and other sensitive locales; many of these reconstructions are now in progress. As longer records are developed, spectrally-based techniques for quantitative comparison between records will allow more precise analysis of the degree of correlation between records.

*Acknowledgments* For constructive reviews and discussion of this manuscript, we thank Chris Charles, Rosanne D'Arrigo, Robert Dunbar, Jonathan Overpeck, and an anonymous reviewer. For support for fieldwork in Tarawa and the Galapagos, we thank Sigma Xi, the Columbia University Department of Geological Sciences and Climate Center, and the Charles Darwin Research Station. Fieldwork in Bali was supported by the Harvard Traveller's Club, the American Association of Petroleum Geologists, the Geological Society of America, the Indonesian Insititute of Sciences, and the Government of Indonesia. We gratefully acknowledge the tireless field assistance of Rene Espinosa and LeAnne Adams and the trace-metal expertise of Laura Linn. Minze Stuiver and Ted McConnaughey provided Galapagos core material. Special thanks are due to Bambang Widoyoko Suwargadi, in his capacity as Indonesian counterpart scientist, and to James Uan and Teekabu Tiikai of the fisheries division of the Kiribati Ministry of Natural Resources Development. This work was supported by the National Science Foundation and the Paleoclimate program of the National Oceanic and Atmospheric Administration. Lamont-Doherty Geological Observatory Contribution No. 4976.

## References

AHARON, P., 1985: Carbon isotope record of late Quaternary coral reefs: possible index of sea surface paleoproductivity. *In* Sundquist, E. T. and Broecker, W. S. (eds.), *The*

*Carbon Cycle and Atmospheric CO₂: Natural Variations Archean to Present.* Geophysical Monograph 32. Washington, D.C.: American Geophysical Union, 343–355.

ALLAN, R. and PARIWONO, J.I., 1989: Ocean-atmosphere interactions in low-latitude Australasia. *International Journal of Climatology*, 10: 145–178.

BARD, E., HAMELIN, B., FAIRBANKS, R.G., and ZINDLER, A., 1990: Calibration of the 14-C timescale over the past 30,000 years using mass spectrometric U-Th ages from Barbados corals. *Nature*, 345: 405–410.

BARNETT, T.P., GRAHAM, N., CANE, M.A., ZEBIAK, S.E., DOLAN, S., O'BRIEN, J., and LEGLER, D., 1988: On the prediction of El Niño of 1986–87. *Science*, 241: 192–196.

BAUMGARTNER, T.R., MICHAELSEN, J., THOMPSON, L.G., SHEN, G.T., and CASEY, R.E., 1989: The recording of interannual climatic change by by high-resolution natural systems: tree-rings, coral bands, glacial ice layers, and marine varves. *In* Peterson, D.H. (ed.), *Aspects of Climate Variability in the Pacific and the Western Americas.* Geophysical Monograph 55. Washington, D.C.: American Geophysical Union, 1–14.

BOYLE, E.A., 1990: Quaternary deepwater paleoceanography. *Science*, 249: 863–870.

BOYLE, E.A., SCLATER, F., and EDMOND, J.M., 1976: On the marine geochemistry of cadmium. *Nature*, 263: 42–44.

CANE, M.A., 1986: El Niño. *Annual Review of Earth and Planetary Sciences*, 14: 43–70.

CARRIQUIRY, J.D., RISK, M.J., and SCHWARCZ, H.P., 1988: Timing and temperature record from the stable isotopes of the 1982–1983 El Niño warming event in eastern Pacific corals. *Palaios*, 3: 359–364.

CHAN, L.H., DRUMMOND, D., EDMOND, J.M., and GRANT, B., 1977: On the barium data from the Atlantic GEOSECS Expedition. *Deep-Sea Research*, 24: 613–649.

COLE, J.E. and FAIRBANKS, R.G., 1990: The Southern Oscillation recorded in the oxygen isotopes of corals from Tarawa Atoll. *Paleoceanography*, 5: 669–683.

COLGAN, M.W., 1990: El Niño and the history of eastern Pacific reef building. *In* Glynn, P.W. (ed.), *Global Ecological Consequences of the 1982-83 El Niño-Southern Oscillation.* New York: Elsevier, 183–232.

COLINVAUX, P.A., 1972: Climate and the Galapagos Islands. *Nature*, 240: 17–20.

COLINVAUX, P.A., 1982: The Galapagos climate: past and present. *In* Perry, R. (ed.), *Key Environments, Galapagos.* Oxford: Pergamon Press, 55–70.

D'ARRIGO, R.D. and JACOBY, G.C., 1992: A tree-ring reconstruction of New Mexico winter precipitation and its relation to El Niño/Southern Oscillation events. *In* Diaz, H.F. and Markgraf, V. (eds.), *El Niño: Historical and Paleoclimatic Aspects of the Southern Oscillation.* Cambridge: Cambridge University Press, 243–257.

DESER, C. and WALLACE, J.M., 1987: El Niño events and their relation to the Southern Oscillation. *Journal of Geophysical Research*, 92: 14,189–14,196.

DESER, C. and WALLACE, J.M., 1990: Large-scale atmospheric circulation features of warm and cold episodes in the tropical Pacific. *Journal of Climate*, 3: 1254–1281.

DONGUY, J.-R. and DESSIER, A., 1983: El Niño-like events observed in the tropical Pacific. *Monthly Weather Review*, 111: 2136–2139.

DRUFFEL, E.M., 1981: Radiocarbon in annual coral rings from the eastern tropical Pacific Ocean. *Geophysical Research Letters*, 8: 59–62.

DRUFFEL, E.R.M., 1985: Detection of El Niño and decade time scale variations of sea surface temperature from banded coral records: implications for the carbon dioxide cycle. *In* Broecker, W.S. and Sundquist, E.T. (eds.), *The Carbon Cycle and Atmo-*

*spheric CO2: Natural Variations Archean to Present.* Geophysical Monograph 32. Washington, D.C.: American Geophysical Union, 111–122.

DRUFFEL, E. R. M., DUNBAR, R. B., WELLINGTON, G. M., and MINNIS, S. A., 1990: Reef-building corals and identification of warming episodes. *In* Glynn, P. W. (ed.), *Global Ecological Consequences of the 1982–83 El Niño-Southern Oscillation.* New York: Elsevier, 233–254.

DUNBAR, R. B., WELLINGTON, G. M., COLGAN, M. W., and GLYNN, P. W., 1991: Eastern tropical Pacific corals monitor low-latitude climate of the past 400 years. *In* Betancourt, J. L. (ed.), *Proceedings of the Seventh Annual Pacific Climate (PACLIM) Workshop.* Sacramento: California Department of Water Resources, 183–198.

DUNBAR, R. B. and WELLINGTON, G. M., 1981: Stable isotopes in a branching coral monitor seasonal temperature variation. *Nature,* 293: 453–455.

EDWARDS, R. L., CHEN, J. H., KU, T.-L., and WASSERBURG, G. J., 1987: Precise timing of the last interglacial period from mass spectrometric determination of thorium-230 in corals. *Science,* 236: 1547–1553.

EDWARDS, R. L., TAYLOR, F. W., and WASSERBURG, G. J., 1988: Dating earthquakes with high-precision thorium-230 ages of very young corals. *Earth and Planetary Science Letters,* 90: 371–381.

ENFIELD, D. B., 1989: El Niño, past and present. *Reviews of Geophysics,* 27: 159–187.

EPSTEIN, S., BUCHSBAUM, R., LOWENSTAM, H. A., and UREY, H. C., 1953: Revised carbonate-water isotopic temperature scale. *Bulletin of the Geological Society of America,* 64: 1315–1326.

FAIRBANKS, R. G., 1989: A 17,000-yr glacio-eustatic sea level record: influence of glacial melting rates on the Younger Dryas event and deep-ocean circulation. *Nature,* 342: 637–643.

FAIRBANKS, R. G. and DODGE, R. E., 1979: Annual periodicity of the O-18/O-16 and C-13/C-12 ratios in the coral *Montastrea annularis. Geochimica Cosmochimica Acta,* 43: 1009–1020.

FAIRBANKS, R. G. and MATTHEWS, R. K., 1979: The marine oxygen isotope record in Pleistocene coral, Barbados, West Indies. *Quaternary Research,* 10: 181–196.

FU, C., DIAZ, H. F., and FLETCHER, J. O., 1986: Characteristics of the response of sea surface temperature in the central Pacific associated with warm episodes of the Southern Oscillation. *Monthly Weather Review,* 114: 1716–1738.

GLANTZ, M., 1984: Floods, fires, and famine: Is El Niño to blame? *Oceanus,* 14–19.

GLYNN, P. W. (ed.), 1990: *Global Ecological Consequences of the 1982–83 El Niño-Southern Oscillation.* New York: Elsevier. 563 pp.

GORDON, A. L., 1986: Interocean exchange of thermocline water. *Journal of Geophysical Research,* 91: 5037–5046.

GOREAU, T. J., 1977: Coral skeletal chemistry: physiological and environmental regulation of stable isotopes and trace metals in *Montastrea annularis. Proceedings of the Royal Society of London, B,* 196: 291–315.

GROVE, J. M., 1988: *The Little Ice Age.* London: Methuen. 498 pp.

HACKERT, E. C. and HASTENRATH, S., 1986: Mechanisms of Java rainfall anomalies. *Monthly Weather Review,* 114: 745–757.

HAMILTON, K., 1988: A detailed examination of the extratropical response to tropical El Niño/Southern Oscillation events. *Journal of Climatology,* 8: 67–86.

HASTENRATH, S., 1988: *Climate and Circulation of the Tropics.* Boston: Reidel. 455 pp.

HO TONG, Y., 1975: *The ASEAN Compendium of Climatic Statistics.* Jakarta: ASEAN Sub-Committee of Climatology, ASEAN Secretariat. 551 pp.

HOREL, J.D. and WALLACE, J.M., 1981: Planetary-scale atmospheric phenomena associated with the Southern Oscillation. *Monthly Weather Review*, 109: 813–829.

KILADIS, G. and DIAZ, H.F., 1989: Global climatic anomalies associated with extremes in the Southern Oscillation. *Journal of Climate*, 2: 1069–1090.

ISDALE, P., 1984: Fluorescent bands in massive corals record centuries of coastal rainfall. *Nature*, 310: 578–579.

LAND, L.S., LANG, J.C., and BARNES, D.J., 1975: Extension rate: a primary control on the isotopic composition of West Indian (Jamaican) scleractinian reef coral skeletons. *Marine Biology*, 33: 221–233.

LEA, D.W., BOYLE, E.A., and SHEN, G.T., 1989: Coralline barium records temporal variability in equatorial Pacific upwelling. *Nature*, 340: 373–376.

LEVITUS, S., 1982: *Climatological Atlas of the World Ocean.* Washington, D.C.: NOAA.

LINN, L.J., DELANEY, M.L., and DRUFFEL, E.R.M., 1990: Trace metals in contemporary and seventeenth-century Galapagos coral: Records of seasonal and annual variations. *Geochimica et Cosmochimica Acta*, 54: 387–394.

LUKAS, R. and LINDSTROM, E., 1991: The mixed layer of the western equatorial Pacific ocean. *Journal of Geophysical Research,* 96 (supplement): 3343–3357.

MARTIN, J.H., BRULAND, K.B., and BROENKOW, W.W., 1976: Cadmium transport in the California Current. *In* Windom, H.L. and Duce, R.A. (eds.), *Marine Pollutant Transfer.* Heath, 159–184.

McCONNAUGHEY, T.A., 1989: C-13 and O-18 isotopic disequilibria in biological carbonates: I. Patterns. *Geochimica et Cosmochimica Acta*, 53: 151–162.

McGLONE, M.S., KERSHAW, A.P., and MARKGRAF, V., 1992: El Niño/Southern Oscillation and climatic variability in Australasian and South American paleoenvironmental records. *In* Diaz, H.F. and Markgraf, (eds.), *El Niño: Historical and Paleoclimatic Aspects of the Southern Oscillation.* Cambridge: Cambridge University Press, 435–462.

MICHAELSEN, J., 1989: Long-period fluctuations in El Niño amplitude and frequency reconstructed from tree-rings. *In* Peterson, D.H. (ed.), *Aspects of Climate Variability in the Pacific and the western Americas.* Geophysical Monograph 55. Washington, D.C.: American Geophysical Union, 69–74.

NAMIAS, J. and CAYAN, D.R., 1984: El Niño: Implications for forecasting. *Oceanus*, 27: 41–47.

NICHOLLS, N., 1984: The Southern Oscillation and Indonesian sea surface temperature. *Monthly Weather Review*, 112: 424–432.

NOZAKI, Y., RYE, D.M., TUREKIAN, K.K., and DODGE, R.E., 1978: A 200 year record of carbon-13 and carbon-14 variations in a Bermuda coral. *Geophysical Research Letters*, 5: 825–828.

PÄTZOLD, J., 1984: Growth rhythms recorded in stable isotopes and density bands in the reef coral *Porites lobata* (Cebu, Philippines). *Coral Reefs*, 3: 87–90.

PHILANDER, S.G.H., 1983: El Niño Southern Oscillation phenomena. *Nature*, 302: 295–301.

PHILANDER, S.G.H., 1985: El Niño and La Niña. *Journal of the Atmospheric Sciences*, 42: 2652–2662.

PHILANDER, S.G.H., 1990: *El Niño, La Niña, and the Southern Oscillation.* San Diego: Academic Press. 293 pp.

POTTS, D.C., DONE, T.J., ISDALE, P.J., and FISK, D.A., 1985: Dominance of a coral community by the genus *Porites* (Scleractinia). *Marine Ecology Progress Series*, 23: 79–84.

QUINN, W.H., NEAL, V.T., and ANTUNEZ de MAYOLO, S.E., 1987: El Niño occurrences over the past four and a half centuries. *Journal of Geophysical Research*, 92: 14,449–14,461.

QUINN, W.H., ZOPF, D.O., SHORT, K.S., and YANG, K.T.W., 1978: Historical trends and statistics of the Southern Oscillation, El Niño, and Indonesian droughts. *Fisheries Bulletin*, 76: 663–678.

RASMUSSON, E.M. and CARPENTER, T.H., 1982: Variations in tropical sea surface temperature and surface wind fields associated with the Southern Oscillation/El Niño. *Monthly Weather Review*, 110: 354–383.

RASMUSSON, E.M. and WALLACE, J.M., 1983: Meteorological aspects of the El Niño/Southern Oscillation. *Science*, 222: 1195–1202.

ROPELEWSKI, C.F. and HALPERT, M.S., 1989: Precipitation patterns associated with the high-index phase of the Southern Oscillation. *Journal of Climate*, 2: 268–284.

SCOFFIN, T.P., TUDHOPE, A.W., and BROWN, B.E., 1989: Fluorescent and skeletal density banding in *Porites lutea* from Papua New Guinea and Indonesia. *Coral Reefs*, 7: 169–178.

SHEN, G.T. and BOYLE, E.A., 1987: Lead in corals: reconstruction of historical industrial fluxes to the surface ocean. *Earth and Planetary Science Letters*, 82: 289–304.

SHEN, G.T. and BOYLE, E.A., 1988: Determination of lead, cadmium, and other trace metals in annually banded corals. *Chemical Geology*, 67: 47–62.

SHEN, G.T., BOYLE, E.A., and LEA, D.W., 1987: Cadmium in corals as a tracer of historical upwelling and industrial fallout. *Nature*, 328: 794–796.

SHEN, G.T., CAMPBELL, T.M., DUNBAR, R.B., WELLINGTON, G.M., COLGAN, M.W., and GLYNN P.W., 1991: Paleochemistry of manganese in corals from the Galapagos Islands, *Coral Reefs*, 10: 91–100 .

SHEN, G.T., LINN, L.J., CAMPBELL, T.M., COLE, J.E., and FAIRBANKS, R.G., 1992a: A chemical indicator of trade wind reversal in corals from the western tropical Pacific. *Journal of Geophysical Research* (in press).

SHEN, G.T., COLE, J.E., LEA, D.W., LINN, L.J., McCONNAUGHEY, T.A., and FAIRBANKS, R.G., 1992b: Surface ocean variability at Galapagos from 1936–1982: Calibration of geochemical tracers in corals. *Paleoceanography* (in press).

SHEN, G.T. and SANFORD, C.L., 1990: Trace element indicators of climate variability in reef-building corals. *In* Glynn, P.W. (ed.), *Global Ecological Consequences of the 1982-83 El Niño-Southern Oscillation*. New York: Elsevier, 255–284.

SWART, P.K., 1983: Carbon and oxygen isotope fractionation in scleractinian corals: a review. *Earth-Science Reviews*, 19: 51–80.

SWART, P.K. and COLEMAN, M.L., 1980: Isotopic data for scleractinian corals explain their palaeotemperature uncertainties. *Nature*, 283: 557–559.

SWETNAM, T.W. and BETANCOURT, J.L., 1990: Fire-Southern Oscillation relations in the Southwestern United States. *Science*, 249: 1017–1020.

THOMPSON, L.G., MOSLEY-THOMPSON, E., and ARNAO, B.M., 1984: El Niño-Southern Oscillation events recorded in the stratigraphy of the tropical Quelccaya ice cap, Peru. *Science*, 226: 50–53.

TOGGWEILER, J.R., 1983: A Six-zone Regionalized Model for Bomb Radiotracers and $CO_2$ in the upper kilometer of the Pacific Ocean. Ph.D. dissertation, Columbia University.

TOGGWEILER, J.R. and TRUMBORE, S.E., 1985: Bomb-test Sr-90 in Pacific and Indian Ocean surface water as recorded by banded corals. *Earth and Planetary Science Letters*, 74: 306–314.

van LOON, H. and ROGERS, J.C., 1981: The Southern Oscillation. Part II: Associations with changes in the middle troposphere in the northern winter. *Monthly Weather Review*, 109: 1163–1168.

WEBER, J.N., 1974: C-13/C-12 ratios as structural tracers elucidating calcification processes in reef-building and non-reef-building corals. *In: Proceedings of the Second International Coral Reef Symposium 2*. Great Barrier Reef Commission, 289–298.

WEBER, J.N. and WOODHEAD, P.M.J., 1970: Carbon and oxygen isotope fractionation in the skeletal carbonate of reef-building corals. *Chemical Geology*, 6: 93-117.

WEBER, J.N. and WOODHEAD, P.M.J., 1972: Temperature dependence of oxygen-18 concentration in reef coral carbonates. *Journal of Geophysical Research*, 77: 463–473.

WELLINGTON, G.M. and GLYNN, P.W., 1983: Environmental influences on skeletal banding in eastern Pacific (Panama) corals. *Coral Reefs*, 1: 215–222.

WRIGHT, P.B., 1989: Homogenized long-period Southern Oscillation indices. *International Journal of Climatology*, 9: 33–54.

WRIGHT, P.B., WALLACE, J.M., MITCHELL, T.M., and DESER, C., 1988: Correlation structure of the Southern Oscillation phenomenon. *Journal of Climate*, 1: 609–625.

WYRTKI, K., 1987: Indonesian through flow and the associated pressure gradients. *Journal of Geophysical Research*, 92: 12,941–12,946.

ZEBIAK, S.E., 1989: On the 30–60 day oscillation and the prediction of El Niño. *Journal of Climate*, 2: 1381–1387.

ZEBIAK, S.E. and CANE, M.A., 1987: A model El Niño/Southern Oscillation. *Monthly Weather Review*, 115: 2262–2278.

# Low-resolution paleoclimate reconstruction of El Niño/ Southern Oscillation: marine and terrestrial proxy indicators

# 19

# Fishery catch records, El Niño/ Southern Oscillation, and longer-term climate change as inferred from fish remains in marine sediments

GARY D. SHARP

*NOAA Center for Ocean Analysis and Prediction, 2560 Garden Road, Monterey, California 93940, U.S.A. and Center for Climate-Ocean Resources Studies, P.O. Box 2223, Monterey, California 93940, U.S.A.*

## Abstract

The causes of rise and decline cycles of many pelagic fisheries have been debated for centuries. It is only within the last several decades that known or potential causal mechanisms have been studied in depth and, where possible, documented. Among the most compelling myths about fisheries collapses are those which invoke single events, such as the famous El Niños of 1972–73, or 1982–83, as the ultimate causes of long-term changes in marine ecosystems in the eastern Pacific Ocean or in other places. Close examination of the facts has shown that the effects of individual El Niño events can be devastating, but within a short time, the majority of the effects fade, and life goes on. El Niño/Southern Oscillation (ENSO) cycles are perturbations of the most important global climate pattern, the seasonal cycle. Ocean and atmosphere are sufficiently interactive on daily, seasonal, and interannual bases that ocean inhabitants have had to adopt responsive survival strategies into their life histories to cope with these frequent environmental processes to persist. The concept that climatic events (e.g., floods and ocean current changes) are important mechanisms for resetting, and even enriching, the ecological systems has yet to be fully appreciated. Examples from regional fisheries provide many lessons that are helpful in this regard. In order to understand changes in local processes it is imperative that one look 'upstream' in time and space for insights about the sources of these changes. Forecasting climate consequences from mechanistic type models is always a goal, but forecasting by analogy will usually be most effective, at least until the physics of climate change and their 'downstream' ecological consequences are better understood.

## Introduction

El Niño/Southern Oscillation (ENSO) comprises a family of natural climatic processes, defined by events, that undoubtedly have bases in global and regional climate patterns and trends. ENSO frequency and intensities apparently also pass through longer distinguishable epochal changes (Anderson et al. 1992, this volume). However, each event that has been studied is dissimilar to some degree to others that have been studied before. The role of analogies in these cases remains elusive, except when contrasting the extreme events with lesser ones. Understanding the distinctions between ENSO events, long-term climate variation, and their respective ecological and societal consequences is a major science issue, and responsibility.

Climatic epochs and extreme events each have left enduring signatures on society and subsequent behaviors in response to perceived threats from climate change and events. Historical compendia that identify important stages in the development of European society, along with the evolution of the global economic system, have long been studied and most recently summarized by Braudel (1981, 1984) to show the interrelations of harvests of natural and cultivated goods, with climate. These history lessons include important insights into the relationships between climate, microclimate related geography, society, and the continuing evolution of world economics, from sales of grains, wines, oil, to fish trading (Braudel 1984, 1990).

An important period in the evolution of European society occurred from the end of the Medieval Warm Period, around A.D. 1200, and makes a good case in point. Northern European feudal economies were forced into a new mode by the rapid decline in crop productions as the post-medieval period of cooling ensued. The fundamental dependence of the Hanseatic League, a powerful coalition of northern European traders, on the herring fisheries of the North Sea and Baltic is underscored by the dates of the League's rise and decline, about 1250 to the late 1500s. The products traded to various markets by the Hansa throughout Europe, Scandanavia, and into the East changed continuously, depending upon local productions of crops, fisheries, and commodities, and serviced specific market needs.

The Hanseatic empire grew and collapsed with the herring fisheries. For example, herring disappeared from the Baltic between the 14th and 15th centuries, opening a special market place. At the same time rich fishing grounds to supply the Hansa were discovered by Dutch and German fishermen off the Dogger Bank in the North Sea, and in the open ocean off the English and Scottish coasts. Herring were traded to eastern Europe and beyond. Human contributions to the evolution from locally focused marketing to global trading included innovations such as onboard herring processing and conservation methods that were discovered in the mid-14th century and served as a major source of trade items, as well as ready nutrition for the Hansa's armed forces for over a century. As the herring declined in the North Sea during the late 1500s, so did the Hanseatic empire.

Lindquist (1984) showed that the Swedes fishing the Bohuslan archipelago off southern Sweden appeared to benefit from abundances of herring resulting from epochs of high abundance and northeastward expansion of North Sea herring. These Bohuslan abundance periods tend to come and go at intervals of 80 or more years, while herring abundances in the nearby open sea occur about every 20 yr. These two facts when brought together suggest that there are at least two processes involved in which a near centennial-scale warming and cooling environment, and the associated ecological changes, interspersed with shorter, decadal scale processes of the sort described by Southward (1974a, 1974b), Southward et al. (1975, 1988), and Colebrook (1991), and many others (see review by Sharp and Csirke 1984).

We might ponder the sort of fisheries science that might have evolved if the North Sea was subject to El Niño-like warming events. It is counter-intuitive that modern fisheries management has evolved from explicit applications of equilibrium mathematical approaches, considering how long there has been recognition of signals from fisheries observations that such equilibria are so unlikely (Hjort 1914, 1926). The relative decadal stability of systems such as the North Sea, where western fisheries science originated, compared to systems where both cool and warm ENSO events dominate local fisheries responses, has provided useful examples of where mathematically convenient theory and observation failed to correspond.

It is ironic that it was exactly these fisheries that provided the initial forum and basis for modern fisheries science to evolve. The lack of success of most modern fisheries management is partially due to an underlying 'myth' of population stability. The problems stem from ecological perturbations in response to both human and climate-induced externalities, as well as from the poverty of the information base from which to apply conventional population dynamics models. The assumption that forms the basis for most modern mathematical modeling methods, whether for population biology or global climate purposes, is the concept of long-term equilibrium. Nowhere is it so obvious that this simplification provides little of practical use as in these two research arenas – climate and fisheries. It is, in fact, the lack of equilibrium that promotes ecological production and climate variation, and the latter that stimulates the former.

The noted economist Schumpeter (1988) provided several bridging concepts between natural lessons as empirical bases for logical, 'scientific' decisions, as a result of human innovations and applied theory. Braudel (1984) and others, have been able to integrate observations and theories into plausible historical interpretations of the evolution of western society, in the context of local and global opportunities. Schumpeter's underlying conceptual basis was his argument that historical, empirical information was necessary to make sense from purely theoretical, basic mathematical approaches to uncertainty about the future. He advocated that business cycles, and therefore economies, are driven by both innovations and evolving societal needs. A complement to his thinking that one can derive from reading historians such as Braudel is the additional premise that climate change provides the continuous backdrop for changing societal needs.

These changes, along with continuous human population growth, i.e., increasing demand, lead to the premise that climate cycles and patterns provide the engine for global economic patterns. It is also germane that since the initial organic processes began on Earth, biology has influenced the global environment and has provided the means to both destabilize and exploit the entire system.

It is important to recognize that the global hydrological cycle – from the oceans into the atmosphere, to mountains and through streams and rivers back into the sea – is the ultimate ecological climate signal. Hydrologic seasonality is a fact of life for terrestrial flora and fauna. The analogies of terrestrial hydrological effects to seasonal upper ocean thermohaline structure, and terrestrial plant biomes to plankton biomes – both forming zoogeographic provinces – provides a basis for interpreting local and regional oceanic ecological changes within a context of continuously changing climate. This context includes events such as ENSO warming or cooling phases as perturbations of seasonal patterns within larger climate processes. The resulting alternative physical environmental states and the responses of those species that are adapted to these states will continue to provide new insights and eventually provide key climate forecast capabilities.

## ENSO and fisheries as an information source

In fisheries contexts, it is important to resolve and distinguish between the different ecological consequences attributable to human harvests, human-induced stressors, and natural climate induced population responses. For example, at the April 1983 Food and Agricultural Organization (FAO) Expert Consultation to Examine Changes in Abundance and Species Composition of Neritic Fish Resources an array of presentations pointed out the significance of longer-term climatic trends and processes that have direct implications to regional ocean fishery production (Sharp and Csirke 1984). This conference took place right in the middle of the 1982–83 ENSO warming event. The proceedings provide examples of the emergent understanding of the relative contributions to marine population variations due to natural and human processes.

No one would argue that man has little or no effect on the sorts of catch data from which we also infer population responses. It is also unreasonable to deny that there are also climate-related ocean environmental causes of the blooms and subsequent collapses of marine populations. For example, many effects of the 1982–83 El Niño on eastern Pacific fisheries were well described: the northeastern Pacific by Wooster and Fluharty (1985); the Chilean experience in IFOP (1985); the Washington and Oregon coast by Pearcy (1984); and the Peruvian coast by Arntz (1984).

Although some still marvel at the magnitude of the greatest 'observed' ENSO event (1982–83), I would like to steer away from this period of tightly focused observations, particularly concerning coastal fisheries. I will suggest that it was merely an apex event within a long epoch of changes. The decade-long warming period within the eastern Pacific region had already set the scene for an array of rapid and dramatic changes as the entire global ocean thermal budget changed in the subsequent years. For example, there are different patterns of species expansion and contraction that have been documented (e.g., Southward 1974a, 1974b, Kawai and Isibasi 1984; Loeb and Rojas 1988; Smith and Moser 1988). Their relative synchrony across ocean basins and around the world is compelling.

Crawford et al. (1991) provide some insights into the variations of *Sardinops* spp. blooms and collapses at the International Symposium on Long-term Variability of Pelagic Fish Populations and Their Environment, held in Sendai, Japan, November 1989. Crawford and his colleagues compared the population response patterns of temporal variability from the Atlantic Basin, the Benguela Current, and the Pacific Basin. They showed that there were some similar regional, ocean basin patterns, and some differences that suggest various lags and levels of independence. However, Pacific Basin catch records suggest that there is a common stimulus within the context of a general ocean/atmosphere regime in which ocean warming periods provide *Sardinops* spp. with unique conditions that promote the survival of early life stages. These conditions result in enhanced population growth and geographic expansion. Sardine populations decline during opposite, cooling ocean environmental conditions.

Consider also the reasons why we study these processes. The economic implications of longer term climate-induced resource changes experienced over the last two decades to, for example, a nation such as Peru, have been devastating. Compare those processes outlined in Valdivia's (1978) description of the effects of the 1972–73 El Niño on the Peruvian fisheries system with Kondo's (1980) initial description of the recent two decades of re-establishment of the far east sardine, *Sardinops melanosticta*, around Japan and into the Pacific Ocean. In contrast to those event-scale, anecdotal regional experiences cited above, the longer term changes in both the Peruvian and Japanese societies have been remarkable, if opposite.

For example, Peru's economy has come to a near halt since the loss of the anchoveta fishery. Japan's recently emergent fish culture industry is reliant upon the low-cost protein from their sardine fisheries for fish feed. The implications of a climate-induced shift in sardine distribution and abundance will have profound effects. Kawasaki's (1984) portrayal of the nearly coherent patterns of catches of sardine (*Sardinops*) fisheries around the Pacific Basin (Fig. 19.1), and globally (Kawasaki 1991) compels consideration of the existence of a basin-wide climate/ocean process that stimulates the bloom and demise of local populations. There are many noncommercial indicator species which undergo parallel changes to these and other climate/ocean changes (Loeb and Rojas 1988).

These latter perspectives offer useful points of reference. The usual or standard collected volumes on ENSO are mostly anecdotal, short-term perspectives related to a single identifiable perturbation of normal seasonal, climatological processes. The basic issues needing discussion are those related to the likely persistence and the distributions of the effects of these perturbations, within larger

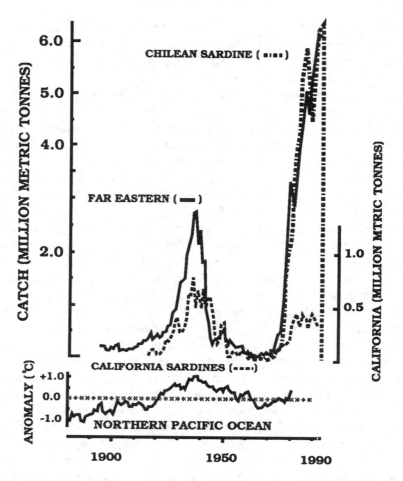

Fig. 19.1 Catch statistic records from three isolated *Sardinops* spp. fisheries around the Pacific Basin are plotted for the periods for which each fishery has been documented. The SST anomaly time series is for the Pacific Basin. Note that the sardines bloom during warming periods, and decline during cooling periods. (Modified from Kawasaki, 1984.)

spatial and temporal contexts. In fact, the basic issues that need to be focused upon are the societal implications, the direct and indirect social consequences of climate variability versus perturbing events. This is why fisheries science has begun evolving away from overfocused studies of purely numerical population dynamics. It was clearly irrational to study fishery dynamics in isolation from climate-driven environmental studies. Fish populations vary with both the effects of human interventions and perturbations of natural hydrological patterns and processes. Hollowed et al. (1987) provide insights into the geographic and temporal mosaic of fish population variations in the eastern Pacific Ocean, which suggests the strong influences of annual and decadal environmental variations. The direct causal processes are obscure, in most cases, but low river flows or extreme ocean temperatures have well-known consequences.

### Fishing and other sources of records

The ocean information bases that provide the most coherence and longest time series are those from coastal fisheries around the globe, particularly eastern boundary currents. These regions produce up to 50% of the world ocean fish catches. Each of the eastern boundary currents has a dominant species array in common, a sardine (*Sardinops* spp.), an anchovy (*Engraulis* spp.), a mackerel (*Scomber* spp.), a bonito (*Sarda* spp.), a jack mackerel (*Trachurus* spp.), and a hake (*Merluccius* spp.) (Bakun et al. 1982), which vary in proportions under different climate/ocean regimes (for reviews of mechanisms and geographic examples see also Sharp 1981, 1988).

The sardine (*Sardinops* spp.) is adapted to the subtropical transition zone, and sardine populations bloom under ocean-warming conditions that do not favor other coastal pelagics such as engraulids (anchovies). It is not the warm water *per se* that induces their blooms, but the shifts from a cooler, upwelling-dominated climate/ocean regime, and the accompanying ecological patterns that promote their successes, or failures, from year to year. Anchovies, alternatively, are apparently adapted to cooler, periodic upwelling conditions (Sharp 1991), and anchovy predators, hakes (Merlucidae), *Scomber scomber, Sarda* spp., and seabirds each appear to fluctuate along with them (Pauly and Tsukayama 1987; Wyatt and Larrañeta 1988; Pauly et al. 1989). The converse situation occurs among sardines, jack mackerels (*Trachurus* spp.), chub mackerels (other *Scomber* spp.), as well as other transition zone predators (Pauly and Tsukayama 1987).

Perhaps the single greatest challenge to successful fisheries research was to change the way fisheries populations were viewed. Over the last few decades the logic of fisheries studies has turned from studies of natural mortality to focus on those conditions necessary for each successive developmental stage to succeed. Historically, fisheries studies have primarily kept accounts of mortality, from time step to time step. It is not mortality that is hard to explain. Well-defined survival conditions and mechanisms are most elusive. It is neither 'mystical' nor difficult to explain what possible factors control the successes and failures of pelagic species from year to year. There are so many factors that can affect each life-history phase that it is difficult to forecast or determine in advance which will dominate (Bakun et al. 1982; Csirke et al. 1989). ENSO heating and cooling cycles are certainly strong affectors of year-to-year variability, but the longer term patterns of aquatic population rise and fall are tuned to greater scale climate-driven environmental variabilities (Soutar and Isaacs, 1969, 1974; Fig. 19.2).

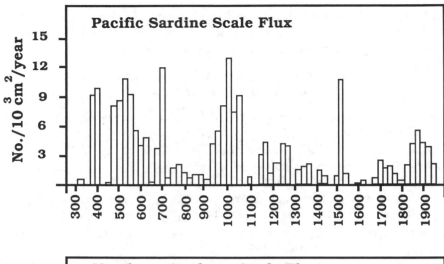

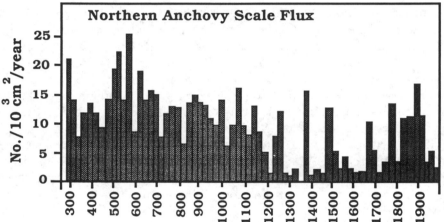

Fig. 19.2  Annually laminated, varved sediment records of the Santa Barbara Basin have been sorted and fish debris (e.g., scales, otoliths, hard parts) identified, and enumerated for contiguous segments for the recent two thousand year period. These records have been the inspiration for much speculation about the periodic ebb and flow of pelagic species, and the climatic regimes that dominated each period (cf., Moser et al., 1987, Smith and Moser, 1988). The time periods of variation range from decadal through centennial, and on to millennial scales. (Data from Soutar, see Anderson et al. 1992, this volume.)

### *Paleosedimentary records of fish population variations*

The studies of anoxic ocean basin sediment cores by Soutar and colleagues (Soutar and Isaacs 1974; DeVries and Pearcy 1982; Baumgartner et al. 1985; and others) have documented the waxing and waning of many species in the Santa Monica and Santa Barbara basins off Southern California, and the Guaymas

Basin in the Gulf of California. Apparently they have been cycling from high to low abundances for millennia. The *Sardinops* spp., engraulids (anchovies and herrings), and other fished species each have their own stories, and many have received much attention, a few of which will be discussed later.

## Physical variability and physiological ecology

Thermohaline properties separate the ocean into specific latitudinal, longitudinal, and depth-related compartments, much as the terrestrial hydrological and thermal climate regimes organize the terrestrial ecology into ecological provinces. Specific physical-chemical habitats form, each of which has provided opportunity over time for biological organisms to evolve within the constraints of these properties and their long- and short-term variations.

The primary forcing of the ocean's physical milieu is by the atmosphere. Changes and gradients in light levels, hydrology, microchemistry, and biological interactions provide individuals and species unique challenges as well as opportunities for survival to reproduction of a next generation. The measure of ecological success is persistence. Perturbation and relaxation of these systems take many forms, but are driven primarily by solar and lunar energies that affect changes on hourly to millennial time scales. Cascades of ecological responses, similarly, occur continuously on diurnal through evolutionary (millennial) time scales. Man has only relatively recently become a major perturber of these systems.

The temporal responses of some organisms to dynamic physical forcing are stretched over many time scales by their adaptations to various processes, depending upon their relative mobilities, or their abilities to resist transport due to currents and turbulence structures in the ocean. Smaller, more passive organisms, have evolved encysting stages that allow them to persist, even when they are carried well below the life-giving photic zone, into the cooler, respiration-conserving abyssal realm. Smetacek (1985, 1988) described the evolutionary and ecological significance of these adaptations for an array of smaller planktonic species. Larger, more mobile species operate on shorter time scales, and seek their shorter-term physiological requirements within the ambit allowed by their somatic energy contents. Sharp (1984), Sharp and Vlymen (1978), and Olson and Boggs (1986) reviewed these features for scombroid fishes and other oceanic predators. These species contrast markedly with smaller engraulid (herrings and anchovies) or small piscivorous species that comprise the major fish catches within coastal oceans (e.g., Vlymen 1977; Muck and Sanchez 1987).

What biological characteristics or environmental properties would promote relative population stability? Among marine species population resilience depends on age classes (longevity) and adaptations to local food deprivation (low energy requirements, or great mobility – a property of larger, nomadic opportunist species). Only strongly (seasonally) patterned oceanographical and ecological processes, therefore habitat dependability is required.

However, population responses to periodic environmental events such as warm (El Niño) phases of the ENSO cycle are often harsh, if not particularly long-lasting. These responses are generally characterized as being typical of thermal stress, e.g., induced physiological responses such as higher metabolic/respiration rates. Particularly affected are the sessile or immobile species that are unable to remove themselves from the local warming, or that become trapped by incursions of warmer waters. Also, as aquarists and florists know well, thermal-stress induced spawning is a chacteristic response by all organisms. Thermal stress induces onset of reproduction, in which most of the stored somatic energy is transferred to the gonads, and gamete production ensues (or flowering is 'forced').

This latter response is perhaps the oldest generalized response in living organisms, from bacteria to man, and plants, ensuring one last effort from the threatened parent organisms to reproduce. This basic response also results in lower growth rates, early onset, i.e., lower size/age, at first reproduction, and a correspondent decreased size at age, hence reduced reproductive potential at age. These processes were well documented for situations around the globe by Sharp and Csirke (1984), and for Peruvian anchoveta from 1964 to 1986 by Tsukayama (1989). Concurrent ecological responses, primarily due to thermal physiological responses include the rapid increase in primary production, enhanced nocturnal respiration and severe anoxic events, and the down stream consequences of these, which can include death by asphyxia of fishes and many invertebrates. All these effects can cause increased sedimentation rates, bottom anoxia, and associated enhancement of the organic load in the bottom sediments.

Conversely, ENSO-cycle cooling periods tend to promote longer, if slower somatic growth; later onset of first maturation; and more efficient use of available foods, resulting in more fat storage and better quality gametes when maturation finally occurs. The entire aquatic ecosystem tends to move upward in size at age, and rates of reproduction slow and better quality gametes are produced. The results are greater biomasses, longer lives, and all the characteristics that we associate with both deep water and highly seasonal higher latitude ecosystems.

The longer patterns of climate-related changes in fish distributions and abundance have been long recognized. The responses of most fisheries to shorter term ENSO events must be recognized as significant system perturbations, although not necessarily more or less important to local societal settings than the longer-term changes. Colebrook (1991) recently emphasized the fact that longer-term variations dominate the biophysical character of the North Atlantic, with some 3–4 yr patterns emerging from the long-term changes.

In fact, the effects of any individual El Niño event on presence or absence of individual coastal fishes and birds are mostly gone within a year or 18 months. Yet changed population characteristics will persist, i.e., those affected individuals will remain small at age, and the resulting population reproduction potential will be lowered. ENSO cool phases tend to slow onset of reproductive development, and enhance somatic growth, yielding larger, later maturing individuals compared to the intermediate or warm phases. The warm ENSO phases tend to stress increasingly each of the size/age groups of aquatic organisms, and younger, smaller fish tend to shift from somatic growth to gonadal development, which results in many, smaller individuals producing more, and smaller, less likely to survive eggs (Ware et al. 1981). The larger fish tend to relocate by moving into deeper, cooler, or more offshore and higher latitude portions of their habitat, or suffer reproductive stress responses that debilitate these individuals and lead to lower viability.

With these diverse responses to environmental conditions, it seems that we could and should be monitoring natural populations and mining that information for clues about status and trends of regional and global climate. For example, during January to March 1985, along the western South American coastline, there was a 'cooler' than usual period that implies more upwelling, and generally higher productivity, without any severely strong winds. The coastal pelagic fish resources enjoyed major reproductive successes. The anchoveta had the largest reproduction since the initial decline period in the early 1970s. The question was, then, whether there will be a continuation of this level of success over an extended period, and what the effects might be on the presently dominant sardine populations?

I have tracked the developments off South America's Pacific coast up through the recent years (Sharp 1981, 1987, 1988). The previously dominant subtropical-oceanic fauna, characterized by many species, including sardines, has moved off-shore and equatorward, as their subtropical, oceanic habitat was displaced by the cooler upwelled water. Another species, the Chilean jack mackerel, (or jurel, *Trachurus murphii*) is presently accounting for as much of the Chilean pelagic fish landings (over 2 million tonnes per year) as were the sardines in previous years. The analytical approach taken by Yañez (1991) provides one means for using both catch and effort data from fisheries, and environmental information to interpret, and, even to forecast to some degree, the behavior of coastal fisheries off Peru and Chile. It cannot, yet, account for species blooms and changes in species targeted by fisheries of the sort just described.

It is certain that the entire Humboldt Current system is under transition. The upwelling foci are reactivating, and anchoveta are blooming, along with an array of coastal upwelling species, including the jack mackerel. Sardines are in decline, in response to both reduced reproduction and fisheries harvests, and the biomass centroid for the coastal pelagic complex is shifting back toward the north as the coastal habitat becomes cooler, in response to the increased upwelling, and lesser influence of the offshore, oceanic environment. This is in nearly complete opposition to the changes that have occurred off the Californias over the same period, during which coastal upwelling has been reduced, and onshore advances of the oceanic environment have transported novel, open ocean associated species into the coastal regime and bays.

Likewise, we should not forget the longer climate/ocean processes and their effects in the far North Pacific. The prices of canned and fresh Pacific salmon increased from 1980 to 1984 in the world marketplace due to the sequential displacement of the usual salmon migratory pathways and complex ecological displacements and perturbations that resulted from warm water insurgences along the Pacific coast. The 1985–86 North Pacific oceanographic picture returned to a more 'typical' cool epoch pattern, only to become again influenced by the subtropical ocean 4 yr later.

Coastal wetlands, spawning streams, and local populations of salmon throughout the western continental United States have suffered recently (1976 to present) from drought. The only respites from this drought have been from anomalous storms associated with ENSO warming events. So, although the coastal pelagic species may suffer habitat warming, individual breeding groups of salmonids may have benefitted as consequences of some of the smaller of these events. The large-scale warming events take their toll on coastal fishes, but resulting rainfall can stimulate entire drainage systems (Cayan and Webb 1992, this volume), providing opportunities for some species.

On longer time scales, the Santa Barbara Basin scale deposition records illustrated in Figure 19.2 suggest that there was a major shift in climate/ocean dynamics from an earlier regime to the present patterns about A.D. 1050 to 1100, or near the end of the Medieval warm period. The previous epoch was much more productive for both indicator species, suggesting both warmer and wetter conditions, along with greater upwelling productivity that is consistent with records of the lake level stands in and about southwestern North America (Hubbs 1960), and with Bakun's (1990) consequence scenarios for general global warming on eastern boundary current systems. There is also, apparently a rather characteristic 250- to 350-yr ebb and flow during the last thousand years, which is evident in data from high resolution tree ring, river flow, and hydrological studies from the region as well as with geological evidence described by Anderson et al. (1992, this volume).

The 1989–90 ocean surface warming and onshore movement of the oceanic regime off the Californias heralded the onset of conditions that are conducive to local *Sardinops* blooms, along with other oceanic and subtropical species, and the suppression of the coastal upwelling species. The numbers of partial data sets and analogous studies that have been started that show these expected patterns of temporal changes are remarkable (Flegal 1990; Ralston and Lenarz 1990; Ramp 1990; Tisch et al. 1990; and Whipple 1990). The nested nature of the problems needing to be examined makes the direct – one hypothesis, one experiment, one variable – approach to climate and climate consequences studies very tenuous, perhaps even impossible. The 'wheels within wheels' or clockworks analogy emerges. Multiple processes, all interacting on many time scales, lead to the kinds of epoch to epoch, year to year, and season to season 'surprises' that evolutionary biologists and ecologists must deal with.

Demands on fisheries production have increased in parallel with the growth of the human population. Where societal pressures may have been minimized by acculturation as the human population has grown, and standards of living increased over the last several decades, the productivity of the land and sea margins has also become increasingly stressed in response to increased human population density. For example, building flood control dikes and dams has too often brought only short-term, local relief in the form of decreased incidence of inundation and hardship arising from seasonal flood cycles. Some of the longer term effects are truly ecological tragedies.

For example, the High Aswan Dam so thoroughly regulates the flows of the Nile that the Egyptian Nile delta has become an ecological nightmare, with the salinity of the soil rising at such a pace that it is difficult to see how the region can ever be self-sustaining again. The once important eastern Mediterranean sardine population, which thrived at the mouth of the Nile and provided nutritional protein food for the region, has collapsed with the decreased Nile flow and the subsequent decline in local primary production. Historically, the peak production from that population resulted from the strong surges of Nile flood waters and pulses of nutrients into the Mediterranean Sea.

Decadal time series from marine biological sampling are available from regional studies, and fisheries catch/research records. These data sets are all much like the instrumental period in meteorology, especially in oceanography in that the records are often so short or inconsistent that their interpretations are complicated by perennial statistical problems associated with too few observations and changing measurement techniques. Therefore, the conclusions drawn are often either misleading, or worse, simply false. The best-studied historical records are the scale counts in sediment laminae (annual varves) for the Southern California Bight and the Gulf of California (Soutar and Isaacs 1974; Soutar and Crill 1977; Baumgartner et al. 1985, 1989; Fig. 19.2) of the relative abundances of anchovies and sardines. Similar studies, of upwelling areas off Peru and Walvis Bay, South Africa, have shown similar patterns of change (DeVries and Pearcy 1982; Shackleton 1987).

The message(s) within these figures certainly suggest climate and ocean interactions and their ecological consequences occur over what appears to be relatively long periods (decades to century and longer scales). The status of fish and plankton populations that reside within, in this case, the Subtropical Transition Zones of the Pacific Ocean Basin offer useful indicators of climate/ocean states.

### Atmospheric processes, and ocean ecological responses

There is continuous heat flux between the ocean and atmosphere, and from the tropics poleward within the ocean and atmosphere. Thermal exchange is manifest in many seasonal and climate event records, and on long time scales in

(a)

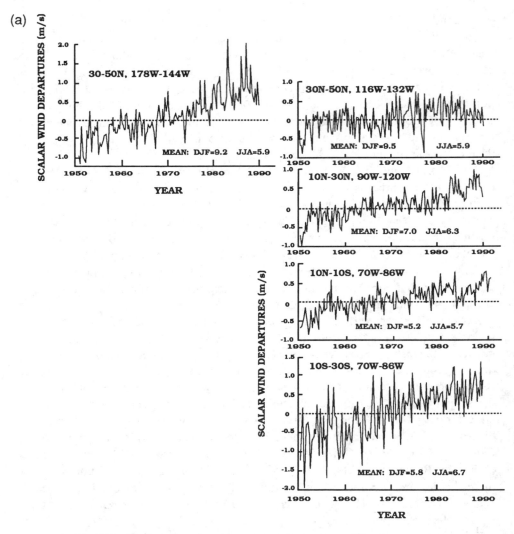

Fig. 19.3 (a) Variations in uncorrected scalar surface winds (in m/s) are compared on a common scale for the western coastline of the Americas from about Vancouver, B.C., down the coast by 20° latitudinal increments to Valparaiso, Chile. The offshore region for the northernmost block is shown at the upper left, and the other blocks are arrayed north to south, from top to bottom. Note the mean wind speeds for each region, which are given for December through February and June through August, the seasonal peak and low periods, respectively. (These data are from the Comprehensive Ocean and Atmosphere Data Set (COADS), and were summarized and provided by R. Pyle of the NOAA Environmental Research Laboratories, Climate Research Division in Boulder, Colorado.

many environments (see for example contributions by Cayan and Webb 1992, this volume; Anderson 1992, this volume). The variability expressed in records in terrestrial environments arise mostly from local, seasonal hydrological and thermal cycles, and their distortions caused by larger scale regional and global

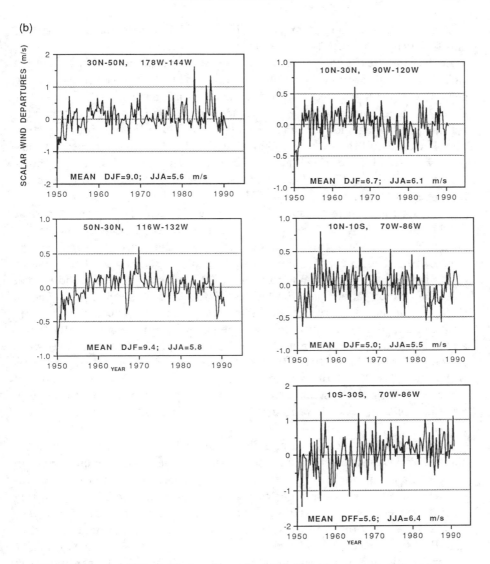

Fig. 19.3 (*cont.*)    (b) Variations in scalar surface winds (in m/s) corrected for assumed proportions of anemometer records and Beaufort Scale estimates are compared, as in (a). Note the rather significant changes over time in the trends due to the corrections. (Corrected data were provided by H. Diaz, of NOAA/ERL, Boulder.)

forcing. These terrestrial consequences are clearly analogous to the processes that induce the upper ocean thermohaline dynamics, as forced by surface winds, changes in cloudiness, storm tracks, frontal interactions, etc. (see, Gray 1990), and the cycling of water.

The significance of Gray's (1990) studies of frequency of hurricane landfall in the southeastern United States and Sahel rainfall should not be ignored in the contexts of other analyses that have recently been reported. For example, Bakun (1990) identified a set of likely scenarios that would have persistent effects on coastal fog and heat budgets, and upwelling, and therefore on coastal ecosystems and fisheries, as a general response to climate warming.

Let us examine the coherences of recent and long-term instrumental data for various locations around the globe. A more immediate local ecological forcing agent than those associated with changes in either mean sea surface or mean terrestrial surface temperatures are the surface winds. Comparison of these records within and among regions around the globe, stratified by latitude for the eastern Pacific Ocean and the western Indian Ocean, exhibit similar if somewhat temporally offset shifts during that period, with some latitudinal and onshore-offshore variations. Figures 19.3 and 19.4 show these patterns for the recent instrumental record period (1950–1991).

Compare the surface scalar winds along the western coast of the Americas (30°N–30°S) to those for the western boundary of the Indian Ocean (26°N–26°S, Fig. 19.4). In Figure 19.4 three climate regimes are shown: from the equator to the north coast of the Arabian Sea (0°–26°N); south of the equator to Madagascar (0°–16°S); and from 16° to 32°S (to the Cape of Good Hope). Note the various patterns and similarities to the records from the eastern Pacific Ocean. In the northern tropics and subtropics surface winds peaked in the early 1960s, and then declined until they became extremely erratic in the early 1970s, a phenomenon that recurs in many areas. In the early 1970s the surface winds peak again and decline until the 1982–83 ENSO period, and since then most areas within the 30°N–30°S latitudes show increases in progress. Despite fundamental climatic differences between the Indian Ocean subregions, and the eastern Pacific, similar increases in wind speeds are also found at higher latitudes in both areas.

Having noted both coherent and distinctive regional aspects of many global climatic variations, it is important to recognize some of the nonuniformity in climate change around the globe. There are, however, some remarkable synchroneities, although signs and degrees of response are modified by shifting seasonal and epochal patterns of such features as the Inter-Tropical Convergence Zone (ITCZ), and the various warm and cool land areas and ocean pools, as described by Ropelewski and Halpert (1987), Gray (1990), and many others.

Comparative methods prove invaluable for studies in which longer time series are not available (cf., Bakun and Parrish, 1990; 1991). Gray's (1990) summary figures provide clear evidence of a relation between the Sahel epochal rainfall patterns and eastern Atlantic cyclogenesis, as can be seen from Figures 19.5 and 19.6. By comparing these regional climatic records with fishery and coastal environmental information striking epochal changes can be identified. For example, Figures 19.5, 19.6, and 19.7 show that the period from about 1968 to 1972 was one of change from high rainfall to low rainfall; decreased hurricane activity; low to high surface wind speeds off Senegal; and a parallel shift from low to high pelagic fish catches off Morocco.

The problem is to find long enough time series to identify the most useful signals and thereby provide diagnostic tools of use to both agriculturists and

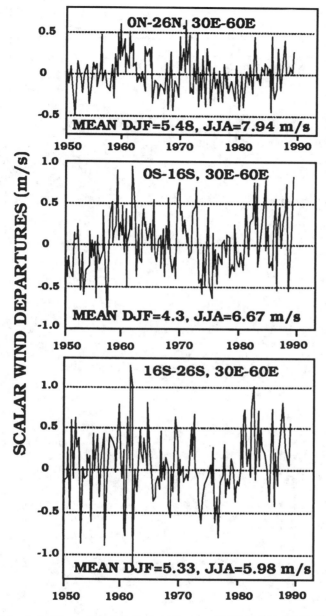

Fig. 19.4 As in Figure 19.3, but for western area of the Indian Ocean which was divided into three ocean/climate regimes. The scalar wind departures are plotted about the mean of this time series. However, the two mean values given here are for the two peak monsoon periods, December through February, and June through August. The long-term patterns are not different from those from the eastern Pacific Ocean. The period of record is from 1950 to 1989.

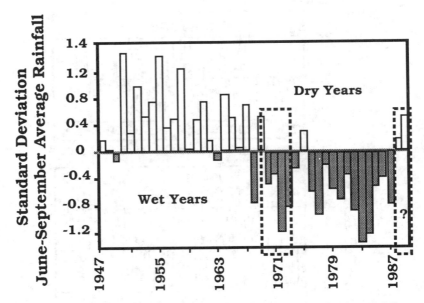

Fig. 19.5   The plots of standard deviations of the June through September rainfall as measured at 38 stations in the Sahel show that about 1968 there was a shift from sporadic but plentiful rains to a two-decade-long drought which ameliorated in 1988. This rainfall is associated with the warming of the southern Atlantic Ocean equatorial water mass, compared to the northern areas, and is a symptom of the southerly extension of the ITCZ in the region. (Figure modified from Gray 1990.)

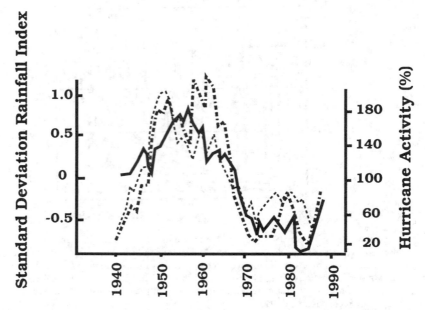

Fig. 19.6   The relationship between seasonal Sahel rainfall standard deviations and the Atlantic Ocean hurrican activity, as measured by the numbers of named hurricanes. The solid line represents the rainfall deviation, and the light dashed line represents the stronger hurricanes; the dotted line plots the number of active hurrican days from these stronger hurricanes. (From Gray 1990.)

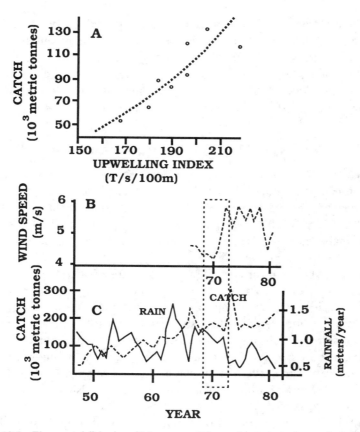

Fig. 19.7   The coastal fisheries of Morocco and Senegal experienced very clear shifts in patterns of fish production, as described in these three panels. In panel A the relation between upwelling intensity (in T/s/100 m) and fish production is shown for the Moroccan sardine fishery (from Belveze and Erzini 1984). In Panel B the coastal wind speed for Senegal is shown. Freon (1984) showed that Senegalese fishery production was affected by the climate changes that occured in the early 1970s. In Panel C the Moroccan sardine catch is plotted along with rainfall, to show that periods of peak catches are negatively correlated with periods of lowest rainfall. (From Belveze and Erzini 1984.)

fisheries for use in their planning and management. We will revisit these topics when the issues of winners and losers in climate change are discussed in the next section. It is imperative that a broad perspective be maintained in discussing the relative merits of one climate epoch versus another, as the consequences vary greatly in time and space, and are important to the long-term health of ecosystems.

We have learned from empirical and logical explorations that the causal links between the sporadic rises and declines among the various regional fish resources around the globe are ultimately very 'local' responses by individuals that produce abundances of viable young, which survive to reproduce again, or not, in response to local opportunities, even when the major forcing may be from quite distant climatic processes. The array of processes and optional perturbations that can dominate these local conditions have been described many times, separately, but the combination of geography, climatology (oceanographical and meteorological), geophysics, and ecology set the scenes for myriad responses (cf. reviews by Sharp 1981, 1988, 1987, 1991; Bakun et al., 1982; Csirke and Sharp 1984; Cury and Roy 1989; Tomoda and Hironaga 1991).

### Winners and losers in changing climate regimes

The effects of the long-and short-term north-south excursions of the Inter-Tropical Convergence Zone (ITCZ) are evident in the time series for scalar winds and sea surface temperatures (SST) in the Panama Bight (0°–10°N and 90°–100°W; Figure 19.8), which is the area in the eastern Pacific Ocean of maximum seasonal excursion of the ITCZ, and an area during which El Niño warm events tend to produce large rainfalls.

An important concern for coastal (and fresh water) fisheries, particularly for anadromous species (e.g., the salmonids, shads, and many other estuarine-dependent species such as shrimp) are changes in magnitudes and patterns of seasonal rainfall. For example, the lush north coast of Honduras is an environment with year-round rainfall. The south coast is a drier environment with strong seasonal rains bordered by a hypersaline estuary and the highly saline Gulf of Fonseca, indicating that over the annual cycle there is a net fresh water deficit for the entire region. Shrimp stocks and associated fisheries are known to respond to these patterns (Garcia and LaReste 1981). The patterns of rainfall variation for the south coastal region are quite different from those of the north coast reflecting the year-round rainfall from the unstable Gulf of Mexico and Caribbean marine layer that dominates in the north. The seasonal swings of the ITCZ dictate the rains along the near-desert Pacific perimeter of Honduras. The seasonal and interannual patterns for each coastal region are shown in Figure 19.9. Note also the strong periodic nature of the annual Pacific record, which reflects the extents and durations of northerly or southerly seasonal ITCZ excursions across the Central American region.

During periods when the Central America Pacific coast has lowest rainfall values, the north coast of Ecuador and northern Peru often have strong positive anomalies. For example, both Ecuador and Peru had severe floods from the 1982-83 ENSO warming, rerouting rivers, and creating new alluvial deposits. There have been several such floods within recent centuries. These are recognizable major events that have occurred sporadically within the last several centuries, as determined from pluvial stratigraphic studies (Wells 1990).

Along the western coast, shrimp fisheries from Mexico south to Panama have declined since the onset of the 1970s eastern tropical Pacific Ocean warming. Due to the increased rainfall, and local flooding associated with the decadal scale warming within the region, and the associated southerly location of the ITCZ, northern Ecuador benefited from an explosion of natural shrimp production, all

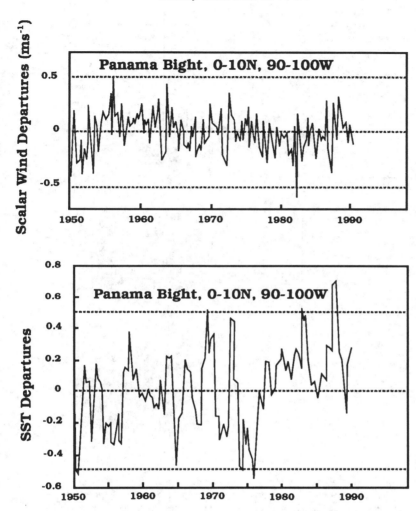

Fig. 19.8 COADS time series for the Panama Bight show recent (1950 to present) changes in both corrected surface wind speeds and sea surface temperature (SST). The latter have been more irregular, although it should be noted that lower wind speeds and increasing SSTs occured during similar time periods. The SST time series display strong interannual variability, and is likely associated with both seasonal and interannual fluctuations of the ITCZ.

of which contributed to their newfound shrimp culture industry. This industry will continue to rise and fall as a function of the shifts in the location of the ITCZ, as illustrated in the time series of rainfall in southern Honduras. These longer term patterns of north and south shift of the atmospheric and oceanic equator, and their ecological consequences are also evident from secondary indicators based on fishery records such as those in Figure 19.10.

Similar effects are evident from Bakun's (1990) study of eastern boundary current upwelling regimes since 1947 pointing out several important issues (see Fig.

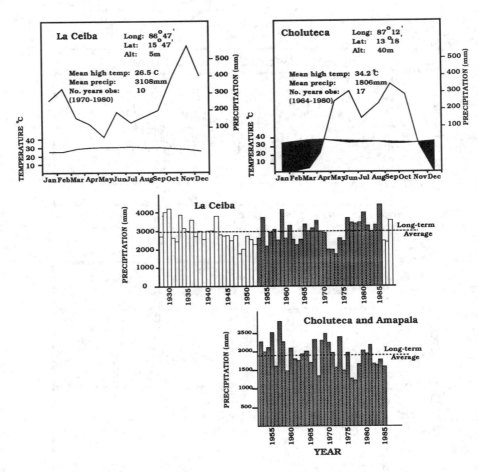

Fig. 19.9 Separated only by a mountainous ridge, the north and south coasts of Honduras are completely different climate regimes. Moisture from the Gulf of Mexico and Caribbean provides the northern area with year-round rainfall, while the arid southern coast exhibits a strong seasonal cycle and only receives rainfall from passing tropical storms and the seasonal transits of the Inter-Tropical Convergence Zone. The shaded area in the right upper climate diagram indicates periods when evaporation exceeds precipitation. The rainfall records are shaded to indicate the overlapped years. Different cycles and trends dominate long-term patterns of each region.

19.11). The long-term trend reported by Bakun is clearly in parallel with the scalar wind information that is offered in Figures 19.3 and 19.4. However, the fact that the trend line does not describe the surface wind behavior of the last 15 yr or so, these data suggests that the relatively short instrumental climate record that is available (1947 to present) is too short to describe potentially important low-frequency oscillations on the scale of decades to centuries. The large excursions, outlined for all regions by the dotted rectangle, are not unprecedented changes, but we do see rather consistent downward trends following this period. This 15-yr trend is in marked contrast to expectation from the

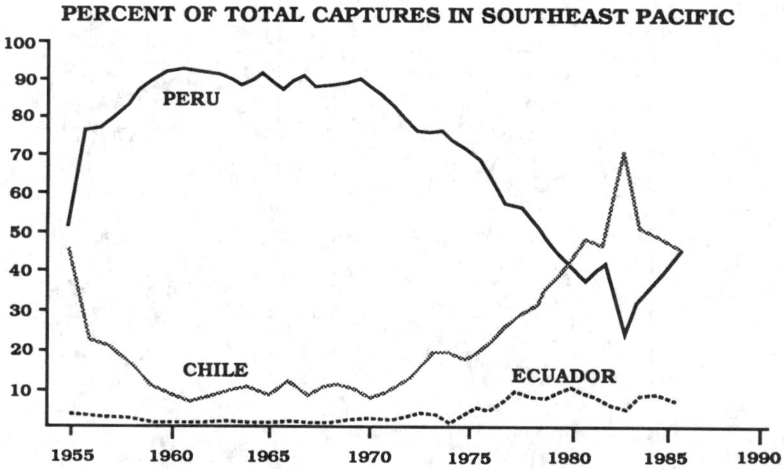

**PERCENT OF TOTAL CAPTURES IN SOUTHEAST PACIFIC**

Fig. 19.10  The proportion of catch of coastal pelagic fishes by Ecuador, Peru, and Chile has shifted over the period from the mid 1950s to present, but the major shift of record related to species composition, distributions and biomass changes began about 1968–69, reaching greatest extent by 1983, and shifting back toward the equator since. This indicates a general north-south shift of the habitat, and the fish biomass in the eastern Pacific Ocean, reflecting significant changes in the atmospheric circulation patterns. (Updated from Avaria 1985.)

regional trends in upper ocean warming that transpired in each system for that period. From Figure 19.3 it is possible to see that the general downward trend from the late 1960s to about 1983–84 reversed for the western coast of South America and heralded the significant changes in the coastal ecology that are underway.

It is clear from Bakun's (1990) discussion that he remains skeptical that the available records are unbiased physical evidence of global warming trends, particularly given the short time periods for which the instrumental records have been collected, and the changes in technique that have occurred throughout the period. Other issues emerge from close examination of the data set, beyond the questions about instrumental and observational biases, and the consequent changes in upwelling intensity due to the processes outlined by Bakun.

For example, while the long-term wind-stress trend analyses for each of Bakun's five upwelling regimes (California, the Iberian Peninsula, Morocco, Peru summer, and Peru winter) each result in similar positive forcing trends for the past 50 or so years, it is also intriguing that commencing at around 1968–1972 each regime experienced a rather rapid and dramatic excursion about the long-term mean, followed by a step well above the mean. Then, with the exception of Morocco, the wind stress for each regime trends downward, with increasing amplitude oscillations. By contrast, there was ocean warming as well as declining upwelling frequency and intensity over most of this period in each area, even though this occurred at different rates in each area.

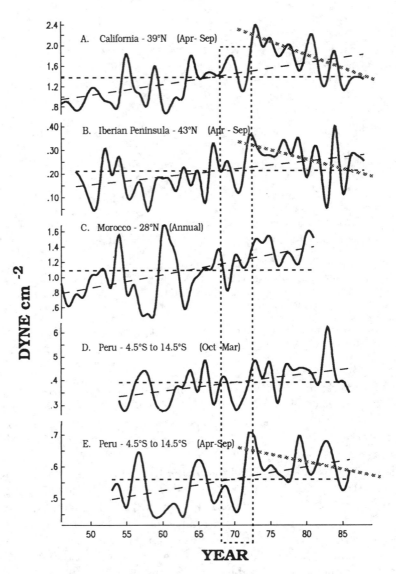

Fig. 19.11   Yearly averages of monthly estimates of along shore wind stress in five upwelling regimes (after Bakun 1990) are compared to show their relative coherences. The period from 1968–1972, outlined by the dotted box, indicates that each regime was subject to a large excursion of at least 0.2 dynes $cm^{-2}$, and about 1.2 dynes $cm^{-2}$ for the California example.

There has been a series of indications that the upwelling intensity and the ecological changes associated with a suite of environmental changes that occurred during the 1968–1972 period were concurrent with, if not the cause of, these dramatic changes. For example, Loeb and Rojas (1988) also described this period as the start in declines of coastal upwelling fish populations off northern and central Chile, as well as the incursion of the offshore oceanic species as the coastal upwelling declined. Entire marine populations, not only commercially exploited pelagic fishes, failed; other populations blossomed within each of the study areas defined by Bakun during the last decades of these series (Sharp and Csirke 1984; Sharp 1987, 1988). Clearly further study is needed to clarify some of the linkages and to improve our understanding of natural physical processes and their ecological consequences.

### *Transfers of climate consequences from the tropics: higher latitude response*

The southwestern United States is a complex, remotely forced hydrological regime with strongly seasonal patterns of climatic forcing (Cayan and Peterson 1989). The late spring shift of the Bermuda high westward from the offshore Atlantic Ocean onto the central plains, provides the low-level moisture necessary for the development of summer thunderstorms that range initially from the southeastern states, and later in summer to Colorado where the Rocky Mountains act as a barrier to farther western extension. In late June or early July, the terrestrial heat balance over the Sonora desert and Baja California generates a true monsoon flow of moist oceanic air from the eastern Pacific and the lower Gulf of California, which also spreads northeastward to include the southern Rocky Mountains and much of the southwestern desert region. The dominance of these two moist air masses shifts from year to year in response to the changes in sea surface temperature (SST) in the Gulf of Mexico and in the eastern Pacific Ocean, and to the upper atmospheric dynamics. The precipitation record exhibits a strong biennial signature, perhaps tied to the quasi-biennial oscillation (QBO) in the tropical stratosphere.

In El Niño years the late summer cyclones that either cross into the eastern Pacific from the Gulf of Mexico, or form off the west coast due to the increased ocean heat content and resulting latent heat fluxes are often steered northeastward, off their normal northwestern course, such that they recurve over Baja California, or across mainland Mexico from about Acapulco, northward. These storms bring deluges and major floods to this region. By contrast during times of cool, lesser heat content periods, in the eastern Pacific SST, when the Caribbean and Gulf of Mexico tend to be warmer than usual, tropical storms, although more abundant, tend to form farther east, and make fewer crossings into the Pacific. Those that do cross into or form in the eastern Pacific tend to be steered north and west, away from the Mexico, and more toward Hawaii. Those that form in the Atlantic off Africa are sometimes guided up the eastern United States seaboard by the combined effects of the suitably warm waters, and favorable upper-level wind patterns (Gray 1990).

These differences in tropical storm activity have important local ecological consequences. Domes form in the thermocline since the upper ocean loses heat, introducing nutrients and organisms into the photic zone, leading to local production, which propagates through the food web. Cyclones and deep convection processes are equivalent to 'plows' in terrestrial contexts, turning the upper ocean, inducing ecological production by bringing nutrients and organisms into the photic zone.

Cayan and Webb (1992, this volume; and Fig. 19.12) and colleagues at the USGS laboratory in Tucson, Arizona, have amassed time series from drainage basins throughout the southwestern United States and Sonora desert regions. Highest river flows have been associated with two distinctively different processes: (1) periods when cool, 'La Niña' phases of the ENSO cycle dominate the eastern tropical Pacific region, which tends to bring large low pressure systems into the Pacific from the Caribbean; and (2) those when tropical cyclones originating from within the eastern Tropical Pacific are drawn northeastward, rather than following their more usual northwestward pathways, during warm 'El Niño' phases of the ENSO cycle. Fish and invertebrates, from lakes, rivers, and coastal wetlands are affected by hydrological flow patterns and associated local temperature patterns. It is not at all surprising that fisheries throughout the Gulf of Mexico, the Gulf of California, and the western watersheds of North America fluctuate with both long-and short-term hydrologic patterns.

Hollowed et al. (1987) have performed studies to compare eastern Pacific Ocean regional responses by assemblages of species, in order to find which groups are associated with which climate regimes. Bakun (1988) and colleagues

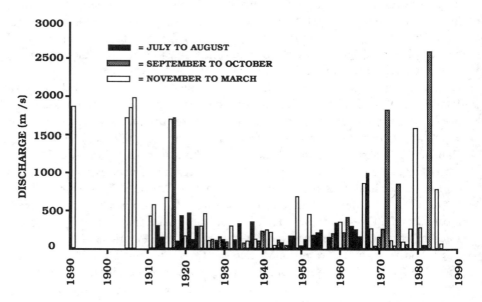

Fig. 19.12  Webb and colleagues (Hjalmarson, 1990, Webb and Betancourt, 1990) have compiled records from several drainage basins, and collated the peak discharge rates by season and year so that the various components of the seasonal cycle and atmospheric states might be inferred. There is a low-frequency pattern of extreme events within these records, which may be linked with the north-south oscillations of the ITCZ.

(Bakun and Parrish 1990a, b) have been following comparative approaches to studies of climate and fisheries as part of a decade-long effort to formalize such methods. Long-term patterns and trends such as are evident from records like those illustrated in Fig. 19.12 suggest that the problems associated with traditional, one hypothesis, one experiment methods may not be fruitful in either climate research, or climate consequences studies such as applied fisheries oceanography research. The problem lies with moving baselines. Characteristically, short-term (using only decade-long data sets) statistical studies of natural ecosystems continue to fail to provide consistent, credible forecasts, simply because environmentally dependent ecosystems change response patterns with changing environmental regimes. The fish and environmental records shown here suggest that several climatic clocks are ticking, and each nurtures different ecological opportunities within the changing environment.

### El Niño warming as a perturbation of marine ecosystems

Extinctions of kelp stands that were harvested for food, providing economic support for entire villages and gross perturbations of coastal populations of neritic (shelf) and intertidal organisms from the southern Galapagos Islands to central Chile (IFOP 1985) got little coverage in the world's newspapers. In contrast, there were frequent articles on such topics as the abandonment of central Pacific nesting islands for the 1982–83 breeding season by seabirds. Adult sea birds are well adapted to such short-term events, and when they found conditions to be unsuitable for raising their young, they simply abandoned the sites. Most sea birds have long lives in which to resolve their reproductive requirements in the face of climatic events. Humans have long lives too, but disruptions of resources for entire communities are catastrophic economic events.

The recent (1989–90) Pacific Ocean conditions provide another lesson about ENSO and longer term changes. The upper ocean heat loss in the Pacific Ocean since the 1968–1983 warming phase ended has been remarkable, punctuated by moderate warming during the 1986–87 and 1989–90 warm El Niño event. This recent 'El Niño' was measured and monitored throughout its development by (among many others) ocean modelers at the Navy Oceanography and Atmospheric Research Laboratories (NOARL) in Mississippi and Monterey. Their near real-time models of sea-level heights and analyses of differences between 1988–89 and 1989–90 observations showed the clear progression of an 'El Niño' Kelvin wave from west to east across the Pacific and then a poleward coastal progression of this Kelvin wave into the higher latitude coastal areas off the Americas. The process began in late fall of 1989 and was mostly over by mid-March of 1990.

The baseline ocean surface temperatures against which this series of events occurred were low relative to the previous decade's. Meanwhile, during the early fall 1989 ground fish survey cruises off central California found that the *Sebastes* (rockfishes) species had depleted fat stores. They also had low fertility or had resorbed their gametes (Whipple 1990). Depleted fat stores and lowered fecundities are classic thermal stress symptoms in aquatic species. Fish respond to habitat temperature shifts about their present thermal adaptation temperatures. These temperatures change continuously, usually associated with seasonal temperature patterns, and at any period a fish may have a thermal set point as much as 3°C or more different from another time of year, or even another location within the species' range.

Within the same region, the Point Reyes Bird Observatory staff had recorded the failure of sea birds nesting at the Farallon Islands and elsewhere along the central California coast (Ainley 1990). In early February 1990 many birds were found along the seashore that apparently had died of starvation. These birds feed heavily on the young, surface-swimming stages of some of the rockfish species. Fish which, along with the lowered abundances due to physiological debilities experienced by the adult rockfish, were simply not found at the surface due to onshore movement of warmer, oceanic waters. This onshore motion caps the usual coastal ocean regime, and precludes the smaller food fishes from being at the surface for birds to feed on.

Later in February, several tons of dead anchoveta were found along the northern shores of Peru, also causing some stir in the news about El Niño. The deaths were associated with lowered oxygen levels in the nearshore surface waters off northern Peru. Oxygen depletion in such environments is due to increased physiological rates associated with increased environmental temperatures, and increased night-time respiration, hence use of oxygen by phytoplankton. This is one of the nearly immediate results of the warm tropical surface water incursions associated with El Niño in the region. In extreme anoxic events, sulfides are produced, and mass fish die-offs are observed. The continuous monitoring of physiological properties, and results from extreme events such as described above, are justifiable means for making inferences about ENSO stages. Although oceanographers would like to have continuous ocean monitoring capabilities at many sensitive sites, fish have little recourse. They are subject to and responsive to daily events in the ocean, and provide a unique array of environmental indicators, particularly of extreme events. It is important that these information sources be integrated into the existing ocean monitoring system, as a source of information that can be used to supplement the available physical sensing systems.

### Insights into the ocean dynamics through ecological consequences

The implications of the many processes and societal effects that are linked to the ENSO events have yet to be fully appreciated. Certainly, the longer-term aspects of the oceanic heating process in which the 1982–83 El Niño happened to be embedded (Dorsey et al. 1991) have yet to be fully understood. For example, the principal eastern Pacific warm event of 1982 occurred against the backdrop of a continuous tropical Pacific warming period from about the late 1960s, which was manifested primarily by a poleward expansion of the southern subtropics. In 1982, when the equatorial Pacific easterly winds collapsed and released a series of Kelvin waves from the western Pacific, it was superimposed on an already very warm eastern Pacific Ocean, with a deeper than usual upper mixed layer.

The physiological set-temperatures of most marine organisms living in the upper subtropical eastern Pacific would have been much higher due the several years incremental warming in the region. Many of the species, particularly those with sessile stages, were devastated by the additional physiological stress imposed by the 1982–83 warm event. On the other hand, some residual tropical populations living in refuges along the South American coast since previous warm period(s), bloomed in response to the extreme warming (Arntz 1984; Avaria 1985; Illanes 1985).

Longer-term warming and cooling of the upper ocean, reflecting changes in heat content, are well known (cf. Dorsey et al. 1991), although SST alone may not reflect these changes very well. The subsurface thermal structure is extremely dynamic, and is often ignored in favor of assumptions that the upper ocean heat content is directly reflected by SST alone. The upper ocean heat content is orders of magnitude greater than that of the lower atmosphere with which it interacts, and SST, therefore, is a complicated variable, resulting from interactive upper ocean and atmospheric dynamics. Reliance on SST alone as a direct measure of thermal interactions with the atmosphere can be quite misleading. It is the depth-integrated heat variations that is needed to account for the long-term changes in the coupled ocean-atmosphere system. It is the volume of water with specific physiological properties (temperature, oxygen salinity, etc.) that aquatic species respond to, not merely SST, and exclusive use of this parameter has led to many prediction failures.

A major complication for ocean ecologists is the fact that most of the regions that have high biological activity are characterized by steep thermal gradients and a highly interactive ocean/atmosphere interface, e.g., eastern boundary currents, western boundary currents, and transition zones. The resulting cloud cover poses several problems too. For example, clouds limit the utility of irradiance-based remote-sensing technologies for monitoring upper ocean dynamics. Most of the important environmental processes related to nonboreal fisheries usually involve structures and changes in structure that occur in specific upper ocean environments.

It is important that the underlying concepts and requirements for accomplishing credible cause and effect research in fisheries contexts be clear in everyone's mind. It is useful to think of marine fishes, for example, as inhabiting a porous bounded volume with specific temperature and oxygen properties. They must constantly search this volume for overlapping regions where their food organisms' habitats intersect their own habitat. Most fish species avoid the upper few meters of the ocean, simply because predators attack them from both below and from the air. Fish schools that congregate at the surface are often held there by either low oxygen levels at shallow depths or by the concentration of their food at the surface, for whatever reasons. The two prevalent habitat boundary conditions are temperature, and oxygen, and salinity plays a role in specific environments, e.g., estuaries and lagoons. Light is another important variable. The habitat is best considered to be a continuously changing bounded volume, defined by several somewhat independent physiological gradients.

Principal perturbations of these volumes are the seasonal current patterns, and physical events on all time scales, from daily light levels, tidal effects, and variously forced mixing periods, through the seasons, and on to the ENSO time scales and beyond. Epochal shifts in habitat location and patterns are common, and the behaviors of specific populations and their physiological adaptations tell us exactly what sorts of climatic perturbations that they experience and survive. Sardine habitat expansion occurred off South America from 1969 to 1989. Kondo's (1980; Fig. 19.13) updated observations about the far eastern sardine exemplify the extremes of these changes. Great ongoing behavioral and ecological changes imply another shift in these systems is well underway (for recent reviews, see Kawasaki et al. 1991).

While dynamic habitat shifts over long time periods are dramatic, and important, the perturbations caused by either warm or cool phases of ENSO-scale events are also important, and affect many species, on all time scales. The best examples are from the eastern boundary currents, particularly off South America and the Californias. In the California Current, prominent upwelling regimes off Baja California provide a steady source of potential colonizers. Also from the south, the subtropical convergence, which can range seasonally from well south of Baja California to mid-California over the long term, affects the southern California Bight on different time scales during longterm epochs of equatorial warming and cooling, and on the shorter term during ENSO warm events. These epochs clearly enhance or suppress the success of *Sardinops*, and many other species, and thereby the sediment deposition records provide insights into relevant climatic regimes for each of the stages that can be discriminated.

From the north, the California Current flow and along-shore wind and upwelling regime characterizes this and other eastern boundary current upwelling regimes, and supports the associated species, i.e., *Engraulis, Sardinops, Sarda, Scomber,* and *Merluccius* spp. (Parrish et al. 1981). From the west, the open oceanic habitat can dominate the coastal regimes during periods of lower alongshore winds (which happened off central California for several months, October 1989 to summer 1990). During this recent period unusual sightings were reported and strandings of oceanic species such as Risso's dolphin and leatherback turtles also occurred. Also, the previously mentioned suppression of usual forage species for bird colonies in the Farallon Islands resulted in population stresses and lowered nesting success for these and other species (Ainley 1990). In Monterey Bay, fishermen have recently reported enormous (in their limited experience) schools of mackerel and sardines, as well as several larger predator species from the adjacent subtropical and oceanic habitats usually found to the south and well offshore to the west.

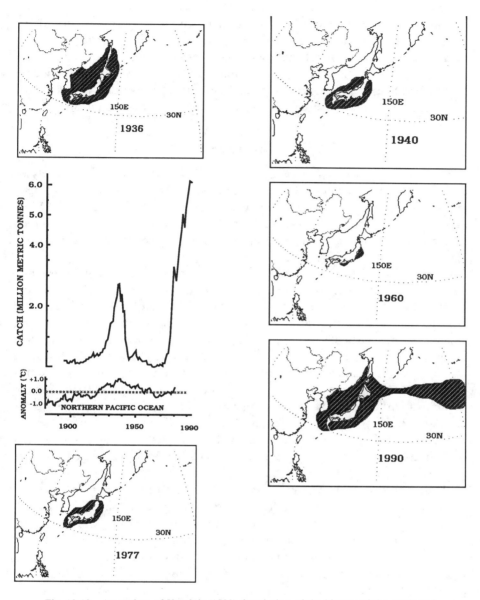

Fig. 19.13 An update of Kondo's (1980) description of the history of the catch and distribution of the far east sardine, *Sardinops melanosticta*, in and around the Sea of Japan and the western Pacific Ocean. Note that the shaded areas represent known distributions from fisheries records.

We need to recognize that the rises and falls of the great sardine fisheries of the world are simply responses – first by the sardines through reproductive successes and colonization of new habitat; then by fishermen, as they harvest those species with the highest yield in value per cost to catch them – all within a context of continuous climatic variations and system-wide trends. The catches of the world's sardines may appear an unlikely signal to watch, or to make predictions from. I would also point out that there is no reason to make wagers from these data, but I would be loath to ignore the emerging pattern, given the knowledge that the sardines of the eastern boundary currents of the world, as well as off Japan are particularly responsive to epochal environmental opportunities.

## Conclusions

Critical questions needing to be answered, deal with the impacts on society of climate change to be answers that define future expectations in globally averaged terms, such as 'an average annual global surface temperature rise of 3–5°C' is uninterpretable in societal terms. Such a statement is not useful in any but purely theoretical terms since in almost every practical application climate change is measured in terms of changes in seasonal patterns.

Avoidance of efforts to place climate into a social context is counterproductive, yet it is a characteristic human defensive behavior. Empirical studies and improved understanding about the cascade of climate-driven processes are necessary to validate and verify the available tool-kit for defining climate and climate trends, particularly the projections of likely consequences of climate change. In spite of the fact that there is a perception that somehow human activities have become a dominant force, in practical terms, it makes little difference what the source of climate variation might be. The long-term problem needing attention now is to develop coping strategies in the face of ever-increasing human population demands.

There are many potential benefits to the immediate implementation of a coherent, *in situ* ocean observing system, including biological indicators, that can provide needed insights into global climate. For example, studies of laminated or stratified sediments, collected along pole to pole transects, particularly fish debris and other indicator species from ocean margin, along with this century's catch information, could provide many useful insights into characteristics of the ocean climate, including ENSO-related processes. Sediments from meromictic lakes from areas within climate transition zones can provide similar information about terrestrial climates and hydrology. This is so, because it is possible to track temperature-related and water mass specific zoogeographic provinces and their expansion and contraction rates in response to climate forcing from such sediments. Records such as described in Figures 19.1, 19.2, 19.10, and 19.13 provide insights into both rates, frequencies and general magnitudes of important climate/ocean climate changes and events. Thorough, global scale studies of other such records are needed.

By tying together *in situ* ocean measurements, satellite observations, and studies of ocean fisheries, sediments, and other sources with similar studies of tropical corals, tree rings, meromictic lakes, and hydrology a great inferential data series and climate monitoring network could evolve. Patterns of coherence of climate consequences (or responses) might then be determined. This would be of great practical application as the patterns of seasonal hydrology and ocean dynamics might be traceable from local through regional scales, and their atmospheric analogs can be inferred, as has been done for some terrestrial systems (e.g., Enzel et al. 1989; Gray 1990).

Lastly, we need to communicate better our perceptions in terms that can be applied to real world decisions. Dealing with societal issues, particularly with the consequences of the climate/environment processes that each of us studies, in relevant and timely fashion, is a responsibility more of us should undertake. ENSO warming events provide frequent reminders, to both scientists and laymen, about important societal impacts of climate that have relevance to biophysical systems on a broad range of time and space scales.

## References

AINLEY, D.G., 1990: Investigations of ecological interactions in the Gulf of the 'Farallones: the unprecedented groundwork and the tantalizing potential. *EOS, Transactions American Geophysical Union*. 71(43): 1356.

ANDERSON, R.Y., 1992: Long-term changes in the frequency of occurrence of El Niño events. *In* Diaz, H.F. and Markgraf, V. (eds.), *El Niño: Historical and Paleoclimatic Aspects of the Southern Oscillation*. Cambridge: Cambridge University Press, 193–200.

ANDERSON, R.Y., SOUTAR, A., and JOHNSON, T.C., 1992: Long-term Changes in El Niño/Southern Oscillation: evidence from marine and lacustrine sediments. *In* DIAZ, H.F. and MARKGRAF, V. (eds.), *El Niño: Historical and Paleoclimatic Aspects of the Southern Oscillation*. Cambridge: Cambridge University Press, 419–433.

ARNTZ, W.E., 1984: El Niño and Peru: positive aspects. *Oceanus*, 27: 36–39.

AVARIA, S., 1985: Efectos de El Niño en las pesquerias del Pacifico Sureste. *Investigaciones Pesquera* (Chile), 32: 101–116.

BAKUN, A., 1990: Global climate change and intensification of coastal upwelling. *Science*, 247: 198–201.

BAKUN, A., BEYER, J., PAULY, D., POPE, J.G., and SHARP, G.D., 1982: Ocean sciences in support of living marine resources: a report. *Canadian Journal of Fisheries and Aquatic Science*, 39: 1059–1070.

BAKUN, A. and PARRISH, R.H., 1990: Comparative studies of coastal pelagic fish reproductive habits: the Brazilian sardine (*Sardinella aurita*). *Journal de le Conseil international pour le Exploracion du Mer*, 46: 269–283.

BAKUN, A. and PARRISH, R.H., 1991: Comparative studies of coastal pelagic fish reproductive habits – a subtropical western ocean boundary example: the anchovy (*Engraulis anchoita*) of the southwestern Pacific. *ICES Journal of Marine Science*, 48: 343–361.

BAUMGARTNER, T.R., FERREIRA-BARTRINA, V., SCHRADER, H., and SOUTAR, A., 1985: A 20 year (varve record of siliceous phytoplankton variability in the Gulf of California. *Marine Geology*, 64: 825–848.

BAUMGARTNER, T.R., MICHAELSEN, J., THOMPSON, L.G., SHEN, G.T., SOUTAR, A., and CASEY, R.E., 1989: The recording of interannual climatic change by highresolution natural systems: tree-rings, coral bands, glacial ice layers, and marine varves. *In*

D. Peterson, (ed.), *Aspects of Climate Variability in the Pacific and Western Americas.* Geophysical Monograph 55. Washington D.C.: American Geophysical Union, 1–15.

BELEVZE, H. and ERZINI, K., 1984: The influence of hydroclimatic 2 factors on the 'availability of the sardine (*Sardina pilchardus*, Walbaum). *In* Sharp, G. D. and Csirke, J. (eds.), *Proceedings of the Expert Consultation to Examine the Changes in Abundance and Species Composition of Nerittc Fish Resources, San Jose, Costa Rica, 18–29 April 1983.* FAO Fish Rep. Ser. 291, Vol. 2, 285–327.

BRAUDEL, F., 1981: *Civilization and Capitalism, 15th–18th Century: Volume 1, The Structures of Everyday Life.* New York: Harper and Row. 623 pp.

BRAUDEL, F., 1984: *Civilization and Capitalism, 15th–18th Century: Volume 3, The Perspective of the World.* New York: Harper and Row. 699 pp.

BRAUDEL, F., 1990: *The Identity of France. Volume 1, History and Environment.* New York: Harper and Row. 423 pp.

CAYAN, D. and WEBB, R., 1992: El Niño/Southern Oscillation and its relationship to terrestrial flood frequency. *In* Diaz, H. F. and Markgraf, V. (eds.), *El Niño: Historical and Paleoclimatic Aspects of the Southern Oscillation.* Cambridge: Cambridge University Press, 29–68.

CAYAN, D. R. and PETERSON, D., 1989: The influence of North Pacific atmospheric circulation on streamflow in the west. *In* Peterson, D. H., (ed.), *Aspects of Climate Variability in the Pacific and Western Americas.* Geophysical Union Monograph 55. Washington, D.C.: American Geophysical Union, 325–398.

COLEBROOK, J. M., 1991: Continuous plankton records: from seasons to decades in the 'plankton of the North-east Atlantic. *In* Kawasaki, T., Tanaka, S., Toba, Y., and Taniguchi, A., (eds.), *Long-term Variability in Pelagic Fish Populations and Their Environment. Proceedings of the International Symposium, Sendai, Japan, 14–18 November 1989.* Tokyo, London: Pergamon Press, 29–46.

CURY, P. and ROY, C., 1989: Optimal environmental window and pelagic fish recruitment success in upwelling areas. *Canadian Journal of Fisheries and Aquatic Science*, 46: 60-680.

CRAWFORD, R. J. M., UNDERHILL, L. G., SHANNON, L. V., LLUCH BELDA, D., SIEGFRIED, W. R., and VILLACASTIN-HERERO, C. A, 1991: An empirical investigation of trans-oceanic linkages between areas of high abundance of sardine. *In* Kawasaki, T., Tanaka, S., Toba, Y., and Taniguchi, A., (eds.), *Long-term Variability in Pelagic Fish Populations and Their Environment. Proceedings of the International Symposium, Sendai, Japan, 14–18 November 1989.* Tokyo, London: Pergamon Press, 319–332.

CSIRKE, J., MENDO, J., ZUZUNAGA, J., CARDENAS, G., MUCK, P., and CHAVEZ, F. 1989: 'Informe de taller de trabajo sobre modelos para la prediccion del rendimiento en el sistema de afloramiento Peruano. Imarpe, Callao, 24–28 Agosto 1987. *In The Peruvian Upwelling Ecosystem: Dynamics and Interactions.* Instituto del Mar del Peru (IMARPE), Callao, Peru; Deutsche Gesellschaft for Technische Zusammenarbeit (GTZ), GmbH, Eschbom, Federal Republic of Germany; and International Center for Living Aquatic Resources Management, Manila, Philippines. ICLARM Conference Proceedings 18, 1–13.

DORSEY, K. T., SHARP, G. D., and MACLAIN, D. R., 1991: Climate and Fisheries: cause and effect, ENSO and longer cycles. *In* Kawasaki, T., Tanaka, S., Toba, Y., and Taniguchi, A., (eds.), *Longterm Variability of Pelagic Fish Populations and Their Environment. Proceedings of the International Symposium, Sendai, Japan, 14–18 November 1989.* Tokyo, London: Pergamon Press, 387–390.

DEVRIES, T.J. and PEARCY, W.G., 1982: Fish debris in sediments of the upwelling zone off central Peru: a late Quaternary record. *Deep Sea Research*, 28(1A): 87–109.

ENZEL, Y., CAYAN, D.R., ANDERSON, R.Y., and WELLS, S.G., 1989: Atmospheric circulation during Holocene lake stands in the Mojave Desert: evidence of regional climatic change. *Nature*, 341: 44–47.

FLEGAL, A.R., 1990: Lead isotopic disequilibria among planktonic assemblages and surface waters within a northern California upwelling region. *EOS, Transactions American Geophysical Union*, 71: 1350.

FREON, P., 1984: Production models as applied to sub-stocks depending on upwelling fluctuations. *In* Sharp, G. and Csirke, J., (eds.), *Proceedings of the Expert Consultation to Examine Changes in Abundance and Species Composition of Neritic Fish Resources. FAO Fisheries Report*, 291(3): 1047–1064.

GARCIA, S. and LERESTE, L., 1981: Life cycles, dynamics, exploitation and management of coastal penaeid shrimp stocks. *FAO Fisheries Technical Paper* 203: 215 pp. (Original in French, same ref.)

GLANTZ, M.H., KATZ, R., and KRENZ, M., 1987: *The Societal Impacts Associated with the 1982–83 Worldwide Climate Anomalies.* NCAR/ESIG, Boulder, Colorado. 105 pp.

GRAY, W.M., 1990: Strong association between west African rainfall and U.S. landfall of intense hurricanes. *Science*, 249: 1251–1256.

HOLLOWED, A.B., BAILEY, K.M., and WOOSTER, W.S., 1987: Patterns in recruitment of marine fishes in the northeast Pacific Ocean. *Biological Oceanography*, 5: 99–131.

HJALMARSON, H.W., 1990: Flood of October 1983 and history of flooding along San 'Francisco River, Clifton, Arizona. *U.S. Geological Survey Water-Resources Investigations Report*, 85–4225-B. 42 pp.

HJORT, J., 1914: The fluctuations in the great fisheries of northern Europe viewed in the light of biological research. *Rapport Pour-vous Reunion de le Conseil permanente internacionale de Exploracion de le Mer*, 20. 284 pp.

HJORT, J., 1926: Fluctuations in the year classes of important food fishes. *Journal de le Conseil permanente internacionale de Exploracion de le Mer*, 1(1): 5–38.

HUBBS, C.L., 1960: Quaternary paleoclimatology of the Pacific coast of North America. *CalCOFI Reports*, VII: 105–112.

ILLANES, J.E., AKABOSHI, S., and URIBE, E., 1985: Efectos de la temperatura en la 'reproduccion del ostion del norte *Chlamys (Argopecten) purpuratus* en la Bahia Tongoy, durante el fenomeno El Niño 1982–83. *In Taller Nacional: fenomeno El Niño 1982–83. Investigaciones Pesquera* (Chile), 32: 167–50 Instituto de Fomento Pesquero (IFOP), 1985: Taller Nacional: fenomeno El Niño 1982-83. *Investigaciones Pesquera* (Chile) 32. Numero especial. 256 pp.

KAWASAKI, T., 1984: Why do some fishes have wide fluctuations in their number? A biological basis of fluctuation from the viewpoint of evolutionary ecology. *In* Sharp, G. D. and Csirke, J., (eds.), *Proceedings of the Expert Consultation to Examine Changes in Abundance and Species Composition of Neritic Fish Resources. FAO Fisheries Report*, 291(3): 1065–1080.

KAWASAKI, T., 1991: Long-term variability in the pelagic fish populations. *In* Kawasaki, T., Tanaka, S., Toba, Y., and Taniguchi, A., (eds.), *Longterm Variability in Pelagic Fish Populations and Their Environment. Proceedings of the International Symposium, Sendai, Japan, 14–18 November 1989.* Tokyo, London: Pergamon Press, 47–59.

KAWAI, T. and ISIBASI, K., 1984: Changes in abundance and species composition of neritic pelagic fish stocks in connection with larval mortality caused by cannibalism and predatory loss by carnivorous plankton. *In* Sharp, G. D. and Csirke, J., (eds.), *Proceedings of the Expert Consultation to Examine Changes in Abundance and Species Composition of Neritic Fish Resources. FAO Fisheries Report*, 291(3): 1081–1112.

KONDO, K., 1980: The recovery of the Japanese sardine – the biology basis of stock size fluctuations. *Rapports Pour-vous Reunion. de le Conseil Internacionale de Exploracion du Mer*, 177: 332–352.

LINDQUIST, A., 1981: Herring and sprat: Fishery independent variations in abundance. *In* Sharp, G. D. and Csirke, J., (eds.), *Proceedings of the Expert Consultation to Examine Changes in Abundance and Species Composition of Neritic Fish Resources. FAO Fisheries Report*, 291(3): 813–821.

LOEB, V. and ROJAS, O., 1988: Interannual variation of ichthyo-plankton composition and abundance relations off northern Chile. *Fisheries Bulletin, U.S.*, 86(1): 1–24.

MOSER, H.G., SMITH, P.E., and EBER, L.E., 1987: Larval fish assemblages in the 'California Current, 1951–1960, a period of dynamic environmental change. *CalCOFI Reports*, XXVIII: 97–127.

MUCK, P. and SANCHEZ, G., 1987: The importance of mackerel and horse mackerel predation for the Peruvian anchoveta stock (a population and feeding model). *In* Pauly, D. and Tsukayama, I., (eds.), *The Peruvian Anchoveta and Its Upwelling Ecosystem: Three Decades of Change*. Instituto del Mar del Peru (IMARPE), Callao, Peru; Deutsche Gesellschaft für Technische Zusammenarbeit (GTZ), GmbH, Eschbom, Federal Republic of Germany; and International Center for Living Aquatic Resources Management, Manila, Philippines, *ICLARM Studies and Reviews*, 15: 276–293.

OLSON, R.J. and BOGGS, C.H., 1986: Apex predation by yellowfin tuna: Independent estimates from gastric evacuation and stomach contents, bioenergetics and caesium concentration. *Canadian Journal of Fisheries and Aquatic Science*, 43: 1760–1775.

PARRISH. R.H., NELSON, C.S., and BAKUN, A., 1981: Transport mechanisms and reproductive success of fishes in the California Current. *Biological Oceanography*, 1: 175–203.

PAULY, D. and TSUKAYAMA, I. (eds.), 1987: *The Peruvian Anchoveta and Its Upwelling Ecosystem: Three Decades of Change*. Instituto del Mar del Peru (IMARPE), Callao, Peru; Deutsche Gesellschaft for Technische Zusammenarbeit (GTZ), GmbH, Eschbom, Federal Republic of Germany; and International Center for Living Aquatic Resources Management, Manila, Philippines. *ICLARM Studies and Reviews*, 15: 351 pp.

PAULY, D., MUCK, P., MENDO, J., and TSUKAYAMA, I, 1989: *The Peruvian Upwelling Ecosystem: Dynamics and Interactions*. Instituto del Mar del Peru (IMARPE), Callao, Peru; Deutsche Gesellschaft for Technische Zusammenarbeit (GTZ), GmbH, Eschbom, Federal Republic of Germany; and International Center for Living Aquatic Resources Management, Manila, Philippines. *ICLARM Conference Proceedings* 18: 438 pp.

PEARCY, W.G., (ed.), 1984: *The Influence of Ocean Conditions on the Production of Salmonids In the North Pacific: a workshop. 8–10 November 1983, Newport, Oregon.* Oregon State University Sea Grant College Program. 327 pp.

RALSTON, S. and LENARZ, W.H., 1990: Interspecific synchrony in the dynamics of pelagic juvenile rockfish from the Gulf of the Farallones (1983–90). *EOS, Transactions American Geophysical Union*, 71(43): 1356.

RAMP, S.R., 1990: Preliminary results from the Point Sur Transect (POST) moored array: low fequency current, temperature, and conductivity variability. *EOS, Transactions American Geophysical Union*, 71(43): 1351.

ROPELEWSKI, C.F. and HALPERT, M.S., 1987: Global and regional scale precipitation patterns associated with El Niño/Southern Oscillation, *Monthly Weather Review*, 115: 1606–1626.

SCHUMPETER, J., 1988: *Essays on Entrepreneurs, Innovations, Business Cycles, and the Evolution of Capitalism.* (Richard V. Clemence, ed.). New Brunswick and Oxford: Transaction Publishers. 341 pp.

SHACKLETON, L.Y., 1987: A comparative study of fossil fish scales from three upwelling regions. *In* Payne, A.I.L., Gulard, J.A., and Brink, K.H., (eds.), *The Benguela and Comparable Ecosystems. South African Journal of Marine Science*, 5: 79–84.

SHARP, G.D., 1981: Report of the Workshop on Effects of Environmental Variation on the Survival of Larval Pelagic Fishes. *In* Sharp, G.D., (convenor, ed.), *Report and Documentation of the Workshop on the Effects of Environmental Variation on the Survival of Larval Pelagic Fishes. IOC Workshop Report Series*, No. 28. Unesco/IOC, Paris, 1–47.

SHARP, G.D., 1984: Ecological efficiency and activity metabolism. *In* Fasham, M.J.R. (ed.), *Flows of Energy and Materials in Marine Ecosystems: Theory and Practice.* NATO conference series. IV Marine sciences v.13. New York: Plenum Press, 459–474.

SHARP, G.D., 1987. Climate and Fisheries: Cause and Effect or managing the long and short of it all. *In* Payne, A.I.L., Gullard, J.A., and Brink, K.H., (eds.), *The Benguela and Comparable Ecosystems. South African Journal of Marine Science*, 5: 811–838.

SHARP, G.D., 1988: Neritic systems and fisheries: their perturbations, natural and man h-induced. *In* Postma, H. and Zijlstra, J.J., (eds.), *Ecosystems of the World. Vol 27. Continental Shelves.* Amsterdam: Elsevier, 155–202.

SHARP, G.D., 1991: Climate and Fisheries: cause and effect a system review. *In* Kawasaki, T., Tanala, S., Toba, Y., and Taniguchi, A., (eds.), *Long-term Variability of Pelagic Fish Populations and Their Environment. Proceedings of the International Symposium, Sendai, Japan, 14–18 November 1989.* Tokyo, London: Pergamon Press, 239–258.

SHARP, G.D. and CSIRKE, J., (eds.), 1984: Proceedings of the Expert Consultation to Examine the Changes in Abundance and Species Composition of Neritic Fish Resources, San Jose, Costa Rica, 18–29 April 1983. *FAO Fisheries Report Series* 291, vols. 2–3. 1294 pp.

SHARP, G.D. and VLYMEN, III, W.J., 1978: The relation between heat generation, conservation, activity and swimming energetics of tunas. *In* Sharp, G.D. and Dizon, A.E., (eds.), *The Physiological Ecology of Tunas.* San Francisco: Academic Press, 213–232.

SMETACEK, V., 1988: Plankton Characteristics. *In* Postma, H. and Zijlstra, J.J., (eds.), *Ecosystems of the World, Vol. 27, Continental Shelves.* Amsterdam: Elsevier, 93–130.

SMETACEK, V.S., 1985: Role of sinking in diatom life-history cycles: ecological, evolutionary and geological significance. *Marine Biology*, 84: 239–251.

SMITH, P.E. and MOSER, G.H., 1988: CalCOFI time series: an overview of fishes. CalCOFI Report, XXIX: 66–90.

SOUTAR, A. and CRILL, P.A., 1977: Sedimentation and climatic patterns in the Santa Barbara Basin during the 19th and 20th centuries. *Bulletin of the Geological Society of America*, 88: 1161–1172.

SOUTAR, A. and ISAACS, J.D., 1969: History of fish populations inferred from fish scales in (anaerobic sediments off California. *CalCOFI Report*, XIII: 63–70.

SOUTAR, A. and ISAACS, J.D., 1974: Abundance of pelagic fish during the 19th and 20th centuries as recorded in anaerobic sediments of the Californias. *Fisheries Bulletin, U.S.*, 72: 257–273.

SOUTHWARD, A.J., 1974a: Changes in the plankton community in the western English Channel. *Nature*, 259: 5433.

SOUTHWARD, A.J., 1974b: Long term changes in abundance of eggs Pin the Cornish pilchard (*Sardina pilchardus* Walbaum) off Plymouth. *Journal of the Marine Biological Association* UK NS, 47: 81–95.

SOUTHWARD, A.J., BOALCH, G.T., and MATTOCK, L., 1988: Fluctuations in the herring and pilchard fisheries of Devon and Cornwall linked to change in climate since the 16th century. *Journal of the Marine Biological Association*, U.K., 68: 423–445.

SOUTHWARD, A.J., BUTLER, E.I., and PENNYCUICK, L., 1975: Recent cyclic changes in climate and abundance of marine life. *Nature*, 253: 714–717.

TISCH, T.D., RAMP, S.R., and COLLINS, C.A., 1990: A study of seasonal variability of the geostrophic velocities and water mass structure off Pt. Sur, California along 36° 20' N. *EOS, Transactions American Geophysical Union*, 71(43): 1351.

TOMODA, Y. and HIRONAGA, S., 1991: The effect of geophysical phenomena on long-term variation of the catch of pelagic fish. *In* Kawasaki, T., Tanaka, S., Toba, Y., and Taniguchi, A., (eds.), *Long-term Variability in Pelagic Fish Populations and Their Environment. Proceedings of the International Symposium, Sendai, Japan, 14–18 November 1989*. Tokyo, London: Pergamon Press, 359–366.

TSUKAYAMA, I., 1989: Dynamics of fat content of Peruvian anchoveta (*Engraulis ringens*). *In* Pauly, D., Muck, P., Mendo, J., and Tsukayama, I. *The Peruvian Upwelling Ecosystem: Dynamics and Interactions*. Instituto del Mar del Peru (IMARPE), Callao, Peru; Deutsche Gesellschaft for Technische Zusammenarbeit (GTZ), GmbH, Eschbom, Federal Republic of Germany; and International Center for Living Aquatic Resources Management, Manila, Philippines. *ICLARM Conference Proceedings 18*, 125–131.

VALDIVIA, J., 1978: Biological aspects of the 1972-73 El Niño. *Rapports Pour-vou Reunion de Conseil internacionale Exploracion du Mer*, 173: 196–202.

VLYMEN, W.J., III., 1977: A mathematical model of the relationship between larval anchovy (*E. mordax*) growth, prey microdistribution and larval behavior. *Environmental Biology of Fishes*, 2: 211–217.

WARE, D.M., de MENDIOLA, B.R., and NEWHOUSE, D.S., 1981: Behavior of first-feeding Peruvian anchoveta larvae (*Eugraulis vingens* J). *In* Boletín del Instituto del Mar del Peru, Volumen Extraordinario, *Investigación Cooperativa de la Anchoveta y su Ecosistema* – ICANE –entre Peru y Canada, Callao, Peru, pp. 80–87.

WEBB, R.H. and BETANCOURT, J.L., 1990: Climate effects on flood frequency: an example from southern Arizona. *In* Betancourt, J.L. and Mackay, A.M., (eds.), *Proceedings of the Sixth Annual Pacific Climate (PACLIM) Workshop, March 5-8, 1989*. California Department of Water Resources Interagency Ecological Studies Technical Report 23, 61–66.

WELLS, L., 1990: Holocene history of the El Niño phenomenon as recorded in flood sediments. *In* Betancourt, J. L. and Mackay, A. M., (eds.), *The Proceedings of the Sixth Annual Pacific Climate (PACLIM) Workshop, March 5-8, 1989*. California Department of Water Resources Interagency Ecological Studies Program Technical Report 23, 141-144.

WHIPPLE, J. A., 1990: Fish as harbingers of environmental change. *EOS, Transactions 'American Geophysical Union*, 71(43): 1356.

WOOSTER, W. W. and FLUTARTY, D. L., (eds.), 1985: *El Niño North: Niño Effects in the Eastern Subarctic Pacific Ocean*. Seattle: Washington Sea Grant Program. 312 pp.

WYATT, T. and LARRAÑETA, M. G., (eds.), 1988: *Long Term Changes in Marine Fish Populations*. Proceedings of a Symposium in Vigo, Spain 18-21 Nov. 1986. Imprento REAL, Bayona.

YAÑEZ, E., 1991: Relationships between environmental variations and fluctuating major pelagic resources exploited in Chile (1959-1988). *In* Kawasaki, T., Tanaka, S., Toba, Y., and Taniguchi, A., (eds.), *Long-term Variability in Pelagic Fish Populations and Their Environment. Proceedings of the International Symposium, Sendai, Japan, 14-18 November 1989*. Tokyo, London: Pergamon Press, 301-310.

# 20

# Long-term changes in El Niño/ Southern Oscillation: evidence from marine and lacustrine sediments

ROGER Y. ANDERSON

*Department of Geology, University of New Mexico, Albuquerque, New Mexico 87131, U.S.A.*

ANDY SOUTAR

*Scripps Institution of Oceanography, La Jolla, California 92037, U.S.A.*

THOMAS C. JOHNSON

*Duke University Marine Laboratory, Beaufort, North Carolina 28516, U.S.A.*

## Abstract

The primary ~4-yr oscillation of the El Niño/Southern Oscillation (ENSO) appears to be recorded and preserved in varved marine sediments and also may be recorded in some lacustrine sediments. In addition, more than 400 yr of historical record, the tree-ring record, and varved marine sediments contain evidence for long-term changes in the frequency and amplitude of ENSO's primary oscillation or in effects related to ENSO. Preferred periodicity, in addition to ENSO's 3- to 7-yr cycle, appears to be expressed between 80 and 100 yr, at ~50 yr, and possibly near 22 yr in both historical data and in proxy data of climatic variability. The origin of long-term ENSO variability is not known, and cyclicity at lower-than-ENSO frequencies may be the result of direct forcing of the natural ENSO oscillator (e.g. from solar forcing), or ENSO may be a more passive agent that transfers the effects of long-term changes in climate into the sedimentary record.

## Introduction

Climatic effects from the El Niño/Southern Oscillation (ENSO) reach far beyond their place of origin in the tropical Pacific. ENSO is now recognized as a global phenomenon (Ropelewski and Halpert 1986; Barnett et al. 1991; Kiladis and Diaz 1989). A single El Niño event persists for months and may temporarily and dramatically change patterns of wind, precipitation, and temperature in some geographic regions. ENSO's strong and far-reaching influence suggests that the ENSO system (natural ENSO oscillator) also may have the capacity to affect regional and global climate more permanently. If, for example, El Niño events should become stronger and more persistent, and if the short-term effects from such events should accumulate over several decades, ENSO itself could be an important agent of long-term climatic change. The instrumental and historical record is too brief to supply evidence for cumulative climatic effects related to ENSO phenomena and such information must be sought in longer, paleoclimatic records. In this chapter we briefly review historical evidence that El Niño is variable over decades and we examine some evidence that suggests that the effects of ENSO may accumulate over time. Also, we briefly consider possible mechanisms leading to changes in ENSO and whether the El Niño phenomenon is an active or a passive agent of long-term climatic change.

## ENSO in the historic period

### *Primary period of ENSO*

El Niño events are recorded in instrumental and historical records as departures from average conditions in certain ocean-atmosphere systems related to ENSO. For example, an El Niño event in the tropical Pacific is marked by a strong positive correlation with mean annual sea-level pressure in the western Pacific (Trenberth and Shea 1987), and by weakened circulation in the California Current system along the eastern margin of the Pacific (Chelton and Davis 1982). El Niños are generally believed to be part of a natural (ENSO) climatic oscillator (Graham and White 1988) that is strongly phase-locked with the annual cycle. Average recurrence interval for El Niño events is slightly less than ~4 yr in the historical record since A.D. 1520 (Quinn et al. 1987) and El Niños are generally separated by 3 to 7 yr. Several decades may elapse between exceptionally strong events. El Niños often are followed by anti-El Niño events, or La Niñas, and a biennial tendency in ENSO indices has been documented (Meehl 1987; Rasmusson et al. 1990).

The effects of ENSO's primary ~4-yr cycle, expressed as changes associated with El Niños and La Niñas, in addition to being recorded by meteorological and oceanographic instruments, have been recognized in biologic systems, in ice cores, corals, and tree-rings (see review in Enfield 1989) and in sediments derived from limnologic and oceanographic systems. Geologic records, in addition to recording a response to ENSO's primary oscillation, are capable of recording long-term changes associated with the ENSO system and evidence exists in historical records that such long-term changes in the frequency and amplitude of ENSO do occur.

### *Long-term variations in the historical ENSO record*

A significant change in global temperature and other climate variables has taken place since 1880. Diaz and Kiladis (1992, this volume) show that ENSO is a factor in defining the regional character of these long-term changes in regional and global climate. The global temperature anomaly has been interpreted as both a long-term 'trend' toward higher regional and global temperature and as a climatic 'step function,' with the shift to a warmer regime taking place about mid-way in the ~100-yr interval of accurate records, or after ~1925 (Fig. 20.1; also see figures in Diaz and Kiladis 1992, this volume). Changes in ENSO accompany the long-term change in temperature. For example, amplitude of the Tahiti-Darwin Southern Oscillation Index (SOI) shifts toward lower values about mid-way through the period of instrumental observation and the frequency of occurrence of El Niño events has a similar trend (Fig. 20.1).

The instrumental record is too short to establish a long-term pattern of variability for ENSO but the longer historical record suggests that changes in variables related to El Niño are organized into deviations that are longer than ENSO's primary oscillation. For example, Quinn et al. (1987) presented tabular data for the recurrence of El Niño events since A.D. 1522 and noted that El Niño events are not evenly distributed over time. Enfield (1988) illustrated the changes in the Quinn et al. (1987) data and recognized a long-term trend of about 100 yr (also see Enfield 1992, this volume). Quinn (1992, this volume) extended the record of ENSO events to A.D. 622 by using Nile data, and Anderson (1992a, this volume) transformed these same data into a plot that depicts changes in the frequency of occurrence of ENSO events. A plot ot the recurrence of El Niños (or,

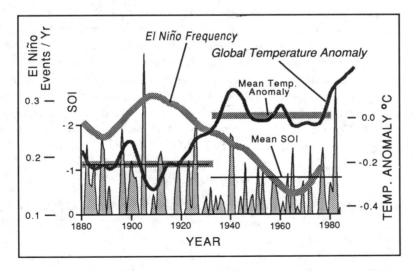

Fig. 20.1 Changes in global temperature, frequency of El Niño events, and amplitude of the Southern Oscillation since 1880. SOI illustrated is the largest negative minimum of the monthly SOI (Ropelewski and Jones 1987). Temperature anomaly from University of East Anglia Climatic Research Unit. A compilation of El Niño frequency is from data of Quinn et al. (1987) and modified by Anderson (1990; also see Anderson 1992a, this volume).

more properly, Southern Oscillation-related activity) since A.D. 622 and power spectra for the same time-series display periods of 22–24, ~50 and ~90 yr (see Figs. 9.1 and 9.2 in Anderson, 1992a, this volume; Diaz and Pulwarty 1992, this volume). Michaelsen (1989) reconstructed changes in ENSO over the past 400 yr by means of climatic indices obtained from tree-rings and identified a response in tree-rings to the primary ENSO cycle as well as a longer response with a period between 80 and 100 yr. Michaelsen also noted that frequency and amplitude of ENSO events were positively associated and Anderson (1990, 1992a, this volume) also noted a positive association between frequency of events and amplitude of the SOI, as depicted (Fig. 20.1) and between frequency of events and Quinn's (1992, this volume) ranking of events in the Nile record.

The findings noted above indicate that changes in both frequency and amplitude of El Niño events take place over time-scales of decades. In addition to changes in the frequency of occurrence of ENSO events, climatic and related effects associated with ENSO's primary oscillation appear to be organized into longer cycles that exhibit periodicity between 80 to 100 yr, and possibly at lower frequencies as well. Such long-term changes can be expected to be preserved in lacustrine and marine sediments.

## ENSO Variability in marine and lacustrine sediments

### Expression of El Niño events in varved sediments

The temporal resolution provided by varves permits identification of ENSO's primary cycle (El Niño events) in some marine and lacustrine environments. Such a connection has been recognized in varved marine sediments by Baumgartner et al. (1985), Casey et al. (1989), and Schimmelmann et al. (1990). The link between ENSO events and sedimentation is less direct in remote lacustrine environments but has been suggested for laminated sediments in East African Rift lakes (Halfman and Johnson 1988). These, and other studies for the historical period (see Enfield 1989), demonstrate that biologic and geologic systems are responsive to ENSO. However, only a few examples are presently available to define long-term variability related to ENSO.

### Santa Barbara Basin

Dunbar (1983), with the aid of A. Soutar, reconstructed a 225-yr history of oxygen isotopic variability from *Globigerina bulloides* in the Santa Barbara Basin, California (Fig. 20.2). Dunbar concluded that isotopic changes in *G. bulloides* reflected changes in surface water; possibly temperature. The isotopic record contains several ~50-yr cycles as well as a longer cycle of about double that period (Fig. 20.3). Although the two time-series appear to track quite closely (Fig. 20.2), other data suggests that intensive study will be required to demonstrate a clear relationship to El Niño. For example, a comparison of isotopic ratios in *G. bulloides* with ENSO-related changes in sea level measured

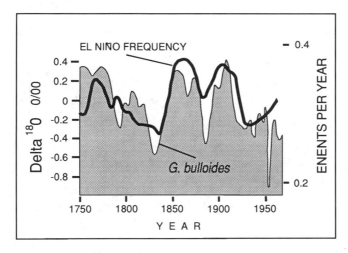

Fig. 20.2 Changes in $\delta^{18}O$ in *G. bulloides* in the Santa Barbara Basin since 1750 compared with changes in frequency of El Niño events in the historic record. (Isotopic data from Dunbar, 1983; El Niño frequency based on data of Quinn et al. 1987 and method of Anderson 1990; also see Anderson 1992a, this volume.)

at San Diego (Fig. 20.3) suggests that interannual sea-level changes are submerged in the longer-term geochemical change. Part of the difficulty in directly comparing $\delta^{18}O$ to instrumental records is an artifact of a sampling strategy for sediments that results in sampling intervals ranging from 2 to 6 yr. Thus, although low $\delta^{18}O$ values around 1960 appear to associate with the high stand of sea level that occurred in connection with the 1957/58 El Niño event, previous sea-level events of nearly the same magnitude are not represented.

Direct micropaleontological comparison to local ENSO expression indicates that there have been significant changes in both the qualitative and quantitative flux of microorganisms to the sediments. Lange et al. (1990) report an increase in the flux of warm-water affinity diatoms during the 1957/58 and 1982/83 El Niño events, but a decrease in total diatom flux during these times. Previously, Weinheimer et al. (1986) and Casey et al. (1989) had shown that radiolarian flux increased during recent El Niño events, and although warm-water affinity forms showed the most response, cold-water forms also apparently had an increased flux. Considering a somewhat longer record (Fig. 20.3), the flux of both warm-water affinity radiolarians and cold-water affinity planktonic foraminifera responded directly to local El Niño conditions in 1957/58, 1940/41, and 1930. Only the 1914/15 high sea-level stand does not appear to have an effect on either radiolarian or foraminiferal flux.

While the above direct comparison of instrumental and micropaleontological data suggests the ability of laminated sediments in the Santa Barbara Basin to record ENSO-related events, it is also evident that recording mechanisms are not entirely faithful. Furthermore, the response (higher flux) is inconsistent with the low productivity in the California Current that is believed to be characteristic of

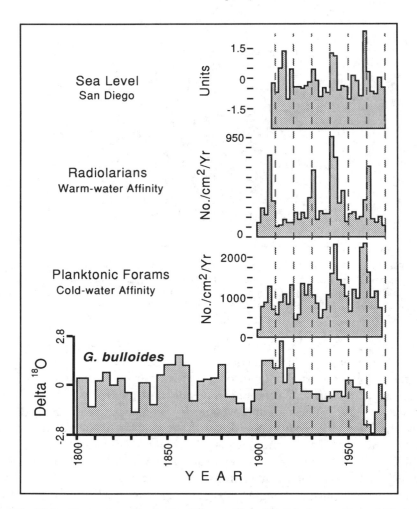

Fig. 20.3   Changes in sea level at San Diego and flux of radiolarians and foraminifera in the Santa Barbara Basin since 1900, and changes in $\delta^{18}$O in *G. bulloides* in the Santa Barbara Basin since 1800.

El Niño periods (McGowan 1985). Even so, oscillations of about 20-yr that are expressed in both the radiolarian and foraminiferal records (Fig. 20.3), as well as the longer ~50-yr, and 80- to 100-yr oscillations in $\delta^{18}$O (Fig. 20.2), appear in spectra of the frequency of El Niño events in the historical record (Anderson 1992a, this volume) and in other geologic records from the eastern Pacific.

### Varve-bioturbation cycles off California

*Decadal cycles* Anderson et al. (1987, 1989, 1990) described varved and bio-turbated late Pleistocene marine sediments deposited along the upper continental slope off the coast of northern and central California. Units of varved sediments alternate with zones of bioturbation on several spatial and time-scales. Bioturbated zones are believed to represent conditions of slower marine circulation in the California Current, warmer surface water, and increased dissolved oxygen in the water column as found during El Niño events. Varved zones represent the cooler La Niña oceanic regime, more intense upwelling and productivity, and lower concentrations of dissolved oxygen (see rationale in Anderson et al. 1990). Decimeter-scale alternations of varved and bioturbated zones, assuming that the associations with El Niño and La Niña apply for long periods (Anderson et al. 1990), reflect changes in El Niño-like and La Niña-like conditions over one to a few millennia. Millimeter to centimeter-scale alternations of varves and bioturbation reflect changes in the California Current and the ENSO system over a few years to decades. For example, millimeter-scale varves are organized into groups of about 10 varve couplets and separated by thin zones of bioturbation (Fig. 20.4) that are about the same thickness as a unit of varved sediment.

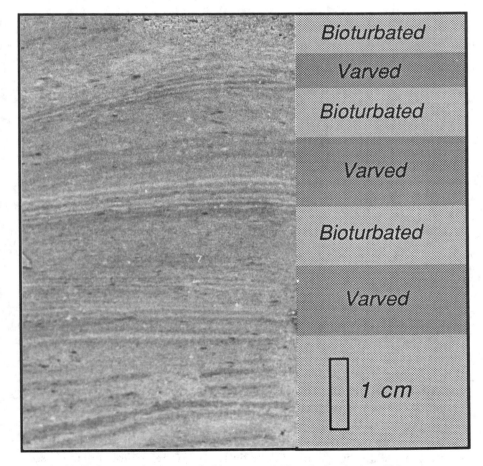

Fig. 20.4 Groups of varves and thin zones of bioturbation in late Pleistocene sediments off coast of California. Note that varve-bioturbation cycles occur in groups of about 10-varve couplets separated by a bioturbated zone of similar thickness. Core G145.

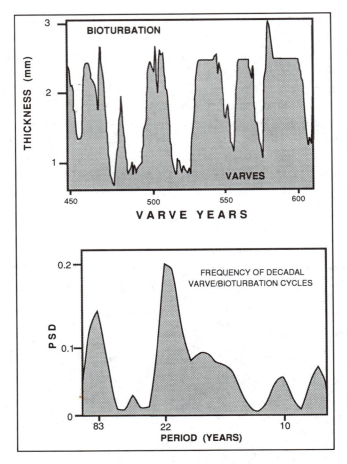

Fig. 20.5  Plot showing alternation of zones of varves and zones of bioturbation in late Pleistocene sediments off coast of California. Note that increased spectral density (lower graphs) corresponds approximately to temporal spacing of varve-bioturbation cycles in Fig. 20.4 and that a longer-period cycle also is present. Core G117.

A temporal pattern for varved and bioturbated zones has been compiled for sediments from another core from the Russian River area of northern California (Fig. 20.5). The pattern of alternation shows a tendency for groupings of about 10 varve couplets separated by an equal spacing of bioturbated sediment. Bioturbated zones have a like composition, suggesting a similar rate of accumulation (Anderson et al. 1990). The period for a complete varve-bioturbation cycle, assuming constant accumulation, is near 20 yr (Fig. 20.5). Part of a longer cycle is also present in the sequence.

A limited amount of core material has been recovered from localities off California and it is not yet possible to define ENSO-related periodicity based on the few examples available. However, the temporal spacing of the varved and bioturbated zones, if they reflect El Niño-like and La Niña-like regimes of ocean circulation, shows that conditions associated with these regimes persist for decades. At other times circulation regimes associated with ENSO effects persist for centuries.

*Millennial cycles* In the western Pacific El Niño and La Niña circulation is linked to and phase-locked with the seasonal monsoonal circulation system. High-resolution varved sediment records have yet to be described from the western Pacific, but high sediment accumulation rates in the Sulu Sea provide enough resolution to define millennial variability in the late Pleistocene and Holocene. Linsley (1990) reports a strong 2500-yr cycle in oxygen isotope ratios that also is manifested as changes in the mass accumulation rate (MAR) of carbonate. Profound, long-term changes in monsoonal circulation are implicated and such changes also may be linked to ENSO.

The decadal alternations of varved and bioturbated sediments found off the Russian River (see *Decadal cycles* above) occur within decimeter-thick zones of varved sediments that represent accumulations of several hundred to about a thousand years. These decimeter-scale cycles, as for the shorter cycles, are believed to represent conditions of increased wind stress, upwelling, and productivity (La Niña-like conditions) followed by equally long episodes of more sluggish circulation and lower productivity (El Niño-like conditions, see Anderson et al. 1989). These prolonged episodes in marine circulation mimic the circulation associated with El Niño and La Niña. The best development of these millennial-scale alternations in varved sediments has been found in the late Pleistocene between about 45,000 and 12,000 yr BP. Their period of expression in the late Pleistocene is not yet known but the few available dates for cores off northern California indicate that a complete cycle lasts for about 1000 to 3000 yr.

Well-developed, millennial varve-bioturbation cycles have not been found along the continental slope off California in the Holocene. However, a varved record for the last 8000 yr in the Santa Barbara Basin, compiled by Pisias (1978), contains at least three well-defined ∼2000-yr cycles in the winter temperature of surface waters. Santa Barbara Basin is strongly influenced by the modern ENSO system and its Holocene millennial cycle, the Holocene and late Pleistocene millennial carbonate cycles in the western Pacific (Linsley, 1990), and the the late Pleistocene cycles off the Russian River, can be defined within a framework of marine circulation changes possibly related to long-term changes in the ENSO system.

### Climatic variability in East Africa

Halfman and Johnson (1988) reported strong cyclicity in the paleoclimate record preserved in sediments of Lake Turkana in northern Kenya. These sediments are laminated but not varved. Alternating light-dark layers contain more or less calcite, respectively, and average about 10 mm in thickness. Each light-dark couplet represents, on average, about 4 yr and may reflect an ENSO signal in rainfall on the Ethiopian Plateau north of the lake. Nicholson (1989) cites several studies that have found quasi-periodic fluctuations in historical rainfall records throughout equatorial and southern Africa that may relate to the Southern Oscillation.

Time-series analysis of lamination thickness and carbonate abundance data from a 12 m piston core representing the last 4000 yr at Lake Turkana showed periodicities of 100, 165, 200, and 270 yr that were common to both the carbonate and thickness records. Lamination thickness data, because of its higher resolution, showed higher-frequency cycles as well, at 25, 31, 44, and 78 yr (Halfman and Johnson 1988). Several of the cycles in lamination thickness found in the original core subsequently have been documented in carbonate and lamination thickness records of other cores from Lake Turkana (Halfman and Hearty 1990), and in a stable isotope profile, also from Lake Turkana (Johnson et al. 1991).

The 4000-yr record from Lake Turkana is evidence for cyclic climatic change in tropical Africa that may be related to ENSO on a time-scale of decades to centuries. Additional proxy records and an improved [14]C geochronology are being assembled in order to refine information about the period and strength of the climatic oscillations and their association with ENSO.

### Expression of ENSO's effects in long-term climatic change

It remains to be determined how ENSO's primary oscillation and the climatic, biological and physical responses that result become organized into longer cycles. Other factors being equal, conditions associated with both El Niño and La Niña during ENSO cycles would cancel each other, leaving no long-term effect. However, if El Niños or La Niñas are more prevalent at some times than others, and if their relative amplitude should differentially increase or decrease along with their frequency of occurrence, then long-term changes in climatic and related variables would follow. Quinn's (1992, this volume) ranking of the events recorded in Nile River data suggests that ENSO events are stronger when more frequent, as do the results of Michaelsen (1989) and changes in the SOI (Fig.20.1; Anderson 1990). One interpretation is that ENSO's primary oscillation is modulated to alter its frequency and/or amplitude. If this is the case, it raises the question of a forcing mechanism.

ENSO has been viewed as a natural oscillator (Graham and White 1988), with its period of oscillation determined by the fixed boundary conditions of the tropical Pacific Ocean coupled with variability that is inherent in the tropical atmosphere. Models of the coupled ENSO system (e.g., Zebiak and Cane 1987; Meehl 1990) generate complex patterns of interannual variability, including changes in frequency and amplitude over several decades. If ENSO's changes reflect only the nonlinear dynamics of the ocean-atmosphere system then the long-term El Niño-like and La Niña-like associations of organisms and geochemistry found in the geologic record (Anderson et al. 1990) simply represent the 'imprinting' of conditions that accompany El Niño and La Niña, phase-locked with the annual cycle. In this explanation, stochastic changes in ENSO's primary oscillation are incidental to long-term variability.

Alternatively, changes in the frequency or amplitude of ENSO's primary oscillation may accompany decade and longer oscillations at low latitudes that originate as extratropical disturbances, heat advection in the ocean's mixed layer, biogeochemical cycling, interaction with monsoon regions, etc. (see review in Ghil et al. 1991). No forcing that is external to the ocean-atmosphere is required but ENSO, through relative changes in amplitude or frequency, may be a vehicle for expressing or transferring climatic changes that originate outside the tropical Pacific.

Indirect evidence supports an hypothesis that long-term changes in ENSO may be indirectly linked to solar-geomagnetic forcing, possibly through the effects of winds. Currently, attention is being focused on solar-geomagnetic events and changes in tropospheric wind fields at middle latitudes that appear to be related to the 11-yr solar cycle, to solar modulation of the geomagnetic field, and changes in the flux of cosmic particles (Tinsley and Deen 1991). The cosmic ray association also may include ~200-yr, Maunder-scale changes in surface winds at middle latitudes (Anderson 1992b). These examples of a solar effect on tropospheric winds are not related to the geographic region of ENSO's primary expression but so little is known about controls on long-term ENSO variability that they are included. An association also is reported for the 11-yr solar cycle and the quasi-biennial oscillation (QBO) (van Loon and Labitzke 1988), as expressed in stratospheric winds above regions where the effects of ENSO are first recognized.

Examples of solar associations rely on known solar periodicities in climatic and paleoclimatic data. For example, cycles associated with ENSO have been observed near the period of the Hale (22-yr) solar cycle and the Gleissberg (86–88 yr) solar cycle (Michaelsen 1989; Anderson 1990, 1992a, this volume; Enfield 1992, this volume). In addition, an association has been identified between the 11-yr solar cycle, stratospheric winds, and sea surface temperature (Barnett 1989) at tropical latitudes where ENSO also occurs.

Solar-geomagnetic-climatic effects appear to be most clearly recorded in cycles of ~200 and ~2400 yr (Thomson 1990), with the 86- to 88-yr Gleissberg cycle less strongly recorded (Stuiver et al. 1991). Paleoclimatic records of ENSO with sufficient length and resolution to characterize century-scale or millennial associations are rare (Anderson et al. 1990) and most evidence so far is for decadal associations. Until mechanisms are found to link solar-geomagnetic changes to winds (Dickinson 1975), and the affected wind fields to ENSO, solar periods of 11, 22, or ~90 yr in ENSO phenomena will be considered fortuitous. Additional evidence that ENSO may be related to these or longer solar periods, however, might stimulate a search for causal mechanisms.

## Age of ENSO

How long has ENSO been a feature of Pacific circulation? Evidence for a response to ENSO in tree-ring records (Michaelsen 1989) shows that El Niños have occurred for at least 400 yr. Millennial-scale oscillations in El Niño-like and La Niña-like conditions are recorded over the last 8000 yr in the Santa Barbara Basin (Pisias 1978), suggesting that a measure of ENSO's teleconnections to mid-latitudes has persisted through the Holocene. Off the coast of California, El Niño-like conditions alternate with La Niña-like conditions during the late Pleistocene ($<$ 45,000 yr, Anderson et al. 1989). Whatever the origin of ENSO's long-term variations, these examples suggest that ENSO has been with us for some time, at least as a quasi-permanent feature of the climate of the tropical Pacific. Nevertheless, McGlone et al. (1992, this volume) suggest that before about 4 KBP, the Southern Hemisphere circulation may have lacked a similar response to ENSO as observed at present. Additional research is needed to understand these differences further.

*Acknowledgment* Anderson's studies of El Niño were supported by National Science Foundation Grant ATM8707462.

## References

ANDERSON, R. Y., 1990: Solar-cycle modulation of ENSO: A possible mechanism for Pacific and global climatic change. *In* Betancourt, J. L. and Mackay, A. M. (eds.), *Proceedings of the Sixth Annual Pacific Climate (PACLIM) Workshop, March 5-9, 1989,* California Department of Water Resources Interagency Ecological Studias Program Technical Report 23, 77-81.

ANDERSON, R. Y., 1992a: Long-term changes in the trequency of occurrence of El Niño events. *In* Diaz, H. F. and Markgraf, V. (eds.), *El Niño: Historical and Paleoclimatic Aspects of the Southern Oscillation.* Cambridge: Cambridge University Press, 193-200.

ANDERSON, R. Y., 1992b: Possible connection between surface winds, solar activity and the earth's magnetic field. *Nature*, 358: 51-53.

ANDERSON, R. Y., GARDNER, J. V., and HEMPHILL-HALEY, E., 1987: Persistent late Pleistocene-Holocene upwelling and varves off the coast of California. *Quaternary Research*, 28: 307-313.

ANDERSON, R. Y., GARDNER, J. V., and HEMPHILL-HALEY, E., 1989: Variability of the late Pleistocene-Holocene oxygen-minimum zone off northern California. *In* Peterson, D. H., (ed.), *Aspects of Climate Variability in the Pacific and Western Americas.* Geophysical Monograph 55. Washington, D.C.: American Geophysical Union, 75-84.

ANDERSON, R. Y., LINSLEY, B. K., and GARDNER, J. V., 1990: Expression of seasonal and ENSO forcing in climatic variability at lower than ENSO frequencies: Evidence from marine varves off California. *In* Meyers, P. A. and Benson, L. V. (eds.), *Paleoclimates: The Record from Lakes, Ocean, and Land: Palaeogeography, Palaeoclimatology, Palaeoecology,* 78: 287-300.

BAUMGARTNER, T. R., FERREIRA-BARTINA, V., SCHRADER, H., and SOUTAR, A., 1985: A 20-year varve record of siliceous phytoplankton variability in the central Gulf of California. Marine *Geology*, 64: 113-129.

BARNETT, T. P., LATIF, M., KIRK, E., and ROECKNER, E., 1991: On ENSO physics. *Journal of Climate*, 4: 487–515.

BARNETT, T. P., 1989: Solar-ocean relation: fact or fiction? *Geophysical Research Letters*, 16(8): 803–806.

CASEY, R. E., WEINHEIMER, A. L., and NELSON, C. O., 1989: California El Niños and related changes in the California Current system from recent and fossil radiolarian records. *In* Peterson, D. H., (ed.), *Aspects of Climate Variability in the Pacific and Western Americas*. Geophysical Monograph 55. Washington, D.C.: American Geophysical Union, 85–92.

CHELTON, D. B. and DAVIS, R. E., 1982: Monthly mean sea-level variability along the west coast of North America. *Journal of Physical Oceanography*, 9: 757–784.

DIAZ, H. F. and KILADIS, G. N., 1992: Atmospheric teleconnections associated with the extreme phases of the Southern Oscillation. *In* Diaz, H. F. and Markgraf, V. (eds.), *El Niño: Historical and Paleoclimatic Aspects of the Southern Oscillation*. Cambridge: Cambridge University Press, 7–28.

DICKINSON, R. E., 1975: Solar variability and the lower troposphere. *Bulletin of the American Meteorological Society*, 56: 1240–1248.

DUNBAR, R. B., 1983: Stable isotope record of upwelling and climate from the Santa Barbara Basin, California. *In* Theide, J. and Suess, E. (eds.), *Coastal Upwelling*, New York: Plenum Press, 217–245.

ENFIELD, D. B., 1988: Is El Niño becoming more common? *Oceanography*, 1(2): 23–27.

ENFIELD, D. B., 1992: Historical and prehistorical overview of El Niño/Southern Oscillation. *In* Diaz, H. F., and Markgraf, V. (eds.). El Niño: Historical Paleoclimatic Aspects of the Southern Oscillation. Cambridge: Cambridge University Press, 95–117.

GHIL, M., KIMOTOA, M., and NEELIN, J. D., 1991: Nonlinear dynamics and predictability in the atmospheric sciences. U.S. National Report to International Union of Geodesy and Geophysics 1987–1990, American Geophysical Union, Reviews of Geophysics, Supplement, 46–55.

GRAHAM, N. E. and WHITE, W. B., 1988: The El Niño cycle: A natural oscillator of the Pacific ocean-atmosphere system. *Science*, 240: 1293–1302.

HALFMAN, J. D. and HEARTY, P., 1990: Cyclical sedimentation in Lake Turkana, Kenya. *In* Katz, B. J. (ed.) *Lacustrine Basin Exploration: Case Studies and Modern Analogs*. Tulsa, Oklahoma, American Association Petroleum Geologists, Memoir 50, 187–196.

HALFMAN, J. D. and JOHNSON, T. C., 1988: High-resolution record of cyclic climatic change during the past 4 ka from Lake Turkana, Kenya. *Geology*, 16: 496–500.

JOHNSON, T. C. and HALFMAN, J. D., and SHOWERS, W. J., 1991: Paleoclimate of the past 4000 years at Lake Turkana, Kenya based on isotopic composition of authigenic calcite. *Palaeogeography, Palaeoclimatology, Palaeoecology*, 85: 189–198.

KILADIS, G. N. and DIAZ, H. F., 1989: Global climatic anomalies associated with extremes in the Southern Oscillation. *Journal of Climate*, 2: 1069–1090.

LANGE, C. B., BURKE, S. K., and BERGER, W. H., 1990: Biological production off southern California is linked to climate change. *Climate Change*, 16: 319–329.

LINSLEY, B. K., 1990: Temporal variability of Pacific ocean climate as recorded at continental margin sites in the Sulu Sea and northern California slope. PhD dissertation, University of New Mexico 196 pp.

McGLONE, M. S., KERSHAW, A. P., and MARKGRAF, V., 1992: El Niño/Southern Oscillation climatic variability in Australasian and South American paleoenvironmental records. *In* Diaz, H. F. and Markgraf, V. (eds.), *El Niño: Historical and*

*Paleoclimatic Aspects of the Southern Oscillation*. Cambridge: Cambridge University Press, 435–462.

McGOWAN, J. A., 1985: El Niño 1983 in southern California Bight. *In* Wooster, W. S. and Fluharty, D. L. (eds.), *El Niño North – Niño Effects in the Eastern Subarctic Pacific Ocean*. University of Washigton Sea Grant Program, Seattle, 166–184.

MEEHL, G. A. 1987: The annual cycle and interannual variability in the tropical Pacific and Indian Ocean regions. *Monthly Weather Review*, 115: 27–50.

MEEHL, G. A. 1990: Seasonal cycle forcing of El Niño-Southern Oscillation in a global coupled ocean-atmosphere GCM. *Journal of Climate*, 3: 72–98.

MICHAELSEN, J., 1989: Long-period fluctuations in El Niño amplitude and frequency reconstructed from tree-rings. *In* Peterson, D. H. (ed.), *Aspects of Climate Variability in the Pacific and Western Americas* Geophysical. Monograph 55. Washington, D.C.: American Geophysical Union, 69–74.

NICHOLSON, S. E., 1989: African drought: Characteristics, causal theories, and global teleconnections. *In* Berger, A., Dickinson, R. E., and Kidson, J. W. (eds.), *Understanding Climatic Change*. Geophysical Monograph 52. Washington, D.C.: American Geophysical Union, 79–100.

PISIAS, N. G., 1978: Paleoceanography of the Santa Barbara Basin during the last 8000 years. *Quaternary Research*, 10: 366–384.

QUINN, W. H. 1992: A Study of Southern Oscillation-related climatic activity for A.D. 622–1990 incorporating Nile River flood data. *In* Diaz, H. F. and Markgraf, V. (eds.), *El Niño: Historical and Paleoclimatic Aspects of the Southern Oscillation*. Cambridge: Cambridge University Press, 119–149.

QUINN, W. H., NEAL, V. T., and ANTUNEZ de MAYOLO, S. E., 1987: El Niño occurrences over the past four and a half centuries. *Journal of Geophysical Research*, 92: 14,449–14,461.

RASMUSSON, E. M., WANG, X., and ROPELEWSKI, C. F., 1990. The biennial component of ENSO variability. *Journal of Marine Systems*, 1: 71–96.

ROPELEWSKI, C. F. and HALPERT, M. S., 1986: North American precipitation and temperature patterns associated with El Niño/Southern Oscillation (ENSO). *Monthly Weather Review*, 115: 2352–2362.

ROPELEWSKI, C. F. and JONES, P. D., 1987: An extension of the Tahiti-Darwin southern oscillation index. *Monthly Weather Review*, 115: 2161–2165.

SCHIMMELMANN, A., LANGE, C. B., MICHAELSEN, J., and BERGER, W., 1990: Climatic changes reflected in laminated Santa Barbara Basin sediments. *In* Betancourt, J. L., and MacKay, A. M. (eds.), *Proceedings of the Sixth Annual Pacific Climate (PACLIM) Workshop, March 5–9, 1989*. California Department of Water Resources Interagency Ecological Studies Program Technical Report, 23, 97–99.

STUIVER, M., BRAZIUNUS, T. F., BECKER, B., and KROMER, B., 1991: Climatic, solar, oceanic, and geomagnetic influences on late-glacial and Holocene atmospheric $^{14}C/^{12}C$ Change. *Quaternary Research*, 35: 1–24.

THOMSON, D. J. 1990: Time series analysis of Holocene climate data. *Philosophical Transactions of the Royal Society of London*, 330: 601–616.

TINSLEY, B. A. and DEEN, G. W., 1991: Apparent tropospheric response to MeV-GeV particle flux variations: a connection via electrofreezing of supercooled water in high-level clouds? *Journal of Geophysical Research*, 96: 22,283–22,296.

TRENBERTH, K. E. and SHEA, D. J., 1987: On the evolution of the Southern Oscillation. *Monthly Weather Review*, 115, 3078–3096.

VANLOON, H. and LABITZKE, K, 1988: Association of the 11-year solar cycle, the QBO, and the atmosphere. Part II: surface and 700 mb on the northern hemisphere in winter. *Journal of Climate*, 1, 905.

WEINHEIMER, A. L., CARSON, T. L., WIGLEY, C. R., and CASEY, R. E., 1986: Radiolarian responses to Recent and Neogene California El Niño and anti El Niño events. *Palaeogeography, Palaeoclimatology, Palaeoecology*, 53, 3–25.

ZEBIAK, S. E. and CANE, M., 1987: A model ENSO. *Monthly Weather Review*, 115, 2262–2278.

# 21

# El Niño/Southern Oscillation climatic variability in Australasian and South American paleoenvironmental records

M.S. McGLONE

*DSIR Land Resources, Christchurch, New Zealand*

A. P. KERSHAW

*Department of Geography and Environmental Science, Monash University, Clayton, Victoria 3168, Australia*

VERA MARKGRAF

*Institute of Arctic and Alpine Research, University of Colorado, Boulder, Colorado 80309-0450, U.S.A.*

## Abstract

The dominant effect of El Niño/Southern Oscillation (ENSO) on amphi-South Pacific climates is to increase the variability of precipitation. The characteristic climatic patterns which accompany the alternating phases of ENSO are opposed in the eastern and western Pacific. Broadly speaking, in Australasia droughts accompany El Niño events, and wetter than average conditions accompany La Niña events, whereas in western South America south of the equator, wet conditions characterize El Niño and drier conditions characterize La Niña events. New Zealand and southern South American climates are somewhat cooler during El Niño events. Paleoenvironmental records from Australasia show a drier and less variable climatic regime than that of the present during the early Holocene. From around 7000 BP effective moisture increased, and the most consistently warm and mild climates were experienced in Australia between 7000 and 4000 BP. In New Zealand, mild, reliably moist but drier climates than present prevailed until about 7000 BP, after which the climate gradually cooled and became wetter overall. In South America the early Holocene was warm and dry along the South Pacific side (central Chile), but cool and moist along the Atlantic side (southeastern Brazil and Venezuela). After 8000 BP moisture in the southern part of South America increased, both in the Pacific and Atlantic sector, whereas northern South America became drier.

From 5000 BP to 3000 BP colder climates with drier summers similar to the present became established in both Australia and New Zealand; evidence of drought, fire, erosion, and the spread of stress- and disturbance-tolerant vegetation indicate more variable climates. In South America during the last 5000 to

3000 yr conditions were overall wetter in the southern sector and drier in the northern sector of the continent, but accompanied throughout by greater variability.

The generally more stable climates of the early Holocene period and lack of comparable precipitation patterns in the amphi-South Pacific land areas point to either a much reduced amplitude of the ENSO fluctuations, or to a change in the extratropical expression of ENSO due to different climatic boundary conditions. It is unlikely that typical ENSO cycles were a major factor in Australasian and South American climates before about 7000 BP, and they only began to exercise their present strong influence beginning at 5000 and fully developed by 3000 BP.

## Introduction

The Southern Oscillation (SO) pressure anomalies in the tropical Pacific have been linked to a wide range of oceanic and climatic anomaly patterns, not only within the circum-Pacific regions but also globally (Rasmusson and Carpenter 1982; Kiladis and Diaz 1989). The primary characteristic of the SO is the high temporal and spatial variability of its related climate patterns. This is exemplified by the high-frequency alternation between its extreme warm (El Niño) versus cold (La Niña) phases.

Oceanic and atmospheric circulation patterns of the SO phases are characterized by a trans-Pacific modulation of pressure in the tropics, which are in turn related to changes in intensity of the trades and hence a shift of tropical convection. These are both related to a trans-Pacific swing of sea-surface temperature anomalies. In the eastern Pacific, the El Niño (the negative SOI–Tahiti minus Darwin Southern Oscillation Index), is characterized by anomalously warm sea surface temperatures (SSTs), reduced upwelling along the Peruvian coast, low surface pressure, anomalously southward displaced near-equatorial trough, weaker subtropical high pressure, and weaker mid-latitudes westerlies. During La Niña, the positive SOI phase, the opposite holds true (see Diaz and Kiladis 1992, this volume).

In an attempt to identify the past strength and existence of ENSO, we looked in paleoclimatic records for evidence of environmental variability, and for the existence of spatially opposing climate patterns in the amphi-South Pacific land regions and also between northern and southern South America. In this paper we compare present-day ENSO climate patterns with those reconstructed from paleoenvironmental records in Australasia and South America to determine the character and timing of past ENSO occurrences.

## ENSO in Australasia

A major component of climatic variability in the Australasian region is linked to fluctuations in the SOI. Changes in sea surface temperatures and pressure to the north of Australia, which accompany fluctuations in the SOI, have marked effects on the Australian monsoon and other aspects of the climate such as the frequency and strength of tropical cyclones (see Nicholls 1992, this volume). In the far south of Australia, alterations in the strength and direction of the westerly winds which accompany fluctuations in the SOI have a pronounced effect on weather patterns (Harris et al. 1988). Because of the northeast-southwest-trending spine of high mountains, which runs the length of the southern two-thirds of the New Zealand archipelago, and the strong mid-latitude westerly windflow over the country, SOI variations produce their effect largely through varying the strength and direction of the mean airflow.

During an El Niño (or negative SOI) positive pressure anomalies develop over most of Australia, and negative pressure anomalies develop to the south and southeast of New Zealand (Fig. 21.1). Over New Zealand surface level windflow tends more towards the southwest, and westerlies strengthen, increasing snowfall in the west (Gordon 1985). Temperatures tend to be lower over southernmost Australia and New Zealand and warmer in the north and northeast of Australia (Gordon 1985; Diaz and Kiladis 1992, this volume). In New Zealand the colder conditions brought by increased southwesterly airflow are offset to some extent in eastern districts by increased westerly windflow across the major mountain ranges producing warming föhn winds. Australian droughts generally coincide

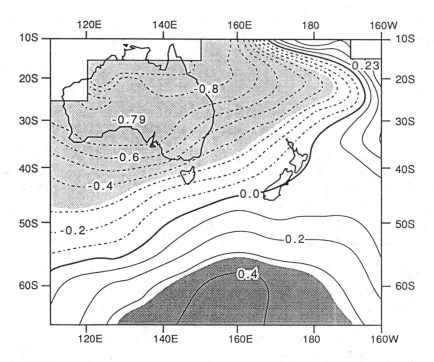

Fig. 21.1 Correlation coefficients between average mean sea-level pressure and the contemporary average (Tahiti minus Darwin) Southern Oscillation Index for September–November 1957–1984 for the Australasian region (after Gordon 1985). This pattern resembles that for the year as a whole.

with El Niño events (Fig. 21.2), a relationship which is most consistent over eastern and northern Australia (Whetton et al. 1990). Streamflow from the Darling River, which is a good index of rainfall over most of eastern Australia, shows an association between low or falling river levels and El Niño events. Rainfall increases in the south and west of the South Island of New Zealand and decreases in the east and over much of the North Island during El Niño events (Pittock and Salinger 1982). Negative SOIs also tend to bring higher pressures and more settled weather to the northern North Island. Droughts in New Zealand tend to coincide with El Niño events (Burrows and Greenland 1979). Of the 14 'strong' and 'very strong' El Niño events recorded by Quinn and Neal (1992), all but one (1911–12) were accompanied by droughts in New Zealand, and the 1877–1878, 1957–1958, 1972–1973, 1982–1983, and 1987–1988 events were accompanied by severe or very widespread droughts.

Positive SOI episodes (La Niña) reverse the pressure anomaly pattern of the El Niño. Pressure tends to be lower than normal over most of Australia and the central Tasman Sea, and higher to the south and east of New Zealand. Prolonged, heavy rainfall and consequent flooding accompanies La Niña in northern and eastern Australia. In New Zealand, La Niña reverses the El Niño windflow pattern anomaly, and there is more northerly and northeasterly windflow. As a consequence, temperatures are higher. The northeast is wetter and warmer, while the southwest is drier. La Niña events can also bring drought conditions to some areas of the country and the 1988–89 La Niña prolonged the drought of 1987–88 in the eastern South Island. La Niña can bring southerly polar outbreaks behind the low pressure systems as they cross the south of the country.

## ENSO in South America

The most recent review on climatic patterns in South America related to ENSO by Aceituno (1988) gives the following picture. The best-documented climate anomaly in South America linked to the negative (El Niño) phase of the SOI is the occurrence of high precipitation periods during the austral summer along the otherwise extremely arid coast of northern Peru and southern Ecuador. The anomalously high precipitation leads to flooding along South America's western coastal plains, south of the equator, where flood deposits have been analysed to reconstruct past El Niño events (Rollins et al. 1986; Wells 1987). During these times of coastal flooding in western South America, northeastern Brazil, the Caribbean, and the Andes north of the equator show a tendency towards conditions drier than normal. Changes in net accumulation, particulates, and stable oxygen isotope ratios, in annually laminated ice cores from the Quelccaya Ice Cap in southern Peru (Thompson et al. 1984; Thompson et al. 1992, this volume), record periods of these dry conditions in northeastern South America during the last 1500 yr that are related to El Niño events. During the positive SOI phase, La Niña, the opposite precipitation signal characterizes these regions: no

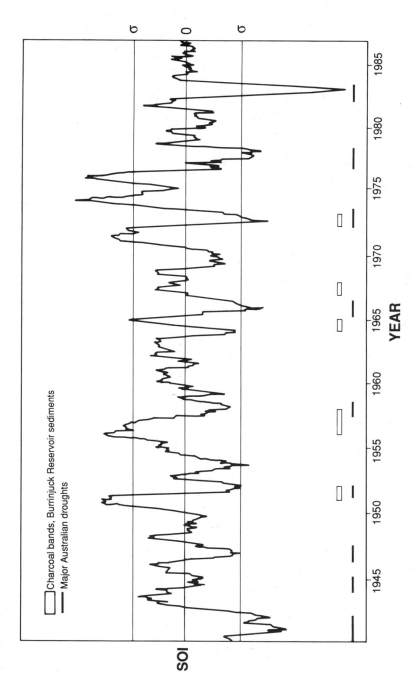

Fig. 21.2   The relationship between the Southern Oscillation Index (SOI) shown with the mean and one standard deviation (Nicholls 1991) and major east Australian droughts, as determined from documentary records (Nicholls 1991) and charcoal bands in Burrinjuck Reservoir sediments indicative of major fires since 1946 (Clark 1986).

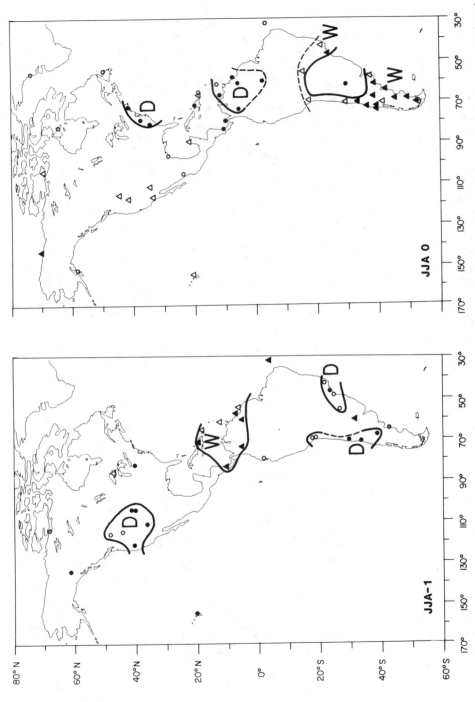

Fig. 21.3  The spatial relationship between precipitation events in the Americas during ENSO (D, drier; W, wetter). Pattern for JJA-1 is representative of La Niña conditions; JJA 0 of El Niño conditions. (After Kiladis and Diaz 1989, by permission of the American Meteorological Society.)

rainfall in coastal Peru and southern Ecuador (the normal pattern), and as a general rule, higher rainfall in northeastern South America.

Rainfall anomalies in extratropical regions of South America have also been related to the SOI (Fig. 21.3). A correlation exists between the negative SOI phase (El Niño) and high rainfall in the semi-arid region of central Chile and the subtropics of southern Brazil/northern Argentina. During La Niña events these regions are anomalously dry. This intercontinental precipitation contrast between the northern Andes (dry during El Niño) and southern Brazil/Uruguay (wet during El Niño) has been documented also by river discharge records of the Magdalena River and the Paraná River, respectively (Aceituno 1988).

## Detection of ENSO events

The sort of climatic variability that we should expect in the circum-South Pacific region would depend on exactly how the ENSO system operated. If, for instance, in the tropical Pacific the fluctuations of the ENSO system were dampened, we would expect less extreme events in more southerly regions but other than that no major climatic change. In contrast, if the ENSO system as a whole were to move to either a higher or lower average index, we might expect substantial climatic change resulting in markedly different climatic and vegetation patterns, but perhaps no overall change in variability. On the other hand, Meehl and Branstator (1992, this volume) suggest on the basis of GCM experiments that a change in the basic climate state could alter the response of climatic systems to ENSO in the extratropics, and this implies the possibility of a modified ENSO signature in the paleoclimatic record of the extratropics. If this is the case, detecting an ENSO signal in the extratropics during periods when the boundary conditions have been different is problematical. For that reason, we confine our discussion to the Holocene period during which Australasian and South American climates were similar to those of the present.

If there have been substantial changes in the structure of ENSO, the main way in which this could be detected in the amphi-South Pacific region is by changes in the variability in precipitation. If the ENSO fluctuations were dampened, we would expect a decline in the number and severity of droughts, and severe flooding episodes. This could then lead to a more zonal atmospheric circulation with reduced climate variability than at present.

However, if the SO as a whole were to become more positive, that is if the average atmospheric circulation were to resemble La Niña episodes, we might expect a substantially wetter eastern Australia, a warmer and wetter eastern and northeastern New Zealand, a drier southern South Island, and a permanently drier southwestern and wetter northeastern South America. A more negative SO would result in much drier conditions over most of eastern and northern Australia, and eastern and northeastern New Zealand, but a wetter southern South Island, northern coast of Peru, central Chile, southern Brazil and northern Argentina. While we compare modern ENSO-induced climatic patterns with reconstructed paleoclimatic patterns, and discuss this possibility, it is difficult to make a convincing case that changes in ENSO alone altered broad-scale climatic patterns.

The key to recognizing ENSO fluctuations is the establishment of the presence of relatively high variability in precipitation (Nicholls 1988). Prolonged or recurrent drought events probably leave the most easily interpretable paleoenvironmental records. How then is drought recognized in the paleoclimatic record?

Drought is often accompanied both in Australia and New Zealand by widespread fire. Fire can be detected in two main ways: charcoal and disturbance to the vegetation structure. A limited, fine-resolution study of reservoir sediments in southeastern Australia (Fig. 21.2) demonstrates some relationship between charcoal from fires and drought-related ENSO events, although not all droughts appear to have been accompanied by fires within the area (see also Swetnam and Betancourt 1992, this volume). However, other Australian paleoenvironmental studies have not permitted resolution of individual fires from a sedimentary charcoal record (Clark 1983). Inwash of charcoal is highly dependent on rainfall events, and this can confuse the relationship between fire and charcoal peaks in a sedimentary site. On the other hand, sustained increases in fire frequency can be detected over long periods as an increased influx of charcoal fragments. Charcoal in sediments and soil is difficult to date precisely because trunk and branch charcoal can have a considerable age at deposition. Small charcoal fragments absorb humic acids from the surrounding matrix and may date younger than their true age. It is not usually possible to say whether a single fire event or a series of events produced a given single layer of soil or sedimentary charcoal. Charcoal dates, under most circumstances, are therefore capable of a resolution of centuries only.

Destruction of vegetation by fire is not only registered as an increased influx of charcoal particles to soils and sediments, but also by its direct effect on vegetation community composition and hence on fossil assemblages. In Australia, increase in fire frequency may be accompanied by decreases in the more fire-sensitive taxa such as *Casuarina*, *Callitris*, and almost all rainforest plants, and increases in fire-tolerant taxa such as *Eucalyptus* and myrtaceous shrubs. In the wetter areas of Australia where the majority of vegetation histories have been constructed, fire-sensitive rainforest tends to be replaced by fire-tolerant and fire-promoting sclerophyll vegetation. In New Zealand and temperate South America there are few forest taxa which can tolerate repeated fire, and fire frequency during the Holocene seems, in most areas of New Zealand and temperate South America, to have remained at levels at which no major change occurred in the forest vegetation. However, in the semiarid regions of South America most woody taxa (trees and shrubs) are adapted to fires. Their importance in the vegetation decreases towards the temperate zones of higher latitudes and fire-adapted taxa are limited to the very driest regions, such as along the steppe/forest limit in Patagonia. There, fire can both substantially destroy the forest cover and prevent effective regeneration (Veblen and Lorenz 1987, 1988).

The ENSO phenomenon is likely to increase the severity of outbreaks of fire because of the alternation of very wet years with very dry ones. Excessive rainfall in normally dry areas encourages lush vegetation and this provides the fuel for outbreaks of fire in dry years. The La Niña events may therefore be an important part of the drought-fire cycle.

## High-resolution records

### *Australasia*

There have been no Australasian studies specifically designed to detect ENSO signals in the paleoenvironmental record. A dendrochronological study on *Nothofagus* in the Southern Alps of New Zealand (Norton et al. 1989) yielded a record of summer temperatures back to A.D. 1730. There was a concentration of variance in the reconstructed temperature record at around 3 yr, thus matching time series of variables associated with the modern ENSO record in New Zealand. The dendrochronological record therefore suggests that New Zealand temperatures have been affected by ENSO fluctuations for at least 260 yr. This is an entirely expected result because there is no reason to suspect that New Zealand climates have altered markedly over this period from the pattern experienced in the recent past, and the historical global record of ENSO shows it to have been a dominant global climate signal on time scales of months to a few years (Thompson et al. 1984; Quinn et al. 1987). In both Australia and New Zealand there are longer dendrochronological records which should be analysed to see if there is strong periodicity which could be connected to ENSO. Growth rings in coral from northern Australia, small icefields in the Southern Alps of New Zealand, and perhaps a few Australian lakes with finely laminated sediments may eventually be used to supplement and extend the dendrochronological record. However, there are no long paleoenvironmental records in Australasia which document climatic variability at anything approaching annual resolution. We therefore have to depend on low-resolution indicators of overall climatic variability.

### *South America*

Several different types of paleoenvironmental data have been analysed for the occurrence of ENSO patterns in South America. Records with annual resolution include the Quelccaya ice core, that dates back 1500 yr and tree-ring records from the central and southern Andes. In the ice-core record from southern Peru, El Niño events are characterized by decreased accumulation, high particle amounts, high conductivity, and less negative $\delta^{18}O$, because El Niño events are associated with periods of greater than normal aridity (Thompson et al. 1984; Thompson et al. 1992, this volume). Statistical analysis of the $\delta^{18}O$ data in the ice record suggests that back to A.D. 1630, ENSO explains 20 to 25% of the variance in the data (Michaelsen and Thompson 1992, this volume). The tree-ring records from the central and southern Andes on the other hand so far have not been analysed for ENSO occurrences. The dendroclimatic analysis instead related the precipitation events found in the records with a periodicity of 3 to 4 yr to anomalous positions of the subtropical high pressure zone (Villalba 1989). However, because the high pressure is linked to SO as well, the tree-ring records eventually may be interpreted also as a record of past ENSO occurrences.

### Low-resolution records: the austral Holocene climatic pattern

Records of climatic variability at an annual resolution are scarce in Australasia and South America, and none extends farther back than about 1500 yr. These tree-ring and ice-core records are consistent with the hypothesis that the ENSO system has been operating essentially in its present mode over that period. For the greater part of the Holocene we must use low-resolution records, primarily derived from pollen and charcoal analyses, to infer the presence or absence of ENSO type climatic fluctuations. However, austral climates have changed considerably in the course of the Holocene, and possible alterations to the ENSO regime are only one contributing factor to these changes (Markgraf et al. 1992). We therefore discuss the broad features of Holocene climates in this section as well as possible ENSO changes. In the concluding section we attempt to distinguish the general sweep of climatic change during the Holocene from evidence for specific ENSO effects.

### *Australia*

The majority of continuous records of paleoclimate have been inferred from pollen diagrams. Most of these diagrams are restricted to the wetter coastal margins of eastern Australia (Fig. 21.4) where a number of regional syntheses have been attempted (Fig. 21.5). Unfortunately, few data are available for the arid part of the continent where ENSO-related variability might be recorded more sensitively than in coastal areas.

Within all documented areas, there was a general climatic amelioration from the height of the last glacial period, about 17,000 BP, to the beginning of the Holocene, around 10,000 BP. This is indicated in the vegetation by the replacement of steppe in southeastern Australia and open savanna woodland in northeastern Australia by eucalypt- and particularly *Casuarina*-dominated forests and woodlands. Temperatures appear to have increased more rapidly than precipitation resulting in an effective reduction in moisture in some areas between about 15,000 and 12,000 to 11,000 BP. Precipitation continued to increase in the early Holocene and the present extent of forest vegetation was achieved by about 9000 BP. *Casuarina* rather than *Eucalyptus* continued to form the canopy in many areas and it is likely that conditions were somewhat drier and warmer than present.

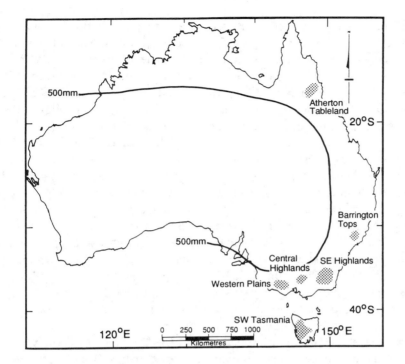

Fig. 21.4 Regional concentrations of pollen diagrams in Australia: Southwest Tasmania (Macphail 1979; Markgraf et al. 1986), Central Highlands (Mc Kenzie 1989), Southeastern Highlands (Kenyon 1989; Kershaw and Strickland 1989; Martin 1986; Pittock 1989), Barrington Tops (Dodson 1987; Dodson et al. 1986); Western Plains (D'Costa et al. 1989), Atherton Tablelands (Kershaw 1975, 1983; Kershaw and Nix 1988).

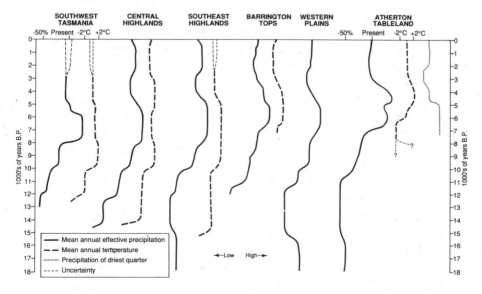

Fig. 21.5 Regional temperature and precipitation curves inferred from late Quaternary pollen records from eastern Australia.

Maximum levels of effective precipitation (precipitation – evaporation) occurred between about 8000 and 4000 BP although timing varies from place to place. There may have been a latitudinal shift in the moisture optimum from the earliest, between 8000 and 6000 BP in Tasmania, to the latest, 6000 to 4000 BP on the Atherton Tableland. During this period forest, including rainforest in northeastern Australia and Tasmania and wet sclerophyll forest dominated by eucalypts in all wetter areas, reached its maximum Holocene extent. Bioclimatic analyses, based on the known present day climatic ranges of recorded indicator taxa (Fig. 21.5), suggest that this precipitation maximum was accompanied by a reduction in temperature in southwest Tasmania (Markgraf et al. 1986) and the Central Highlands of Victoria (McKenzie 1989), and perhaps also, during the early part of the period, on the Atherton Tableland (Kershaw and Nix 1988). This temperature reduction may have been an effect of locally more constant cloud cover effectively reducing solar radiation, rather than a broad regional temperature decline, as temperatures subsequently increased again. The sharp reduction in dry season rainfall at about 5000 BP, corresponding with a late attainment of a temperature maximum suggested by bioclimatic analysis results from the Atherton Tableland, adds some support for the proposed importance of cloud cover (see Fig. 21.5).

Lowest levels of precipitation occurred in the late Holocene around 3500 to 2500 BP and may have been accompanied by lower temperatures as suggested by the return to periglacial conditions in the southeastern highlands (Costin and Polach 1971). At, or slightly after, this time there was a substantial increase in the rate of growth of many existing swamps and bogs, and initiation of others, primarily in the highlands of southeastern Australia, including Tasmania (Macphail and Hope 1985). Their preferred explanation is that evaporation rates were reduced, due to an increase in effective precipitation with increased summer rainfall. However, this pattern is not consistent through the region.

Cooler and drier conditions generally are indicated after 6000 to 4000 BP by a shift downslope in the treeline in Tasmania (Markgraf et al. 1986) and the contraction of wet sclerophyll forest and rainforest on the Australian mainland (Kershaw 1975; Dodson et al. 1986; D'Costa et al. 1989). However, the subsequent climatic changes responsible for the surge in bog growth did not significantly affect regional vegetation patterns.

There is also evidence for increased environmental variability after 5000 BP. This is expressed areally in marked differentiation of vegetation along the present west-east precipitation gradient in Tasmania (Markgraf et al. 1986), but other indicators record significant temporal variability.

On the Atherton Tableland, there are increases in pollen of secondary successional plants (especially *Macaranga* and *Mallotus*) and the relatively light-demanding conifer, *Agathis*, in pollen records. These increases, dating either from ca. 5000 BP or 3000 BP, are illustrated in the pollen diagram from Lake Euramoo (Fig. 21.6) and suggest disturbance to the forest or an opening up of the forest canopy. One explanation is increased seasonality, as implied

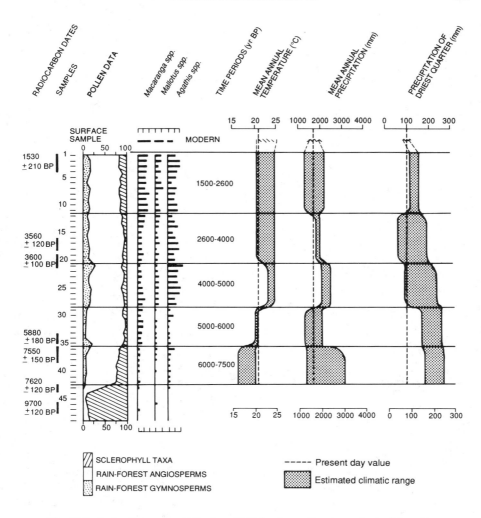

Fig. 21.6 Lake Euromoo (Atherton Tableland, northeastern Australia) pollen records and bioclimatic estimates.

previously from the bioclimatic information, but temporally coarser climatic fluctuations are another possibility. This latter suggestion may be supported by apparently interannual, finely-banded sediments deposited in nearby Lake Barrine which are strongly represented from 5000 BP (Kershaw and Walker 1988).

At a number of sites, charcoal records indicate an increase in burning after 5000 to 4500 BP. These include Lynch's Crater on the Atherton Tableland (Kershaw 1983), Rotten Swamp in the southeastern highlands (Clark 1987), and West Basin in the western plains of Victoria (see Fig. 21.7). The latter site, one of the few meromictic lakes besides Lake Barrine examined in Australia, also displays variability in the form of increased sediment banding including some fine laminations within the last 5000 yr. The increased burning is likely to have

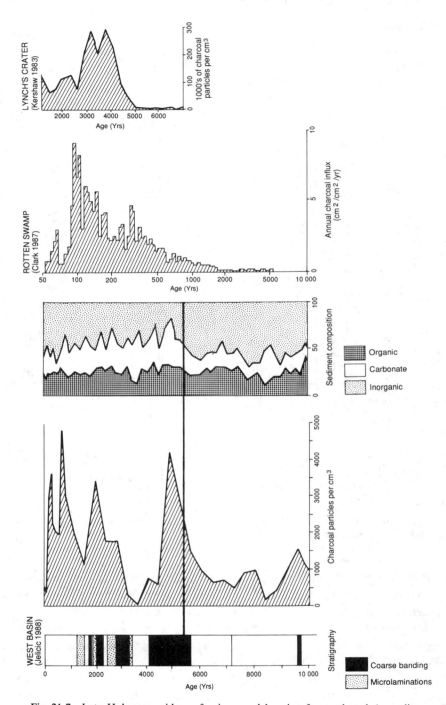

Fig. 21.7   Late Holocene evidence for increased burning from selected Australian charcoal records from Victoria (West Basin; Jelicic 1988), from the southeastern highlands (Rotten Swamp; Clark 1987), and the Atherton Tablelands (Lynch's Crater; Kershaw 1983) and for increased sedimentation variability in the West Basin record.

contributed substantially to widespread slope instability documented on the southern tablelands of New South Wales and elsewhere in the late Holocene (Williams 1978; Hughes and Sullivan 1981).

Much of this increased disturbance and burning could be anthropogenic rather than climatic in origin. The evidence for burning and disturbance on the Atherton Tableland corresponds with the earliest archaeological evidence for rainforest occupation (Horsfall and Hall 1990), and it is unlikely that fires on Lynch's Crater swamp surface could have been caused by climate alone as the site has been surrounded by complex rainforest that would have effectively excluded major fires through this period. Athough it is known that Aborigines had colonized most other parts of Australia over the last 30,000 to 40,000 yr (Allen 1989), archaeologists have argued for intensification of occupation with increased use of fire within the last 4000 to 5000 yr (Hughes and Sullivan 1981; Head 1989).

### New Zealand

Shortly after 10,000 BP New Zealand glaciers had shrunk back to their minimum Holocene extent, and they did not undergo major readvance until about 5000 BP (Gellatly et al. 1988). During this early Holocene period isotopic evidence shows that annual temperatures were apparently at their highest (Hendy and Wilson 1968), although there is no evidence that treelines have ever been higher than at present. McGlone (1988) has shown that during the early Holocene interval (10,000 to 7000 BP) plants characteristic of mild, moist climates, in particular *Ascarina lucida* and tree ferns (*Cyathea*), were widespread (see Fig. 21.8C). Subalpine scrub and low forest was dominant where subalpine *Nothofagus* forests are now both widespread and abundant. Rogers and McGlone (1989) suggested that annual rainfall during early Holocene was somewhat lower than at present, as a result of lesser incursion of winter rainfall, but that summers were cloudy and moist. Cloudy conditions during summer at high-altitude sites may have counteracted the generally higher annual temperatures experienced in the early Holocene. Many sites which now have permanent water or bog vegetation were dry in the early Holocene (McGlone 1990), which supports the argument for lower precipitation overall, while the presence of moisture-demanding vegetation over large areas of New Zealand favors summers with little or no moisture stress.

From about 5000 BP, glaciers in the Southern Alps of the South Island began to advance, and there have been at least 11 major periods of advance since then, the last period having terminated about 100 yr ago (Gellatly et al. 1988). Conditions suitable for glacier advance occur when south to southwest airflow and cooler conditions persist over the country (Burrows and Greenland 1979).

The late Holocene in New Zealand is characterized by the spread of vegetation tolerant of edaphic and climatic extremes. *Nothofagus* forest spread rapidly in central and southern North Island and northern South Island from about

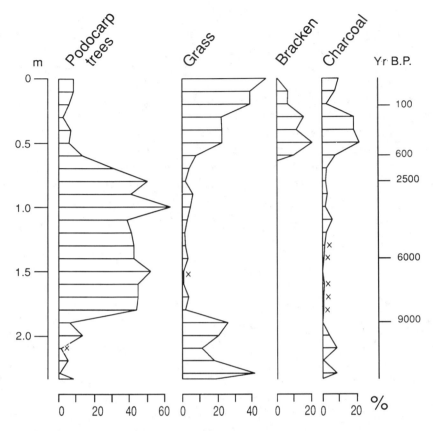

Fig. 21.8   Pollen and charcoal record from the Garvie Mountains, Central Otago, South Island, New Zealand (McGlone and Mark, unpublished data). Increased burning is apparent from about 6000 BP, but widespread forest destruction (illustrated by decline of podocarp trees, *Podocarpus* ssp., *Dacrydium* ssp., and *Dacrycarpus* ssp.) occurred only after 2500 BP. Bracken (*Pteridium aquilinum*), grass (Poaceae), and charcoal increase after 600 BP is a result of Polynesian settlement.

7000 BP (McGlone 1988; Rogers and McGlone 1989). *Nothofagus* forest in general can tolerate more variable climates, poorer soils, and more disturbance than conifer-broadleaved forests. In the northern North Island, *Agathis australis* forest spread from around 6000 BP, and expanded vigorously from about 3000 BP (McGlone et al. 1984; McGlone 1988; Newnham et al. 1989). *Agathis australis* is favored by dryish, warm summers (Ogden and Ahmed 1989) and its rapid increase in the late Holocene probably represents the regular occurrence of summer moisture deficits. This pattern of increase is similar to that recorded for *Agathis* on the Atherton Tableland in northeastern Australia. Other trees which spread at the same time (*Knightia excelsa*, *Phyllocladus trichomanioides*; see Fig. 21.8) are also favored by high solar radiation and relatively drier summer conditions.

There is abundant evidence in the late Holocene for increasing fire frequency (Fig. 21.8). In Northland (northern North Island) fire levels increased from about 2600 BP (Enright et al. 1988), and Kershaw and Strickland (1988) report a Northland raised bog which was burned after 3000 BP. Increasing levels of charcoal are recorded from Waikato sites (central North Island) from 6000 BP but more particularly post-3000 BP. Burrows and Russell (1990) show that fire was more prominent from about 7000 BP in central eastern South Island, with outbreaks occurring at irregular intervals up until recent times. In central Otago (southeastern South Island) destruction of a large part of the forest cover occurred between about 3000 and 1000 BP in a series of widespread outbreaks (McGlone 1988; Fig. 21.9). Although charcoal and disturbance by fire had occurred in the early Holocene in central Otago, the scale and widespread distribution of fire in the late Holocene is clearly of a different order of magnitude.

It therefore appears paradoxical that during the late Holocene – at the same time that fire frequency and drought-tolerant vegetation spread – numerous bogs, swamps, and lakes were initiated in many areas of New Zealand, and particularly in the drier eastern districts (Rogers and McGlone 1989; McGlone 1990; Fig. 21.10). Rogers and McGlone (1989) suggest that this is explicable if winters

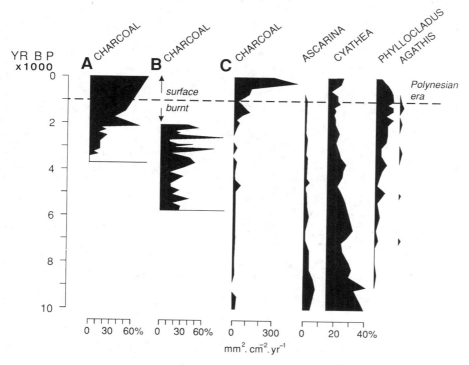

Fig. 21.9  Charcoal and pollen records from the North Island of New Zealand. A, charcoal as percentage of pollen count, North Cape (Enright et al. 1988); B, charcoal percentages, Ohinewai, Waikato Basin (McGlone et al. 1984); C, pollen and charcoal, Lake Rotomanuka, Waikato Basin (Newnham et al. 1989).

became both cooler and wetter during the late Holocene, which would permit recharging of groundwater, and thus maintain or expand wetland areas. Consistently drier summers would, at the same time, promote fires and lead to the expansion of moisture-stress tolerant vegetation.

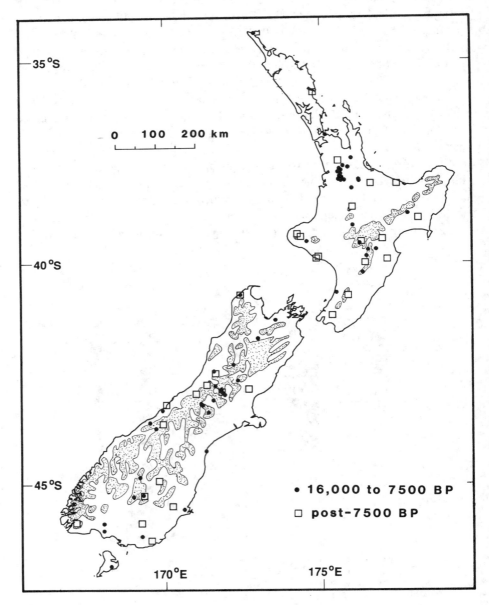

Fig. 21.10   Dated peat bog and lake initiations. Dots: initiations between 16,000 and 7500 BP. Squares: initiations post-7500 BP. Note the concentration of post-7500 BP sites on the axial ranges and in eastern districts. Shaded area land above 1000 m elevation.

## South America

Description of a detailed Holocene paleoclimate history from regions with a significant ENSO signal is based on pollen and lake level records (Fig. 21.11) from Venezuela (Lake Valencia, 10°N: Bradbury et al. 1981; Leyden 1985), from southeastern Brazil (Salitre, 19°S: Ledru 1991), the Galapagos (Los Juncos, 0° S: Colinvaux 1972; Colinvaux and Schofield 1976), north-central Chile (Quereo/ Quintero, 31°S: Villagran and Varela 1990), central Chile (Laguna Tagua Tagua, 34°S: Heusser 1990), and west-central Argentina (Gruta del Indio, 35°S: D'Antoni 1983; Vaca Lauquen, 37°S: Markgraf 1987; Lago Morenito, 41°S: Markgraf 1984).

Between 10,000 and 8000 BP in both records from northeastern and southeastern South America, Lake Valencia and Salitre, respectively, vegetation and lake level changes were interpreted to indicate conditions cooler than today. Lake levels were high and montane forests expanded into the lowlands. On the

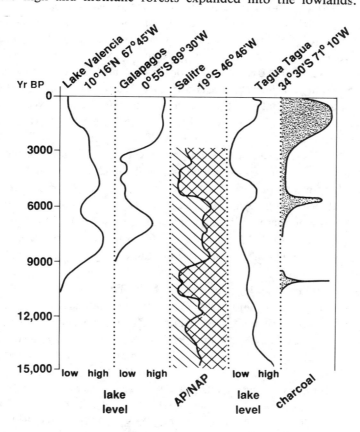

Fig. 21.11  South American lake level records (Lake Valencia, Bradbury et al. 1981; Galapagos, Colinvaux and Schofield 1976; Tagua Tagua, Heusser 1990), interpreted from diatom and aquatic plant stratigraphy, and paleoclimate record based on Arboreal (AP) versus Nonarboreal (NAP) pollen proportion changes (Salitre, Ledru 1991) and charcoal (particles/cm$^{-3}$ × 10$^6$) from Tagua Tagua.

Galapagos, in central Chile and Argentina, on the other hand, climates were very dry at that time. Lake levels were low, the swamp forest islands of north-central Chile were dominated by semiarid shrubland, the mixed *Nothofagus* forest islands along the equatorward limit of their distribution were species-poor woodlands, and the Monte desert lacked even the herbaceous therophytes.

After 8000 BP the climatic pattern in South America became regionally more diverse. Lake levels of Lake Valencia fell substantially and in the surrounding vegetation savanna elements became important. Paleoclimatic interpretation calls for conditions warmer and drier than before. At Salitre, in contrast, meso-phytic forest replaced the *Araucaria* forest indicating higher temperatures as well as increased moisture. Lake levels in the Galapagos and Laguna Tagua Tagua rose, but remained substantially below modern levels. *Nothofagus* forests in west-central Argentina became denser and more species rich, suggesting increased moisture.

Between 5000 and 3000 BP Lake Valencia became a freshwater lake and the surrounding vegetation of mesic semievergreen forest suggests more mesic con-ditions than before. Salitre showed expansion of savanna elements instead, suggesting drier conditions. Galapagos lake levels were lower than before between 5000 and 3000 BP and also north-central Chile continued dry. By 3000 BP essentially modern climate conditions became established. Lake Valen-cia turned into a saline lake and the vegetation assemblages showed great varia-bility from 3000 BP onwards. At Salitre semideciduous vegetation expanded suggesting increased moisture, although the levels never reached the amount of those prior to 6000 BP. On the Galapagos, in central Chile and western Argen-tina high lake levels and forest expansion indicate the onset of the relatively highest levels of moisture during the Holocene. The only charcoal record from semiarid southern South America, Laguna Tagua Tagua in central Chile, documents a drastic increase in charcoal particles by 3000 BP, suggesting high fire frequency. At latitudes 37 to 41°S conifers such as *Austrocedrus* began to increase proportionally in the vegetation, indicating expansion of drought-tolerant taxa.

In historical times, lake level changes in Lake Titicaca show a weak positive correlation with SO (Aceituno 1988). The paleolimnological record shows dif-ferences in amplitude of lake level change. During the last 4000 yr, the time during which lake levels rose, the amplitude of lake level change also increased, which could indicate the onset of the ENSO pattern (Ybert 1987; Wirrmann et al. in press).

Lower resolution analysis of past El Niño events was obtained from flood deposits in the Peruvian coastal plains. El Niño events dated back to about 7500 BP (Wells 1987), and 5000 BP (Rollins et al. 1986), respectively, with 11 events alone during the last 3500 yr (Wells 1990). Some of these events apparently were of far greater amplitude than those witnessed in historical times. Undated, but highly weathered flood deposits have also been found, indicating that El Niño events were not restricted to the Holocene. Sediment analysis of strandlines along the coast of southern Brazil have been taken as another proxy for El Niño events. Radiocarbon-dated sequences of strandlines dating back to about 5000 BP indicate that there were frequent El Niño-like events between 5000 and 4000 BP and during the last 2500 yr (Martin et al. 1984; Martin and Suguio 1989).

## The history of ENSO from the Holocene paleoclimatic record

As outlined above, several hypotheses can be proposed for the state of the ENSO phenomenon during the early Holocene in Australasia and South America: (1) The SO may have become more negative or positive; (2) its characteristic fluctuations may have had a reduced amplitude; (3) changed baseline conditions, in Australasia especially, may have caused it to be expressed differently.

The generally drier conditions which are suggested by paleoclimatic records for Australasia for the early Holocene (Fig. 21.12) are consistent with the SO being in a more negative phase. Earlier suggestions on the basis of Australasian paleoclimatic data that the early Holocene (or Hypsithermal) was in a more positive phase (Pittock and Salinger 1982) are not supported by this finding. However, drier conditions in eastern Australia and New Zealand are the only evidence for a more negative SO. Lower precipitation may have resulted from reduced tropical sources of moisture (as is suggested by drier Queensland conditions in particular), and greater continentality due to lower sea levels, in many areas, and therefore is in itself not conclusive evidence. New Zealand, on the other hand, seems to have been somewhat warmer during the early Holocene, total precipitation lower (Fig. 21.12), especially in the southeast, and glaciers at their minimum extent. This is not consistent with increased southwesterly airflow which occurs during negative phases of the SO. A striking feature of early Holocene climates in Australasia is the lack of variability. Although conditions were drier, fire was by no means as common as in the late Holocene. The vegetation of the early Holocene is more typical of situations with low amounts of stress or disturbance, in that trees and shrubs typical of stressed sites or secondary vegetation tend to be much less common than they are now. Sediments in lakes and soil profiles in Australia reflect the relative lack of widespread or intense disturbance of the landscape. As the sharp oscillations between the negative and positive phases of the ENSO phenomenon tend to create disturbance of the vegetation and landscape at present, and the drought which accompanies the negative phase promotes fire, it seems improbable that the ENSO system was operating as it is now. Either the cycles were dampened or the different base state of Australasian climates resulted in the ENSO fluctuations not being expressed.

Increasing fire and disturbance of vegetation and landscape in Australia has been attributed to increasing Aboriginal populations and intensification of land use (Hughes and Sullivan 1981; Head 1989). However, in New Zealand the late arrival of the Maori settlers (at or after 1000 BP) rules out an anthropogenic explanation. Given the close similarity between Australian and New Zealand climatic evolution, the Australian late Holocene increase in fire and disturbance

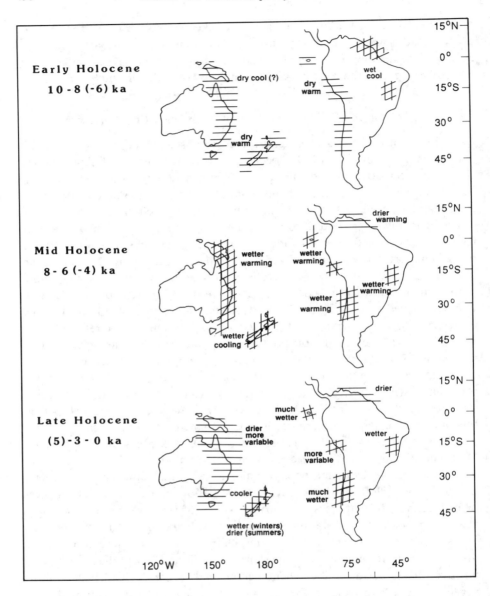

Fig. 21.12   Paleoclimatic patterns of Australasia and South America for the early,
mid- and late Holocene (for explanation see text).

was probably a result of increasing climatic variability and drought. Aboriginal
land-use practices, which included frequent and widespread use of fire, may have
accentuated a trend which would have been apparent in the absence of humans.
Vegetation types – such as tropical forest – previously of little value may have
become more accessible as a result of climatic change.

   The patterns of early Holocene climates in South America (Fig. 21.12) lead
to similar conclusions as those based on Australasian evidence. In the early

Holocene paleoenvironmental conditions suggest fairly stable climates. Absence of flood deposits in northern Peru, and opposition of paleoclimate patterns between northeastern and southwestern South America, the former with moister and cooler climates, the latter with drier and warmer climates is suggestive of La Niña conditions (or positive SOI). However, high moisture in southeastern Brazil at that time and drier conditions in Australasia that would be reflective of El Niño patterns instead contradict this interpretation. This, and the fact that overall environmental and hence climate variability was low at that time suggests that ENSO fluctuations were dampened or differently expressed. The evidence also refutes the idea that ENSO could have been 'stuck' in either extreme, a proposition questioned earlier on climatological grounds (Nicholls 1979; Barnett 1991).

Between 8000 and 5000 BP climate patterns in South America shifted (Fig. 21.12). The eastern Pacific (Galapagos) and western South America became moister, without reaching todays moisture levels, while northeastern South America became drier. This pattern resembles the El Niño precipitation anomalies and one flood deposit in coastal Peru dates to that interval (Wells 1990). Records from southeastern Brazil and Australasia, however, showing high moisture levels at that time contradict the El Niño pattern, suggesting again that ENSO's role was not comparable to today's.

After 5000 BP variability in the records appears to increase to become a dominant feature after 3000 BP. This is expressed by enhancement of the amplitude of change, especially well documented in records of lake level change, and by an 'anomalous' combination of plant assemblages, indicating simultaneouly both increased moisture and increased drought and fire frequency. Repeated occurrences of ENSO events could explain the paleoclimatic evidence. Numerous flood deposits in coastal Peru and El Niño-type strandlines in southeastern Brazil indicate increased frequency of ENSO events. The precipitation patterns over South America, showing generally much higher moisture levels in southwestern and southeastern South America, but greater aridity in northern South America (Fig. 21.12) suggests predominance of the El Niño signal and/or high frequency of 'strong' El Niño events.

The evidence from Australasia and South America therefore points to much reduced climatic variability in the early and mid-Holocene. There is no evidence that during that time ENSO remained in one extreme position or the other. Our preferred explanation is a lower amplitude ENSO system in which marked El Niño/La Niña events were not strongly expressed in the circum-South Pacific. A possible explanation for this modification of the ENSO system could be the different boundary conditions of the early Holocene which altered the development of monsoonal circulation in the Southern Hemisphere. These altered boundary conditions were (1) the reduced amount of ocean to the north of Australia, and continuing land connection between Australia and New Guinea because of lower sea level; and (2) the reduced seasonality in the Southern Hemisphere as a probable result of lower summer insolation and higher winter insolation at the top of the atmosphere.

After 3000 BP environmental variability, as demonstrated by increased fire frequence, erosion, or expansion of disturbance-favored taxa, is high in nearly all circum-South Pacific paleoenvironmental records. By that time the vegetation and fire regimes of Australia, New Zealand, and South America became nearly identical to those of the present. In particular, the vegetation in all three landmasses shows increases in taxa that depend on higher levels of moisture simultaneously with increase in taxa tolerant of drought. Fire, flood, and disturbance were common throughout, and cold southwesterly airflow became a regular feature of both New Zealand and southernmost South America. Our preferred explanation is that El Niño/La Niña events became more frequent and more intense only during the latter part of the Holocene.

In conclusion, the paleoenvironmental evidence suggests that the early Holocene (10,000 to 7000 BP) was a time of equable climates in circum-South Pacific areas outside of the tropics. Extreme climatic events, such as the prolonged droughts or devastating floods now commonly associated with El Niño and La Niña, were much rarer. Since 7000 BP there has been increasing indications in the paleorecord of such events, and by 3000 BP variable climatic regimes similar to those of the present were well established. It is therefore unlikely that ENSO cycles were a major influence before 7000 BP in this region, and it is only over the last 3000 yr that they have played their current important role.

*Acknowledgment* Vera Markgraf acknowledges the support of NSF grants ATM–86–18217 and DPP–86–1398.

# References

ACEITUNO, P., 1988: On the functioning of the Southern Oscillation in the South American sector. Part I: surface climate. *Monthly Weather Review*, 116: 505–524.

ALLEN, J., 1989: When did humans first colonize Australia? *Search*, 20: 149–154.

BARNETT, T.P., 1991: The interaction of multiple time scales in the tropical climate system. *Journal of Climate*, 4: 269–285.

BRADBURY, J.P., LEYDEN, B., SALGADO-LABOURIAU M.L., LEWIS, W.M. Jr., SCHUBERT, C., BINFORD, M.W., FREY, D.G., WHITEHEAD, D.R., and WEIBEZAHN, F.H., 1981: Late Quaternary environmental history of Lake Valencia, Venezuela. *Science*, 214: 1299–1305.

BURROWS, C.J. and GREENLAND, D.E., 1979: An analysis of the evidence for climatic change in New Zealand in the last thousand years: evidence from diverse natural phenomena and from instrumental records. *Journal of the Royal Society of New Zealand*, 9: 321–373.

BURROWS, C.J. and RUSSEL, J.B., 1990: Aranuian vegetation history of the Arrowsmith Range, Canterbury I. Pollen diagrams, plant macrofossils, and buried soils from Prospect Hill. *New Zealand Journal of Botany*, 28: 323–345.

CLARK, R.L., 1983: Pollen and charcoal evidence for the effects of Aboriginal burning on the vegetation of Australia. *Archaeology in Oceania*, 18: 32–37.

CLARK, R.L., 1987: Fire history from charcoal: effects of sampling on interpretation. *In* Ambrose, W.R. and Mummery, J.M.J. (eds), *Archaeometry: Further Australasian*

*Studies.* Department of Prehistory, Australian National University, Canberra, 135–142.

CLARK, R.L., 1986: Pollen as a chronometer and sediment tracer, Burrinjuck Reservoir, Australia. *Hydrobiologia*, 143: 63–69.

COLINVAUX, P.A., 1972: Climate and the Galapagos Islands. *Nature*, 240: 17–20.

COLINVAUX, P.A. and SCHOFIELD, E.K., 1976: Historical ecology in the Galapagos Islands. *Journal of Ecology*, 64: 989–1012.

COSTIN, A.B. and POLACH, H.A., 1971: Slope deposits in the Snowy Mountains, southeastern Australia. *Quaternary Research*, 1: 228–235.

D'ANTONI, H.L., 1983: Pollen analysis of Gruta del Indio. *Quaternary of South America and Antarctic Peninsula*, 1: 83–104.

D'COSTA, D.M., EDNEY, P., KERSHAW, A.P., and DE DECKER, P., 1989: Late Quaternary palaeoecology of Tower Hill, Victoria, Australia. *Journal of Biogeography*, 16: 461–482.

DIAZ, H.F. and KILADIS, G.N., 1992: Atmospheric teleconnections associated with the extreme phases of the Southern Oscillation. *In* Diaz, H.F. and Markgraf, V. (eds.), *El Niño: Historical and Paleoclimatic Aspects of the Southern Oscillation.* Cambridge: Cambridge University Press, 7–28.

DODSON, J.R., 1987: Mire development and environmental change, Barrington Tops, New South Wales, Australia. *Quaternary Research*, 27: 73–81.

DODSON, J.R., GREENWOOD, P.W., and JONES, R.L., 1986: Holocene forest and wetland vegetation dynamics at Barrington Tops, New South Wales. *Journal of Biogeography*, 13: 561–585.

ENRIGHT, N.J., McLEAN, R.F., and DODSON, J.R., 1988: Late Holocene development of two wetlands in the Te Paki region, far northern New Zealand. *Journal of the Royal Society of New Zealand*, 18: 369–382.

GELLATLY, A.F., CHINN, T.F. and RÖTHLISBERGER, F., 1988: Holocene glacier variations in New Zealand: a review. *Quaternary Science Reviews*, 7: 227–242.

GORDON, N.D., 1985: The Southern Oscillation: a New Zealand perspective. *Journal of the Royal Society of New Zealand*, 15: 137–155.

HARRIS, G.P., DAVIES, P., NUNEZ, M., and MEYERS, G., 1988: Interannual variability in climate and fisheries in Tasmania. *Nature*, 333: 754–757.

HEAD, L., 1989: Prehistoric Aboriginal impacts on Australian vegetation: an assessment of the evidence. *Australian Geographer*, 20: 37–46.

HENDY, C.H. and WILSON, A.T., 1968: Palaeoclimatic data from speleothems. *Nature*, 219: 48–51.

HEUSSER, C.J., 1990: Ice age vegetation and climate of subtropical Chile. *Palaeogeography, Palaeoclimatology, Palaeoecology*, 80: 107–127.

HORSFALL, N. and HALL, J., 1990: People and the rainforest: an archaeological perspective. *In* Webb, L.J. and Kikkawa, J. (eds.), *Australian Tropical Rainforests.* Melbourne: CSIRO, 33–39.

HUGHES, P.J. and SULLIVAN, M.E., 1981: Aboriginal burning and late Holocene geomorphic events in eastern NSW. *Search*, 12: 277–278.

JELICIC, L., 1988: A palaeoecological investigation of Holocene sediments from West Basin, western plains of Victoria. Unpubl. B.A. (Hons) thesis, Monash University. 105 pp.

KENYON, C.E., 1989. A late Pleistocene and Holocene palaeoecological record from Boulder Flat, East Gippsland. Unpubl. B.Sc. (Hons.) thesis, Monash University. 117 pp.

KERSHAW, A.P., 1975: Late Quaternary vegetation and climate in northeastern Australia. *In* Suggate, R.P. and Creswell, M.M. (eds), *Quaternary Studies*. Wellington: Royal Society of New Zealand, 181–187.

KERSHAW, A.P., 1983: A Holocene pollen diagram from Lynch's Crater, North-eastern Queensland, Australia. *New Phytologist*, 94: 669–682.

KERSHAW, A.P. and NIX, H.A., 1988: Quantitative palaeoclimatic estimates from pollen data using bioclimatic profiles of extant taxa. *Journal of Biogeography*, 15: 589–602.

KERSHAW, A.P. and STRICKLAND, K.M., 1988: A Holocene pollen diagram from Northland, New Zealand. *New Zealand Journal of Botany*, 26: 145–152.

KERSHAW, A.P. and STRICKLAND, K.M., 1989. The development of alpine vegetation on the Australian mainland. *In* Good, R. (ed), *The Scientific Significance of the Australian Alps*. Canberra: Australian Academy of Science, 113–126.

KERSHAW, A.P. and WALKER, D., 1988: Vegetation, historical plant geography and late Cainozoic vegetation history of the northeast Queensland humid tropics. *Excursion LBG Guide, 7th International Palynological Congress*, Brisbane, 1988.

KILADIS, G.N. and DIAZ, H.F., 1989: Global climatic anomalies associated with extremes in the Southern Oscillation. *Journal of Climate*, 2: 1069–1090.

LEDRU, M.P., 1991: Etude de la pluie actuelle des forêts du Brésil central: climat, végétation, application a l'étude de l'évolution paléoclimatique des 30.000 dernières années. Thèse de Doctorat, Paris, 1991. 193 pp.

LEYDEN, B.W., 1985: Late Quaternary aridity and Holocene moisture fluctuations in the Lake Valencia Basin, Venezuela. *Ecology*, 66: 1279–1295.

MacPHAIL, M.K., 1979: Vegetation and climates in southern Tasmania since the last glaciation. *Quaternary Research*, 11: 306–341.

MacPHAIL, M.K. and HOPE, G.S., 1985: Late Holocene mire development in montane southeastern Australia: a sensitive climatic indicator. *Search*, 15: 344–349.

MARKGRAF, V., 1984: Late Pleistocene and Holocene vegetation history of temperate Argentina: Lago Morenito, Bariloche. *Dissertationes Botanicae*, 72: 235–254.

MARKGRAF, V., 1987: Paleoenvironmental changes at the northern limit of the subantarctic *Nothofagus* forest, lat 37°S, Argentina. *Quaternary Research*, 28: 119–129.

MARKGRAF, V., BRADBURY, J.P. and BUSBY, J.R., 1986: Paleoclimates in Southwestern Tasmania during the last 13,000 years. *Palaios*, 1: 368–380.

MARKGRAF, V., DODSON, J.R., KERSHAW, A.P., McGLONE, M.S., and NICHOLLS, N., 1992. Evolution of late Pleistocene and Holocene climates in the circum South Pacific land areas. *Climate Dynamics*, 6: 193–211.

MARTIN, A.R.H., 1986: Late glacial and Holocene alpine pollen diagrams from the Kosciusko National Park, New South Wales, Australia. *Review of Palaeobotany and Palynology*, 47: 367–409.

MARTIN, L., FLEXOR, J.M., KOUSKY, V., FONSECA de ALBUQUERQUE, CAVALCANTIA, I., 1984: Inversions du sens du transport littoral enregistrées dans les cordons littoraux de la plaine côtière du Rio Doce (Brésil): possible liaison avec des modifications de la circulation atmosphérique. *Compte Rendue Academie Sciences, Paris*, 298: 25–27.

MARTIN, L. and SUIGUIO, K., 1989: Paleoclimatic changes during the Holocene, registered in littoral deposits. *In*: *International Symposium on Global Changes in South America during the Quaternary, Sao Paolo, Brazil*. 1989. ABEQUA/INQUA, Special Publication No. 2: 44–58.

McGLONE, M.S., 1988: New Zealand. *In* Huntley, B. J. and Webb, T., III, (eds.) *Vegetation History*. Handbook of Vegetation Science 7. Dordrecht: Kluwer, 557–599.

McGLONE, M.S., 1990: The early Holocene as an analogue for a $2 \times CO_2$ New Zealand. Unpublished report, New Zealand, DSIR Land Resources, Christchurch.

McGLONE, M.S., NELSON, C.S., and TODD, A.J., 1984: Vegetation history and environmental significance of pre-peat and surficial peat deposits at Ohinewai, Lower Waikato lowland. *Journal of the Royal Society of New Zealand*, 14: 233–244.

McKENZIE, G.M., 1989: Late Quaternary vegetation and climate of the Central Highlands of Victoria with particular reference to *Nothofagus cunninghamii* (Hook) Oerst. rainforest. Ph.D thesis, Monash University, 248 pp.

MEEHL, G. and BRANSTATOR, G.W., 1992: Coupled climate model simulation of El Niño/Southern Oscillation: implications for paleoclimate. *In* Diaz, H.F. and Markgraf, V. (eds.), *El Niño: Historical and Paleoclimatic Aspects of the Southern Oscillation*. Cambridge: Cambridge University Press, 69–91.

MICHAELSEN J. and THOMPSON, L.G. Jr., 1990: A comparison of proxy records of El Niño/Southern Oscillation. *In* Diaz, H.F. and Markgraf, V. (eds.), *El Niño: Historical and Paleoclimatic Aspects of the Southern Oscillation*. Cambridge: Cambridge University Press, 323–348.

NEWNHAM, R.M., LOWE, D.J., and GREEN, J.D., 1989: Palynology, vegetation and climate of the Waikato lowlands, North Island, New Zealand, since c. 18,000 years ago. *Journal of the Royal Society of New Zealand*, 19: 127–150.

NICHOLLS, N., 1979. A simple air-sea interaction model. *Quarterly Journal of the Royal Meteorological Society*, 105: 93–105.

NICHOLLS, N., 1988. El Niño – Southern Oscillation and rainfall variability. *Journal of Climate*, 1: 418–421.

NICHOLLS, N., 1991: The El Niño/Southern Oscillation and Australian vegetation. *Vegetatio*, 91: 23–36.

NICHOLLS, N., 1992: Historical El Niño/Southern Oscillation variability in the Australasian region. *In* Diaz, H.F. and Markgraf, V. (eds.), *El Niño: Historical and Paleoclimatic Aspects of the Southern Oscillation*. Cambridge: Cambridge University Press, 151–173.

NORTON, D.A., BRIFFA, K.R., and SALINGER, M.J., 1989: Reconstruction of New Zealand summer temperatures to 1730 AD using dendroclimatic techniques. *International Journal of Climatology*, 9: 633–644.

OGDEN, J. and AHMED, M., 1989: Climate response function analysis of kauri (*Agathis australis*) tree ring chronologies in northern New Zealand. *Journal of the Royal Society of New Zealand*, 19: 205–221.

PITTOCK, A.B. and SALINGER, M.J., 1982: Towards regional scenarios for a $CO_2$ - warmed Earth. *Climatic Change*, 4: 23–40.

PITTOCK, J., 1989: Palaeoenvironments of the Mt. Disappointment Plateau (Kinglake West, Victoria) from the late Pleistocene. Unpubl. B.Sc. (Hons.) thesis, Monash University. 154 pp.

QUINN, W.H. and NEAL, V.T., 1992: The historical record of El Niño events. *In* Bradley, R.S. and Jones, P.D. (eds.), *Climate Since A.D. 1500*. London: Routledge, 623–648.

QUINN, W.H., NEAL, V.T., and ANTUNEZ de MAYOLO, S.E., 1987: El Niño occurrences over the past four and a half centuries. *Journal of Geophysical Research*, 92: 14,449–14,461.

RASMUSSON, E.M. and CARPENTER, T.H., 1982: Variations in tropical sea surface temperature and surface wind fields associated with the Southern Oscillation/El Niño. *Monthly Weather Review*, 110: 354-384.

ROGERS, G. and McGLONE, M.S. 1989: A postglacial vegetation history of the southern-central uplands of North Island, New Zealand. *Journal of the Royal Society of New Zealand*, 19: 229-248.

ROLLINS, H.B., RICHARDSON, J.B., III, and SANDWEISS, D.H., 1986: The birth of El Niño: geoarchaeological evidence and implications. *Geoarchaeology*, 1: 3-16.

SWETNAM, T.W., and BETANCOURT, J.L., 1992: Secular variability of the Southern Oscillation detected in tree-ring data from Mexico and the southern United States. *In* Diaz, H.F. and Markgraf, V. (eds.), *El Niño: Historical and Paleoclimatic Aspects of the Southern Oscillation*. Cambridge: Cambridge University Press, 259-269.

THOMPSON, L.G., MOSLEY-THOMPSON, E., and THOMPSON, P.A. 1992: Reconstructing interannual climate variability from tropical and subtropical ice-core records. *In* Diaz, H.F. and Markgraf, V. (eds.), *El Niño: Historical and Paleoclimatic Aspects of the Southern Oscillation*. Cambridge University Press, 295-322.

THOMPSON, L.G. Jr., MOSLEY-THOMPSON, E., and MORALES, A.B., 1984: El Niño-Southern Oscillation events recorded in the stratigraphy of the tropical Quelccaya ice cap, Peru. *Science*, 229: 971-973.

VEBLEN, T.T. and LORENZ, D.C., 1987: Post-fire stand development of *Austrocedrus-Nothofagus* forests in northern Patagonia. *Vegetatio*, 71: 113-126.

VEBLEN, T.T. and LORENZ, D.C., 1988: Recent vegetation changes along the forest/steppe ecotone of northern Patagonia. *Annals of the Association of American Geographers*, 78: 93-111.

VILLAGRAN, C. and VARELA, J., 1990: Palynological evidence for increased aridity on the Central Chilean coast during the Holocene. *Quaternary Research*, 34: 198-207.

VILLALBA, R., 1989: Latitude of the surface high pressure belt over western South America during the last 500 years as inferred from tree-ring analysis. *Quaternary of South America and Antarctic Peninsula*, 7: 273-304.

WELLS, L.E., 1987: An alluvial record of El Niño events from northern coastal Peru. *Journal of Geophysical Research*, 92: 14,463-14,470.

WELLS, L.E., 1990: Holocene history of the El Niño phenomenon as recorded in flood sediments of northern coastal Peru. *In: Proceedings Sixth Annual Pacific Climate (PACLIM) Workshop*, March 5-8, 1989. California Department of Water Resources, Interagency Ecological Studies Program Technical Report 23: 141-144.

WHETTON, P., ADAMSON, D., and WILLIAMS, M., 1990: Rainfall and river flow variability in Africa, Australia and East Asia, linked to El Niño-Southern Oscillation events. *In* Bishop, P. (ed), *Lessons for Human Survival: Nature's Record from the Quaternary*. Geological Society of Australia, Symposium Proceedings 1: 71-82.

WILLIAMS, M.A.J., 1978: Late Holocene hillslope mantles and stream aggradation in the Southern Tablelands, NSW. *Search*, 9: 96-97.

WIRRMANN, D., YBERT, J.-P., and MOURGIART, P., in press: A 20000 years paleohydrological record from Lake Titicaca. *In* Dejoue, C. and Iltis, A. (eds.), *Lake Titicaca, Synthesis of the Knowledge*. Dordrecht: Kluwer.

YBERT, J.-P., 1987: Spectres palynologiques de tourbières et de sédiments de la fin du Pléistocène et de l'Holocène des Andes de Bolivie. *Géodynamique*, 2: 108-109.

# 22

# Synthesis and future prospects

HENRY F. DIAZ

*NOAA/ERL, 325 Broadway, Boulder, Colorado 80303, U.S.A.*

VERA MARKGRAF

*Institute of Arctic and Apline Research, University of Colorado, Boulder, Colorado 80309, U.S.A.*

MALCOLM K. HUGHES

*Laboratory of Tree Ring Research, University of Arizona, Tucson, Arizona 85721, U.S.A.*

## ENSO in the climate system

The importance of the global-scale climatic feature known as El Niño/Southern Oscillation (ENSO) to regional and hemispheric-scale climatic variability is now well documented. This book provides a broad overview of some of the fundamental components of the ENSO system, underscores a few of the significant advances that have been made over the last couple of decades in understanding this phenomenon, and summarizes much of the recent work in applying paleo-climatic methods to the study of long-term changes in ENSO. Questions regarding climate and global environmental change arising from human activities are increasingly drawing the attention of scientists and policymakers worldwide. Hence, documenting the sources of natural variability of the climate system, and understanding how potential changes to intrinsic global-scale features of the present climate, such as ENSO, may evolve have become important topics of scientific research.

The ENSO phenomenon has long been known to be a critical feature of the climate of the tropical Pacific region. As such, it has a profound influence on the terrestrial and marine fauna and flora of this region, and hence on its human population. Nicholls (Chapter 7) provides a comprehensive review of the literature which illustrates how the fauna and flora of much of Australasia are well adapted to high climatic variability – frequent droughts interspersed with high rainfall periods. Such high variability is characteristic of the climate of this region; indeed, it is a principal effect of the ENSO cycle on regional precipitation in areas most strongly affected by the oscillation. The influence of ENSO on the interannual and possibly longer-term climatic variability of many continental regions outside the immediate ENSO core areas of the Indo-Pacific Ocean, for example in Africa and North America, has also been thoroughly documented and shown to be highly significant.

Due to the large-scale nature of ENSO teleconnections, a long-term chronology of ENSO-related activity was derived by Quinn (Chapter 6) from recorded changes in the annual flood level of the Nile River near Cairo, Egypt. This important record, based on a number of documentary sources, reconstructs an element of the Southern Oscillation that is related to climatic variability in the western portion of the nominal ENSO core-region, namely the summer monsoon rainfall over the highlands of Ethiopia. The record is available back to the beginning of the Muslim Era in A.D. 622.

Analysis of changes in the frequency of occurrence of warm ENSO events (El Niño) during the period of record, which starts with the beginning of the Spanish colonial period in the Americas, indicates that the processes associated with El Niño development are not stationary through time. There is some indication that changes in solar activity, as measured by changes in sunspot activity, may modulate the frequency of warm events, and in particular, that of the strongest categories (Enfield and Cid 1991; Anderson, Chapter 9; Enfield, Chapter 5). On the basis of the record of ENSO occurrences derived from a variety of instrumental and documentary sources, Enfield (Chapter 5) illustrates how ENSO statistics have varied during the past three centuries. In a separate contribution, Diaz and Pulwarty (Chapter 8) evaluate the relationship among several different measures of the overall ENSO system, documenting the degree of temporal synchrony present during ENSO development. These authors also provide evidence to indicate that ENSO recurrence has varied over time, although the characteristic recurrence interval of 3 to 4 yr for moderate and stronger El Niño events is expressed throughout most of the record (since the early 1500s).

Given the potentially large impact on the welfare of regions affected by ENSO that could arise from changes in the intensity and/or frequency of occurrence of ENSO extremes as a result of changes in global climate, great efforts have been made to use paleoclimatic techniques to reconstruct elements of the ENSO record for the past millennium or so. These efforts, illustrated in a number of contributions to this book, have yielded considerable insight into the nature of low frequency changes in the ENSO phenomenon.

## ENSO in the paleoclimate record

From the outset, our aim was to give the reader a sense of both the richness of the available paleoenvironmental indicators of ENSO activity, as well as some measure of the limitations of each of the various proxy records. We emphasize that the various ENSO proxy indices may respond to different aspects of the ENSO phenomenon (and at different times in its evolution) and that all contain a certain degree of local variability (or noise) unconnected to ENSO signals.

The challenge in interpreting paleoclimatic records within the context of the ENSO phenomenon, therefore, is to isolate from a range of climatic forcing parameters those that are strictly related to ENSO events (Baumgartner et al. 1989). This varies for different indicators and different geographic regions, and also for records of different temporal and spatial resolution. In recognition of this fact, evaluation of how each indicator records the ENSO signal constitutes a critical aspect that is addressed throughout the book.

There are several types of environmental indicators that can record changes in interannual scale climatic variability associated with ENSO, which allow us to extend the record of ENSO back beyond the instrumental and historical period. Other indicators have temporal resolution that is on decadal to century scale, and these records enable us to extend the record of ENSO climatic variability much farther back in time. One of the aims of this book was to examine both, high and low frequency behavior of different ENSO indices over different periods of time, in order to shed light on issues related to low frequency variations in teleconnections and their possible mechanisms.

Marine sediment records, including time series of sediment types and their characteristics (Anderson et al., Chapter 20), fisheries data, including the distribution of fish scales in coastal marine sediments (Sharp, Chapter 19), and records from corals (Cole et al., Chapter 18), all sense different types of ENSO-related changes in the ocean environments, such as sea surface temperatures, ocean currents, wind velocity, upwelling intensity, changes in nutrients and trace metals, etc. The same paleoenvironmental indicator may sense different parameters, depending on the geographic location. For example, in the western Pacific, corals will primarily sense ENSO events in terms of changes in ocean salinity and temperature, which are related to changes in atmospheric forcing associated with ENSO (e.g., monsoonal precipitation). In the eastern Pacific, on the other hand, the signal in the coral record is related to upwelling intensity as it pertains to ENSO perturbations.

In the terrestrial setting, the proxy signals related to ENSO are as diverse as those in the ocean setting. Ice-core records from southern Peru (Thompson et al., Chapter 16), where El Niño events are characterized by droughts, measure changes in snow accumulation, particle amounts, and temperature as seen by oxygen isotope changes. But because not every drought event is necessarily due to an El Niño event, other records need to be linked to the ice record for verification and validation. This aspect is discussed by Michaelsen and Thompson (Chapter 17), who used various statistical tests to examine the interrelationships between the Peruvian ice core record and tree-ring records from southwestern United States. By doing so, ambiguities in interpretation of past ENSO events can be eliminated and the overall interpretation is thereby greatly strengthened.

With exception of one record from Java, the tree-ring records discussed here are primarily from the southwestern United States, Texas, and northern Mexico, where the ENSO signal has been shown to have a strong climatic effect on cool season precipitation (Diaz and Kiladis, Chapter 2). Although the signal weakens from northern New Mexico (D'Arrigo and Jacoby, Chapter 13) towards the Great Plains (Cleaveland et al., Chapter 15), and shifts from an October through February time frame to a December through May one, this seasonal precipitation signal is well captured by the trees because it affects summer (growing season) soil moisture conditions. Past summer drought occurrences and their relation to a history of wildfires is discussed by Swetnam and Betancourt (Chapter 14). They were able to document changing periods of high temporal variability during times when the Southern Oscillation exhibits strong variability near the biennial period. In essence, when the characteristic ENSO rhythm is dominated by lower frequency components, the correlation between fire history and the tree-ring (precipitation) record is lower than during times with high interannual variability.

Another type of approach in reconstructing past ENSO events is discussed by Lough (Chapter 11) and Meko (Chapter 12), who analyse geographical patterns of tree-ring (precipitation) records and their relation to known ENSO teleconnection patterns. Especially at 3- to 5-yr time-scale, the ENSO-induced climate signal exhibits an opposition between northwestern and southwestern North American tree-ring chronologies, consistent with observed ENSO teleconnection patterns.

The same geographic region of western North America is examined for evidence of ENSO teleconnections using precipitation and streamflow data, and major flood records in relation to atmospheric circulation patterns (Cayan and Webb, Chapter 3). With such information at hand on the modern relation between ENSO events and floods, it is clear that future steps should involve the study of past flood events in those regions, similar to the successful ENSO flood history developed for Peru by Wells (1990).

Analysis of large-scale climate patterns appears to be the most promising method for reconstructing past ENSO behavior over long time scales (on 1000-yr time scales), and over periods where climate boundary conditions were substantially different from today. Comparison of past changes in vegetation patterns in Australia, New Zealand, and South America (McGlone et al., Chapter 21) points to the possibility that during early Holocene times, when insolation-related seasonality contrasts were lower in the Southern Hemisphere than today, ENSO events do not appear to have manifested themselves as they are today. Only during late Holocene times (after about 5000 BP), do ENSO-related climate patterns develop, as reflected by great environmental variability expressed on decadal to century time scales, the presence of a mixture of fire-adapted and mesic plant communities, and geographical moisture patterns that would suggest high frequency alternations of both ENSO phases.

We have summarized, in Figure 22.1, the type and location of the various paleoenvironmental ENSO indicators discussed in the book. In the future, we hope to see an expansion in the number of sites and types of records available for analysis. Clearly, greater coverage of high-resolution proxy records in the more sensitive tropical regions would be valuable, for example, in testing

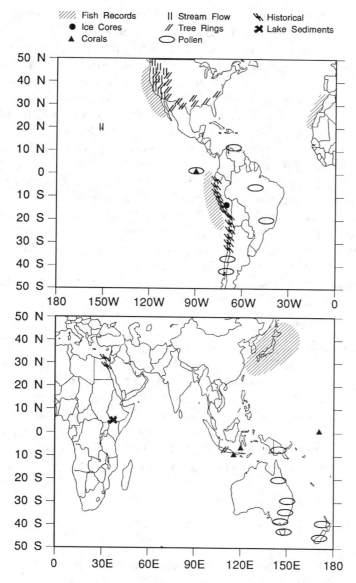

Fig. 22.1 Location of historical and paleoenvironmental records of ENSO.

inferences which have been made using historical sources with regard to changes in ENSO frequency.

## ENSO and future climate

Another important question addressed in the book concerns the sensitivity of the ENSO phenomenon to the overall climate in which it operates. For example, questions have been raised as to whether the characteristic ENSO features of today were similar to those during the cold episodes of the Little Ice Age (from approximately A.D. 1500 to the end of the 19th century) and the warm periods associated with the Medieval Warm Period (from approximately A.D. 900 to 1300). In this regard, the contribution by Meehl and Branstator (Chapter 4) is of interest. Using modelling results with a coupled ocean-atmosphere general circulation model (GCM) with both normal and doubled atmospheric $CO_2$ concentrations, they show that the extratropical atmospheric response in the warm (or El Niño) phase of the oscillation in the experiment with doubled $CO_2$ concentrations is quite different from the response obtained for the present (control) climate. The implication that the nature of extratropical ENSO signals may be different under significantly altered boundary conditions compared to the present climate is also considered by McGlone and colleagues (Chapter 21), primarily on the basis of evidence from pollen, charcoal sediment loads, and other biological paleoindicators for regions surrounding the South Pacific Ocean (South America, Australia, and New Zealand).

As such studies on similar time scales are carried out for other regions, it would be profitable to explore the possible causes of such sustained mode shifts in tropical-extratropical interaction, should the different extratropical response to ENSO turn out to be a confirmed feature of Holocene climate. If the behavior of the ENSO differed from present in the early Holocene ($\sim$ 9000 to 6000 BP), that period could provide indications of the effects of a warmer Earth on ENSO. If changes in Holocene atmospheric patterns can be linked physically to changes in boundary conditions, they could guide the design of models of ENSO applicable beyond present climate conditions. High temporal resolution records from the core ENSO regions from such earlier times would be invaluable in examining the relationship between the expression of ENSO (for example in the frequency domain) and the background climate, even if these records were not absolutely dated.

## Prospects and opportunities

Given these various lines of evidence, what are some potentially fruitful strategies for future paleoclimatic research as applied to ENSO reconstruction? Enfield (Chapter 5) discusses a set of conceptual alternatives for the state of the ENSO system over long time scales ($> 10^4$ yr). He suggests that one important objective of paleoclimatic research in this area should be aimed at resolving the issue of the robustness of the ENSO phenomenon to major changes in boundary conditions (ice ages, greenhouse warming, etc.). Enfield also points out the importance of determining the sensitivity of the ENSO system to large external (solar variability, volcanoes) or internally induced (variable modes of the ocean thermohaline circulation) changes in forcing.

These research questions which deal with the possible instability of the extratropical effects of ENSO suggest that future progress in the study of paleoclimatic aspects of ENSO would be enhanced by the adoption of different strategies for the 'core region' and for the extratropics. The issue of robustness of the basic phenomenon may be answerable by the further development of promising records close to the Equator, particularly carefully chosen analyses of long-lived corals as described by Cole et al. (Chapter 18). As a better understanding is gained of the mechanisms generating interannual variability in the various coral measurements, it should become possible to develop records of interannual variability in the core region by using even older (or fossil) corals, either exposed naturally or obtained by deep drilling of atolls. This might provide valuable direct information on the robustness of the ENSO phenomenon. Direct fisheries records for the historical period and sedimentary records of marine biotic change, as discussed by Sharp (Chapter 19), could add information to that obtainable from corals, from large critical regions of the oceans along the continental margins. Land-based records within the core region, such as analyses of tropical tree rings (see D'Arrigo and Jacoby, Chapter 13) and historical and paleoflood records (see Cayan and Webb, Chapter 3) would further strengthen the record.

Several chapters in this book furnish evidence that combining proxy records from multiple extratropical regions (e.g., McGlone et al., Chapter 21), or extratropical regions combined with parts of the core region (Michaelsen and Thompson and D'Arrigo and Jacoby, Chapters 17 and 13, respectively); would produce informative records of paleo-ENSO activity. Diaz and Kiladis (Chapter 2) have identified regions of strong precipitation or temperature anomalies associated with extreme positive or negative values of the Southern Oscillation. An investigation of extratropical records of ENSO based on carefully chosen records from within such regions offers the possibility of reconstructing large-scale extratropical responses, assuming no fundamental change in the teleconnections. This assumption would probably be most reasonable if only applied to the late Holocene up to the turn of this century, when climate system boundary conditions can be assumed to have changed relatively little. Even within those regions with strong modern ENSO teleconnections, records should be chosen carefully, bearing in mind the limitations discussed in this volume. In particular, extratropical local climates contain much non-ENSO-related 'noise,' that partially conceals any ENSO 'signal.' In addition, each individual proxy record itself contains nonclimatic 'noise,' and responds only partially to climate, usually seasonally (e.g., winter precipitation or summer temperature), and may not necessarily respond to that in a linear fashion. In fact, threshold effects may render some proxy records sensitive only to extreme events in local climate. Further, proxy records are often asymmetrical in their response. For example, semiarid region tree rings are likely to record drought more faithfully than flood. Despite all these limitations, the exploratory analyses presented here indicate that sets of well-chosen proxy records may be almost as strongly teleconnected with the Southern Oscillation as instrumental records from the same region.

Nonetheless, variable correlations in time among proxy records from different regions indicate that an integrated network of annual or better resolution records of regional-scale climatic changes, with global coverage, would be needed for an exhaustive search for extratropical ENSO signals differing from those found in the instrumental record. Establishing such a network for the past 2000 yr is a main objective of a number of interdisciplinary climate and paleoclimate programs in the United States and elsewhere. This is a major task that could well take at least a decade to perform. In the meantime, the possibility of different extratropical modes of expression of ENSO should be borne in mind.

Whatever strategies are adopted, their success will depend as much on the quality of the reconstructions derived from proxy records as on the soundness of the experimental design. Particularly when dealing with records of annual or seasonal temporal resolution, it is vital that there be precise calendrical control. It is really important, for example, in studying the relative timing of strong ENSO events and the possible climatic influence of major volcanic eruptions, that the dates ascribed to both should be precise and accurate. In the case of episodic records, for example paleofloods, chronology is a more complex question, depending not only on chronometric technique, but also on contextual interpretation. Similarly, the development of reconstructions of past environmental conditions such as climate from natural records rests on models describing the relationship between the environment and the sediment or organism. In many cases these models assume time-invariance in these interactions, particularly the statistical rather than mechanistic models. Rigorous statistical cross-validation of such records is a very important test of such models, but it should not be considered sufficient. A clear and comprehensive understanding of the mechanisms producing each natural record is essential to reconstructing the remote past (i.e., before the Industrial Revolution).

Clearly, increased interaction between the paleoclimate community and the ocean and atmospheric sciences community regarding scientific research on ENSO is desirable. The fact that the Workshop on Paleoclimate Aspects of El Niño/Southern Oscillation was held, which provided the impetus for this volume, is proof that such interaction is underway. The time is right, the issues are important and scientifically challenging. We hope that publication of this interdisciplinary work on historical and paleoclimatic aspects of ENSO will spur many of our readers to establish or enhance such cooperative activities.

## References

BAUMGARTNER, T.R., MICHAELSEN, J., THOMPSON, L.G., SHEN, G.T., SOUTAR, A., and CASEY, R.E., 1989: The recording of interannual climatic change by high-resolution natural system: tree-rings, coral bands, glacial ice layers, and marine varves. *In* Peterson, D. H. (ed.), *Aspects of Climate Variability in the Pacific and the Western Americas.* Geophysical Monograph 55. Washington, D.C.: American Geophysical Union, 1–14.

ENFIELD, D.B. and CID S., L., 1991: Low-frequency changes in El Niño-Southern Oscillation. *Journal of Climate*, 4: 1137–1146.

WELLS, L., 1990: Holocene history of the El Niño phenomenon as recorded in flood sediments. *In* Betancourt, J.L. and MacKay, A.M. (eds.), *Proceedings of the Sixth Annual Pacific Climate (PACLIM) Workshop, March 5–8, 1989.* California Department of Water Resources, Interagency Ecological Studies Program Technical Report 23, 141–144.

# Index